Mushrooms
Nutraceuticals and Functional Foods

Editors:

Deepu Pandita
Senior Lecturer
Government Department of School Education, Jammu
Union Territory of Jammu and Kashmir, India

Anu Pandita
Dietician, Vatsalya Clinic
New Delhi, India

CRC Press
Taylor & Francis Group
Boca Raton London New York

CRC Press is an imprint of the
Taylor & Francis Group, an **informa** business

A SCIENCE PUBLISHERS BOOK

First edition published 2023
by CRC Press
6000 Broken Sound Parkway NW, Suite 300, Boca Raton, FL 33487-2742

and by CRC Press
4 Park Square, Milton Park, Abingdon, Oxon, OX14 4RN

CRC Press is an imprint of Taylor & Francis Group, LLC

Library of Congress Cataloging-in-Publication Data (applied for)

ISBN: 978-1-032-34453-9 (hbk)
ISBN: 978-1-032-34456-0 (pbk)
ISBN: 978-1-003-32223-8 (ebk)

DOI: 10.1201/9781003322238

Typeset in Times New Roman
by Radiant Productions

Preface

Worldwide 1.5 million fungi are assessed, out of which around 14,000 species produce the fruiting bodies known as mushrooms and around 2000 species are edible and 270 are reported to be medicinal. The fungal kingdom is phylogenetically closer to man than plant kingdom. So, it is believed that fungi may give better health benefits to human systems. These mushrooms with nutraceutical and pharmacological properties may help in designing of health- enhancing myco-foods and myco-pharmaceuticals. The nutraceutical word is a juxtaposition of "nutrition" and "pharmaceuticals" and means any substance which may be considered a food or part of food and provides health-improving effects, prevents and cures some human disorders.

Mushrooms are known as poor people's protein 'elixir of life', 'the gift from God Osiris' and 'the food of the Gods'. Mushrooms are labelled as the "little miniature chemical factories" that produce a gigantic collection of novel components and active secondary metabolites and nutraceuticals. Due to their nutritional profile, edible and medicinal mushrooms are considered healthy dietary food, functional food or nutraceuticals such as phenolics, polyketides, flavonoids, α-tocopherol, terpenoids, immunostimulatory glucans, lectins and steroids, ergothioneine, polysaccharides, Polyunsaturated fatty acids (PUFA) especially the n-3-fatty acid family, lysine and leucin amino acids and other bioactive molecules which possess more than 130 medicinal properties and act as immune enhancers, anti-carcinogenic, antitumor, hepatoprotective, anti-HIV anti-diabetic, anti-obesity, anti-hypercholesterolemic, antimicrobial, anti-inflammatory, anti-nephritis, immunomodulatory, cardio protective, anti-aging, anti-arthritis, reduce or prevent age-related neurodegenerative processes, such as Alzheimer's and Parkinson's diseases, etc. This book contains 23 chapters with the introductory chapter discussing the antioxidant and secondary metabolite profiling of mushrooms as sources of Nutraceuticals and Functional foods. The other chapters deliberate the nutraceutical and functional foods aspects of *Agaricus bisporus* (Button Mushroom), *Agaricus subrufescens* (God's mushroom), *Armillaria mellea* (Honey Fungus), *Auricularia auricula-judae* (Jelly Ear), *Auricularia polytricha* (Cloud Ear Fungus), *Calvatia gigantean* (Giant Puffball), *Cantarellus cibarius* (Girolle), *Craterellus cornucopioides* (Black Trumpet), Flammulina velutipes (Velvet Shank), *Ganoderma lucidum* (Reishi mushroom), *Grifola frondosa* (Maitake), *Hericium erinaceus* (Lion's Mane), *Lentinula edodes* (Shiitake), *Lignosus rhinocerotis* (Tiger Milk Mushroom), *Morchella esculenta* (Gucchi), *Ophiocordyceps sinensis* (Cordyceps), *Pleurotus ostreatus* (Oyster Mushroom), *Polyporus confluens* (Sheep Polypore), *Trametes versicolor* (Turkey Tail), *Tremella fuciformis* (Snow Ear), *Volvariella volvacea* (Paddy Straw Mushroom), and *Wolfiporia extensa* (Peck) Ginns. (China Root).

Contents

CHAPTER 1

Antioxidants and Secondary Metabolites of Mushrooms as a Source of Nutraceuticals and Functional Food

Ankur Singh and *Aryadeep Roychoudhury**

1. Introduction

Since origin, man has mostly lived in a hunter-gatherer society, depending on the biodiversity of their surroundings for their sustenance. Mushrooms have probably been eaten for as long as people have walked on the surface of the Earth. The history of mushrooms dates back to 300 BC when Theophrastus first mentioned it as 'truffles'. Romans regarded mushrooms as 'food of the Gods' whereas Greek warriors used to consume them during battles to acquire strength. Basidiomycete fungi, during sexual reproduction, produce fruiting bodies that are commonly known as 'mushrooms'. Spores released from the fruiting bodies germinate to form mycelia which in the presence of suitable substrate (dead branches, fallen leaves, wood, etc.) further develop into primoridia. Finally, primoridia gives rise to the fruiting bodies, i.e., the mushrooms. According to Royse et al. (2017), the production of mushrooms was 30 million tons at a cost of about $63 billion in 2013 which shows a significant rise in the production as compared to that of 1978 when only one million ton of mushrooms was produced worldwide. Of all the cultivated mushrooms, around 54% (~ $34 billion) was edible. Most of the cultivated edible mushrooms consist of five genera: *Flammulina* (11%), *Agaricus* (15%), *Auricularia* (17%), *Pleurotus* (19%) and *Lentinula* (22%) (Raut, 2019). For centuries, mushrooms have been widely used as traditional medicines and foods. However, in spite of wide use, the pharmacological and nutritional property of mushrooms has been recently recognised worldwide. In recent times, various scientific reports have described the pharmacological properties (antifungal,

Post-Graduate Department of Biotechnology, St. Xavier's College (Autonomous), 30 Mother Teresa Sarani, Kolkata – 700016, West Bengal, India.
* Corresponding author: aryadeep.rc@gmail.com

antibacterial, antiviral, antitumour and immune-stimulant property) of mushrooms and components extracted from them (Ma et al., 2018). Both secondary metabolites and cellular components of a wide variety of mushrooms have scientifically proved to boost the immune system of the host. Secondary metabolites are of great interest to the medical industry due to their unusual structure and active formation. Along with medicinal interest, other fields, such as agricultural and cosmetic industries, also uses metabolites extracted from mushrooms as their raw material. In spite of such a wide role, the identification and characterisation of compounds showing pharmacological response mostly rotates around higher molecular weight compounds, i.e., protein, polysaccharides and peptides, with only a few reports showing the efficacy of low molecular-weight compounds. Thus, this chapter mostly emphasises the importance of bioactive secondary metabolites derived from edible mushrooms in the nutraceutical industries.

2. Nutritional Value of Mushrooms

Mushrooms are widely popular due to their high protein content and unsaturated fatty acids. Additionally, mushrooms are also known to be a good source of soluble fibre, particularly β-glucan (Cheung, 2013). Fruiting bodies of mushrooms are mostly composed of carbohydrates (50–60%) and sugars (11%) on dry weight basis (Temesgen, 2018), whereas freshly harvested mushrooms have moisture content of about 70–95% depending mostly on the time of harvest and environmental conditions. Khan and Tania (2012) reported that the carbohydrate content of *Pleurotus ostreatus* was 56%. Mannitol is a major sugar found in mushrooms, constituting around 80% of total free sugar (Wannet et al., 2000). Another important component that is widely found in fruiting bodies of mushrooms is proteins. As compared to that of other vegetables and wild plants, the protein level in mushrooms is appreciably high. According to a data published by the USDA (2019), the nutrients present in mushrooms are fat (0.34 g/100 g dry weight), fibre (1.0 g/100 g dry weight), protein (3.09 g/100 g dry weight) and carbohydrate (3.23 g/100 g dry weight) which provides 22 kcal of energy upon consuming 100 g raw mushrooms. According to the report published by U.S. Department of Agriculture (2018), the glycaemic index of mushrooms is quite low which does not significantly affect the glucose level and insulin response of consumers. Along with this, mushrooms are also reported to be a good source of micronutrients and mineral elements that are important for maintaining proper health of humans. The major elements found in mushrooms are potassium (K), calcium (Ca), magnesium (Mg), phosphorus (P), iron (Fe), copper (Cu), zinc (Zn) and sodium (Na) (Malinowska et al., 2004). According to Thongbai et al. (2015), mushrooms are also a good source of vitamins, such as vitamin B complex, including vitamin B_5 that plays a major role in the release of energy from fat, carbohydrate and proteins. However, the minerals and nutritional content of the mushrooms mostly depend on the species, age, size of fruiting bodies and the mineral content of the surrounding or substrate supplied at the time of cultivation (Rudawska and Leski, 2005).

3. Economic Importance of Micro Fungal Nutraceuticals

In the last decade, the severity of health-related problems has significantly increased due to which the demand for nutraceuticals and dietary supplements has also parallelly enhanced due to their potential role in improving the life expectancy and preventing health-related disorders in human beings. The nutraceuticals industry first came into existence in 1990s (Ali and Nizar, 2018). From the very beginning, the nutraceuticals market became more competitive due to large investments by higher food and pharmaceutical companies. According to a report published by PMMI Business Intelligence, the global nutraceutical market is expected to grow up to $373 billion by 2025 as against around $241 billion in 2019, i.e., nutraceutical market is expected to grow at a rate of 7.5% compound growth annually (PMMI Business Intelligence, 2019; Chrzan, 2019).

Of late, various reports have shown the medicinal importance of consuming mushrooms that acts as a driving force for its increased market demand, thus paving the path for its inclusion in the nutraceutical market. According to a study conducted by Datam Intelligence, the global demand for edible mushrooms is expected to grow at 7.9% compound annual growth rate. Edible mushrooms had a global market value of $42.42 billion in the year 2018, $45.3 billion in 2020, and are forecasted to increase up to $62.19 billion in 2023 and $72.5 billion by 2027 (Research and Markets Global Edible Mushrooms Market). According to a report by Technavio (2018–2022), the market value of major medicinal mushrooms, such as Chaga and Reishi mushrooms, will enhance by $13.88 billion by 2018–2022 in European, American, Middle Eastern and African markets. They further assume that the inclination of people toward vegan diet will boost the market of mushrooms, further increasing the market by 9% compounded annual growth rate during 2018–2022. Various mushroom-based pharmaceuticals, i.e., krestin, lentinan, coriolan, schizophyllan, etc. are already available in the world market (Badalyan, 2014). The market value of soluble fibre glucan, which is highly present in mushrooms, is expected to grow significantly in food, cosmetic and medicinal industries (Venkatachalam et al., 2020). Dietary habits, positive outlook towards nutrient-rich food and change in lifestyle are the major causes of the rise of nutraceutical industry and in coming times, consumer preference for more affordable products will also be a major factor contributing towards the growth of the mushroom industry.

4. Nutraceutical Metabolites of Mushrooms

It is obvious that people have been always in search of new substances which can improvise biological significance to make them healthier and fitter. Today science depends on herbs and plants as enhancers of the food sources. These products are regarded as variously functional foods, vitamins, dietary supplements, nutraceuticals, etc. (Chen and Miles, 1996; Brower, 1998; Zeisel, 1999). In this regard, mushrooms are now receiving much appreciation for their health benefits, medicinal and nutritional attributes (Roychoudhury, 2021). Human beings consume mushrooms as dietary ingredients, supplementing and complementing deficiency in food items from animal and plant origin. Unique chemical composition of mushrooms makes them preferable for certain groups of physiological disorders or ailments. Such attributes help mushrooms to be treated as a healthy food. The term 'nutraceutical', derived from a combination of two words: 'nutrition' and 'pharmaceutical', refers to any substance with immense health benefits along with the treatment and prevention of diseases (Biesalski, 2001). Nutraceuticals can be herbal products, isolated nutrients, dietary supplements or engineered and processed food products, like cereals, beverages and soups. These nutraceuticals show both remarkable biological activities as well as physiological functions (Andlauer and Furst, 2002). Regular consumption of nutraceuticals can cause a regression state in disease as well as increase the immune response. The mushrooms, a good source of nutraceuticals, are important for enhancing the immune system, balancing the nutrients, increasing natural body resistance and preventing diseases. Some species of mushrooms have valuable nutraceutical functions, especially the varieties like *Lentinus (Lentinula), Flammulina, Hericium, Tremella, Auricularia, Agaricus, Lactarius, Grifola, Russula, Pisolithus*, etc. Mushroom nutraceuticals extracted from mycelium or fruiting bodies can be consumed as tablets or capsules as per the requirements in the diet. There is a small chance of toxicity from an overdose of mushroom nutraceuticals, though it has a higher nutritional value (almost equal to meat and milk) and low calorific value. Recent studies prove that mushrooms are considered culinary delicacies as also being well known for their medicinal and nutritional qualities. Palatability of mushrooms can be exercised by flavour, texture, colour and taste, whereas nutritional value is determined by the composition of vitamins, proteins, amino acids, minerals, nucleic acids, fibres, fatty acids and so on. Nutritional composition depends upon the environmental condition and genetic diversity. Mushrooms contain essential amino acids, like methionine, leucine, valine and lysine that are deficient commonly in other plant nutrients for human use (Rahi et al., 2002). Most of the mushrooms are free from cholesterol with greater protein content, around 20–30%, which helps the vegetarians to increase the amount of plant protein in their balanced diet. The fibrous component of mushroom is also helpful in the human digestive system and in lowering cholesterol. Generally, vitamins B and D are animal vitamins, not found in plants, but present in mushrooms to provide a good diet for the vegetarians. Mushrooms also have a variety of oxidising and hydrolysing enzymes (Rajarathnam and Bano, 1993). On the basis of dry weight, mushrooms contain 32% protein, 55% carbohydrate, 2% fat and the rest contain minerals (Crisan and Sands, 1978). Therefore, the cultivation of mushrooms has a commercial value in increasing the exploitation of their metabolic by-products for health benefits.

4.1 Essential Amino Acids

Among the 20 amino acids, nine cannot be synthesised within the human body, but are essential for biological and physiological functions and are called essential amino acids (valine, lysine, leucine, tryptophan, methionine, threonine, histidine, isoleucine and phenylalanine). The proteins from common edible mushrooms contain all the nine amino acids essential for humans. Lysine as well as methionine and tryptophan are the most abundant and lowest, respectively among the essential amino acids present in edible mushrooms.

4.2 Carbohydrates

Mushrooms possess a wide variety of carbohydrates, such as pentoses, hexoses, methylpentoses, disaccharides, sugar alcohols, amino sugars and sugar acids (Crisan and Sands, 1978). *Pleurotus* among edible mushroom contain carbohydrates in the range of 46.6–81.8% in comparison to 60% in *Agaricus bisporus* on the basis of their dry weight.

4.3 *Proteins*

The protein content in edible mushrooms cultivated in various countries ranges between 1.75–5.9% and 19–35% of their fresh weight and dry weight, respectively which are about twice those of asparagus, cabbage and four to 12 times than that of orange and apple (Chang, 1980). On the basis of dry weight, mushrooms contain higher proteins, in comparison to 13.2% in wheat, 7.3% in rice, 25.2% in milk and 39.1% in soybean. Crude protein content in mushrooms ranks lower than most animal meats, but is greater than in most other foods, including milk (Chang and Miles, 1993). *Morchella esculenta* (1.62 g/100 g), *Pleurotus sajor-caju* (1.60 g/100 g), *Agaricus bisporus* (1.80 g/100 g), *Pleurotus ostreatus* (1.68 g/100 g) and *Boletus edulis* (2.20 g/100 g) also contain ample amounts of proteins among the mushrooms (Boda et al., 2012).

4.4 *Fat*

The crude fat constituents of mushrooms constitute all classes of lipid compounds consisting of phospholipids, free fatty acids, sterols, triglycerides, monoglycerides, diglycerides and sterol esters. Saturated fatty acids are harmful for health, whereas unsaturated fatty acids make food healthier to consume and almost 72% of mushroom fatty acids are known to be unsaturated (Mattila et al., 2002). On the basis of dry weight, the range of fat content varies between 1.1–8.3% which on an average appears to be of 4% in different species of edible mushrooms.

4.5 *Vitamins*

Mushrooms are an ideal source of various vitamins. The B complex vitamins, like thiamine (vitamin B_1), riboflavin (vitamin B_2), biotin, niacin and other vitamins like ascorbic acid (vitamin C) are abundant in mushrooms (Mattila et al., 2001). The thiamine content is 1.14 mg in *Agaricus bisporus*, 0.35 mg in *Volvariella volvacea*, 4.80 mg in *Pleurotus* spp., and 7.8 mg in *Lentinus edodes* on the basis of dry weight in mg/100 g mushroom. The niacin content varies between 46–108.7 mg in *Pleurotus* spp., 55.7 mg in *Agaricus bisporus*, 54.9 mg in *Lentinus edodes* and 64.88 mg in *Volvariella volvacea*. The ascorbic acid content is the highest in *Lentinus edodes* (9.4 mg/100 g dry sample) and 1.4 mg, 1.8 mg, 7.4 mg in *Volvariella volvacea*, *Agaricus bisporus* and sajor-caju, respectively (Chang and Miles, 1993). *Volvariella volvacea* (1.63–2.98 mg) has lower riboflavin range than that of *Agaricus bisporus* (5.0 mg) and *Lentinus edodes* (4.9 mg) per 100 g of dry weight.

4.6 *Minerals*

Mushrooms are also a heavy source of minerals. The calculated concentration of the total ash content is about 56–70% of sodium (Na), potassium (K), calcium (Ca), magnesium (Mg) and phosphorus (P) (Li and Chang, 1982). Potassium, being an extremely vital mineral for regulating blood pressure, keeps the cells functional and is abundantly present to near about 45% of total ash content. A large portobello mushroom has a higher potassium content than a banana. Na and Ca are equally present in all the mushroom species except for *Lentinus edodes* where Ca is present in excessively greater amounts. Copper (Cu) is higher in all species of *Pleurotus* ranging between 12.2–21.9 mg L^{-1}. Cu participates in the formation of red blood cells and helps the body in oxygen absorption. The amount of Pb and Ca vary between 1.5–3.2 and 0.3–0.5 ppm, respectively, in *Pleurotus*. The zinc content among all the heavy metals is generally the greatest in *Pleurotus*. Selenium is a key content of mushrooms as an antioxidant that helps to neutralise free radicals for the prevention of cell damage and reduces the risk factor for cancer. Mushrooms also contain zinc as another important mineral.

4.7 Fibre

Fibre plays a pivotal role in a balanced diet. Diets containing high fibre for diabetic patients reduce the daily requirements for insulin, stabilising the glucose profile and possibly decreasing the glucose absorption rate in blood (Anderson and Ward, 1979). Most of the water insoluble or soluble active polysaccharides from mushrooms are identified as dietary fibres having high molecular weight and are absorbed directly without digestion (Vahouny and Kritchevsky, 1986). Mushrooms in dried matter constitutes 10–50% dietary fibres, whereas the fibre content range is 7.4–27.6% in *Pleurotus* species in comparison to 4–20% in *Volvariella volvacea* and 10.4% in *Agaricus bisporus*.

4.8 Nucleic Acids and Nucleotide Derivatives

Nucleotides are the pillars to form nucleic acids, such as DNA and RNA. In the form of nucleoside triphosphates (ATP, GTP, TTP, CTP and UTP), nucleotides play a crucial role in metabolism and constitute a reservoir of energy. Basidiomycetous fungi can produce nucleosides and also exhibit interesting bioactive potential. For example, in Japan, shiitake mushrooms (*Lentinus edodes*) are cultivated for their edible delicacies and nutritional value on fallen logs and constitute a good source of lentinacin, a hypoglycaemic activity containing nucleoside compound (cholesterol lessening). Viikari and Linko (1977) reported that nucleic acid content of mushrooms is 8–25%, whereas other fungi and algae contain 3.2–4.7% and 3–8% nucleic acids, respectively on the basis of dry weight (Viikari, 1972). *Clitocybe inversa* has been reported to produce an insecticidal metabolite named clitocine. The Protein Advisory Group of the United Nations System has reported that the safe nucleic acid intake limit for adults is 4 g at the maximum. Recent reports state that among the four highest edible mushrooms, *Pleurotus sajor-caju* contains the highest nucleic acid content of 4.06% and *Agaricus bisporus* contains the lowest nucleic acid content of 2.66% on the basis of their dry weight. People can consume 392.5 g of fresh *Pleurotus sajor-caju* as it is safe and healthy, whereas this limit can be changed for lower nucleic acid-containing mushrooms.

4.9 Phenolic Compounds

Phenolic compounds are one of the major groups of secondary metabolites found in fruiting bodies of mushrooms. They possess a single aromatic ring with one or more hydroxyl groups. Phenolic compounds found in mushrooms exhibit a wide range of physiological properties, such as anti-inflammation, anti-allergenic, anti-microbial, anti-antherogenic, vasodilator, anti-thrombotic and cardioprotective. Additionally, mushrooms are a rich source of antioxidants due to the presence of phenolic compounds that are well-known radical scavengers, metal ion chelators and singlet oxygen quenchers (Heleno et al., 2015; Ferreira et al., 2009). Major phenolic compounds identified and isolated from mushrooms include p-coumaric acid, protocatechuic acid, cinnamic acid and p-hydroxybenzoic acid that play a pivotal role in lessening the severity of cancer, neurodegenerative disorders and cardiovascular diseases (Heleno et al., 2015; Valverde et al., 2015).

5. Bioactivity of Secondary Metabolites

It is impossible to completely estimate the number of secondary metabolites produced in mushrooms due to their complex chemical structure and small size. Chadwick and Whelan (1992) have stated that the secondary metabolites formed in mushrooms are mainly cellular waste generated due to metabolic activity and are significantly important for mankind. Additionally, Berdy (2005) demonstrated that secondary metabolites produced by mushrooms act as chemical communicators between microbes or other non-microbial systems, including plants and mammals. Thus, it can be undoubtedly concluded that secondary metabolites formed in fungi possess a vast array of biological activities. In spite of this huge significance, the acting mechanism and chemical structure of some known and unknown compounds of mushrooms still need to be investigated. In other words, all the secondary metabolites found in mushrooms possess inherent activities that are yet to be discovered. This section of the chapter is mainly devoted to summarising the major biological activities of mushroom-derived secondary metabolites.

5.1 Anti-microbial Activity

Following the discovery of penicillin by Alexander Fleming, tremendous progress has been made in the field of antibiotics that reduce the severity of several diseases. Due to simple screening system and higher availability, investigation on higher fungi has led to the identification of a wide range of anti-microbial activities of secondary metabolites. According to the report of Zjawiony (2004), around 75% of polypore fungi have been proved to have strong anti-microbial activity that might play a major role in developing new antibiotics in the near future. Around 22,000 bioactive compounds have been isolated and identified from mushrooms of which most of the compounds, i.e., approximately 20,000 bioactive compounds show antibiotic activity (Berdy, 2005). Compounds extracted from mushrooms have antifungal, antiviral and/ or antibacterial activities. Garcia et al. (2012) proved that the compounds extracted from *Laetiporus sulphurous* showed anti-microbial activity as compared to that of some of the commonly used mycotic and antibiotics. Tuberculosis is one of the major health-related problems that affects nearly 2–3 million people worldwide. Rifampicin and isoniazid are two well-known drugs that are commonly used during the treatment of tuberculosis; however, various reports have shown that these two compounds have hepatotoxic activity. Arpha et al. (2012) isolated the chemical constituents from *Astraeus odoratus* and showed their antibacterial activity that also lowers the cytotoxicity caused due to *Mycobacterium tuberculosis*. Thus, the compounds isolated from *Astraeus odoratus* can be used as an alternative to the most commonly used anti-cancerous and antibacterial drugs.

5.2 Anticancer Activity

Cancer chemotherapy mostly depends on cytotoxic drugs that can cause cell death by inhibiting tumour cell proliferation and unnatural cell growth. On the basis of studies conducted in the past years, it has been proved that secondary metabolites of higher fungi show cytotoxic activities against several tumour cell lines. In China and Japan, an ascomycete parasite of bamboo twigs, *Shiraia bambusicola*, is used in traditional Chinese medicine for the treatment of pneumonia and rheumatism. A cytotoxic fungal pigment, named hypocrellin D from *Shiraia bambusicola*, inhibits the growth of tumour cell lines, Anip-973, Bel-7721 and A-549 significantly with 38.4, 1.8 and 8.8 mg mL^{-1} IC$_{50}$ values, respectively (Fang et al., 2006). Several compounds were extracted through separation protocol guided by specific bioassay from *Phellinus igniarius*, a medicinal mushroom by Shi and co-workers followed by revealing their cytotoxicity *in vitro* against few human cancer cell lines (Mo et al., 2004; Wang et al., 2005a; Wang et al., 2007). Among these compounds, some exhibited potent selectivity against Bel7402 and A549 and another compound exhibited moderate activity against human colon cancer HCT-8 and ovarian cancer A 2,780 cell line (Wang et al., 2005a). Several compounds extracted from *Phellinus igniarius* were tested inactive against human cancer cell line, but could inhibit protein tyrosine phosphate 1B, whereas the other exhibited exact reverse activities from above (Wang et al., 2007). Hsp90 is discovered as a vital role player in cancer phenotype to provide an effective drug target for cancer chemotherapy as it is crucial in maintaining the conformational and functional stability of the key oncogenic client proteins (Whitesell and Lindquist, 2005; Chen and Ding, 2004). 17-allylamio-17-desmethoxygeldanamycin, the Hsp90 inhibitor is in clinical trial stage as this inhibitor exhibited remarkable antitumour activity in both *in vivo* and *in vitro* model systems (Chen and Ding, 2004). Natural product-based Hsp90 inhibitors have currently been developed by Gunatilaka and co-workers through fractionation guided by bioassay with cytotoxic compound and monocillin I from *Paraphaeosphaeria quadriseptata* and *Chaetomium chiversii* (Turbyville et al., 2006). Some compounds are applied as antitumour drugs to block the microtubule depolymerisation in ovarian and breast cancer treatment (Demain, 1999). Cancer cell line proliferation is remarkably inhibited by metabolites identified from *Albatrellus confluens* which are also involved in inducing apoptosis by activating caspases-3, 8 and 9, increasing Bax level, decreasing Bcl-2 level and releasing cytochrome c from mitochondria (Ye et al., 2005). Cell motility, a crucial cause for tissue invasion and responsible for 90% death of cancer patients, leads to *in vitro* primary tumours to cell migration and metastasis and is related to cell invasion and angiogenesis. Therefore, cell motility appears as an alternative drug target in cancer therapy (Fenteany and Zhu, 2003; Carmeliet, 2003). Future investigation is focused upon natural products as angiogenesis and cell motility inhibitors. Cyclohexadepsipeptide, discovered by Zhan et al. from *Fusarium oxysporum*, an endophytic fungi can inhibit metastatic breast cancer (MDA-MB-231) and prostate cancer (PC-3M) cell migration, exhibit antiangiogenic activity in HUVEC-2 cell line and also possesses insecticidal activities, cytohomeostasis action and induces apoptosis of tumour cells (Zhan et al., 2007; Jow et al., 2004; Wu et al., 2002).

5.3 Anti-inflammatory Activity

Inflammation is the first response to irritation by outside invaders with the five significant characteristic signs of pain, redness, heat, swelling and loss of function. Anti-inflammation or anti-inflammatory function is to reduce inflammation, as uncontrolled inflammation leads to greater damage to tissue and organ dysfunction. Hence the importance of discovering anti-inflammatory agents from natural resources is increasing day by day. *Cycas pruinosa* metabolic extracts (CPME) can cause down regulation of gene expression responsible for inflammation mediators, including TNF-α, IL-1b, COX-II and inducible nitric oxide synthase (iNOS) by inhibiting the activation of NF-kB in RAW 264.7 cells. Lipopolysaccharide (LPS) stimulated mice showed a downstream suppression of inflammatory mediators both *in vitro* and *in vivo*, such as PGE2, IL-1b, NO and TNF-α (Kim et al., 2003). Hence, CPME exhibited a remarkable anti-inflammatory activity by inhibition of NF-kB-dependent inflammatory gene expression, which leads to the beneficial function of CPME to treat sepsis or endotoxin shock. Further observation stated that LPS-induced nitric oxide production can be inhibited by iNOS mRNA synthesis inhibition (Quang et al., 2006a). Several cytotoxic compounds, produced from *Albatrellus caeruleoporus*, an inedible mushroom, also exhibited inhibition to LPS-stimulated production of nitric oxide in RAW 264. Similarly, down regulation of iNOS gene expression suppressed the LPS-stimulated production of nitric oxide in RAW cells (Quang et al., 2006b). Moreover, in the process of inflammation, both COX-I and COX-II are involved in converting arachidonic acid in prostaglandins (Lipsky et al., 1998), where COX-I is responsible for cytoprotective effects of prostaglandin and COX-II for inducing inflammatory conditions (Smith et al., 1998). Metabolites produced from the fruiting body of an edible mushroom *Agrocybe aegerita* reduce COX-I and COX-II enzyme activity (Zhang et al., 2003).

5.4 Antioxidant Activity

Active oxygen, and specifically free radicals, can induce oxidative stress in cells which leads to atherosclerosis, cancer, inflammatory diseases, aging and cardiovascular diseases, while antioxidant helps to protect healthy cells from oxidative damage caused by active oxygen and free radicals. Natural antioxidants, derived from various plant and higher fungi sources, are preferred over synthetic antioxidants that have toxicity and risk factors for health that eventually enhanced the importance of natural antioxidants over decades (Sun and Fukuhara, 1997; Quang et al., 2006c). Recent studies showed that *Ganoderma lucidum* has no or little side effects (Yen and Wu, 1999). Sun et al. (2004) observed that *Ganoderma lucidum* peptide (GLP) has greater antioxidant activity than butylated hydroxytoluene in soybean oil. GLP can block the activity of soybean lipoxygenase in a dose-dependent manner with 27.1 mg mL^{-1} IC$_{50}$ value. GLP quenched superoxide radical produced in pyrogallol autoxidation, showing scavenging activity towards hydroxyl radicals. GLP blocked auto-haemolysis of rat's red blood cells and exhibited substantial antioxidant property in mitochondrial membrane peroxidation system and rat liver tissue homogenates. GLP plays a vital role in inhibiting the peroxidation of lipid in cell by its metal chelating and antioxidant and scavenging activities. Polyporaceae family fungi, named *Phellinus igniarius*, have been used to treat bloody gonorrhoea, abdominalgia and fester. Almost 20 metabolites are discovered from the *Phellinus igniarius* fruiting bodies, out of which some have potent antioxidant activity to inhibit microsomal lipid peroxidation in rat liver (Wang et al., 2005b). In recent studies, four antioxidants were identified and isolated by Kim and colleagues from *Cyathus stercoreus* fermented mushroom. Those antioxidants showed higher antioxidant activity than that of Trolox and BHA, reference antioxidants and was evaluated through ABTS and DPPH radical scavenging activity assays (Kang et al., 2007).

6. Conclusion and Future Perspectives

Mushrooms have gained significant importance and attention from scientific and industrial community due to their high nutritional and medicinal values. From previous studies, it is widely proved that mushrooms are an abundant source of a wide range of useful natural products with diverse chemical structure and bioactivity. In spite of such an immense role in various industries, the commercial products of mushrooms are still very limited due to higher cost value and labour-intensive process. Increasing product yields and development of highly efficient production and purification of secondary metabolites in the near future is required to make the availability of these important compounds more feasible. Additionally, more studies are required to identify and isolate significantly important secondary metabolites that will help in lessening the severity of life-threatening diseases that will eventually increase the life expectancy of human beings. Further studies in coming years will be helpful in deciphering the genes and enzymes involved in the biosynthetic pathway of these metabolites and will greatly contribute in manipulation of their biosynthetic route for enhanced production. The bioactive compounds from mushrooms hold a promise for future innovations in drug development and as a supplement for preventing human diseases. Nutritionally-rich mushrooms are a valuable health food and have been used for medicinal purposes for centuries across various parts of the world. However, limited awareness and high cost limit their market demand. Additionally, lack of quality control is also a major cause of concern for people involved in this industry and its related fields. In the coming time, further research regarding the nutraceutical applications of mushrooms should be conducted to establish their efficacy and awareness among the common people.

Acknowledgements

Financial assistance from Science and Engineering Research Board, Government of India through the grant [EMR/2016/004799] and Department of Higher Education, Science and Technology and Biotechnology, Government of West Bengal, through the grant [264(Sanc.)/ST/P/S&T/1G-80/2017] to Dr. Aryadeep Roychoudhury is gratefully acknowledged.

References

Ali, M.E. and Nizar, N.N.A. (2018). *Preparation and Processing of Religious and Cultural Foods.* Elsevier: Amsterdam, The Netherlands, ISBN 9780081018927.
Anderson, J.W. and Ward, K. (1979). High-carbohydrate, high-fibre diets for insulin-treated men with diabetes mellitus. *Am. J. Clin. Nutr.*, 32: 2312–2321.

Andlauer, W. and Furst, P. (2002). Nutraceuticals: A piece of history, present status and outlook. *Food Res. Int.*, 35: 171–176.

Arpha, K., Phosri, C., Suwannasai, N., Mongkolthanaruk, W. and Sodngam, S. (2012). Astraodoric Acids A–D: New lanostane triterpenes from edible mushroom *Astraeus odoratus* and their anti-mycobacterium tuberculosis H37Ra and cytotoxic activity. *J. Agric. Food Chem.*, 60: 9834–9841.

Badalyan, S.M. (2014). Potential of mushroom bioactive molecules to develop healthcare biotech products. pp. 373–378. *In: Proceedings of the 8th International Conference on Mushroom Biology and Mushroom Products*, New Delhi, India.

Berdy, J. (2005). Bioactive microbial metabolites: A personal view. *J. Antibiotics*, 58: 1–26.

Biesalski, H.K. (2001). Nutraceuticals: Link between nutrition and medicine. pp. 1–26. *In:* Kramer, K., Hoppe, P.P. and Packer, L. (eds.). *Nutraceuticals in Health and Disease Prevention*, Marcel Dekker, New York, NY, USA.

Boda, R.H., Wani, A.H., Zargar, M.A., Ganie, B.A., Wani, B.A. and Ganie, S.A. (2012). Nutritional values and antioxidant potential of some edible mushrooms of Kashmir valley. *Pak. J. Pharm. Sci.*, 25: 623–627.

Brower, V. (1998). Nutraceuticals: Poised for a healthy slice of healthcare market. *Nat. Biotechnol.*, 16: 728–730.

Carmeliet, P. (2003). Angiogenesis in health and disease. *Nat. Med.*, 9: 653–660.

Chadwick, D.J. and Whelan, J. (1992). Secondary metabolites: Their function and evolution. *Ciba Foundation Symposium 171*, Wiley, Chichester, UK, pp. 1–328.

Chang, S.T. (1980). Mushrooms as human food. *BioScience*, 30: 399–401.

Chang, S.T. and Miles, P.G. (1993). The nutritional attributes and medicinal value of edible mushrooms. pp. 27–39. *In: Edible Mushrooms and their Cultivation*. CRC Press.

Chen, A.W. and Miles, P.G. (1996). Biomedical research and the application of mushroom nutraceuticals from *Ganoderma lucidum*. pp. 153–159. *In:* Royes, D.L. (ed.). *Mushroom Biology and Mushroom Products*. Pennsylvania State University, USA.

Chen, Y. and Ding, J. (2004). Heat shock protein 90: Novel target for cancer therapy. *Chin. J. Cancer*, 23: 968–974 (in Chinese).

Cheung, P.C.K. (2013). Mini-review on edible mushrooms as source of dietary fibre: Preparation and health benefits. *Food Sci. Hum. Wellness*, 2: 162–166.

Chrzan, J. (2019). *The Global Market for Nutraceuticals Set for Robust Growth*. https://www.healthcarepackaging.com/markets/neutraceuticalsfunctional/article/13296428/the-global-market-for-nutraceuticals-set-for-robust-growth.

Crisan, E.V. and Sands, A. (1978). Nutritional value. pp. 137. *In:* Chang, S.T. and Hayes, A. (eds.). *The Biology and Cultivation of Edible Mushrooms*. Academic Press, New York, NY, USA.

Demain, A.L. (1999). Pharmaceutically active secondary metabolites of microorganisms. *Appl. Microbiol. Biotechnol.*, 52: 455–463.

Fang, L.Z., Qing, C., Shao, H.J., Yang, Y.D., Dong, Z.J., Wang, F., Zhao, W., Yang, W.Q. and Liu, J.K. (2006). Hypocrellin D, a cytotoxic fungal pigment from fruiting bodies of the ascomycete *Shiraia bambusicola. J. Antibiot.*, 59: 351–354.

Fenteany, G. and Zhu, S. (2003). Small-molecule inhibitors of actin dynamics and cell motility. *Curr. Top Med. Chem.*, 3: 593–616.

Ferreira, I., Barros, L. and Abreu, R. (2009). Antioxidants in wild mushrooms. *Curr. Med. Chem.*, 16: 1543–1560.

Garcia, A., Bocanegra-Garcia, V., Palma-Nicolas, J.P. and Rivera, G. (2012). Recent advances in anti-tubercular natural products. *Eur. J. Med. Chem.*, 49: 1–23.

Heleno, S.A., Martins, A., Queiroz, M.J.R.P. and Ferreira, I.C.F.R. (2015). Bioactivity of phenolic acids: Metabolites versus parent compounds: A review. *Food Chem.*, 173: 501–513.

Jow, G.M., Chou, C.J., Chen, B.F. and Tsai, J.H. (2004). Beauvericin induces cytotoxic effects in humanacute lymphoblastic leukemia cells through cytochrome c release, caspase 3 activation: The causative role of calcium. *Cancer Lett.*, 216: 165–173.

Kang, H.S., Jun, E.M., Park, S.H., Heo, S.J., Lee, T.S., Yoo, I.D. and Kim, J.P. (2007). Cyathusals A, B, and C, antioxidants from the fermented mushroom *Cyathus stercoreus*. *J. Nat. Prod.*, 70: 1043–1045.

Khan, M.A. and Tania, M. (2012). Nutritional and medicinal importance of *Pleurotus* mushrooms: An overview. *Food Rev. Int.*, 28: 313–329.

Kihlberg, R. (1972). The microbe as a source of food. pp. 427–466. *In*: Clifton, L.E., Raffel, S. and Starr, M.P. (eds.). *Annual Review of Microbiology*, vol. 26.

Kim, K.M., Kwon, Y.G., Chung, H.T., Yun, Y.G., Pae, H.O., Han, J.A., Ha, K.S., Kim, T.W. and Kim, Y.M. (2003). Methanol extract of *Cordyceps pruinosa* inhibits *in vitro* and *in vivo* inflammatory mediators by suppressing NF-kappa B activation. *Toxicol. Appl. Pharm.*, 190: 1–8.

Li, G.S.F. and Chang, S.T. (1982). The nucleic acid content of some edible mushrooms. *Eur. J. Appl. Microbiol. Biotechnol.*, 15: 237–240.

Lipsky, L.P., Abramson, S.B., Crofford, L., Dubois, R.N., Simon, L.S. and Van de Putte, L.B. (1998). The classification of cyclooxygenases inhibitors. *J. Rheumatol.*, 25: 2298–2303.

Ma, G., Yang, W., Zhao, L., Pei, F., Fang, D. and Hu, Q. (2018). A critical review on the health promoting effects of mushrooms nutraceuticals. *Food Sci. Hum. Wellness*, 7: 125–33.

Malinowska, E., Szefer, P. and Falandysz, J. (2004). Metals bioaccumulation by bay bolete, *Xerocomus badius*, from selected sites in Poland. *Food Chem.*, 84: 405–416.

Mattila, P., Könkö, K., Eurola, M., Pihlava, J.M., Astola, J., Vahteristo, L., Hietaniemi, V., Kumpulainen, J., Valtonen, M. and Piironen, V. (2001). Contents of vitamins, mineral elements, and some phenolic compounds in cultivated mushrooms. *J. Agric. Food Chem.*, 49: 2343–2348.

Mattila, P., Lampi, A.-M., Ronkainen, R., Toivo, J. and Piironen, V. (2002). Sterol and vitamin D_2 contents in some wild and cultivated mushrooms. *Food Chem.*, 76: 293–298.

Mo, S., Wang, S., Zhou, G., Yang, Y., Li, Y., Chen, X. and Shi, J. (2004). Phelligridins, C–F cytotoxicpyrano[4,3-c][2]benzopyran-1,6-dione and furo[3,2-c]pyran-4-one derivatives from the fungus *Phellinus igniarius*. *J. Nat. Prod.*, 67: 823–828.

PMMI Business Intelligence. (2019). Nutraceuticals Market Assessment. https://www.pmmi.org/report/2019-nutraceuticals-market-assessment.

Protein Advisory Group. (1970). *Single-Cell Protein Guideline No. 4*, FAO/WHO/UNICEF Protein Advisory Group, United Nations, New York, NY, USA.

Quang, D.N., Harinantenaina, L., Nishizawa, T., Hashimoto, T., Kohchi, C., Soma, G.I. and Asakawa, Y. (2006a). Inhibitory activity of nitric oxide production in RAW 264.7 cells of daldinals A–C from the fungus *Daldinia childiae* and other metabolites isolated from inedible mushrooms. *J. Nat. Med.*, 60: 303–307.

Quang, D.N., Hashimoto, T., Arakawa, Y., Kohchi, C., Nishizawa, T., Soma, G. and Asakawa, Y. (2006b). Grifolin derivatives from *Albatrellus caeruleoporus*, new inhibitors of nitric oxide production in RAW 264.7 cells. *Bioorg. Med. Chem.*, 14: 164–168.

Quang, D.N., Hashimoto, T. and Asakawai, Y. (2006c). Inedible mushrooms: A good source of biologically active substances. *Chem. Rec.*, 6: 79–99.

Rahi, D.K., Shukla, K.K., Rajak, R.C. and Pandey, A.K. (2002). Nutritional potential of *Termitomyces heimii*: An important wild edible tribal mushroom species of M.P. *J. Basic Appl. Mycol.*, 1: 36–38.

Rajarathnam, S. and Bano, Z. (1993). Handbook of applied mycology. pp. 241–292. *In*: Arora, D.K., Mukherji, K.G. and Marth, E.H. (eds.). *Foods and Feeds*. Marcel Dekker, New York, NY, USA.

Raut, J. (2019). Current status, challenges and prospects of mushroom industry in Nepal. *Int. J. Agric. Econ.*, 4: 154–160.

Research and Markets Global Edible Mushrooms Market – Industry Trends, Opportunities and Forecasts to 2023. https://www.researchandmarkets.com/reports/4451952/global-edible-mushrooms-market-industry-trends.

Roychoudhury, A. (2021). Mushrooms as medicinal and therapeutic agents. *International Journal on Current Trends in Drug Development & Industrial Pharmacy*, 5: 3–12.

Royse, D.J., Baars, J. and Tan, Q. (2017). Current overview of mushroom production in the world. pp. 5–13. *In*: Zied, D.C. and Pardo-Giménez (eds.). *Edible and Medicinal Mushrooms: Technology and Applications*. John Wiley & Sons Ltd, Singapore.

Rudawska, M. and Leski, T. (2005). Macro- and micro-element contents in fruiting bodies of wild mushrooms from the Notecka forest in west-central Poland. *Food Chem.*, 92: 499–506.

Smith, C.J., Zhang, Y., Koboldt, C.M., Muhammad, J., Zweifel, B.S., Shaffer, A., Talley, J.J., Masferrer, J.L., Seibert, K. and Isakson, P.C. (1998). Pharmacological analysis of cyclooxygenase-1 in inflammation. *Proc. Natl. Acad. Sci.*, USA, 95: 13313–13318.

Sun, B. and Fukuhara, M. (1997). Effects of co-administration of butylated hydroxytoluene, butylated hydroxyanisole and flavonoids on the activation of mutagens and drug-metabolising enzymes in mice. *Toxicology*, 122: 61–72.

Sun, J., He, H. and Xie, B.J. (2004). Novel antioxidant peptides from fermented mushroom *Ganoderma lucidum J. Agric. Food Chem.*, 52: 6646–6652.

Temesgen, T. (2018). Application of mushroom as food and medicine. *Adv. Biotechnol. Microbiol.*, 11: 555817.

Thongbai, B., Rapior, S., Hyde, K.D., Wittstein, K. and Stadler, M. (2015). *Hericiumerinaceus*, an amazing medicinal mushroom. *Mycol. Prog.*, 14: 91.

Turbyville, T.J., Wijeratne, E.M.K., Liu, M.X., Burns, A.M., Seliga, C.J., Luevano, L.A., David, C.L., Faeth, S.H., Whitesell, L. and Gunatilaka, A.A.L. (2006). Search for Hsp90 inhibitors with potential anticancer activity: isolation and SAR studies of radicicol and monocillin I from two plant-associated fungi of the Sonoran desert. *J. Nat. Prod.*, 69: 178–184.

U.S. Department of Agriculture. (2018). https://fdc.nal.usda.gov/fdc-app.html#/food-details/169251/nutrients.

Vahouny, G.V. and Kritchevsky, D. (1986). *Dietary Fibre—Basic and Clinical Aspects*. Plenum Press, New York, NY, USA.

Valverde, M.E., Hernández-Pérez, T. and Paredes-López, O. (2015). Edible mushrooms: Improving human health and promoting quality life. *Int. J. Microbiol.*, 2015: 1–14.

Venkatachalam, G., Arumugam, S. and Doble, M. (2020). Industrial production and applications of α/β linear and branched glucans. *Indian Chem. Eng.*, 1–15.

Viikari, L. and Linko, M. (1977). Reduction of nucleic acid content of SCP. *Process Biochem.*, 12: 17–19.

Wang, Y., Mo, S.Y., Wang, S.J., Li, S., Yang, Y.C. and Shi, J.G. (2005a). A unique highly oxygenated pyrano[4,3-c][2]benzopyran-1,6-dione derivative with antioxidant and cytotoxic activities from the fungus *Phellinus igniarius*. *Org. Lett.*, 7: 1675–1678.

Wang, Y., Wang, S.J., Mo, S.Y., Li, S., Yang, Y.C. and Shi, J.G. (2005b). Phelligridimer A, a highly oxygenatedand unsaturated 26-membered macrocyclic metabolite with antioxidant activity from the fungus *Phellinus igniarius*. *Org. Lett.*, 7: 4733–4736.

Wang, Y., Shang, X.Y., Wang, S.J., Mo, S.Y., Li, S., Yang, Y.C., Ye, F., Shi, J.G. and He, L. (2007). Structures, biogenesis, and biological activities of pyrano[4,3-c]isochromen-4-one derivatives from the fungus *Phellinus igniarius*. *J. Nat. Prod.*, 70: 296–299.

Wannet, W.J.B., Hermans, J.H.M., van der Drift, C. and Op den Camp, H.J.M. (2000). HPLC detection of soluble carbohydrates involved in mannitol and trehalose metabolism in the edible mushroom *Agaricus bisporus. J. Agric. Food Chem.*, 48(2): 287–91.

Whitesell, L. and Lindquist, S.L. (2005). Hsp90 and the chaperoning of cancer. *Nat. Rev. Cancer*, 5: 761–772.

Wu, S.N., Chen, H., Liu, Y.C. and Chiang, H.T. (2002). Block of L-type Ca^{2+} current by bauvericin, atoxic cyclopeptide, in the NG108-15 neuronal cell line. *Chem. Res. Toxicol.*, 15: 854–860.

Ye, M., Liu, J.K., Lu, Z.X., Zhao, Y., Liu, S.F., Li, L.L., Tan, M., Weng, X.X., Li, W. and Cao, Y. (2005). Grifolin, a potential anti-tumour natural product from the mushroom *Albatrellus confluens*, inhibits tumour cell growth by inducing apoptosis *in vitro*. *FEBS Lett.*, 579: 3437–3443.

Yen, G.C. and Wu, J.Y. (1999). Antioxidant and radical scavenging properties of extracts from *Ganoderma tsugae. Food Chem.*, 65: 375–379.

Zeisel, S.H. (1999). Regulation of 'nutraceuticals'. *Science*, 285: 1853–1855.

Zhan, J., Burns, A.M., Liu, M.X., Faeth, S.H. and Gunatilaka, A.A.L. (2007). Search for cell motility and angiogenesis inhibitors with potential anticancer activity: beauvericin and other constituents of two endophytic strains of *Fusarium oxysporum*. *J. Nat. Prod.*, 70: 227–232.

Zhang, Y., Mill, G.L. and Nair, M.G. (2003). Cyclooxygenase inhibitory and antioxidant compounds from the fruiting body of an edible mushroom, *Agrocybe aegerita*. *Phytomedicine*, 10: 386–390.

Zjawiony, J.K. (2004). Biologically active compounds from aphyllophorales (Polypore) fungi. *J. Nat. Prod.*, 67: 300–310.

CHAPTER 2

Button Mushroom
(*Agaricus bisporus*)

Aliza Batool,[1] *Umar Farooq,*[1] *Afshan Shafi,*[1] *Naqi Abbas,*[1]
Zahid Rafiq[1] and *Zulqurnain Khan*[2,*]

1. Introduction

The term 'mushroom' is not a taxonomic classification. It is defined as "a macro fungus with a characteristic fruiting body, which can be seen with naked eyes is also epigeous or hypogeous and is harvested by hand." Mushrooms are classified as basidiomycetes, but some ascomycete species also fall under this category. Mushrooms have at least 14,000 species and maybe as many as 22,000. The world's total number of mushroom species is believed to be 140,000 with just 10% of them known. Estimating that only 5% of the undiscovered and unexamined mushrooms are beneficial, this means that there are 7,000 species yet to be identified that could be of service to mankind (Hawksworth, 2001). Even among the known species, a fraction of mushrooms that have been thoroughly researched is extremely low. This fact, combined with the knowledge of microscopic fungi's (*Claviceps purpurea, Aspergillus, Penicillium, Tolypocladium inflatum, W. Gams*) great potential for producing bioactive metabolites has a strong potential to help in pharmacological, improved genetic and chemical analysis (Lindequist et al., 2005).

Mushrooms are significant components of minor forest products that grow on the biosphere's most abundant biomolecule—cellulose. The mushroom's fruiting body is visible, while the majority of the mushroom remains underground as mycelium. Mushrooms were present in the world long before man arrived, as proven by fossil records from the lower cretaceous period. As a result, anthropologically speaking, it's possible that man ate mushrooms as a food source when he was still a hunter and gatherer in the cultural evolution timeline. Mushrooms have a wide range of

[1] Department of Food Science & Technology, MNS-University of Agriculture, Multan, Pakistan.
[2] Department of Biotechnology, Institute of Plant Breeding and Biotechnology (IPBB), MNS-University of Agriculture, Multan, Pakistan.
* Corresponding author: zulqurnain.khan@mnsuam.edu.pkcccccf

uses, including food and medicinal, in addition to their important ecological roles. Mushrooms are an important source as man's future nutritious and appetising food. Mushrooms are very important in different type of diseases, like cancer, stress, asthma, cholesterol reduction, sleeplessness, allergies and diabetes. They can be used as a protein malnutrition gap due to their high protein content. Mushrooms, as functional foods, are utilised as nutrient supplements in the form of pills to boost immunity. They are suitable for diabetic and cardiac patients due to their minimal starch and cholesterol content. In the mushrooms, one-third of the iron is in usable form. Anticancer drugs are made from their polysaccharide content. They've even been proven to be effective in the fight against HIV. Antifungal, antioxidant, antibacterial and antiviral capabilities have been found in biologically active chemicals derived from mushrooms, which have also been utilised as pesticides and nematicides. In the light of the numerous applications of mushrooms, there are various aspects of mushrooms with human health advantages, like food, minerals, medicine and medications (Wani et al., 2010).

2. Excellent Source of Food

Since antiquity, people have been foraging for wild mushrooms. Due to their chemical composition, which is appealing from the nutritional aspect, fungi have been used as a source of food. Mushrooms were ingested in the early days of civilisation primarily for their flavour and palatability. The modern use of mushrooms is considerably different from the historical use because much research has been done on the chemical makeup of mushrooms, proving that they may be used as a disease-fighting diet. A lot of researchers have looked into the early history of mushroom use in various countries. *Hydnum coralloides, Polyporus mylittae, Morchella esculenta, Agaricus campestris, Helvella crispa* and *Hypoxylon vernicosum* were all used in India much earlier (Mattila et al., 2001).

The composition of mushrooms is around 90% of water and 10% of dry substance; especially the composition of protein is around 27–48%. Carbohydrates make up for less than 60% of the total, whereas fats make up 2–8%. A typical mushroom has

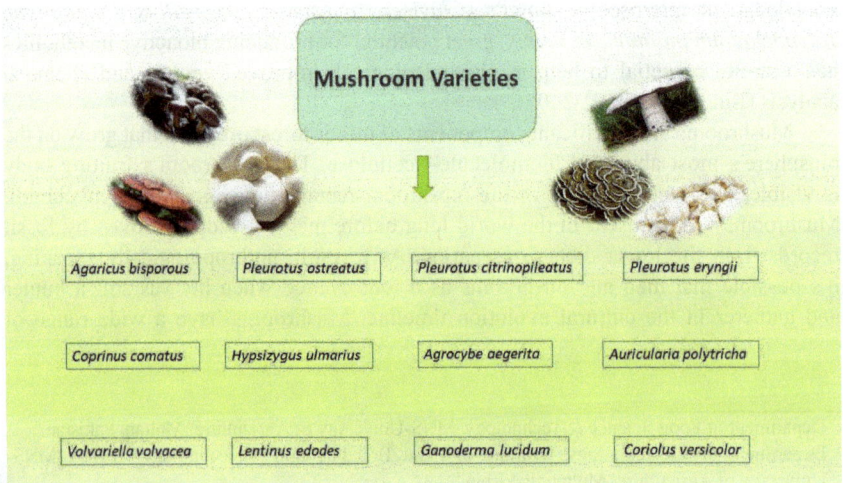

Mushroom Varieties

Agaricus bisporous	Pleurotus ostreatus	Pleurotus citrinopileatus	Pleurotus eryngii
Coprinus comatus	Hypsizygus ulmarius	Agrocybe aegerita	Auricularia polytricha
Volvariella volvacea	Lentinus edodes	Ganoderma lucidum	Coriolus versicolor

Fig. 1. Varieties of mushrooms.

approximately 16.5% dry matter, including 7.4% of crude fiber, 14.6% of crude protein and 4.48% of fat and oil. The edible mushrooms are a high-nutrient food source that outperforms meat, eggs and milk. Only over 2,000 mushroom species are considered edible worldwide, with only about 20 being cultivated professionally and only four to five being used in industrial manufacture. There is also a large nutritional differential between pileus and stalks (Wani et al., 2010).

3. *Agaricus bisporus* (Button Mushroom)

Humans have long sought out *Agaricus bisporus*, sometimes known as white button mushroom, for consumption. Mushrooms were thought to have medicinal properties by the ancient Chinese, who believed that they would strengthen the human body and prolong health. The Romans utilised it as a meal and a medication and they also employed several fungus species to decorate their structures and places of worship (Safwat and Al Kholi, 2006). Most significant advances in mushroom cultivation were made in France, where *Agaricus bisporus* was first cultivated in agriculture media around AD 1600 and particularly developed for the purpose. A Frenchman first commercialised mushroom cultivation in 1780, when he planted *Agaricus bisporus* near Paris, underground in the quarries. For over a century, in United States and Canada, commercial mushroom farming has been the practice. In 1894, in Pennsylvania, the first specialised commission for the development of mushroom culture was established, which has since been known as the world's mushroom capital. In the United States and Canada, the *Agaricus bisporus* mushroom is the most extensively produced commercial mushroom (Chen et al., 2003). In the United States, there are currently about 346 farmers of *Agaricus* sp. and specialty mushrooms, with sales of the mushroom crop totalling about 450 × 106 kg in that year (USDA, 2013). *Agaricus bisporus* is a rich protein source, highly nutritious and also helps in the management of many (Owaid et al., 2017).

3.1 History and Ethno Pharmacology

The edible basidiomycete mushroom *Agaricus bisporus* is found in meadows throughout Europe and North America. *Agaricus bisporus* is a type of mushroom that is widely produced in most nations and accounts for the majority of mushroom consumption in the United States and Australia. In France, it was first cultivated and cultivar strains emerged in Western Europe, according to historical data. Only the Pharaohs were allowed to eat or even touch mushrooms because the ancient Egyptians believed they could give immortality. Mushrooms were often referred to as 'food for the gods' in ancient Rome. Mushrooms were thought to give humans supernatural power in Russia, China, Mexico and in other parts of the world, according to folklore (Jeong et al., 2012). The fungus has traditionally been used to treat heart disease, cerebral stroke and cancer. In addition, it possesses anti-aging properties. Mushroom is the world's most valuable source of nutrients and an appetising future food. Mushrooms are more beneficial than other foods in fighting cancer, lowering cholesterol, stress, sleeplessness, asthma, allergies and diabetes. Because they are high in protein, they can be utilised to fill the gap in protein deficiency. Mushrooms can help boost a person's immunity. Mushrooms are also beneficial for diabetics and people who have heart problems (due to the presence of low cholesterol and starch). Iron is accessible in free form in one-third of the mushrooms. Presence of polysaccharide contents in the mushroom act as an anticancer drug and fight HIV effectively (Prasad et al., 2015). Antifungal, antibacterial, antioxidant and antiviral capabilities have been found in biologically active chemicals from mushrooms, which have also been utilised as pesticides and nematicides. The consumption of button mushrooms daily lower the risk of breast cancer. The mushrooms may suppress the formation of enzymes that affect the creation of estrogen—a hormone that contributes to the development of cancer. The substance's unique effects on various types of cancer are presently being investigated (Bhushan and Kulshreshtha, 2018).

3.2 Botanics

The common mushroom, white mushroom or button mushroom, is a young fungus that is pale yellow or light brown and has an abutted crown. The premature mushroom is sold as Italian mushroom, cremini mushroom, baby Bella, Roman mushroom, Mini Bella, or brown mushroom in strains with darker flesh. The original cap is a light greyish brown tint with broad horizontal scales on a paler background that fades outwards. It has a hemispheric form at first and then flattens out as it matures, with a diameter of 5–10 cms. The dense, precise gills start as pink and then turn brown, then dark brown with a whitish edge from the cheilocystidia. The cylindrical stipe grows up to 6 cm tall and 12 cm broad, with a thick and thin ring on the upper side that may be streaked. The hard flesh is white, but when bruised, it turns a delicate pinkish hue. The print of the spores is dark brown. The spores are oval to spherical and are 4.5–7.5 m in length. Mushrooms lack chlorophyll (Owaid et al., 2017).

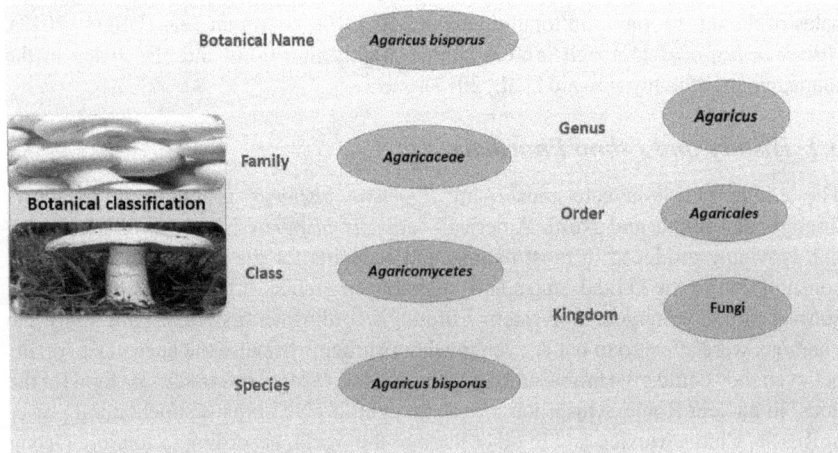

Fig. 2. Biological classification of *Agaricus bisporus.*

3.3 Cultivation of Agaricus bisporus

Mushrooms are heterotrophic (saprophytic) creatures, meaning they lack chlorophyll and rely on the decomposition of organic matter to survive. *Agaricus bisporus* can be grown on a variety of lignocellulosic substrates (Alkaisi et al., 2016). *Agaricus bisporus* plays a vital role in the production of lignin-degrading enzymes (manganese peroxidase and laccase) and is a readily available supply of enzymes, particularly laccase, which is crucial for polyphenol oxidation (Owaid et al., 2017).

The most difficult culture media utilised for edible mushroom production is the substrate for cultivating *Agaricus bisporus* (Baysal et al., 2007). A two-stage method is used to prepare the compost. A mix of raw materials, animal dung (like poultry manure and stable bedding) and gypsum are combined, wetted and moulded into a stack in the first stage (windrow) (Reddy et al., 2013). This species has thrived on a variety of composted organic materials, including manures of chicken, horse, or pigeon, wheat bran, sugarcane bagasse, tea leaves, molasses, Tifton, straw residues from oat, wheat, and feed crops, corn cob, brachiaria (*Brachiaria* sp.), reed plant (*Phragmites australis*) straw, water hyacinth (the casing layer is made of waste paper, wheat straw and discarded tea leaves (Gulser et al., 2003).

The amount of water in the stack is controlled and disassembled and reassembled at regular intervals (Peker et al., 2007). Pasteurisation is performed in the second stage to prepare the compost for inoculation with *Agaricus bisporus* on a selective growth medium (Flores et al., 2009). This two-phase substrate preparation procedure has several drawbacks, like, for cultivation, additional area and time are necessary (Andrade et al., 2013).

The device combines various handling techniques and mechatronic technology. Image analysis and a monochromatic vision system are used to locate and size the mushrooms (Sakae et al., 2006). The sequence in which they should be picked and the picking action are then determined by an expert selection algorithm (bend, twist, or both) (Rehman et al., 2016). The first of two suction cup mechanisms linked to a Cartesian robot's single head is then organised, carefully detaching individual mushrooms and gently placing them into a specially developed, compliant finger conveyer. After high-speed skiving, mushrooms are removed from the conveyor and placed in packs at the machine's side, using a gripper mechanism (Bhushan and Kulshreshtha, 2018).

3.4 Mushroom as a Functional Food

Many components are utilised in the preparation of processed food products; those with intrinsic nutritional and functional features that change the quality of end food products deserve special attention. As a result, ingredients are now regarded as an integral aspect of the formulation of any culinary product. However, natural-source substances or compounds that are generally recognised as harmless are of tremendous interest due to their safety and health properties (Fasseas et al., 2008). According to the Food and Drug Administration, these are compounds that alter numerous qualities and characteristics of any food, both directly and directly. They are used in all stages of downstream processing, such as food manufacture, packing and storage, until the product reaches the customer. By modifying taste and texture, it is possible to improve not only the nutritional quality and safety, but also freshness, attractiveness and general acceptance of food products. Nutraceuticals are additives or parts that have medicinal or beneficial health benefits and play a critical role in the prevention and treatment of many diseases (El Sohaimy et al., 2012).

Whole meals, parts of foods, or even a single food component or extract used as a dietary supplement are all examples of nutraceuticals. These nutraceuticals are only regarded 'functional' when they are added to food or its formulation to achieve a specific intended function, such as improving human well-being and quality of life by lowering the sickness risk (Reis et al., 2017). Mushrooms are low in calories but are highly nutritious, have nutraceutical characteristics and are used as a dietary supplement (Asgar et al., 2010). Furthermore, mushrooms are a high-quality protein source (Kakon et al., 2012).

As a result, adding a high-quality protein source that provides all of the needed amino acids to diets or food products may help to reduce human protein-energy deficiency (Oyetayo et al., 2007). Mushrooms are also used as a bio-therapeutic agent due to the existence of various secondary metabolites, nutraceuticals, or biologically active substances with medicinal value (Prasad et al., 2015). Mushrooms, in general, have all three food functions: nutrition, flavour and physiological functions. They have a unique savoury flavour termed 'umami', which is attributed to the presence of sodium salts of free amino acids, including aspartic and glutamic amino acids, as well as 50-nucleotides (Rathore et al., 2017). The umami flavour, also known as the pleasant flavour, is the overall flavour of the food that has been improved by mono-sodium glutamate (Zhang et al., 2013). As a result of their distinct flavour, mushrooms are preferred and adaptable in most culinary preparations. Sweet, sour, salty, umami and bitter are all taste qualities of peptides with varying structures and lengths. They are normally tasteless in water, but when combined with comparable tastants, they improve the sweet, sour, salty, bitter, or umami flavour (Dunkel et al., 2007). Some peptides and dipeptides (Gly-Leu-Pro-Asp and Gly-His-Gly-Asp), derived from the mushroom *Agaricus bisporus*, are essential compounds in kokumi flavour (Kong et al., 2019). Mouthfulness, complexity and consistency are the finest ways to characterise the taste of kokumi. Kokumi taste compounds have a faint or no flavour of their own and can improve the flavour of basic tastes, like umami, sweet and salty (Feng et al., 2019). When added to a simple chicken broth, these peptides from *Agaricus bisporus* can produce various flavour characteristics, like mouthfulness and complexity (Das et al., 2021).

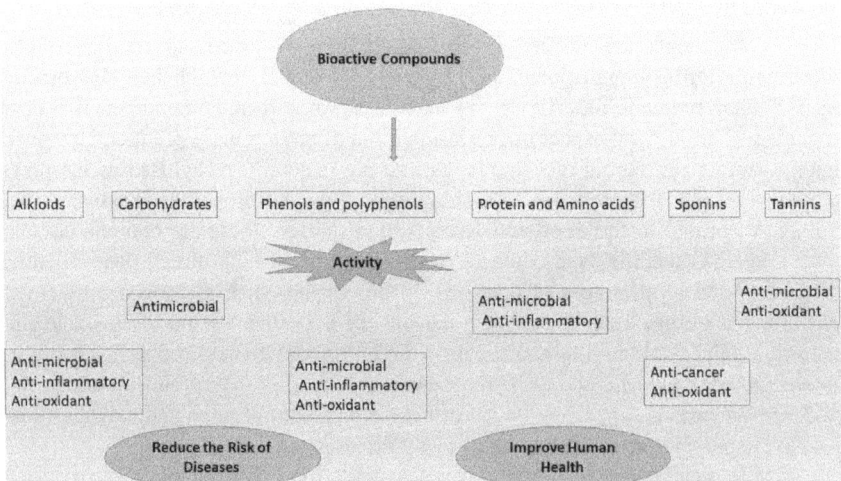

Fig. 3. Biological activity of *Agaricus bisporus*.

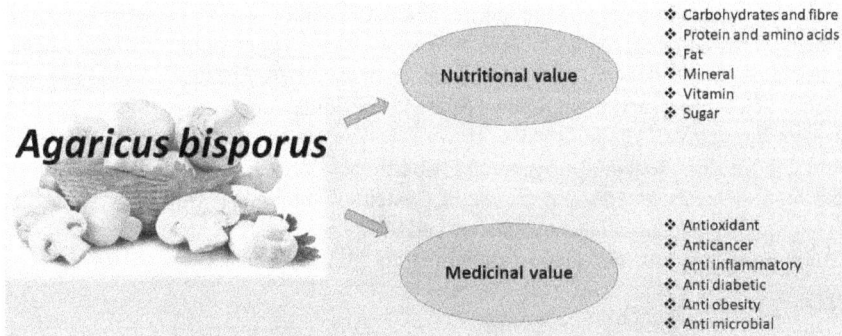

Fig. 4. Nutritional and medicinal value of *Agaricus bisporus.*

3.5 Nutritional Value of Agaricus bisporus

Agaricus bisporus is highly nutritious and contains fats, proteins/amino acids, vitamins, carbohydrates, crude fibres, micronutrients and macronutrients (Owaid, 2015). *Agaricus* sp. contains a significant amount of phosphorus. Protein content ranges from 34–44% in *Agaricus bisporus* fruiting bodies. The concentration of monosodium glutamate is 22.67 g/kg and the concentration of amino acids is 77.92 g/kg. Linoleic, stearic, palmitic and oleic acids are among the unsaturated fatty acids found in *Agaricus bisporus*. In *Agaricus bisporus* the concentration of fructose is 26.2 g/kg, mannitol 236.2 g/kg and other reducing sugars have 57 g/kg. *Agaricus bisporus* is utilised commercially in fresh, dry and canned forms. The dry weight of *Agaricus bisporus* cultivated on wheat straw compost accounts for roughly 8–10% of the total (Colak et al., 2007).

The composition of carbohydrates in *Agaricus bisporus* fruits is 38–48%, crude protein between 21–27%, crude fibre 17–23%, crude ash 8–11% and fat is 3–4% (Tsai et al., 2007). *Agaricus bisporus* is a good source of dietary fibres, which assist to avoid various common illnesses, including obesity, due to its low caloric content (up to 30 calories per 100 gms) (fresh) (Goyal et al., 2015). Vitamins A, B_3 (niacin acid), B_1 (thiamine), D, riboflavin, ascorbic acid, pantothenic acid and folic acid are abundant in *Agaricus bisporus*. It contains threonine, cystine, lysine, isoleucine, methionine, phenylalanine, valine, tyrosine, and leucine, which are all important amino acids for human health (Muszynska et al., 2015).

Micronutrients, like potassium, iron, zinc, sodium, cobalt, selenium, copper and manganese are present in *Agaricus bisporus*. Around 97–98% of *Agaricus bisporus*'s total element content is made up of four elements: potassium, phosphorus, calcium and magnesium. The selenium content of *Agaricus bisporus* is 2.3–2.7 mg kg^{-1}; it is a good source of Se-supplement (Owaid et al., 2017).

3.5.1 Amino Acid and Protein Concentration

After animal sources, mushrooms are a good source of protein in terms of crude protein content (Correa et al., 2016). The chemical composition and nutritional value of grown mushrooms can be influenced by a variety of factors, such as growth substrates, developmental stages and post- and pre-harvest circumstances (Gothwal et al., 2012). The protein content of *Agaricus bisporus* is around 17.7–29% (Ahlavat et al., 2016).

Mushroom proteins have an amino acid composition similar to animal protein, which is essential given the high intake of protein from animal food sources, particularly in developed countries (Guillamon et al., 2010). Because all nine essential amino acids necessary for humans are present in mushrooms, they can be uses as a meat alternative. Alanine, lysine, threonine, phenylalanine, serine, glutamic acid, arginine, proline, leucine, tyrosine and aspartic acid are the amino acids present in the highest levels in *Agaricus bisporus*. Cysteine, methionine, isoleucine, tyrosine, threonine, phenylalanine, lysine, leucine and valine are all essential amino acids found in *Agaricus bisporus*, making it a good meal for humans (Muslat et al., 2014).

3.5.2 Carbohydrate and Fibre

Humans do not use mushroom polysaccharides as a primary source of energy. Non-digestible carbohydrates include oligosaccharides, such as trehalose and non-starch polysaccharides (NSPs), such as chitin, glucans and mannans, which make up the majority of mushroom carbohydrates but are only present in trace amounts (Cheung et al., 2010). Mannitol and glucose, which are normally present in very modest levels, as well as glycogen (5–10%), are digestible carbohydrates. Mannitol and trehalose sugars are abundantly present in the edible mushrooms, with mannitol predominating in *Agaricus bisporus*. Dietary fibre, which comprises components of fungal cell walls like chitin, hemicelluloses, mannans and beta-glucans, is important for some of the mushrooms health benefits. *Agaricus bisporus* has a chitin content of 9.60 g/100 g DM. *Agaricus bisporus* has twice as much chitin as *P. ostreatus*. In comparison to *P. ostreatus* and *L. edodes*, *Agaricus bisporus* contains a higher concentration of chitin (Cherno et al., 2013).

3.5.3 Mineral Content

Agaricus bisporus is an excellent source of minerals, like potassium, iron, zinc, sodium, cobalt, selenium, copper and manganese (Owaid, 2015). Potassium and phosphorus are the most abundant elements in mushroom fruiting bodies, followed by calcium, magnesium, sodium, iron and zinc (Guillamon et al., 2010). The mineral contents in *Agaricus bisporus* fruit bodies range between 0.54–1.58% for potassium, 37.2–61.9 µg/g, 36.6–58.0 µg/g for zinc, for sodium (7.4–7.9 mg/kg), for phosphorus, 56.2–91.1 µg/g for manganese (1.34 mg/kg), for selenium 54.6–163.4 µg/g, for copper, 88.0–76.3 mg/kg for magnesium and 534.2–554.8 mg/kg for calcium (Falandysz and Borovicka, 2013).

Selenium is a micronutrient that both humans and animals need. Dietary supplementation with Se-enriched *Agaricus bisporus* had a strong anticancer effect. Selenium in *Agaricus bisporus* also improves the enzyme activity of GI-specific glutathione peroxidase-2, thioredoxin reductase-1 and cytosolic glutathione peroxidase-1 (Maseko et al., 2014).

3.5.4 Vitamins

Mushrooms are a good source of vitamins. Niacin is the richest nutrient in *Agaricus bisporus*. Vitamin B_1, α-tocopherol, L-ascorbic acid and vitamin B_3 are present in *Agaricus bisporus* (Bernas and Jaworska, 2016). The *Agaricus bisporus* is an excellent source of riboflavin (0.27–0.29 mg/kg), folate (0.09–0.08 mg/kg), thiamine (0.085–0.09 mg/kg) and niacin (3.6–2.9 mg/kg), although less in vitamin C content.

Vitamin B_2 contents in *Agaricus bisporus* are much higher as compared to the many vegetables. Vitamin D is found naturally in mushrooms and *Agaricus bisporus* has a vitamin D level of 984 IU/g. This is significant for bone health and ergosterol is a vitamin D_2 precursor in the body. Composition of ergosterol of *Agaricus bisporus* mushrooms ranges between 39.5–56.7 mg/100 g f.w (Reis et al., 2012).

3.5.5 Fatty Acids

Despite its low fat content, *Agaricus bisporus* contains essential fatty acids, such as linoleic acid. Because of the larger contribution of linoleic acid, *Agaricus bisporus* contains higher polyunsaturated fatty acids and lower monounsaturated fatty acids. The total amount of fatty acids in *Agaricus bisporus* is around 180–5818 mg/kg of dry matter and on an average linoleic acid is nearly about 90% in *Agaricus bisporus* (Barros et al., 2008). In *Agaricus bisporus* different fatty acids are present, like palmitic, linoleic, eicosanoic, oleic, caprylic, stearic and erucic acids; linoleic acid is the main fatty acid (Sadiq et al., 2008). In *Agaricus bisporus* the concentration of linoleic acid is 61.82–67.29% and palmitic acid is 12.67–14.71% (Shao et al., 2010).

For human health, linoleic acid is very necessary and it has several positive effects. By interacting with HDL in the blood, it helps to prevent atherosclerosis. The linoleic acid in *Agaricus bisporus* is 20 and five times higher than in the other mushrooms, like *Pleurotus ostreatus* and *Ganoderma lucidum*, respectively (Hossain et al., 2007).

3.5.6 Soluble Sugar and Volatile Compounds

The most essential qualitative factor contributing to the widespread use of grown mushrooms is flavour and taste. The umami or pleasant flavours of mushrooms, as well as the impression of satisfaction, are a general food flavour sensation created or amplified by monosodium glutamate. In edible mushrooms, the total soluble sugars and polyols of monosodium glutamate (glutamic and aspartic acids) and sweet components like (threonine, alanine and glycine) are high, which may be adequate to suppress and mask the bitter taste caused by bitter components (Tsai et al., 2007). In fresh *Agaricus bisporus* fruit bodies, mannitol is the most important soluble sugar, whereas glucose is the second most prevalent, with concentrations varying between 17.6–28.1 mg/g at various stages of maturity. A high concentration of sugars and polyols, particularly mannitol, would produce a sweet sensation rather than the characteristic mushroom flavour (Smiderle et al., 2011). Both volatile and non-volatile chemicals are associated with mushroom flavour. Their fruiting body and mycelial biomass have a range of valuable odours and flavours due to the terpenes, carbohydrates, lactones and amino acids in their composition (Atila et al., 2017).

3.6 Medicinal Properties of Agaricus bisporus

Medicinal mushrooms, such as *Agaricus bisporus*, have a long history of use in a variety of traditional treatments (Jain et al., 2013). *Agaricus bisporus* extracts are used for bioactivities, like antibacterial activity, anticancer activity, antioxidants activity and anti-inflammation. It fights against different diseases, like cardiovascular disease, cancers, diabetes, immune system disorders, bacterial and fungal infection (Javan et al., 2015; Owaid et al., 2017).

Mushrooms contain natural substances that may enhance human health by lowering the risk of certain diseases or boosting human performance (Ma et al., 2018). Biological active compounds are present in mushrooms and they have potential in the management of many diseases. Mushrooms contain ganoderic acid, phenolics, lectins, triterpenoids, flavonoids, laccase, hispolon, nucleosides, nucleotides, calcaelin, lentinan, proteoglycan and ergosterol, among other nutraceuticals (Patel and Goyal, 2012). In general, depending upon their interactions with biochemical processes and chemical structure, mushrooms' nutraceuticals can have strong biological activities, like antibacterial, anticarcinogenic, antitumour, antidiabetic, antiviral, antimutagenic, anti-obesity, anti-inflammatory and anti-hypercholesterolemic effects (Das et al., 2021).

3.6.1 Anticancer Activity

Cancer is one of the world's deadliest diseases. Polysaccharides, a natural active component found in mushrooms, show substantial anticancer efficacy against a variety of cancer cell lines. Different mushrooms, like *Ganoderma lucidum*, *Grifola frondosa*, *Agaricus blazei* and *Lentinus edodes* have anticancer properties, including cell-growth suppression in the prostate, colon and breast cancer cell lines (Ren et al., 2012). They also induce the process of apoptosis and suppress the process of angiogenesis. They can activate the protein-1 (AP-1) and modify the cell-cycle regulatory protein retinoblastoma (pRb) (Lin et al., 2009). Isolated lectins from *Agaricus bisporus* inhibit the proliferation of colon cancer cell and also enhance the cellular antioxidant defence mechanisms and boost the chemotherapeutic treatment resistance of lung, colon and glioblastoma cancer cells (Adams et al., 2008).

Basidiomycota have therapeutic properties due to the presence of polysaccharide content. Polysaccharides are members of the beta-glucan family of chemicals and they have anti-tumorigenic properties by enhancing cellular immunity. Bioactive chemicals found in *Agaricus bisporus* have been proven to have immunomodulating and anticancer effects (Zhang et al., 2014). The polysaccharide from *Agaricus bisporus* has significant immunostimulatory and anticancer properties (Jeong et al., 2012).

Agaricus bisporus can improve mucosal immunity and health. Secretory immunoglobulin A secretion is greatly accelerated when *Agaricus bisporus* is consumed in the diet (Smiderle et al., 2013). *Agaricus bisporus* extracts boost the formation of interferon-gamma and have an immune-stimulating impact on activated human peripheral blood mononuclear cells (Volman et al., 2010). *Agaricus bisporus* extracts induce the process of apoptosis in HL-60 leukemia cells and also inhibit cell proliferation (McCleary and Draga, 2016).

The arginine in *Agaricus bisporus* slows cancer cell growth and proliferation and it should be taken as a food supplement by cancer patients (Jagadish et al., 2009). Aromatase would be suppressed by *Agaricus bisporus*, lowering the risk of breast cancer (Novaes et al., 2011). Furthermore, *Agaricus bisporus* contains a significant amount of lovastatin with anticancer properties towards breast cancer cell line (Kanaya et al., 2011). *Agaricus bisporus* phytochemicals limit the activity of aromatase, also inhibit the proliferation of breast cancer cell and reduce the development of mammary tumour (Yang et al., 2016).

Unsaturated fatty acids, like linoleic acid, which suppresses aromatase activity, are the most potent chemicals in *Agaricus bisporus* (Roupas et al., 2012). Regular consumption of mushrooms decreases the breast cancer risk. It also reduces the risk of breast cancer in premenopausal women who eat mushrooms (Atila et al., 2017).

3.6.2 Antidiabetics

Agaricus bisporus is high in dietary fibres, antioxidants like vitamin C, D and B_{12}, as well as folates and polyphenols, which may help prevent cardiovascular and diabetic disorders (Jeong et al., 2010). *Agaricus bisporus* has several substances that may provide anti-inflammatory and antioxidant health advantages in adults, who are prone to type-2 diabetes, if consumed often over time (Yamac et al., 2010). High doses of *Agaricus bisporus* extract reduced streptozotocin severity and also decreased the triglyceride (TG) and plasma glucose concentrations to 39.1% and 24.7%. It is also effective against the activities of liver enzymes (alanine aminotransferase and aspartate aminotransferase) and also manage the liver weight gain (Volman et al., 2010). Alpha-glucans from *Agaricus bisporus* are highly effective in the management of diabetes and the production of lipopolysaccharides is reduced by the consumption of alpha-glucans (Calvo et al., 2016). Consumption of *Agaricus bisporus* is a viable dietary option for preventing liver steatosis, which is a reversible condition (Atila et al., 2017).

3.6.3 Anti-hyperlipidemic

In hyperlipidemia, different factors are responsible like high levels of fat or triglycerides. These factors are major casuses of atherosclerosis and heart disease which is one of the most deadly human diseases (Lin et al., 2009). Different phytosterols are present in *Agaricus bisporus*, like ergosta-5,7-dienol, ergosta-7-enol and ergosta-7,22-dienol. These phytosterols can reduce the LDL cholesterol and plasma cholesterol by limiting the cholesterol absorption (Esmaillzadeh and Azatbakth, 2008).

Lovastatin is a statin medicine that is used to minimise the risk of cardiovascular disease by reducing cholesterol in those who have hypercholesterolemia (Xu et al., 2013). In breast cancer cell line, lovastatin has anticancer properties. The lovastatin composition in *Agaricus bisporus* is 565.4 mg/kg, which reduces cholesterol levels in the liver (Yang et al., 2016). Consumption of *Agaricus bisporus* has an antiglycaemic and anticholesterolemic effect, as well as a favourable impact on the liver function and lipid metabolism (Atila et al., 2017).

3.6.4 Antioxidant

Pleurotus eryngii, *Grifola frondosa*, *Pleurotus ostreatus*, and *Lentinula edodes* all have lower antioxidant capacity than *Agaricus bisporus* mushrooms. In the ethanolic extract of *Agaricus bisporus*, different primary phenolic compounds are present, like gallic acid, ferulic acid, protocatechuic acid, myricetin catechin and caffeic acid. The ethanolic extract of *Agaricus bisporus* has a strong antioxidant effect, so it could be used as a new natural antioxidant. Fresh-cut *Agaricus bisporus* mushrooms have phenolic content of 100.78–100.32 mg/100 g. The antioxidant ergothioneine is present in *Agaricus bisporus* with a concentration ranging between 0.21–045 mg/g dw and it has strong antioxidant activity. In mushrooms, phenolic chemicals are the most powerful antioxidants. *Agaricus bisporus* chitosan NPs displayed antioxidant properties (Barros et al., 2008). All of the possible antioxidant properties have an anticancer effect. The feed efficiency, production of adipocytokine, fat deposition in the liver and muscle and fat mass are all reduced by fungal chitosan (Savoie et al., 2008).

Seratonin is a biological molecule with excellent antioxidant properties. Serotonin's antioxidant properties slow the progress of Alzheimer's disease. The concentration of serotonin in *Agaricus bisporus* extracts is high (5.21 mg/100 g) (Liu et al., 2013).

TCP stands for tocopherols, which are fat-soluble antioxidants that also serve a number of other functions in the body. In the fruit bodies of *Agaricus bisporus*, the range of α-tocopherol is between 0.23–0.28 g/100, δ-tocopherol 2.60–2.54 μg/100 g, β-tocopherol 0.85–0.71 g/100 g and γ-tocopherol 1.51–7.63 g/100 (Atila et al., 2017).

3.6.5 Anti-obesity Activity

Mushrooms are extremely nutritive, including large amounts of bioactive substances (alkaloids, flavonoids, polysaccharides, polyphenols, fibres, sterols and terpenes) that have antioxidant properties and can help treat obesity and obesity-related cardiovascular diseases (Gasecka et al., 2018).

Plant sterols, or phytosterols, inhibit cholesterol absorption. LDL cholesterol and plasma cholesterol are reduced by phytosterols isolated from *Agaricus bisporus* (Esmaillzadeh and Azatbakth, 2008). *Agaricus bisporus* also contains a significant amount of lovastatin, a substance that lowers cholesterol levels in the body and reduces the risk of heart disease (Xu et al., 2013). *Agaricus bisporus* basidiocarp consumption modulates anti-cholesterolemic and antiglycaemic effects and *Agaricus bisporus* possesses both anti-hypercholesterolemic and antiglycaemic actions. Furthermore, it has a beneficial effect on lipid metabolism as well as liver function (Usman et al., 2021).

3.6.6 Anti-inflammatory Properties

Agaricus bisporus has strong anti-inflammatory properties. Polysaccharides isolated from *Agaricus bisporus* with analgesic and anti-inflammatory properties include mannogalactan, fucomannogalactan and fucogalactan. Another polysaccharide derived from this species that can help with sepsis is heterogalactan. Sepsis is a serious medical condition that is one of the leading causes of death in intensive care units all over the world and heterogalactan from *Agaricus bisporus* can help in these situations. Because *Agaricus bisporus* has a high anti-inflammatory impact, it may be useful in the fight against sepsis (Usman et al., 2021).

3.6.7 The Genoprotective Effect

The *Agaricus bisporus* edible mushroom is the most widely consumed and farmed. H_2O_2 oxidative damage to cellular DNA was prevented by the fruit bodies of *Agaricus bisporus*. *Agaricus bisporus* is linked to a heat-labile protein found in the fruit body, called FIIb-1, which has been identified as tyrosinase. Cold water extracts of the edible fungus *Agaricus bisporus* have a genoprotective effect that is linked to tyrosinase activity in the mushroom fruit bodies. The enzymatic hydroxylation of tyrosine to L-DOPA and subsequent conversion of this metabolite to dopaquinone is required for *Agaricus bisporus* tyrosinase's genoprotective action (Atila et al., 2017).

Table 1. Medicinal properties of *Agaricus bisporus*.

Diseases/Disorders	Function/Mechanism	References
Bacterial and fungal diseases and cancer	Temperature influenced antibacterial and antifungal action by disrupting the molecule responsible for antimicrobial activity. *Anticancer activity*: Ergosterol, a phenolic molecule, inhibited the aromatase enzyme in a breast cancer cell line *in vitro*. The genoprotective effect of *Agaricus bisporus* tyrosinase is dependent on the enzymatic hydroxylation of tyrosine to L-DOPA and subsequent conversion of this metabolite to dopaquinone.	Dhamodharan and Mirunalini, 2010
Weak immunity	Mushroom polysaccharide fractions increased the production of IL-6, TNF-a and NO which plays important role in immunity.	Jeong et al., 2012
Oxidative stress, diabetes, obesity	Antioxidant due to catechin, ferulic acid, gallic acid, protocatechuic acid and myricetin; anticancer due to polysaccharides; anti-diabetic due to vitamin C, D and B_{12}, as well as polyphenols, folates and dietary fibres; anti-obesity due to phytosterols or plant sterols reducing cholesterol absorption.	Usman et al., 2021
Diabetes, hypercholesterolemia, inflammatory and skin disorders, cancer	Immunomodulating activity, anticholesterolemic and antiglycaemic, anti-inflammatory activity, skin disorders, anticancer activity due to decreased testosterone-induced cell proliferation in MCF-7aro cells.	Bhushan and Kulshreshtha, 2018
Oxidative stress, weak immunity, and inflammatory diseases	Scavenging of superoxide, hydroxyl and DPPH radicals, as well as hydrogen peroxide and increasing antioxidant enzyme activity in mouse serum, liver and heart. Polysaccharides dramatically altered immunological processes by inducing the generation of nitric oxide and cytokines as well as ergosterol, ergosterol peroxide and trametenolic acid.	Zhang et al., 2016
Oxidative stress	Antioxidant activity due to polysaccharides.	Tian et al., 2012
Weak immunity	Through intravenous, intraperitoneal and subcutaneous injection, glucans revealed an immune stimulatory impact mediated by the activation of neutrophils, macrophages, monocytes and natural killer cells.	Yasin et al., 2019
Breast cancer	*Antiproliferative activity*: Suppresses the development of breast cancer cells by inhibiting the function of the aromatase enzyme.	Adams et al., 2008
Tumorigenesis	Aromatase inhibitory activity due to organic-soluble compounds. Polysaccharides inhibit tumorigenesis	Chen et al., 2006
Bacterial & fungal diseases and oxidative stress	Antibacterial, antifungal and antioxidant activities.	Owaid et al., 2017

3.6.8 Antimicrobial

Agaricus bisporus alcoholic extracts have potent antibacterial properties against a variety of bacteria, yeasts and dermatophytes. Microbial inhibition is high in *Agaricus bisporus*. The stipes of *Agaricus bisporus* could be used as natural antimicrobials. The *Agaricus bisporus* aqueous total protein extracts have high antimicrobial activity, especially against *S. aureus* (Ndungutse et al., 2015).

Silver nanoparticles (AgNPs) are a type of metallic nanoparticles with antibacterial and antifungal properties that are widely used. The fungus *Agaricus bisporus* is thought to play a role in the production of silver nanoparticles (AgNPs) (Owaid and Ibraheem, 2017).

After oyster mushroom *Pleurotus* sp., *Agaricus bisporus* has the second-highest level (about 11%) of significant nanoparticle synthesis (Akyuz et al., 2010). The *Agaricus bisporus* extract is used to make AgNPs, which may offer a significant advantage over traditional antibiotics in that they destroy pathogenic bacteria and no organism can build resistance to them. *Pleurotus florida Fomes fomenterieus, Helvella lacunose* and *Ganoderma appalanatum* have lower zones of inhibition against methicillin-resistant *Staphylococcus aureus* strains than AgNPs from the mushroom *Agaricus bisporus* (Abah and Abah, 2010). *Agaricus bisporus*-AgNPs have a good antibacterial activity against Gram-positive bacteria, like *Staphylococcus aureus*, *S. typhi*, *Proteus* sp., *Enterobacter* sp. and *Klebsiella* sp. (Atila et al., 2017).

4. Conclusion

Due to high nutritional value of *Agaricus bisporus*, it can help to prevent malnutrition, especially in emerging and underdeveloped nations. Intake of *Agaricus bisporus* is not only beneficial for nutrition, but it also has many medical benefits, like anticancer, help in the management of cardiovascular disease, antidiabetic, antioxidant and antibacterial properties. Edible mushrooms have become increasingly popular as a source of treatment or as health dietary supplements in the recent decades. Nutraceutical therapy has emerged as a prospective source of new medicines for a variety of life-threatening disorders, according to a majority of studies. Although the isolation of bioactive compounds from *Agaricus bisporus* represents a significant step forward in their characterisation as a source of pharmaceuticals, further clinical evidence is needed to determine the medical effects of *Agaricus bisporus*.

References

Abah, S.E. and Abah, G. (2010). Antimicrobial and antioxidant potentials of *Agaricus bisporus*. *Adv. in Bio. Res.*, 4(5): 277–282.

Adams, L.S., Chen, S., Phung, S., Wu, X. and Ki, L. (2008). White button mushroom (*Agaricus bisporus*) exhibits antiproliferative and proapoptotic properties and inhibits prostate tumour growth in athymic mice. *Nutrition and Cancer*, 60(6): 744–756.

Ahlavat, O.P., Manikandan, K. and Singh, M. (2016). Proximate composition of different mushroom varieties and effect of UV light exposure on vitamin D content in *Agaricus bisporus* and *Volvariella volvacea*. *Mushroom Res.*, 25(1): 1–8.

Akyuz, M., Onganer, A.N. Erecevit, P. and Kirbag, S. (2010). Antimicrobial activity of some edible Mushrooms in the eastern and southeast Anatolia region of Turkey. *Gazi University J. of Sci.*, 23(2): 125–130.

Alkaisi, M.R., Hasan, A.A. and Aljuboori, A.W. (2016). Evaluation of production efficiency for some cultivated mushroom strains *Agaricus bisporus* which was renovated mother culture in multiple methods. *Iraqi J. Sci.*, 57: 38390.

Andrade, M.C.N., de-Jesus, J.P.F., Vieira, F.R., Viana, S.R.F., Spoto, M.H.F. and Minhoni, M.T.A. (2013). Dynamics of the chemical composition and productivity of composts for the cultivation of *Agaricus bisporus* strains. *Braz. J. Micro.*, 44(4): 1139–1146.

Asgar, M.A., Fazilah, A., Huda, N., Bhat, R. and Karim, A.A. (2010). Non-meat protein alternatives as meat extenders and meat analogs. *Compr. Rev. Food Sci. Food Saf.*, 9: 513–529.

Atila, F., Owaid, M.N. and Shariati, M.A. (2017). The nutritional and medical benefits of *Agaricus bisporus*: A review. *J. of Micro. Biotech. Food Sci.*, pp. 281–286.

Barros, L., Cruz, T., Baptista, P., Estevinho, L.M. and Ferreira, I.C.F.R. (2008). Wild and commercial mushrooms as source of nutrients and nutraceuticals. *Food and Chem. Tox.*, 46: 2743–2747.

Baysal, E., Yigitbasi, O.N., Colak, M., Toker, H., Simsek, H. and Yilmaz, F. (2007). Cultivation of *Agaricus bisporus* on some compost formulas and locally available casing materials, Part 1: Wheat straw-based compost formulas and locally available casing materials. *Afr. J. Biotech.*, 6: 222530.

Bernas, E. and Jaworska, G. (2016). Vitamins profile as an indicator of the quality of frozen *Agaricus bisporus* mushrooms. *J. Food Comp. and Analysis*, 49: 1–8.

Bhushan, A. and Kulshreshtha, M. (2018). The medicinal mushroom *Agaricus bisporus*: Review of phytopharmacology and potential role in the treatment of various diseases. *J. of Nature and Sci. of Medi.*, 1(1): 4.

Calvo, M.S., Mehrotra, A., Beelman, R.B., Nadkarni, G., Wang, L., Cai, W., Goh, B.C., Kalaras, M.D. and Uribarri, J. (2016). A retrospective study in adults with metabolic syndrome: Diabetic risk factor response to daily consumption of *Agaricus bisporus* (white button mushrooms). *Plant Foods for Human Nutri.*, 71(3): 245–51.

Chen, R., Chen, L. and Song, S. (2003). Identification of two thermotolerance related genes in *Agaricus bisporus*. *Food Tech. Biotech.*, 41(4): 339–344.

Chen, S., Oh, S.R., Phung, S., Hur, G., Ye, J.J., Kwok, S.L., Shrode, G.E., Belury, M., Adams, L.S. and Williams, D. (2006). Anti-aromatase activity of phytochemicals in white button mushrooms (*Agaricus bisporus*). *Cancer Res.*, 66(24): 12026–12034.

Cherno, N., Osalina, S. and Nikitina, A. (2013). Chemical composition of *Agaricus bisporus* and *Pleurotus ostreatus* fruiting bodies and their morphological parts. *Food and Environ. Safety*, 7(4): 291–299.

Cheung, P.C.K. (2010). The nutritional and health benefits of mushrooms. *Brit. Nutri. Found. Nutri. Bullet.*, 35: 292–299.

Colak, M., Baysal, E., Simsek, H., Toker, H. and Yilmaz, F. (2007). Cultivation of *Agaricus bisporus* on wheat straw and waste tea-leaves-based composts and locally available casing materials, Part 3: Dry matter, protein and carbohydrate contents of *Agaricus bisporus*. *Afr. J. Biotech.*, 6(24): 2855–2859.

Correa, R., Brugnari, T., Bracht, A. and Ferreira, I.C.F.R. (2016). *Pleurotus* spp. (oyster mushroom) related with its chemical composition: A review on the past decade findings. *Trends in Food Sci. and Tech.*, 50: 103–117.

Das, A.K., Nanda, P.K., Dandapat, P., Bandyopadhyay, S., Gullón, P., Sivaraman, G.K., McClements, D.J., Gullon, B. and Lorenzo, J.M. (2021). Edible mushrooms as functional ingredients for development of healthier and more sustainable muscle foods: A flexitarian approach. *Molecules*, 26(9): 2463.

Dhamodharan, G. and Mirunalini, S. (2010). A novel medicinal characterisation of *Agaricus bisporus* (white button mushroom). *Pharmacol. Online*, 2: 456–463.

Dunkel, A., Köster, J. and Hofmann, T. (2007). Molecular and sensory characterisation of Y-glutamyl peptides as key contributors to the kokumi taste of edible beans (*Phaseolus vulgaris* L.). *J. Agric. Food Chem.*, 55: 6712–6719.

El-Sohaimy, S. (2012). Functional foods and nutraceuticals-modern approach to food science. *World Appl. Sci. J.*, 20: 691–708.

Esmaillzadeh, A. and Azatbakth, L. (2008). Food intake patterns may explain the high prevalence of cardiovascular risk factors among Iranian women. *J. Nutri. Epidem.*, 138: 1469–1475.

Falandysz, J. and Borovicka, J. (2013). Macro and trace mineral constituents and radionuclides in mushrooms: Health benefits and risk. *Applied Micro. Biotech.*, 97: 477–501.

Fasseas, M.K.K., Mountzouris, K.C.C., Tarantilis, P.A.A., Polissiou, M. and Zervas, G. (2008). Antioxidant activity in meat treated with oregano and sage essential oils. *Food Chem.*, 106: 1188–1194.

Feng, T., Wu, Y., Zhang, Z., Song, S., Zhuang, H., Xu, Z., Yao L. and Sun M. (2019). Purification, identification, and sensory evaluation of kokumi peptides from *agaricus bisporus* mushroom. *Foods*, 8: 43.

Flores, C., Vidal, C., Trejo-Hernandez, M.R., Galindo, E. and Serrano-Carreon, L. (2009). Selection of *Trichoderma* strains capable of increasing Laccase production by *Pleurotus ostreatus* and *Agaricus bisporus* in dual cultures. *J. Appl. Micro.*, 106: 249–257.

Gasecka, M., Magdziak, Z., Siwulski, M. and Mleczek, M. (2018). Profile of phenolic and organic acids, antioxidant properties and ergosterol content in cultivated and wild-growing species of *Agaricus. Eur. Food Res. Tech.*, 244: 259–268.

Gothwal, R., Gupta, A., Kumar, A., Sharma, S. and Alappat, B.J. (2012). Feasibility of dairy waste water (DWW) and distillery spent wash (DSW) effluents in increasing the yield potential of *Pleurotus flabellatus* (PF 1832) and *Pleurotus sajor-caju* (PS 1610) on bagasse. *3 Biotech.*, 2: 249–257.

Goyal, R., Grewal, R.B. and Goyal, R.K. (2015). Fatty acid composition and dietary fibre constituents of mushrooms of North India. *Emirates J. Food and Agri.*, 27(12): 927–930.

Guillamon, E., García-Lafuente, A., Lozano, M., Arrigo, M.D., Rostagno, M.A., Villares, A. and Martínez, J.A. (2010). Edible mushrooms: Role in the prevention of cardiovascular diseases. *Fitoterapia*, 81: 715–723.

Gulser, C. and Peksen, A. (2003). Using tea waste as a new casing material in mushroom (*Agaricus bisporus* (L.) Sing.) cultivation. *Bio. Tech.*, 88: 153–15.

Hawksworth, D.L. (2001). Mushrooms: The extent of the unexplored potential. *Int. J. Med. Mushrooms*, 3: 333–7.

Hossain, M.S., Alam, N., Amin, S.M.R., Basunia, M.A. and Rahman, A. (2007). Essential fatty acids content of *Pleurotus ostreatus, Ganoderma lucidum* and *Agaricus bisporus*. *Bangladesh J. Mushroom*, 1(1): 1–7.

Jagadish, L.K., Krishnan, V.V., Shenbhagaraman, R. and Kaviyarasan, V. (2009). Comparitive study on the antioxidant, anticancer and antimicrobial property of *Agaricus bisporus* (J.E. Lange) Imbach before and after boiling. *Afr. J. of Biotech.*, 8(4): 654–661.

Jain, N., Karaiya, H., Amrita, K., Tiwari, S., Dubey, V. and Ramalingam, C. (2013). Evaluation of antibacterial properties of the suspension of ginger, black pepper, vinegar, honey and its application in shelf-life extension of *Agaricus bisporus. Int. J. Drug Dev. Res.*, 5(2): 179–186.

Javan, A.J., Nikmanesh, A., Keykhosravy, K., Maftoon, S., Zare, M.A., Bayani, M., Parsaiemehr, M. and Raeisi, M. (2015). Effect of citric acid dipping treatment on bioactive components and antioxidant properties of sliced button mushroom (*Agaricus bisporus*). *J. of Food Quality and Haz. Cont.*, 2: 20–25.

Jeong, S.C., Koyyalamudi, S.R. and Pan, G. (2012). Dietary intake of *Agaricus bisporus* white button mushroom accelerates salivary immunoglobulin A secretion in healthy volunteers. *Nutri.*, 28: 527–531.

Jeong, S.C., Jeong, Y.T., Yang, B.K., Islam, R., Koyyalamudi, S.K., Pang, G., Cho, K.Y. and Song, C.H. (2010). White button mushroom (*Agaricus bisporus*) lowers blood glucose and cholesterol levels in diabetic and hypercholesterolemic rats. *Nutri. Res.*, 30: 49–56.

Jeong, S.C., Koyyalamudi, S.R., Jeong, Y.T., Song, C.H. and Pang, G. (2012). Macrophage immunomodulating and antitumour activities of polysaccharides isolated from *Agaricus bisporus* white button mushrooms. *J. Med. Food.*, 15: 5865.

Kakon, A., Choudhury, M.B.K. and Saha, S. (2012). Mushroom is an ideal food supplement. *J. Dhaka Natl. Med. Coll. Hosp.*, 18: 58–62.

Kanaya, N., Kubo, M., Liu, Z., Chu, P., Wang, C., Yuan, Y.C. and Chen, S. (2011). Protective effects of white button mushroom (*Agaricus bisporus*) against hepatic steatosis in ovariectomised mice as a model of postmenopausal women. *PloS ONE*, 6(10): 1–11.

Kong, Y., Zhang, L.L., Zhao, J., Zhang, Y.Y., Sun, B.G. and Chen, H.T. (2019). Isolation and identification of the umami peptides from shiitake mushroom by consecutive chromatography and LC-Q-TOF-MS. *Food Res. Int.*, 121: 463–470.

Lin, X., Ma, L., Racette, S.B., Spearie, C.L.A. and Ostlund, R.E. (2009). Phytosterol glycosides reduce cholesterol absorption in humans. *American Journal of Physiology, Gastro and Liver Physio.*, 296: 931–935.

Lindequist, U., Niedermeyer, T.H. and Jülich, W.D. (2005). The pharmacological potential of mushrooms. *Evidence-based Complementary and Alternative Medicine*, 2(3): 285–299.

Liu, J., Jia, L., Kan, J. and Jin, C. (2013). *In vitro* and *in vivo* antioxidant activity of ethanolic extract of white button mushroom (*Agaricus bisporus*). *Food and Chem. Tox.*, 51: 310–316.

Ma, G., Yang, W., Zhao, L., Pei, F., Fang, D. and Hu, Q. (2018). A critical review on the health promoting effects of mushrooms nutraceuticals. *Food Sci. Hum. Wellness*, 7: 125–133.

Maseko, T., Howell, K., Dunshea, F.R. and Ng, K. (2014). Selenium-enriched *Agaricus bisporus* increases expression and activity of glutathione peroxidase-1 and expression of glutathione peroxidase-2 in rat colon. *Food Chem.*, 146: 327–333.

Mattila, P., Konko, K., Eurola, M., Pihlawa, J.M., Astola, J., Vahteristo Lietaniemi, V., Kumpulainen, J., Valtonen, M. and Piironen, V. (2001). Contents of vitamins, mineral elements and some phenolic compounds in cultivated mushrooms. *J. Agri. Food Chem.*, 49: 2343–2348.

McCleary, B.V. and Draga, A. (2016). Measurement of β-glucan in mushrooms and mycelial products. *J. of AOAC Int.*, 99(2): 364–373.

Muslat, M.M., Al-Assaffii, I.A.A. and Owaid, M.N. (2014). *Agaricus bisporus* product development by using local substrate with bio-amendment. *Int. J. Environ. Glob. Clim.*, 2(4): 176–188.

Muszynska, B., Krakowska, A., Sułkowska-Ziaja, K., Opoka, W., Reczynski, W. and Bas, B. (2015). *In vitro* cultures and fruiting bodies of culinary-medicinal *Agaricus bisporus* (white button mushroom) as a source of selected biologically-active elements. *J. Food Sci. Tech.*, 52(11): 7337–7344.

Ndungutse, V., Mereddy, R. and Sultanbawa, Y. (2015). Bioactive properities of mushroom (*Agaricus bisporus*) stipe extracts. *J. of Food Process and Pres.*, 39: 2225–2233.

Novaes, M.R.C.G., Fabiana Valadares, M.C.R., Gonçalves, D.R. and Menezes, M.C. (2011). The effects of dietary supplementation with Agaricales mushrooms and other medicinal fungi on breast cancer: Evidence-based medicine. *Clin.*, 66(12): 2133–2139.

Owaid, M.N. (2015). Mineral elements content in two sources of *Agaricus bisporus* in Iraqi market. *J. Adv. App. Sci.*, 3(2): 46–50.

Owaid, M.N. and Ibraheem, I.J. (2017). Mycosynthesis of nanoparticles using edible and medicinal mushrooms. *Eur. J. Nan.*, 9(1): 5–23.

Owaid, M.N., Barish, A. and Shariati, M.A. (2017). Cultivation of *Agaricus bisporus* (button mushroom) and its usages in the biosynthesis of nanoparticles. *Open Agri.*, 2: 537–543.

Oyetayo, F.L., Akindahunsi, A.A. and Oyetayo, V.O. (2007). Chemical profile and amino acids composition of edible mushrooms *Pleurotus sajor-caju*. *Nutr. Health*, 18: 383–389.

Patel, S. and Goyal, A. (2012). Recent developments in mushrooms as anticancer therapeutics: A review. *3 Biotech.*, 2: 1–15.

Peker, H., Baysal, E., Yigitbasi, O.N., Simsek, H., Colak, M. and Toker, H. (2007). Cultivation of *Agaricus bisporus* on wheat straw and waste tea-leaf-based compost formulas using wheat chaff as activator material. *Afr. J. Biotech.*, 6: 4009.

Prasad, S., Rathore, H. and Sharma, S. (2015). Medicinal mushrooms as a source of novel functional food. *IJFS*, 4: 2215.

Rathore, H., Prasad, S. and Sharma, S. (2017). Mushroom nutraceuticals for improved nutrition and better human health: A review. *Pharm. Nutr.*, 5: 35–46.

Reddy, M.T., Reddy, K.A., Reddy, K.A., Reddi, E.U. and Reddi, B. (2013). A study on the production of *Agaricus bisporus* mushrooms using *Eichhornia crassipes* (mart. Solms)—A troublesome exotic aquatic weed of Kolleru Lake. *IJSN*, 4: 1003.

Rehman, M.K., Ali, M.A., Hussain, A., Khan, W.A. and Khan, A.M. (2016). Effect of different casing materials on the production of button mushroom (*Agaricus bisporus* L.). *J. Environ. Agri. Sci.*, 7: 5561.

Reis, F.S., Martins, A., Barros, L. and Ferreira, I.C. (2012). Antioxidant properties and phenolic profile of the most widely appreciated cultivated mushrooms: A comparative study between *in vivo* and *in vitro*. *Food Chem. and Toxicol.*, 50: 1201–1207.

Reis, F.S., Martins, A., Vasconcelos, M.H., Morales, P. and Ferreira, I.C.F.R. (2017). Functional foods based on extracts or compounds derive from mushrooms. *Trends Food Sci. Tech.*, 66: 48–62.

Ren, L., Perera, C. and Hemar, Y. (2012). Antitumour activity of mushroom polysaccharides: A review. *Food Fun.*, 3: 1118–1130.

Roupas, P., Keogh, J., Noakes, M., Margetts, C. and Taylor, P. (2012). The role of edible mushrooms in health: Evaluation of the evidence. *J. Fun. Foods*, 4: 687–709.

Sadiq, S., Bhatti, H.N. and Hanif, M.A. (2008). Studies on chemical composition and nutritive evaluation of wild edible mushrooms. *Iran J. Chem. and Chem. Engineer*, 27(3): 151–154.

Safwat, M.S.A. and Al-Kholi, M.A.J. (2006). *Recent Trends, Reality and Future in the Production, Manufacture and Marketing of Medicinal and Aromatic Plants.* The Egyptian Association for producers, manufacturers and exporters of medicinal and aromatic plants (Asmap.), Giza, Egypt.

Sakae, H., Yutaka, T. and Minoru, T. (2006). Mushroom cultivation using compost produced in the garbage automatic decompose-extinguisher (GADE). *Eurasian J. For. Res.*, 9(2): 61–67.

Savoie, J.M., Minvielle, N. and Largeteau, M.L. (2008). Radical-scavenging properties of extracts from the white button mushroom, *Agaricus bisporus. J. Sci. of Food and Agri.*, 88(6): 970–975.

Shao, S., Hernandez, M., Kramer, J.K.G., Rinke, D.L. and Tsao, R. (2010). Ergosterol profiles, fatty acid composition, and antioxidant activities of button mushrooms as affected by tissue part and developmental stage. *J. Agri. Food Chem.*, 58(22): 11616–11625.

Smiderle, F.R., Ruthes, A.C., Van Arkel, J., Chanput, W., Lacomin, M., Wichers, H.J. and Van-Griensven, L.J.L.D. (2011). Polysaccharides from *Agaricus bisporus* and *Agaricus brasiliensis* show similarities in their structures and their immunomodulatory effects on human monocytic THP-1 cells. *BMC Comp. Alt. Med.*, 11: 58.

Smiderle, F.R., Alquini, G., Tadra-Sfeir, M.Z., Iacomini, M., Wichers, H.J. and Van Griensven, L.J.LD. (2013). *Agaricus bisporus* and *Agaricus brasiliensis* (1 → 6)-β-d-glucans show immunostimulatory activity on human THP-1 derived macrophages. *Carbohydrate Polymers*, 94: 91–99.

Tian, Y., Zeng, H., Xu, Z., Zheng, B., Lin, Y., Gan, C. and Lo, Y.M. (2012). Ultrasonic-assisted extraction and antioxidant activity of polysaccharides recovered from white button mushroom (*Agaricus bisporus*). *Carbohydrate Poly.*, 88(2): 522–529.

Tsai, S.Y., Wu, T.P., Huang, T. and Mau, S.J. (2007). Non-volatile taste components of *Agaricus bisporus* harvested at different stages of maturity. *Food Chem.*, 103. 1457–1464.

Usman, M., Murtaza, G. and Ditta, A. (2021). Nutritional, medicinal, and cosmetic value of bioactive compounds in button mushroom (*Agaricus bisporus*): A review. *Applied Sci.*, 11(13): 5943.

Volman, J.J., Mensink, R.P., van-Griensven, L.J. and Plat, J. (2010). Effects of a-glucans from *Agaricus bisporus* on *ex vivo* cytokine production by LPS and PHA-stimulated PBMCs; a placebo-controlled study in slightly hypercholesterolemic subjects. *Euro. J. of Clin. Nutri.*, 64: 720–726.

Wani, B.A., Bodha, R.H. and Wani, A.H. (2010). Nutritional and medicinal importance of mushrooms. *J. Med. Plants Res.*, 4(24): 2598–2604.

Xu, H., Yang, Y.J., Yang, T. and Qian, H.Y. (2013). Stains and stem cell modulation. *Age. Res. Rev.*, 12: 1–7.

Yamac, M., Kanbak, G., Zeytinoglu, M. and Van-Griensven, L.J.L.D. (2010). Pancreas protective effect of button mushroom *Agaricus bisporus* (J.E. Lange) Imbach (Agaricomycetidae) extract on rats with streptozotocin-induced diabetes. *Int. J. Med. Mushrooms*, 12(4): 379–389.

Yang, T., Yao, H., He, G., Song, L., Liu, N., Wang, Y., Yang, Y., Keller, E.T. and Deng, X. (2016). Effects of Lovastatin on MDA-MB-231 breast cancer cells: An antibody microarray analysis. *J. Cancer*, 7(2): 192–199.

Yasin, H., Zahoor, M., Yousaf, Z., Aftab, A., Saleh, N., Riaz, N. and Shamsheer, B. (2019). Ethnopharmacological exploration of medicinal mushroom from Pakistan. *Phytomedicine*, 54: 43–55.

Zhang, J.J., Li, Y., Zhou, T., Xu, D.P., Zhang, P., Li, S. and Li, H.B. (2016). Bioactivities and health benefits of mushrooms mainly from China. *Molecules*, 21(7): 938.

Zhang, J.J., Ma, Z., Zheng, L., Zhai, G.Y., Wang, L.Q., Jia, M. and Jia, L. (2014). Purification and antioxidant activities of intracellular zinc polysaccharides from *Pleurotus cornucopiae* SS-03. *Carb. Poly.*, 111: 947–954.

Zhang, Y., Venkitasamy, C., Pan, Z. and Wang, W. (2013). Recent developments on umami ingredients of edible mushrooms—A review. *Trends Food Sci. Tech.*, 33: 78–92.

CHAPTER 3

God's Mushroom
(Agaricus subrufescens)

Parthasarathy Seethapathy,[1,] Praveen Thangaraj,[2]*
Anu Pandita,[3] Subbiah Sankaralingam[4] and Deepu Pandita[5,]*

1. Introduction

Mushrooms are macro fungi that are saprophytic and produce prominent sporocarps (fruiting bodies). A mushroom generates big sporocarp, which is regarded as a popular food due to its low caloric content and high mineral, fibre and protein content (Firenzuoli et al., 2008). Around AD 600, historical documentation of the intentional cultivation of edible mushrooms began. However, the identification of new species or strains has yielded little results and mushroom experts are devoting considerable effort to domesticate more strains. Since 1900, *Pleurotus ostreatus*, the first species of edible mushroom, has been cultivated, followed *by Pleurotus sajor-caju* in 1974. Both species have been traditionally utilised around the world as healthy food and as medicinal sources, including antioxidant activity (Kumari et al., 2011). The output of *Agaricus* button mushrooms ranks top in India and globally. More than 3,000 species are recognised as edible mushrooms (Rajaratnam et al., 2012). There are minimal carbohydrates and fats in button mushrooms. Still, they are rich in proteins, vitamins A and D, vitamin B_5, vitamin B_{12}, vitamin C and minerals like phosphorus, potassium, iron, etc. In addition, the mushroom contains vital vitamins, fibres, and amino acids, particularly L-lysine and L-tryptophan, for treating diabetes and cardiac patients. Identifying edible mushrooms correctly is essential for productive exploration. All macro-basidiomycete edible fungi are rich in vitamin B, vitamin C and ergosterol

[1] Department of Plant Pathology, Amrita School of Agricultural Sciences, Amrita Vishwa Vidyapeetham, Coimbatore, 642109, Tamil Nadu, India.
[2] Department of Plant Pathology, Tamil Nadu Agricultural University, Coimbatore, 641003, Tamil Nadu, India.
[3] Vatsalya Clinic, Krishna Nagar, New Delhi, 110051, India.
[4] PG and Research Department of Botany, Saraswathi Narayanan College, Madurai, 625022, India.
[5] Government Department of School Education, Jammu, Jammu and Kashmir 180001, India.
* Corresponding authors: spsarathyagri@gmail.com, deepupandita@gmail.com

(Wisitrassameewong et al., 2012). Recently, scientists from all over the world have placed a greater emphasis on identification of biomolecules from edible mushrooms, as they become attractive for the development of new products for health therapy and are also viewed as a potential alternative for medicinal applications, with a particular emphasis on their significance in anticancer treatment (Ramberg et al., 2010; Sweet et al., 2013). Wasser (2002) examined the antitumour and immunomodulatory effects of bioactive substances extracted from numerous edible mushrooms. In addition, edible mushrooms have attracted many antifungal, antibacterial, antiviral, antioxidant, antidiabetic, anti-allergic, antimitogenic, anti-hypertensive and anti-hypercholesterolemic properties based on the various characteristics of their medicinal value. They have higher applications in antitumour and anti-inflammatory activity. Over the last few decades, numerous new taxa of mushrooms have evolved, mainly for therapeutic and nutritional purposes. One form of this kind of mushroom is the sun mushroom, also known as *Agaricus subrufescens* Peck, *Agaricus blazei* (Murrill) ss. Heinemann, *Agaricus brasiliensis* Wasser et al. All of these names have been attributed to the same species. For a significant number of years, this species has been grown in Brazil on a more modest scale.

Recent emphasis has been placed on the edible mushroom *Agaricus subrufescens* Peck, its usage for medicinal benefits, production practices, nutritional composition and bioactive substances for therapeutic activities, rendering it a matter of interest for research and discussion. *A. subrufescens* culture practices and production follow the same procedure as *Agaricus bisporus* cultivation practices. Despite this production, the button mushroom has substantially shorter cultivation time and a much larger yield than other mushrooms (Giménez et al., 2020). Consequently, the edible *A. subrufescens* mushroom contributes to its research on taxonomy, distribution, growing practices and formulation development of new nutritional and medicinal products based on fungi.

2. History and Significance

A. subrufescens was first reported by the New York state botanist C.H. Peck in 1893; the mushroom was early called the 'God's mushroom' and 'almond mushroom' due to its sweet with almond flavoured taste, and cultivated worldwide to be consumed in the Atlantic region of the United States, from the late 19th to 20th century (Kauffman, 1918; Kerrigan, 2005). After a few decades, the commercial cultivation and production of *A. subrufescens* severely dropped as market trends shifted; moreover, the edible button mushroom species *Agaricus bisporus* appears to have been farmed routinely in the United States (Charles, 1931). In 1960, *A. subrufescens* was re-discovered in Brazil and dubbed as 'Piedade mushroom', which originated in the province of Sao Paolo. The fungus was taken to Japan by Furumoto in 1965 to study the medicinal and therapeutic qualities of *A. subrufescens*. The discovered mushroom was recognised as *Agaricus blazei* Murrill obtained by the Belgian botanist P. Heinemann in 1967. Other names for *A. blazei* mushroom are Himematsutake in Japan, medicinal mushroom or sun mushroom in Brazil, and royal sun *Agaricus* in other countries. Further, the cultivation methods of *A. blazei* were abandoned in Japan and *A. subrufescens* became a crucial edible mushroom for its export of medical importance, thus fetching better prices for Brazil in comparison to other edible button mushrooms (Souza Dias et al., 2004).

3. Taxonomy

A. subrufescens, an edible mushroom, is a gilled sporocarphic macro fungus belonging to the family Agaricaceae, order Agaricales, phylum Basidiomycota (Firenzuoli et al., 2008). It was initially discovered in North America and later in South America. However, the mushroom *A. subrufescens* (= *Psalliota subrufescens* Kauff. = *Agaricus rufotegulis* Nauta) originated in forests throughout Europe, Hawaii and Thailand (Arrillaga and Parra, 2006; Dai et al., 2009; Wisitrassameewong et al., 2011). Kerrigan (2005) was the first to characterise the taxonomic characteristics of *A. subrufescens*, followed by Firenzuoli et al. (2008) and Wisitrassameeewong et al. (2011). Didukh and Wasser rejected the *A. blazei* species designation in 2002, which was subsequently dubbed the Brazilian fungus *A. brasiliensis* (Kerrigan et al., 2005). Richard Kerrigan also conducts genetic and interfertility testing on other fungal strains of *Agaricus* (Wasser et al., 2002) and reports that samples of the Brazilian strains *A. blazei* and *A. brasiliensis* are genetically related to *A. subrufescens*. *A. subrufescens* basidioma exhibited highly varied morphological characteristics. Sporocarps appear robust due to the genotype and the environment (Kerrigan, 2005). During the button stage, the cap's width ranges between 20–70 mm, while in the mature stage, it grows to 60–150 mm. The pileus of *A. subrufescens* is dirty white and occasionally appears brownish-gold or rusty brown. The basidiospores are chestnut brown in colour.

4. Distribution

Over the past many decades, there have been continuous arguments regarding the origin, dispersion and naming of this fungus (da Eira et al., 2005). The cultivation history of *A. subrufescens*, also known as the almond, sun or therapeutic mushroom, can be divided into three different periods. According to Kerrigan (2005), between 1894 and 1918, the *A. subrufescens* species was produced in the United States in 1894. The second era spanned from 1965 to 1997, according to studies by Mizuno (1997). They reported mushroom production in Brazil, China, Japan and Korean countries and provided scientific breakthroughs in mushroom varietal development. The ultimate era started with publishing the use of *A. subrufescens* in Brazil (Colauto et al., 2002), which also permitted a rise in the number of research groups to investigate this mushroom in a wider variety of countries, including Argentina, Canada, France, Mexico, Norway, Slovenia, Taiwan and this continues to this day.

5. Nutritional Properties and Pharmacological Values

Since its identification in 1893, this fungus has been cultivated globally, especially in Brazil, where numerous strains of *A. subrufescens* have been established. They are promoted as a supplemental and beneficial alternative therapy. In nations where vegetarianism predominates, edible mushrooms are chosen as food. Mushrooms, sometimes known as 'white veggies' or 'vegetarian meat without bones', are an excellent source of protein, vitamins and fibre and are helpful in medicinal therapy (Thakur and Singh, 2013). On a wet weight basis, mushrooms typically include 85–95% water, 3% protein, 4% carbohydrates, 0.1% fats and 1% minerals and vitamins (Tewari, 1986; Thakur, 1998). Additionally, 19–35% protein was produced on a dry weight basis, compared to 7.3% for rice, 13.2% for wheat and 25.2% for milk (Chang, 1989). The constitution of *A. subrufescens* shows a low percentage of water, crude fat, crude fibre and ash, but a high percentage of protein, carbohydrate content and the average value of accessible carbohydrates, as well as a high value for energy. The sporocarps of *A. subrufescens* are in high concentrations, packed with beneficial elements, such as phosphorus, potassium, calcium, magnesium and zinc (Györfi et al., 2010). Despite this, a trace amount of cadmium could also be found in the sample. Due to the presence of vital minerals and trace elements, mushrooms are gaining increasing relevance (Chadha and Sharma, 1995), with excellent nutritional and therapeutic value throughout the world, particularly in China (Li, 2012). Nutrient-rich compost is essential for growing *A. subrufescens* and determines the quality of mushroom output (Pardo-Gimenez et al., 2016). However, the nutritional properties of *A. subrufescens* are obscure, particularly in estimating the crude fat and cellulose percentages. The proximate amounts of nutrients produced by of *A. subrufescens* are listed in Table 1. However, the nutritional composition of protein in *A. subrufescens* was very high compared to *A. bisporus*, *A. sylvaticus*, *Pleurotus* sp., *Volvariella volvacea* and *Lentinus edodes* (Fig. 1) (Vinhal Costa Orsine et al., 2012; Machado et al., 2016; Atila et al., 2021; Zied et al., 2017).

The pharmacological benefits of *A. subrufescens* are well-known and are utilised as treatment for multiple disorders in various countries. Additionally, it is regarded as an alternate method of curing diseases. In the past, many authors have made significant contributions in creating a wide variety of pharmaceutical products. They have shown that the functions of biomolecules cover a broad range of activities. It is believed that mushroom polysaccharides suppress tumour genesis, have demonstrated an effective anti-tumour effect against a variety of allogeneic and syngeneic malignancies and limit the progression of tumours to other parts of the body (Ohno et al. 2001). The therapeutic properties associated with *A. subrufescens* (Fig. 2) help to treat patients

Table 1. Proximate macro nutrient of *A. subrufescens*.

Nutrients	Content (g kg^{-1} of Dry Matter)
pH (1:5, w/v)	7.63
Moisture	664.7
Total nitrogen	19.7
Protein	123.1–237.2
Ash	237.2
Organic matter	762.8
Carbon/Nitrogen	22.5
Crude fibre	330
Crude fat	2.5–307.2
N-free extracts	307.2
Total carbohydrates	637.2
Hemicellulose	123.4
Cellulose	204.8
Lignin	208.8
Neutral-detergent sol. fibre	225.8

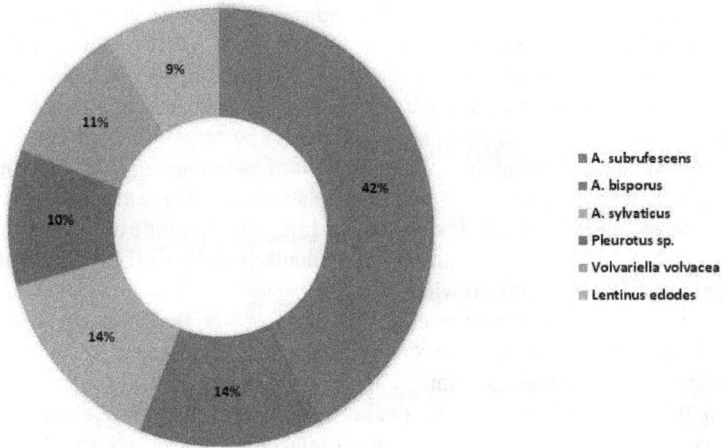

Fig. 1. Nutritional composition of protein in edible mushrooms.

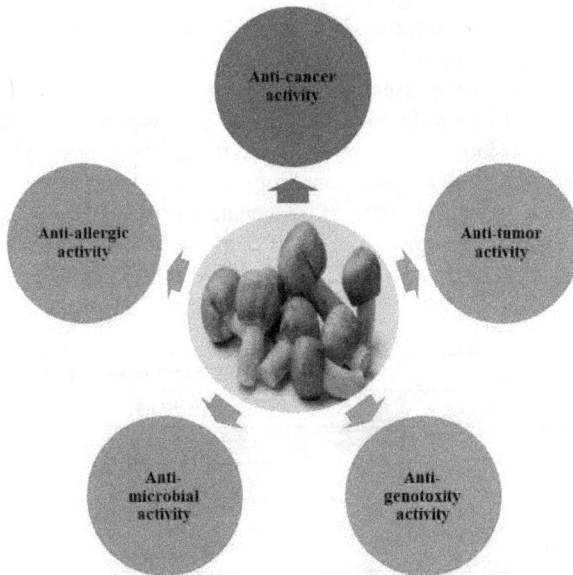

Fig. 2. Therapeutic activity of *A. subrufescens*.

with cancer and immunodeficiency (Firenzuoli et al., 2008). Blazein is a steroid component that was discovered in *A. subrufescens*. It was named after the synonym for *A. subrufescens, A. blazei*. Blazein has an experimental anticancer action on human lung cancer LU99 cells. However, it does not impact the normal lymphocytes (Itoh et al., 2008). Extracts of *A. subrufescens* can potentially be cytotoxic to cancer cells *in vitro*. They can also suppress the growth of human leukaemia cells by inducing apoptosis, which is a process that directly inhibits the growth of cancer cells *in vitro* (Kim et al., 2009). The proteoglycan enriched fraction of *A. subrufescens* which had a ratio of polysaccharides to peptides as 74:26%, triggered apoptotic cell death in leukemic U937 cells. This was accomplished by downregulating Bcl-2, activating caspase-3 and cleaving poly (ADP-ribose) polymerase. Using heterologous U937 cell lines with Bcl-2-comprising plasmid DNA showed that the significant regulatory authorities of the *A. subrufescens* derived designed to induce apoptosis are Bcl-2 and caspace-3 associated with the dephosphorylation of the Akt cascade signalling pathway. This was achieved by enhancing the Bcl-2 levels of expression (Jin et al., 2007). It would appear from this that preparations of polysaccharide-enriched fractions of *A. subrufescens* have a powerful preventive effect on DNA damage brought on by known genotoxic agents. It was shown that patients using high quantities of *A. subrufescens* extracts (11.5 mg/ml) considerably decreased the amount of DNA damage, showing a preventive effect towards varying concentrations of diethylnitrosamine (DEN)-induced liver genotoxicity (Barbisan et al., 2003).

Table 2. Therapeutic effects of *A. subrufescens*.

S. No.	Therapeutic Effects	Mechanisms	References
1.	Antitumour activity	In addition to combating allogeneic and syngeneic malignancies, the mushroom also prevents tumour metastasis.	Kimura et al., 2004
		In vitro, the mushroom is cytotoxic to tumour cells, inhibits the growth of human leukemia cells via causing apoptosis and inhibits tumour cell proliferation directly.	Kim et al., 2009; Jin et al., 2007
2.	Anticancer activity	To strengthen the immune system in order to inhibit or eliminate infections and cancer cells.	Gonzaga et al., 2005; Barbisan et al., 2002; Akiyama et al., 2011
3.	Anti-genotoxicity activity	Reduce DNA damage, demonstrating a protective effect against liver genotoxicity generated by diethylnitrosamine (DEN).	Barbisan et al., 2002
4.	Anti-microbial activity	Enhances their effectiveness against hepatitis B virus (HBV) and foot-and-mouth disease (FMDV).	Gonzaga et al., 2009; Bernardshaw et al., 2006; Faccin et al., 2007; Firenzuoli et al., 2008; Hetland et al., 2008
		Suppresses *Streptococcus pneumoniae* infection in rats, inhibit virus plaque development in cell culture, and display antiviral efficacy against poliovirus type 1 and herpesvirus.	Bernardshaw et al., 2005; Bruggemann et al., 2006; Faccin et al., 2007
5.	Anti-allergic activity	The activation of macrophages by epithelial cells that promote the development of lymphocyte-T cells into Th1 cells in mice appears to have no pro-inflammatory effects.	Bouike et al., 2011
6.	Antidiabetic activity	Anti-hyperglycemic, anti-hypercholesterolemic, anti-hypertriglyceridemic and anti-arteriosclerotic actions are promoted.	Kim et al., 2005a, 2005b

Biologically active substances isolated from *A. subrufescens* cultures are recognised to treat numerous disorders. As a result, the mushroom is known as a medicinal mushroom and is frequently consumed for the prevention of cancer cells, diabetes, hyperlipidemia, atherosclerosis and chronic hepatitis, as well as to activate the immune system (Takaku et al., 2001; Firenzuoli et al., 2008; Levitz, 2010). Bioactive compounds, such as ergosterol and β-glucans, are ubiquitous and considered as biochemical markers. The bioactive compounds isolated and identified from the mushroom *A. subrufescens* are polysaccharides, namely riboglucans, β-glucans and glucomannans (Hikichi et al., 1999; Cho et al., 1999; Fujimiya et al., 1998; Gonzaga et al., 2005). The polysaccharides compositions produced by *A. subrufescens* are represented in Fig. 3, based on the publication (Volman et al., 2010).

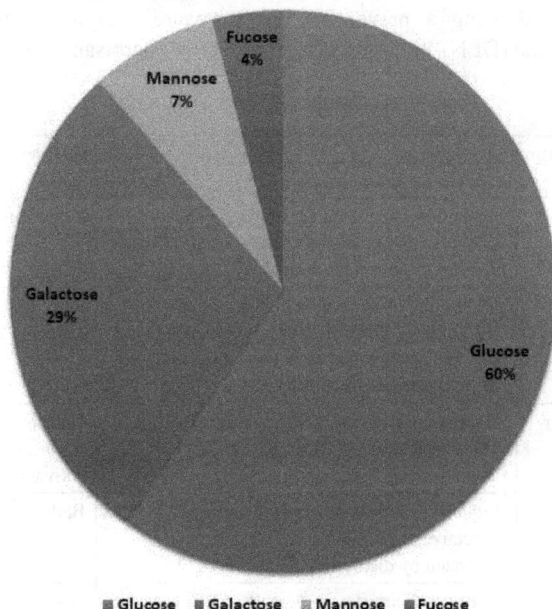

■ Glucose   ■ Galactose   ▨ Mannose   ■ Fucose

Fig. 3. Composition of polysaccharides from *A. subrufescens.*

6. Commercial Cultivation

The cultivation of mushrooms has shown a significant increase in the variety of mushrooms grown and the yield of these mushrooms in recent years. Because they are low in fat and calories while rich in crude fibre, proteins and beneficial vitamins, mushrooms are a very nutritious food that can also be used as a tonic and a form of medicine. In addition to this, they have therapeutic characteristics that are multi-functional and effective against a variety of ailments. Because the raw materials required for mushroom cultivation are inexpensive and easy to obtain, mushroom agriculture has many potentials to expand in Asia. *A. subrufescens* is cultivated commercially, using diverse compost additions, a casing layer and the standardisation of various temperature manipulations for primordia and button initiation. To increase mushroom productivity, agro-waste supplementation involves introducing high nitrogen-containing materials into the compost during spawning or casing. However, compost supplementation offers sufficient nutrients for mushroom germination, resulting in more excellent mushroom production within a shorter cultivation cycle

Table 3. Chemical classes and their health benefits.

Chemical Classes	Compounds	Health Benefits	References
Polysaccharides	β-(1–6)-linked β-D-glucopyranosyl	Treatment of characteristic metabolic syndrome.	Liu et al., 2011; Liu and Sun, 2011
	Sodium pyroglutamate	Immunomodulator action along with anti-angiogenic, anti-tumour and anti-metastatic properties.	Kimura et al., 2004
	Lectin, a glycoprotein with β-(1–6)-glucan	Anti-tumour activity.	Kawagishi et al., 1990
	Agaritine	Anti-tumour activity in leukemic cells.	Endo et al., 2010; Akiyama et al., 2011
	α-(1–4)-; β-(1–6)-glucan	Kill cancer cell activation and apoptosis.	Fujimiya et al., 1998
	α-(1–6)-; α-(1–4)-glucan	Antitumour activity.	Mizuno et al., 1990
	β-(1–6)-; β-(1–3)-glucan	Antitumour activity.	Mizuno et al., 1990
	β-glucan	Antitumour effect against Sarcoma 180 cells.	Ohno et al., 2001; Gonzaga et al., 2009
	Proteoglycan	Induced apoptosis in human leukemic U937 cells.	Jin et al., 2007
	LMPAB	Antitumour activity.	Niu et al., 2009
Sterols	Ergosterol	Anti-tumour activity.	Takaku et al., 2001; Jasinghe and Perera, 2005; Lindequist et al., 2005; Volman et al., 2010
	Blazein	Anticancer efficacy against lung cancer in humans LU99 cells.	Itoh et al., 2008
Proteins	Caspase-3	Anti-proliferative and anti-angiogenic strategies against the development of prostate tumour.	Yu et al., 2009

and enhanced biological efficiency (Arce-Cervantes et al., 2015; Pardo-Giménez et al., 2016). The almond mushroom can only be grown on farms with many crops in various stages of development and enclosed in structures for protection. The cereal straw-produced raw materials, such as wheat, barley or rice are the most common sources of carbohydrates and lignin. These bulk materials have low nitrogen content. Thus, the compost is augmented with animal manure from chicks or turkeys and any other available nitrogen source. Gypsum is used so that nitrogen may be preserved, the pH can be balanced and it can act as a flocculent. These disparate components are combined, given moisture and then composted for two to three weeks. After this interval, the compost is heated to a temperature high enough to pasteurise it and is then continued to be composted under stringent environmental regulations. When there is no longer any ammonia present, the process of composting is terminated. According to the findings presented in this paper's research, the substrate in question now has the potential to have extra nutritious items thoroughly incorporated into it and used as a supplement. This substrate has spawned the mushroom's mycelia that have been colonised on sterile grain and mixed into it. Mycelia invade the substrate for two to three weeks. The substrate can also require some more nutrients when this growth phase is complete. The surface of the substrate that has been colonised is then top-dressed or cased over a substratum that has been buffered with lime to assist in the production of mushroom primordia. Alterations in the substrate, air temperatures and the management of gaseous elements – all play a role in the commencement of the mushroom formation process, which can take anywhere from seven to 15 days to complete. After the mushrooms have formed and are ready to be sold, they are picked by hand. The growth of mushrooms occurs in cycles. The produce is scrapped once it has been grown for the allotted time and assessed before being sold for a profit.

In most cases, the growth of a God's mushroom closely resembles that of a button mushroom. Despite this, the cultivation cycle for button mushrooms is significantly shorter and the yield that can be obtained from them is significantly greater. Therefore, it is possible to cultivate both species using the same methodology; but the final product would not be comparable. In button mushroom production, the hybrid strain is grown preferentially because of the excellent quality and quantity of the basidiocarps. This is because hybrid lines include both parents' and offspring's genetic material (Fukuda et al., 2003). Ruffling in button mushroom production is the process in which the fungus *Agaricus* multiplies in the casing over its entire depth. This culture method permits more precise mushroom production during the first flush and uniform *Agaricus* production. A larger amount of mushrooms was collected, using the ruffing method which contributed to an improvement in the biological efficiency of the process. The optimal temperature for the commencement of sun mushroom fruiting varies (Pardo-Giménez et al., 2017). Several authors lowered the temperature range for primordia initiation to roughly 20–22°C. However, *A. subrufescens*' primordia commencement and harvest intervals are temperature sensitive (Wang et al., 2013). Win et al. (2018) farmed the edible button mushroom *A. subrufescens* commercially according to the mycelial density; compact and slightly compact. In 17 days, in woodchips, compared to 26 days in corncobs, *A. subrufescens* spawned. In the cultivation of *A. subrufescens*, sawdust substrates encouraged longer days for primordial initiation and fruiting body growth than woodchips and corncob substrates. They resulted in a mixture of 100% sawdust and compost. Interestingly, the mushroom size of *A. subrufescens* grown on compost was double that of mushrooms grown on compost that had not been added. Consequently, it was demonstrated conclusively that adding woodchips and corncobs to the primary substrates of compost combined with various agro-residues resulted in higher yields and superior quality almond mushrooms. *A. subrufescens* can also be cultivated commercially on a wheat straw-based substrate, using the same compost mixture as *A. bisporus*. A simple modification of the ruffling of the casing mixture with a constant incubation temperature throughout the crop periods saves energy during the summer. It improves crop yields with greater biological efficiency (Llarena-Hernández et al., 2014), introducing numerous products with added value to the market (Table 4).

Table 4. Products available in Indian markets.

Name of Product	Purpose	Cost (USD)
Agaricus subrufescens Extract and powder 60 capsules	Contains many minerals, such as potassium, zinc, iron, calcium and magnesium, vitamins and provitamins.	22.23
Pure *Agaricus blazei*	To support blood sugar regulation, digestion, and immunity.	130.00
Agaricus blazei powder capsule	Dietary supplements.	55.99
Agaricus extract capsules, organic	Dietary supplements.	33.49
Agaricus blazei extract 30% polysaccharides	Food supplements.	142.00
Agaricus subrufescens pure	Dietary supplements.	–
Agaricus subrufescens extract	Medicinal properties.	23.99
Agaricus blazei 650 mg, ABM Mushroom, *Agaricus Blazei* Murill, Vegan, without magnesium stearate, 120 capsules	Food supplements.	187.62
Agaricus Bio® wellness powder capsules	Promotes regulation of abnormal cell growth, supports strong immune system health, encourages healthy blood sugar and cholesterol levels and antioxidant activity.	–

7. Conclusion

Over many decades, *A. subrufescens* has seen widespread consumption as a protein source. As a result of its possible medical benefits, it has become an essential focus of investigation in a wide variety of fields of study. Scientists have demonstrated that many bioactive chemicals that have been found and characterised have the potential to be turned into therapeutic agents and employed. Furthermore, evidence from drug trials was always required to evaluate whether or not *A. subrufescens* provides any therapeutic effects. There is a potential for toxicological issues as well. Toxicological evidence should be considered, particularly with recognised bioactive chemicals, such as agaritine, blazein and *A. subrufescens* sporocarps; overall epidemiology data is required to connect consumption with any potential adverse effects. Concerns over its therapeutic value must specifically focus on the cytotoxicity and oncogenic potential of agaritine and its metabolites. *A. subrufescens* to be successful and it would have been great if it could have been demonstrated that it does not necessarily lead to potential toxicity. In the long term, *A. subrufescens* has the potential to be an excellent therapeutic medication.

References

Akiyama, H., Endo, M., Matsui, T., Katsuda, I., Emi, N., Kawamoto, Y., Koike, T. and H. Beppu. (2011). Agaritine from *Agaricus blazei* Murrill induces apoptosis in the leukemic cell line U937. *Biochimica et Biophysica Acta*, 1810: 519–525.

Arce-Cervantes, O., Saucedo-García, M., Lara, H.L., Ramírez-Carrillo, R., Cruz-Sosa, F. and Loera, O. (2015). Alternative supplements for *Agaricus bisporus* production and the response on lignocellulolytic enzymes. *Scientia Horticulturae*, 192: 375–380.

Arrillaga, P. and Parra, L.A. (2006). Elgénero Agaricus L. en España. XI. *Agaricus subrufescens,* primera cita para España. *Boletin Sociedad Micologica de Madrid,* 30: 201–207.

Atila, F., Owaid, M.N. and Shariati, M.A. (2021). The nutritional and medical benefits of *Agaricus bisporus*: A review. *Journal of Microbiology, Biotechnology and Food Sciences,* 281–286.

Barbisan, L.F., Miyamoto, M., Scolastici, C., Salvadori, D.M., Ribeiro, L.R., Eira, A.F. and Camargo, de J.L. (2002). Influence of aqueous extract of *Agaricus blazei* on rat liver toxicity induced by different doses of diethylnitrosamine. *Journal of Ethnopharmacology,* 83: 25–32.

Barbisan, L.F., Spinardi-Barbisan, A.L.T., Moreira, E.L.T., Salvadori, D.M.F., Ribeiro, L.R., Da Eira, A.F. and De Camargo, J.L.V. (2003). *Agaricus blazei* (Himematsutake) does not alter the development of rat diethylnitrosamine-initiated hepatic preneoplastic foci. *Cancer Science,* 94(2): 188–192.

Bernardshaw, S., Hetland, G., Grinde, B. and Johnson, E. (2006). An extract of the mushroom *Agaricus blazei* Murrill protects against lethal septicemia in a mouse model of fecal peritonitis. *Shock,* 25(4): 420–425.

Bernardshaw, S., Johnson, E. and Hetland, G. (2005). An extract of the mushroom *Agaricus blazei* Murrill administered orally protects against systemic *Streptpcoccus pneumoniae* infection in mice. *Scandinavian Journal of Immunology,* 62: 393–398.

Bouike, G., Nishitani, Y., Shiomi, H., Yoshida, M., Azuma, T., Hashimoto, T., Kanazawa, K. and Mizuno, M. (2011). Oral treatment with extract of *Agaricus blazei* Murrill enhanced Th1 response through intestinal epithelial cells and suppressed OVA sensitised allergy in mice. *Evidence-based Complementary and Alternative Medicine,* pii: 532180.

Bruggemann, R., Orlandi, J.M., Benati, F.J., Faccin, L.C., Mantovani, M.S., Nozawa, C. and Linhares, R.E.C. (2006). Antiviral activity of *Agaricus blazei* ss. Heinem extract against human and bovine herpes viruses in cell culture. *Brazilian Journal of Microbiology,* 37: 561–56.

Chadha, K.L. and Sharma, S.R. (1995). *Advances in Horticulture (Mushroom)*. Malhotra Publication House, New Delhi, 13: 649.

Chang, S.T. and Quimio, T.H. (1989). *Tropical Mushroom: Biology and Cultivation Methods.* The Chinese University Press, Hong Kong, p. 493.

Charles, V.K. (1931). Some common mushrooms and how to know them, *USDA circular No. 143.* Washington D.C.: Government Printing Office, p. 60.

Cho, S.M., Park, J.S., Kim, K.P., Cha, D.Y., Kim, H.M. and Yoo, I.D. (1999). Chemical features and purification of immunostimulating polysaccharides from the fruiting bodies of *Agaricus blazei. Korean Journal of Microbiology,* 27: 170–174.

Colauto, N.B., Dias, E.S., Gimenes, M.A. and Eira, A.F. (2002). Genetic characterization of isolates of506the basidiomycete *Agaricus blazei* by RAPD. *Brazilian Journal of Microbiology,* 33: 507131–133.

da Eira, A.F., Didukh, M.Y., De Amazonas, M.A.L. and Stamets, P.E. (2005). Is a widely cultivated culinary-medicinal *Royal Sun Agaricus* (Champignon do Brazil, or the Himematsutake mushroom) *Agaricus brasiliensis* Wasser, S. et al., indeed a synonym of *A. subrufescens* Peck? *International Journal of Medicinal Mushrooms,* 7(3): 1–7.

Dai, Y.C., Yang, Z.L., Cui, B.K., Yu, C.J. and Zhou, L.W. (2009). Species diversity and utilization of medicinal mushrooms and fungi in China (review). *International Journal of Medicinal Mushrooms,* 11(3): 287–302.

Dias, E.S., Abe, C. and Schwan, R.F. (2004). Truths and myths about the mushroom *Agaricus blazei. Scientia Agricola,* 61: 545–549.

Endo, M., Beppu, H., Akiyama, H., Wakamatsu, K., Ito, S., Kawamoto, Y., Shimpo, K., Sumiya, T., Koike, T. and Matsui, T. (2010). Agaritine purified from *Agaricus blazei* Murrill exerts antitumour activity against leukemic cells. *Biochimica Biophysica Acta,* 1800: 669–673.

Faccin, L.C., Benati, F., Rincão, V.P., Mantovani, M.S., Soares, S.A., Gonzaga, M.L., Nozaga, C. and Linhares, R.E.C. (2007). Antiviral activity of aqueous and ethanol extracts and of an isolated polysaccharide from *Agaricus brasiliensis* against poliovirus type 1. *Letter in Applied Microbiology,* 45: 24–28.

Firenzuoli, F., Gori, L. and Lombardo, G. (2008). The medicinal mushroom *Agaricus blazei* Murrill: Review of literature and pharmaco-toxicological problems. *Advance Access Publication,* 27: 3–15.

Fujimiya, Y., Suzuki, Y., Oshiman, K.I., Kobori, H., Moriguchi, K., Nakashima, H., Matumoto, Y., Takahara, S., Ebina, T. and Katakura, R. (1998). Selective tumoricidal effect of soluble proteoglucan extracted from the basidiomycetes, *Agaricus blazei* Murrill, mediated via natural kill cell activation and apoptosis. *Cancer Immunology, Immunotherapy*, 46: 147–159.

Fukuda, M., Ohno, S. and Kato, M. (2003). Genetic variation in cultivated strains of *Agaricus blazei*. *Mycoscience*, 44(6): 431–436.

Giménez, P.A., Pardo, J.E., Dias, E.S., Rinker, D.L., Caitano, C.E.C. and Zied, D.C. (2020). Optimization of cultivation techniques improves the agronomic behavior of *Agaricus subrufescens*. *Scientific Reports*, 10(1): 1–9.

Gonzaga, M.L.C., Bezerra, D.P., Alves, A.P.N.N., De Alencar, N.M.N., Mesquita, R.D.O., Lima, M.W., Soares, S.D.A., Pessoa, C., Moraes, M.O.D. and Costa-Lotufo, L.V. (2009). *In vivo* growth inhibition of Sarcoma 180 by an a-(1–4)-glucan-b-(1–6)-glucanprotein complex polysaccharide obtained from *Agaricus blazei* Murrill. *Journal of Natural Medicine*, 63: 32–40.

Gonzaga, M.L.C., Ricardo, N.M.P.S., Heatley, F. and Soares, S.D.A. (2005). Isolation and characterisation of polysaccharides from *Agaricus blazei* Murrill. *Carbohydrate Polymers*, 60: 43–49.

Györfi, J., Geösel, A. and Vetter, J. (2010). Mineral composition of different strains of edible medicinal mushroom *Agaricus subrufescens* Peck. *Journal of Medicinal Food*, 13(6): 1510–1514.

Hetland, G., Johnson, E., Lyberg, T., Bernardshaw, S., Tryggestad, A.M. and Grinde, B. (2008). Effects of the medicinal mushroom *Agaricus blazei* Murrill on immunity, infection and cancer. *Scandinavian Journal of Immunology*, 68: 363–370.

Hikichi, M., Hiroe, E. and Okubo, S. (1999). *Protein Polysaccharide 0041*. European Patent 0939082, 9 January 1999.

Itoh, H., Ito, H. and Hibasami, H. (2008). Blazein of a new steroid isolated from *Agaricus blazei* Murrill (Himematsutake) induces cell death and morphological change indicative of apoptotic chromatin condensation in human lung cancer LU99 and stomach cancer KATO III cells. *Oncology Reports*, 20: 1359 1361.

Jasinghe, V.J. and Perera, C.O. (2005). Distribution of ergosterol in different tissues of mushrooms and its effect on the conversion of ergosterol to vitamin D2 by UV irradiation. *Food Chemistry*, 92(3): 541–546.

Jin, C.Y., Moon, D.O., Choi, Y.H., Lee, J.D. and Kim, G.Y. (2007). Bcl-2 and caspase-3 are major regulators in *Agaricus blazei*-induced human leukemic U937 cell apoptosis through dephoshorylation of Akt. *Biological and Pharmaceutical Bulletin*, 30(8): 1432–1437.

Kauffman, C.H. (1918). *The Agaricaceae of Michigan. Michigan Geological and Biological Survey*. Lansing, Michigan, 1: 918.

Kawagishi, H., Kanao, T., Inagaki, R., Mizuno, T., Shimura, K., Ito, H., Hagiwara, T. and Nakamura, T. (1990). Formolysis of a potent antitumour (1–6)-b-D-glucan–protein complex from *Agaricus blazei* fruiting bodies and antitumour activity of the resulting products. *Carbohydrate Polymer*, 12: 393–403.

Kerrigan, R.W. (2005). *Agaricus subrufescens*, a cultivated edible and medicinal mushroom, and its synonyms. *Mycologia*, 97(1): 12–24.

Kim, C.F., Jiang, J.J., Leung, K.N., Fung, K.P. and Lau, C.B.S. (2009). Inhibitory effect of *Agaricus blazei* extracts on human myelomia cells. *Journal of Ethnopharmacology*, 122: 320–326.

Kim, G.Y., Lee, M.Y., Lee, H.J., Moon, D.O., Lee, C.M., Jin, C.Y., Choi, Y.H., Jeong, Y.K., Chung, K.T., Lee, J.Y., Choi, I.H. and Park, Y.M. (2005a). Effect of water-soluble proteoglycan isolated from *Agaricus blazei* on the maturation of murine bone marrow-derived dendritic cells. *International Immunopharmacology*, 5: 1523–1532.

Kim, Y.W., Kim, K.H., Choi, H.J. and Lee, D.S. (2005b). Antidiabetic activity of b-glucans and their enzymatically hydrolysed oligosaccharides from *Agaricus blazei*. *Biotechnology Letters*, 27: 483–487.

Kimura, Y., Kido, T., Takaku, T., Sumiyoshi, M. and Baba, K. (2004). Isolation of an anti-angiogenic substance from *Agaricus blazei* Murrill: Its antitumour and antimetastatic actions. *Cancer Science*, 95: 758–764.

Kumari, D., Reddy, M.S. and Upadhyay, R.C. (2011). Nutritional composition and antioxidant activities of 18 different wild Cantharellus mushrooms of northwestern Himalayas. *Food Science and Technology International*, 17(6): 557–567.

Levitz, S.M. (2010). Innate recognition of fungal cell walls. *PLoS Pathogens*, 6(4): e1000758.

Lindequist, U., Niedermeyer, T.H.J. and Jülich, W.D. (2005). The pharmacological potential of mushrooms. *eCAM*, 2(3): 285–299.

Liu, J. and Sun, Y. (2011). Structural analysis of an alkali-extractable and water-soluble polysaccharide (ABP-AW1) from the fruiting bodies of *Agaricus blazei* Murrill. *Carbohydrate Polymers*, 86: 429–432.

Liu, J., Zhang, C., Wang, Y., Yu, H., Liu, H., Wang, L., Yang, X., Liu, Z., Wen, X., Sun, Y., Yu, C. and Liu, L. (2011). Structure elucidation of a heteroglycan from the fruiting bodies of *Agaricus blazei* Murrill. *International Journal of Biological Macromolecules*, 07: 003.

Llarena-Hernández, C.R., Largeteau, M.L., Ferrer, N., Regnault-Roger, C. and Savoie, J.M. (2014). Optimisation of the cultivation conditions for mushroom production with European wild strains of *Agaricus subrufescens* and Brazilian cultivars. *Journal of the Science of Food and Agriculture*, 94(1): 77–84.

Machado, A.R.G., Teixeira, M.F.S., de Souza Kirsch, L., Campelo, M.D.C.L. and de Aguiar Oliveira, I.M. (2016). Nutritional value and proteases of *Lentinus citrinus* produced by solid state fermentation of lignocellulosic waste from tropical region. *Saudi Journal of Biological Sciences*, 23(5): 621–627.

Mizuno, T. (1997). Breeding and cultivation of medicinal mushroom. *Food Rev. Int.*, 13: 383–390.

Mizuno, T., Hagiwara, T., Nakamura, T., Ito, H., Shimura, K., Sumiya, T. and Asakura, A. (1990). Antitumour activity and some properties of water-soluble polysaccharides from 'Himematsutake', the fruiting body of *Agaricus blazei* Murrill. *Agricultural Biology and Chemistry*, 54: 2897–2906.

Ohno, N., Furukawa, M., Miura, N.N., Adachi, Y., Motoi, M. and Yadomae, T. (2001). Antitumour-b-glucan from the cultured fruiting body of *Agaricus blazei*. *Biological and Pharmaceutical Bulletin*, 24(7): 820–828.

Orsine, J.V.C., Novaes, M.R.C.G. and Asquieri, E.R. (2012). Nutritional value of *Agaricus sylvaticus*; Mushroom grown in Brazil. *Nutricion Hospitalaria*, 27(2): 449–455.

Pardo-Giménez, A., Catalán, L., Carrasco, J., Álvarez-Ortí, M., Zied, D. and Pardo, J. (2016). Effect of supplementing crop substrate with defatted pistachio meal on *Agaricus bisporus* and *Pleurotus ostreatus* production. *Journal of the Science, Food and Agriculture*, 96: 3838–3845.

Pardo-Giménez, A., Pardo-González, J.E. and Zied, D.C. (2017). Casing materials and techniques in *Agaricus bisporus* cultivation. pp. 149–174. In: Zied, D.C. and Pardo-Gimenez, A. (eds.). *Edible and Medicinal Mushrooms: Technology and Applications*. Wiley-Blackwell, West Sussex, England.

Rajaratnam, S. and Thiagarajan, T. (2012). Molecular characterisation of wild mushroom. *Eur. J. Exp. Biol.*, 2(2): 369–373.

Ramberg, J.E., Nelson, E.D. and Sinnott, R.A. (2010). Immunomodulatory dietary polysaccharides: A systemic review of the literature. *Nutrition Journal*, 9: 54–76.

Sweet, E.S., Standish, L.J., Goff, B. and Andersen, M.R. (2013). Adverse events associated with complementary and alternative medicine use in ovarian cancer patients. *Integrative Cancer Therapies*, 12(6): 508–516.

Takaku, T., Kimura, Y. and Okuda, H. (2001). Isolation of an antitumour compound from *A. blazei* Murrill and its mechanism of action. *American Society for Nutritional Sciences*, 1409–1413.

Tewari, R.P. (1986). Mushroom cultivation. *Extension Bulletin. Indian Institute of Horticulture Research*, Bangalore, India, 8: 36.

Thakur, M.P. (1998). Food and medicinal values of mushrooms. pp. 107–119. In: Puri, S. and William, W.J. (eds.). *Health Care and Development of Medicinal Plants*. Baba Printers, Raipur.

Thakur, M.P. and Singh, H.K. (2013). Advances in the cultivation technology of tropical mushrooms in India. *JNKVV Res J.*, 48(2): 120–135.

Volman, J.J., Helsper, J.P.F.G., Wei, S., Baars, J.J.P., Van Griens-ven, L.J.L.D., Sonnenberg, A.S.M., Mensink, R.P. and Plat, J. (2010). Effect of mushroom-derived b-glucan-rich polysaccharide extracts on nitric oxide production by bone marrow-derived macrophages and nuclear factorjB transactivation in Caco-2 recepter cells: Can effects be explained by structure? *Molecular Nutrition and Food Research*, 54: 268–276.

Wang, J.T., Wang, Q. and Han, J.R. (2013). Yield, polysaccharides content and antioxidant properties of the mushroom *Agaricus subrufescens* produced on different substrates based on selected agricultural wastes. *Scientia Horticulturae*, 157: 84–89.

Wasser, S.P. (2002). Medicinal mushrooms as a source of antitumour and immunomodulating polysaccharides. *Applied Microbiology Biotechnology*, 60: 258–274.

Win, T.T. and Ohga, S. (2018). Study on the cultivation of *Agaricus blazei* (almond mushroom) grown on compost mixed with selected agro-residues. *Advances in Microbiology*, 8: 778–789.

Wisitrassameewong, K., Karunarathna, S.C., Thongklang, N., Zhao, R.L., Callac, P., Chukeatirote, E., Bahkali, A.H. and Hyde, K.D. (2011). *Agaricus subrufescens*: New to Thailand. *Chiang Mai Journal of Science*, 39(2): 281–291.

Wisitrassameewong, K., Karunarathna, S.C., Thongklang, N., Zhao, R., Callac, P., Moukha, S., Ferandon, C., Chukeatirote, E. and Hyde, K.D. (2012). *Agaricus subrufescens*: A review. *Saudi Journal of Biological Sciences*, 19(2): 131–146.

Yu, C.H., Kan, S.F., Shu, C.H., Lu, T.J., Hwang, L.S. and Wang, P.S. (2009). Inhibitory mechanisms of *Agaricus blazei* Murrill on the growth of prostate cancer *in vitro* and *in vivo*. *Journal of Nutritional Biochemistry*, 20: 753–764.

Zied, D.C., Pardo, J.E., Tomaz, R.S., Miasaki, C.T. and Pardo-Giménez, A. (2017). Mycochemical characterisation of *Agaricus subrufescens* considering their morphological and physiological stage of maturity on the traceability process. *BioMed Research International*, 2017.

CHAPTER 4

Honey Fungus
(Armillaria mellea)

Bushra Hafeez Kiani

1. Introduction

The production of free radicals is the main cause of change in environmental factors which are having a strong impact on the health of humans as they can damage the cell structures, like DNA, cell membrane and proteins. Moreover, it may cause diseases like cardiovascular diseases, stroke, or some types of cancers that can prove deadly (Lobo et al., 2010). That is why recent scientific studies are focusing on new products that can have beneficial antioxidant and antimicrobial properties (Souilem et al., 2017). Before attaining scientifically-confirmed empirically-obtained results, using traditional medicines could be a solution for this issue. In present days, mushrooms are considered an important drug source that can have few side effects and can be used as functional foods (Gargano et al., 2017). Moreover, edible mushrooms are rich in nutrition and compounds that have medicinal properties; some of these compounds exhibit antioxidant, antimicrobial and cytotoxic properties (Toledo et al., 2016; Sokovic et al., 2017). This proves that they can be beneficial against new diseases (Roupas et al., 2012).

Armillaria (Basidiomycota, Physalacriaceae) is a genus that has ecological and economic importance. Wood and root rot fungi are distributed worldwide in different climates and affect as many as 500 host species (Hood et al., 1991). Most *Armillaria* species are pathogens that can cause *Armillaria* Root Rot (ARR) disease in hardwood and conifer trees, some herbaceous plants and shrubs. They are facultative necrotrophs that can colonise and kill the host root cambium and switch to the saprobic phase and decompose dead wood tissues. In the saprobic phase, *Armillaria* spp., can affect the white-rot of wood tissues by decomposition of components present in plant cell walls (Kile et al., 1991; Guillaumin et al., 2013).

Department of Biological Sciences (Female Campus), International Islamic University, Islamabad, 44000, Pakistan.
Email: bushrahafeez.kiani@gmail.com

Gastrodia (Tian-ma), the tuber of the orchid, *Gastrodia elata* Blume is being used as treatment for convulsions due to tetanus, stroke, headaches, or epilepsy since past times (Tang and Eisenbrand, 1992). Rhizome of *G. elata* has been in focus to separate active compounds and the main component that has been separated is gastrodin, which is a simple glycoside that contains glucose and 4-hydroxybenzyl alcohol. Moreover, other compounds of gastrodia, like vanillyl alcohol, vanillin (3-methoxy-4-hydroxybenzaldehyde) and 4-hydroxybenzaldehyde exhibit anticonvulsant properties (Ojemann et al., 2006). During clinical treatment of neurasthenia, headache and vertigo, gastrodin – the main component of Gastrodia – is administered through oral, intramuscular injection, or intravenous drip (Lu et al., 2002).

Most important habitat of *G. elata* is East Asia in the mountains of China and Korea. The number of Gastrodia species has decreased and could not fulfil the demands of herbal medication before two symbiotic fungi in the host *G. elata* were discovered in the 1960s (Xu and Guo, 2000). *Mycena osmundicola* (fungus) sprouts seeds and *Armillaria mellea* invades tubers that are below ground, except rootlets to bring up nutrients from the soil. When these two requirements are attained, Gastrodia cultivation increases significantly. It remains an expensive herb that has medicinal properties and is used in China where plant growth is slow and the demand is considerably high.

Medicinal components of Gastrodia are metabolites of *A. mellea* mushroom (Yang et al., 1984). Recent studies expressed the importance of *A. mellea* in different systems, like neural, immune and circulatory systems (Liu et al., 2003). An alternative biotechnological way to control Gastrodia resource issues could be batch fermentation of *Armillaria mycelia*. Much research has been carried out on *A. mellea* to understand its chemical constituents, clinical effects and pharmacology, but the number of review papers on this fungus are few. This review article aims to study the literature of culture, phytochemical and pharmacological aspects of *A. mellea* for the purpose of increasing the development of the application of *A. mellea* metabolites.

Armillaria mellea is a medicinal fungus that has been used for many years in East Asia. Polysaccharides that are separated from *A. mellea* have been seen to exhibit superoxide radical scavenging that has antioxidant (Siu et al., 2016) and anticancer properties due to mitochondrial apoptotic pathway and activation of a caspase cascade (Wu et al., 2012). According to previous data, *A. mellea* can be proved effective against neurodegenerative diseases, like Alzheimer's. It is mainly found in coniferous forests of the Northern Hemisphere (Roberts et al., 2011). *A. mellea* is a polymorphic species complex, comprising 40 described species, ranging from saprotrophic to pathogenic (de Mattos-Shipley et al., 2016). It has been used traditionally for treating headaches, neurasthenia, hypertension, insomnia, epilepsy and many other disorders (Zhang et al., 2015) and hyperglycemic effect is also one of its properties (Zavastin et al., 2015). This chapter summarises the importance of *A. mellea* with reference to functional food and as a nutraceutical (Fig. 1).

2. Distribution

Armillaria mellea is found in excess in Britain, Ireland, mainland Europe, although it is not common in Scandinavia but is common further south. It can also be found easily in parts of North America. The mycelial threads are the means by which *Armillaria*

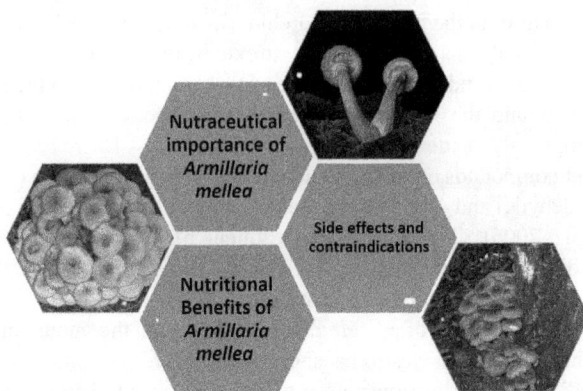

Fig. 1. Importance of *Armillaria mellea* (honey fungus) as a nutraceutical and functional food.

fungi can spread over a tree and move from one tree to another and black bootlace-like rhizomorphs (meaning literally 'root forms') are shown above, i.e., fungal hyphae in the form of parallel bunches.

Firstly, black rhizomorphs grow below the hardwood bark and then the bark falls to show threads of mycelial. Individual rhizomorphs of *Armillaria mellea* are 2 mm in size, but sometimes they combine to form substantial threads of approximately 5 cm in diameter. During their growth in soil, these rhizomorphs link with honey fungus mycelium of the infected tree with a new tree that can act as a host at a great distance. Maximum of 9 m length of rhizomorphs have been investigated (Pegler, 2000).

3. Cultivation

3.1 Solid-state Culture

As compared to different cultivation methods, the morphology of *A. mellea* is different. Rhizomorphs of *A. mellea* can be formed in solid media, growing through agar or over the surface of solid agar. Brown crustose aerial hyphae are formed when white fluffy aerial hyphae slowly turn brown (Hannson and Seifert, 1987). 3.5 pH and 22°C are the optimum conditions for the growth of fungus (Weitz et al., 2001). Biomass and morphology of *A. mellea* depend upon the media. Semi-solid media consisting of 2% glucose, 1% corn powder, 0.6% peptone and 0.5% agar are suitable for growing *A. mellea* (Cheng et al., 2006a).

3.2 Standing Liquid Culture and Shake Flask Culture

Submerged and large clusters of rhizomorphs can be found in standing liquid culture. Mycelium grow in pellets but no growth is observed in the shake flask. According to Tan et al. (2002), optimum initial temperature and nitrogen source in shake flask cultures was 5.0 and soybean cake powder and wheat bran, respectively. Optimum media components were suggested to maintain submerged culture conditions for the mycelial biomass of *A. mellea* (Zhang et al., 2001; Li et al., 2003; Cheng et al., 2007). 1.5% silkworm pupa powder, 1.5% soybean cake power, 1% ethanol, 1.0% glucose,

2.0% sucrose, 0.15% K2HPO4 and 0.075% MgSO4 are optimum values for media constituents of maximum mycelial biomass and polysaccharide production. Elicitors were used to increase the production of mycelial biomass and ethanol can be used to stimulate the growth of mycelia (Weinhold et al., 1963).

3.3 Sporocarp Formation

Armillaria hyphae, after asexually growing, could differentiate into sporocarp. Xue et al. (2004) tissue cultured sporocarp by providing optimum malt agar media and inducing it at 25°C in darkness and later culturing at 18°C with 90% humid environment. Cheng et al. (2006b) cultured sporocarp by inducing the fungus on brain glucose media containing 3% bran, 2% glucose, 0.3% dipotassium phosphate, 0.15% magnesium sulphate and 1.5% bioagar (w/v).

4. Constituents

4.1 Polysaccharides

Polysaccharides make up 4.7% of the *Armillaria* rhizomorphs in the stable phase with 9.24% in the logarithmic growth phase of the overall dry mass (Cheng et al., 2006a). Unrefined polysaccharides from the mycelia culture make up more than 10% (w/w) (Zhang et al., 2001). *Armillaria*-derived polysaccharides have been reported to have antioxidant properties (Yang et al., 2007), reduce motion sickness (Yu et al., 2006a), fight aging (Zhang et al., 2001) and enhance the immune response (Sun et al., 2009). Chen et al. (2001) conducted a detailed study on the polysaccharide content present in *Armillaria* at particular growth periods by extracting, isolating and purifying the rhizomorph, sporocarp, mycelium, along with the fermented media of *Armillaria*. It was reported that the molecular mass of the polysaccharides was 10–70 kg Dalton through size-exclusion chromatography. The polysaccharide obtained from the mycelium and the fermenting media was just glucose, while glucose and xylopyranose were reported in the sporocarp and rhizomorphs. Polysaccharides including D-rhamnose, D-mannopyranoside, D-glucopyranose, D-arabinose, D-galactopyranose and beta-D-fucopyranoside were reported from the fermented media of *Armillaria* with a total mass of 665 Kda in mole ratios of 0.42:1:17.05:0.36:13.67:0.33 respectively (Kong et al., 2003). Another polysaccharide in the rhizomorph was reported to contain uronic acid, D-xylopyranose, D-galactopyranose, D-mannopyranoside and D-glucopyranose with a molecular weight of 138 kilodaltons (Shen and Hong, 1998). Additionally, Zhang et al. (1995) derived two polysaccharides from *Armillaria* mycelia. The former polysaccharide was neutral with hydrophilic containing D-glucopyranose, D-galactopyranose and D-mannopyranoside in a mole ratio of 5:1:1.7; whereas the latter was acidic and hydrophobic in nature, containing D-arabinose, D-xylopyranose, D-rhamnose, beta-D-fucopyranose and D-galactopyranose in the mole ratios of 1:2.7:2:1.7:2.8 (Zhang et al., 1995). Two other hydrophilic polysaccharides had been isolated by a group of researchers where the xylomannan contained 13 D-mannose sugar residues while the other polysaccharide was constituted of Maduro's, D-galactopyranose, 6-deoxy-L-galactopyranose and D-mannopyranoside in mole ratios of 2:6:2:1 (Bouveng et al., 1967).

4.2 Proteins

Armillaria sporocarp is made of a large number of proteins (2.3 g/100 g of dry weight) (Stasiak, 2008). Proteins have over half of the overall nitrogen content of fungi (19–39 g/100 g of dry weight) (Źródłowski 1995; Florczak and Lasota, 1995). Research shows that half of the nitrogenous content of *Armillaria* sporocarp contained albumin, a quarter of globulin, 10% of prolamine, and 5% of glutein (Karkocha, 1964).

4.3 Sesquiterpene Aryl Esters

Sesquiterpene aryl esters are important constituents of *Armillaria*. A novel bacteria-fighting sesquiterpenoids were discovered in cultures of *Armillaria* (Midland et al., 1982). Over 37 sesquiterpene aryl esters have been reported in the fungus since then. After the discovery of *Armillaridin* and *Armillarin* in 1984, other sesquiterpene aryl esters were discovered in *Armillaria*, such as Melleolide H, *Armillaribin*, and *Armillaricin* (Yang et al., 1984; Yang et al., 1989a; Yang et al., 1989b; Yang et al., 1990a; Yang et al., 1990b; Obuchi et al., 1990; Yang et al., 1991a; Yang et al., 1991b). Simultaneously, there were reports of 14 other novel sesquiterpene aryl esters in *Armillaria* that had fungus and bacteria-fighting abilities (Donnelly et al., 1984; Donnelly et al., 1985a; Donnelly et al., 1985b; Donnelly et al., 1986; Donnelly et al., 1987; Donnelly et al., 1990a; Donnelly et al., 1990b).

4.4 Sterols and Sphingolipids

Regardless of the low lipid content of *Armillaria* (1.8%) as compared to average fungal species (5–8%) there are reports of its use for medicinal purposes. Lumisterol is the sterol commonly present in Basidiomyceta organisms and was also reported in *Armillaria* as well as peroxyergosterol, indicating anticancer properties (Florczak et al., 2004; Muszynska et al., 2011). Current research shows the presence of a sphingolipid, armillaramide from *Armillaria* (Muszynska et al., 2011). Sphingolipids have not been thoroughly researched; however, they remain an important aspect of research for their role in the prevention of liver diseases and cancer due to their immunostimulant properties (Gao et al., 2001).

4.5 Fatty Acids

There are relatively low amounts of fats present in fungus; still, unsaturated fats make up two-thirds of the total fat content (Bernaś et al., 2006). Nearly 17 fatty acids were isolated from the sporocarps of *Armillaria* from which four were 18-carbon polyunsaturated fatty acids while 13 were saturated (Muszynska et al., 2011; Cox et al., 2006). Unsaturated fatty acids are crucial for the production of endoperoxide H_2 and epoprostenol—the two components of bile. These fatty acids also have an inhibiting effect on atherogenesis and are necessary for man (Muszynska et al., 2011; Bernaś et al., 2006).

4.6 Indolic Compounds

Research conducted on *Armillaria* sporocarp revealed the presence of indolic compounds, including tryptamine-d4 hydrochloride (2.73 mg per 100 g dry weight),

L-tryptophan (4.47 mg per 100 g dry weight) and serotonin (2.21 mg per 100 g dry weight) (Muszyńska et al., 2001; Zavastin et al., 2018). Such chemicals are commonly present in mycelia or sporocarp extracts of *Basidiomycota* species (Muszyńska et al., 2011; Muszyńska et al., 2009). Tryptamine-d4 hydrochloride functions as a non-selective receptor for enteramine and triple releasing agent (TRA) through interactions with monoamine oxidase inhibitors which could be deadly (Isbister et al., 2004). Some catabolites and anabolises of L-tryptophan have been shown to cause injury to the nervous system by developing disorders and illness over time, which is why caution should be observed while administering such medication (Stone et al., 2003).

Enteramine has an important function in the regulation of temperature, weight, appetite, cellular proliferation and sleep. Abnormality in enteramine concentrations could lead to alcohol addiction, anxiety, depressive disorder, obsessive-compulsive disorder and even suicide, which is why treatment of these problems in the majority of cases requires enteramine reabsorption by nerve cells (Muszynska et al., 2011; Chattopadhyay et al., 1996).

4.7 Biocatalysts

Lee et al. (2005) isolated *A. mellea* metalloproteinase involved in fibrinolysis from cultured *Armillaria* mycelium by using size-exclusion and ion-exchange chromatographic techniques. The metalloprotease could digest the A chain of fibrinogen and hydrolyse it with the help of its metallic co-factor. Additionally, Healy et al. (1999) purified a protease biocatalyst distinct from lysine from the *Armillaria* sporocarp which also had positive effects on fibrinolysis. Colak et al. (2007) reported that an unrefined enzymatic extract of *Armillaria* had polyphenol oxidative properties on homocatechol. A copper-containing oxidoreductase enzyme, laccase, had been reported in *Armillaria* as well (Xiao et al., 2002; Wu et al., 2001).

4.8 Minerals and Vitamins

Fungi contain large amounts of minerals, radionuclides and metallic compounds with high densities. Such compounds can be found in fungi predominantly present alongside busy and dirty roads (Svoboda et al., 2006; Kemp, 2002). The concentration of digestible minerals present in the fungi largely depends on the substrate, cap diameter and age of the fungus as the majority of mineral content is present in the fungal cap (Przybyłowicz and Donoghue, 1988; Muszynska et al., 2011). Research conducted by Falandysz et al., showed that metals, including sodium, potassium, calcium, magnesium, iron, manganese, lead, zinc, copper, silver, cadmium, mercury and selenium were present in *Armillaria mellea* fungus (Falandysz et al., 2002; Falandysz et al., 1992).

Vitamins are very important as they are involved in many biological processes of animals and humans. Fungal organisms are some of the most valuable resources of vitamin content, specifically B complex vitamins (Furlani and Godoy, 2008; Mattila et al., 2001). Some studies indicate that in comparison with *Basidiomycota*, *Armillaria mellea* has larger quantities of vitamin B_3 and thiamin but smaller quantities of riboflavin than other fungal organisms (Podlewsk, 2006; Majdańska, 2007).

4.9 Other Constituents

Watanabe et al., in 1990, derived N 6-substituted adenosine with properties of cerebral protection from *Armillaria mellea*. Gao et al. (2001) secluded sphingolipid armillaramide, 22, 23-dihydro ergosterol and peroxyergosterol from *Armillaria mellea*. Guo et al. (2007) derived triterpenes 3-hydroxyfriedel-3-en-2-one, fridelin and 3 alpha-hydroxy friedel-2-one along with steroids 6, 9-epoxy-ergosta-7, 22-dien-3-ol. peroxyergosterol, and ergosterin from *Armillaria mellea*. Shi et al. (1998) isolated 9, 11-dehydroergosterol peroxide from the sporocarp of *Armillaria mellea*.

5. Advantages of *Armillaria mellea*

It is only after 2018 that there have been *in vitro* experimental findings suggesting *Armillaria mellea* as a therapeutic agent. Although this fungus has been in use in Asian medicine for centuries, impartial trials are needed to verify the efficiency of *Armillaria mellea*. Many pieces of research show the presence of biologically active chemicals and molecules in the sporocarp of *Armillaria mellea* mushroom. These molecules include lipids, amino acids, sugars, minerals, starches and essential nutrients and have shown promising results as medicinal agents. However, opinions from a practicing doctor are needed as self-administration is not advised and full caution needs to be observed for the safety of the patient.

5.1 Antioxidant Properties

Polysaccharides derived from *Armillaria mellea* have been observed to have antioxidative action against free radicals at 1 mg/ml. These free radicals cause damage to genetic information, premature aging and various other maladies. Such *in vitro* antioxidative agents hold the viable solution to these problems but require further research (Chen et al., 2019).

5.2 Immunomodulation

B lymphocytes and T lymphocytes have important roles in acquired immunity, where the former is involved in antibody-mediated immunity while the latter in cell-mediated immunity. The effects of *Armillaria mellea* on these immune cells were studied *in vitro* and the experimental data deduced that this fungus contains a compound involved in inducing cellular division and caused increments in the level of both T cells and B cells with concanavalin A and lipopolysaccharides, respectively. Macrophages were also stimulated along with their phagocytic properties under controlled conditions (Chen et al., 2019). These findings may be used to develop a methodology to manipulate and modulate the defence system as per need.

5.3 Anticancer

The anticancer effects of a biologically active chemical, armillaridin, extracted from *Armillaria mellea*, have been studied on liver cancer cell lines under controlled conditions. This chemical had an inhibitory effect on the tumorous hepatocytes and is essential for preventing tumour growth. A cytotoxicity test deduced that armillaridin has the potential to cause apoptosis, thereby concluding that this *Armillaria mellea*-derived chemical exhibits important anticancer characteristics (Leu et al., 2019).

5.4 Antidepressant

Armillaria mellea-derived sesquiterpene has been employed for countering depressive disorder in mice models. This fungus had shown depression-fighting abilities as well as the ability to alter the levels of chemical transmitters, such as dopamine, involved in mice brains. A decrease in dopamine or its chemoreceptors leads to depressive disorder. Further studies are needed to fully understand the mechanism of action of *Armillaria mellea* against depression (Zhang et al., 2019).

5.5 Diabetes and Lipid Metabolism

Diabetes mellitus is a condition that results in increase of blood sugar concentrations and may damage the vascular epithelium, nerve cells and tissues if left unchecked. Rats, administered dexamethasone and lipid-rich diets, were diagnosed with type 2 diabetes. These rats were administered large quantities of *Armillaria mellea* extracts and showed betterment in resistance to insulin and a decrease in blood sugar concentrations. There was also an increased amount of lipase enzymes that break down lipids and reduce lipid concentrations in blood. Furthermore, fatty acids stopped accumulating in hepatocytes (Yang et al., 2018). This finding serves as an indicator that extracts of *Armillaria mellea* can function as both fat-metabolising and antidiabetic agents.

5.6 Reduce Pro-inflammatory Response

Xylosyl, derived from *Armillaria mellea*, acts as a repressor for mediators and cytokines involved in the pro-inflammatory response under laboratory conditions (Chang et al., 2018). This is an experimental proof that *Armillaria mellea* can be used to decrease inflammation and with additional investigations, it can be potentially consumed to manage inflammatory disorders, such as osteoarthritis and diabetes.

6. Nutraceutical Importance of *Armillaria mellea*

6.1 Impact on the Nervous System

Armillaria mellea has anticonvulsive activity; both its fermented and aqueous gastrodia extracts have been seen to increase resistance to seizures in mice injected with pentylenetetrazol (New Drug Group, 1977). *Armillaria mellea* polysaccharides are effective against motion dizziness (Yu et al., 2006a). AMG-1, derived from *Armillaria mellea* mycelium, is a thousand times stronger than adenosine present in cerebral protection (Watanabe et al., 1990). AMG-1 is effective on the A1 receptor presynapse to stop neurotransmitters from releasing. It is also effective in stopping neurogenic twitch response caused by the electric field stimulants. There was a decrease in the sensitivity to exogenous acetylcholine in the vas deferens of rats, indicating pre-synapse and post-synapse depression (Xiong and Huang, 1998).

Pills that contain mycelia of *Armillaria mellea* have been in use for a long time in Chinese medicine, especially for treating brain-related disorders, such as headaches, dizziness, seizures, sleeplessness, mental exhaustion, high blood pressure and idiopathic endolymphatic hydrops (Ojemann et al., 2006; Wu et al., 2012). Fermented extracts of *A. mellea* have been shown to have an antiepileptic action by increasing the resistance to seizures in mice that had been injected with pentylenetetrazol. On the other hand, adenosine derived from *A. mellea* mycelia stopped the neurogenic twitch response caused by electric field stimulants with both pre- and post-synapse depression. It was also reported to be 1000-fold more powerful than adenosine involved in cerebral protection (Yu et al., 2006). Polysaccharides extracted from *A. mellea* have been reported to improve dizziness caused by motion (Kalyoncu et al., 2010) (Fig. 2).

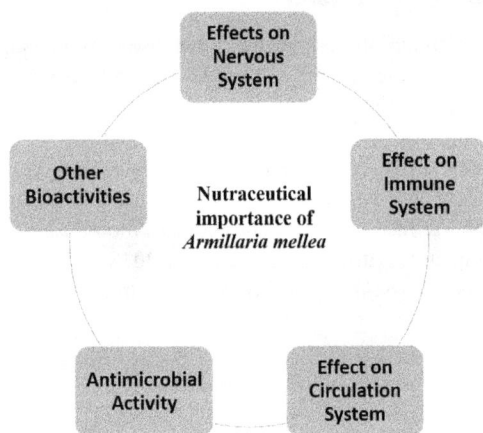

Fig. 2. Nutraceutical importance of *Armillaria mellea.*

6.2 Impact on the Defence System

It was discovered that polysaccharide AP, derived from *A. mellea*, could act as an immunostimulant in mice. AP acted by considerably enhancing the synthesis of serum hemolysin in both normal mice and immune-compromised mice that had been injected cyclophosphamide. AP also had a role in enhancing the immunity of body fluids and increasing the synthesis of spleen plaque-forming cells in mice that were normal. It was observed that in the laboratory, AP increased lymphocytic cell division in the spleen of normal mice; however, there were no effects on delayed cutaneous hypersensitivity to 2, 4-dinitro-1-chlorobenzene in normal mice It was additionally noted that AP led to an increase in the action of phagocytes present in the abdomen as well as the rate of clearance of charcoal in the bloodstream of normal mice (Lin et al., 1988). The findings of Dai et al. (2000) and Yu et al. (2001) are analogous to this experiment. *Armillaria mellea*-derived extracellular polymeric substances have also been reported to act as mice immune-response enhancers (Kong et al., 2007). Another study reported that polysaccharides derived from *Armillaria mellea* showed improvement in the mice's immunity through adjustments in the concentration of cytokines and other immune cells (Wang et al., 2007). Experiments conducted by Li et al. (2005) also showed the safeguarding abilities of polysaccharides against bone marrow destruction caused by cytophosphane in mice.

6.3 Impact on Circulatory System

Vertebrobasilar insufficiency had been affected by capsules containing chemicals derived from *Armillaria mellea* (Wang et al., 2007). These pills had the ability to inhibit platelet aggregation, increase blood flow to the brain and lower the relative viscosity of blood. Yu and Shen (2002) conducted experiments using *Armillaria mellea*-derived polysaccharides AMP-1 and AMP-2. It was observed that both can reduce blood sugar levels in diabetic mice induced with alloxan. AMP-1; on the other hand, it increased sugar tolerance in unaffected mice. There are also reports of using *A. mellea* pills to considerably lower blood cholesterol concentrations during patient trials (Zhang et al., 1983).

6.4 Aseptic Properties

Research shows that *Armillaria mellea* has aseptic properties (Sun, 2004; Yamac and Bilgili, 2006). It is proposed that sesquiterpene aryl esters may be responsible for the antimicrobial properties of the fungus. 4-O-methylmelleolide and judeol were both novel sesquiterpene aryl esters identified by scientists to show aseptic properties against Gram-positive bacteria strains (Donnelly et al., 1985). Obuchi et al. (1990) observed the inhibition of Gram-positive strains and yeast in the presence of armillaric acid. Melleolides K, L and M were discovered to have aseptic effects on Gram-positive bacterial strains, fungal species and yeast. Arnone et al. (1986) isolated antimicrobial sesquiterpenoids, melleolides B and D from *Armillaria mellea.*

6.5 Other Biological Roles

Armillaria mellea-derived glycan has antioxidative properties against hyper oxides (Yang et al., 2007). The Am 1 glycan has been discovered to act as an anti-mutagen in *Drosophila melanogaster* (Zhang et al., 2002). Age-defying properties have also been discovered in extracts of *Armillaria mellea* (Yu et al., 2006b).

7. Nutritional Benefits of *Armillaria mellea*

There is little research available on the nutritional profile of honey fungus. For the most part, the nutritional facts attributed to the species are inferred from knowledge of the nutrition of mushrooms in general (such as good vitamin D_2 content). However, studies performed on *Armillaria mellea* honey fungus specimens from Romania show the different values for various microelements (Zavastin et al., 2018). The same study also revealed the presence of selenium. Another study on *A. mellea* mushrooms sourced from Romanian forests revealed a high content of both iron and chromium. Calcium, magnesium, manganese, phosphorus, potassium and sodium were also identified in the mushroom cap and stipe or stalk (Table 1).

Table 1. Nutritional compounds in *Armillaria mellea.*

S. No.	Nutritional Components	Amount (g)
1	Carbohydrates	83
2	Proteins	2
3	Fats	2
4	Mannitol	6
5	Oxalic acid	0.5
6	Malic acid	7
7	Citric acid	0.6
8	Cinnamic acid	73
9	Fumaric acid	0.5
10	Copper	0.02
11	Iron	0.2
12	Zinc	0.07
13	Energy	7

7.1 Armillaria mellea *Health Benefits*

The following health effects are possible as a result of a reasonable consumption of honey fungus as part of an overall healthy, varied and balanced diet (Zsigmondy et al., 2015):

1. More regular bowel movements and constipation prevention and relief from dietary fibre.
2. Healthier gut flora and an increase in the number and diversity of lactic acid bacteria.
3. Good food for weight loss, thanks to its low energetic value (low-calorie content).
4. Source of vitamin D_2 with benefits for immunity, fertility, bone health and mental health.
5. Source of iron helps boost vitality and combat fatigue.
6. Varied mineral profile contributes to meeting nutritional demands.
7. Source of vitamin D, calcium, magnesium, phosphorus promotes strong bones and teeth.
8. Compounds in the honey fungus have been found to improve insulin sensitivity.
9. Exhibits anti-inflammatory activity and antioxidant and anticancer properties.

7.2 *Food Values of* Armillaria mellea

Armillaria mellea have sufficient nutritional values; however, they are not preferred by some and their hard stalks are normally avoided (Davis et al., 2012). A few people have shown hypersensitivity against *Armillaria mellea* that led to upset stomachs. It has been suggested to avoid collection of mushrooms attached to trees and plants, such as *Aesculus glabra*, Tasmanian blue gum tree, *Robinia pseudoacacia*, and *Conium maculatum.* The flavour profile of these mushrooms has been known to show slight nutty and sweet notes and different cooking methods result in different textures, varying from crispy to chewy. Some mushrooms are commonly blanched to remove the intense flavour prior to eating, which could potentially decrease the stomach-upsetting agents (Phillips et al., 2010). It has been suggested to only consume cooked mushrooms. Preserving mushrooms by drying can add to their flavour; however, they can be hard to consume once rehydrated (Kuo, 2007). Pickling and roasting mushrooms is also an option. Honey mushrooms on being cooked in soup and broth and their natural mucilage will help give body to the soups.

8. Side Effects and Contraindications

The edibility of honey fungus mushrooms is questioned. Some experts consider the species edible, but others classify it as conditionally-edible and warn about its potential for side effects. Some voices recommend honey fungi to be eaten only when cooked or after being parboiled to remove potentially toxic compounds, such as melleolides which have cytotoxic properties. In addition to honey fungi requiring cooking prior to consumption to render them safe for eating, it's sometimes advised to avoid mushrooms growing on certain trees, such as some horse chestnut species of locust trees, or even eucalyptus. The explanation is that the mushrooms may concentrate potentially harmful amounts of certain compounds with toxic properties since they feed off the trees.

8.1 Source of Heavy Metals

All mushrooms accumulate heavy metals from the surrounding environment and honey fungi too have the ability to do so. This makes it important to source them from pristine areas to avoid heavy metal contamination and subsequent side effects. Avoid collecting mushrooms growing by roadsides or in the vicinity of industrial areas or big cities.

8.2 Gastrointestinal Upset

Side effects, such as gastrointestinal upset with symptoms, such as abdominal cramps, nausea, vomiting sensation, loose stools, or diarrhoea have been reported. It is contraindicated to associate some *Armillaria* species with alcohol. The potential for an upset digestive system is also the reason why it is recommended to parboil honey fungi. Still, some people may experience indigestion symptoms following consumption.

9. Conclusion

Armillaria mellea is a common fungus that can be eaten and survives in varying environments all over the world. There are a plethora of uses and characteristics unique to this fungus and have been a focal point for research, ranging from morphological to habitual characteristics of *Armillaria*. There have been various publications indicating the biological and therapeutic benefits of *Armillaria* against numerous pathogens and disorders.

Researchers have centred their research on another use of fungi, known as mycoremediation. This bioremediation process involves the use of fungal species

Basidiomycota, such as *Armillaria mellea*, to degrade waste present in the environment. This area of research has not been thoroughly investigated yet and has a great potential to become a global phenomenon for combating environmental pollution.

10. Current and Future Developments

Armillaria mellea remains to be fully investigated, but present-day data indicate that *Armillaria mellea* is an unmatched organism for developing anticonvulsants as well as for discovering novel biologically-active compounds. The therapeutic components of *Gastrodia elata* host were majorly *Armillaria* catabolites and anabolites. A plausible scenario is that there is an exchange of DNA amongst the symbiont and host that leads to the production of a few commonly present metabolites throughout evolution. A need for genetic examination is present to unveil the mystery of mutualistic relationships amongst *Armillaria* and *Gastrodia elata*. The fungus *Armillaria mellea* is not cultivated in fields as its mycelia may lead to the spoilage of some plant roots (Aguín et al., 2006). Biotechnology is a practical solution to produce the biologically active compounds present in *Armillaria mellea* in large quantities as it allows optimising and improving the fungal strains. Bioreactor culture is crucial for producing bulk quantities of *Armillaria mellea* and furthering its scope on an industrial scale. Compounds derived from *Armillaria mellea* offer new windows for immunostimulant and antiepileptic therapies along with other medicinal uses. These compounds have to be further evaluated in the context of structure and mechanism of action.

References

Aguín, O., Mansilla, J.P. and Sainz, M.J. (2006). *In vitro* selection of an effective fungicide against *Armillaria mellea* and control of white root rot of grapevine in the field. *Pest Manage. Sci.*, 62: 223–228.

Arnone, A., Cardillo, R. and Nasini, G. (1986). Structures of melleolides B-D three antibacterial sesquiterpenoids from *Armillaria mellea*. *Photochemistry*, 25: 471–474.

Bernaś, E.G., Jaworska, Z. and Lisiewska, M. (2006). Edible mushrooms as a source of valuable nutritive constituents Acta *Scientarum polonorum*. *Technologia Alimentaria*, 5(1): 5–20.

Bouveng, H.O., Fraser, R.N. and Lindberg, B. (1967). Polysaccharides elaborated by *Armillaria mellea* tricholomataceae Part II. Water-soluble mycelium polysaccharides. *Carbohydr. Res.*, 4: 20–31.

Chang, C.C. et al. (2018). Purification, structural elucidation and anti-inflammatory activity of xylosyl galactofucan from *Armillaria mellea*. *Int. J. Biol. Macromol.*, 114: 584–591.

Chattopadhyay, A., Rukmini, R. and Mukherjee, S. (1996). Photophysics of a neurotransmitter: Ionization and spectroscopic properties of serotonin. *Biophys. J.*, 71: 1952–1960.

Chen, X.M., Guo, S.X., Wang, Q.Y. and Xiao, P.G. (2001). Studies on polysaccharides in different development stages of *Armillaria mellea*, China. *J. Chin. Mater. Med.*, 26: 381–384.

Chen, R., Ren, X. and Yin, W. (2019). Ultrasonic disruption extraction, characterization and bioactivities of polysaccharides from wild *Armillaria mellea*. *International Journal of Biological Macromolecules*, 156: 1491–1502. Doi: 10.1016/j.ijbiomac.2019.11.196. PMID: 31785299.

Cheng, X.H., Liu, L.D., Dong, H.X., Qu, H.G. and Cai, D.H. (2007). Optimisation of submerged culture condition for production of mycelial biomass by *Armillaria mellea*. *J. Chin. Med. Mater.*, 30: 509–512.

Cheng, X.H.S. and Guo, X. (2006). The inducement and fruiting condition of fruit body of *Armillaria Mellea*. *Microsystem*, 25: 302–307.

Cheng, X.H.S. and Guo, X. (2006). Cultural characteristics of *Armillaria Mellea* on solid media. *Acta Acad. Med. Sin.*, 28: 553–557.

Colak, A., Sahin, E., Yildirim, M. and Sesli, E. (2007). Polyphenol oxidase potentials of three wild mushroom species harvested from Liser high plateau Trabzon. *Food Chem.*, 103: 1426–1433.

Cox, K.D., Scherm, H. and Riley, M.B. (2006). Characterization of *Armillaria* spp. from peach orchards in the southeastern United States using fatty acid methyl ester profiling. *Mycol. Res.*, 110: 414–422.

Dai, L., Wang, H. and Shen, Y.S. (2000). The effects of *Armillaria mellea* isolated polysaccharide (MHG) on murine macrophage. *J. Biol.* 17: 20–21.

Davis, R., Michael, S., Robert, M. and John, A. (2012). *Field Guide to Mushrooms of Western North America*, Berkeley: University of California Press, pp. 134–135. ISBN 978-0-520-953604. OCLC 797915861.

De Mattos-Shipley, K., Ford, M.J.K.L., Alberti, F., Banks, A.M.B., Bailey, A.M. and Foster, G.D. (2016). The good, the bad and the tasty are the many roles of mushrooms. *Stud. Mycol.*, 85: 125–157.

Donnelly, D.M., Polonsky, X.J., Prangé, T., Snatzke, G. and Wagner, U. (1984). The absolute configuration of the orsellinate of armillol application of the coupled oscillator theory. *J. Chem. Soc. Chem. Commun.*, 4: 222–223.

Donnelly, D.M.X., Abe, F., Coveney, D., Fukuda, N. and Reilly, J. (1985). Antibacterial sesquiterpene aryl ester from *Armillaria mellea. J. Nat. Prod.*, 48: 10–16.

Donnelly, D.M.X., Coveney, D.J. and Polonsky, J. (1985). Melledonal and melledonol sesquiterpene esters from *Armillaria mellea. Tetrahedron Lett.*, 26: 5343.

Donnelly, D.M.X., Coveney, D.J., Fukuda, N. and Polonsky, J. (1986). New sesquiterpene aryl esters from *Armillaria mellea. J. Nat. Prod.*, 49: 111–116.

Donnelly, D.M.X., Hutchinson, R.M., Coveney, D. and Yonemitsu, M. (1990). Sesquiterpene aryl esters from *Armillaria mellea. Phytochemistry*, 29: 2569–2572.

Donnelly, D.M.X.R. and Hutchinson, M. (1990). Armillane: A saturated sesquiterpene ester from *Armillaria mellea. Phytochemistry*, 29: 179–182.

Donnelly, D.M.X.R., Quigley, P.F., Coveney, D.J. and Polonsky, J. (1987). Two new sesquiterpene esters from *Armillaria mellea. Phytochemistry*, 26: 3075–3077.

Falandysz, J., Gucia, M., Skwarzec, B., Frankowska, A. and Klawikowska, K. (2002). Total mercury in mushrooms and underlying soil from the Borecka forest, northeastern Poland. *Arch. Environ. Contam. Toxicol.*, 42: 145–154.

Falandysz, J., SicinÂska, B., Bona, H. and Kohnke, D. (1992). Metals content in *Armillariella mellea. Bromatologia i Chemia Toksykologiczna*, 25: 171–176.

Florczak, J. and Lasota, W. (1995). Wchłaniane i wiązanie kadmu przez boczniaka ostrogowatego w warunkach uprawy. *Bromatol. Chem. Toksykol.*, 28: 17–23.

Florczak, J., Karmańska, A. and Wędzisz, A. (2004). Comparison of the chemical contents of selected wild growing mushrooms. *Bromatol. Toxykol. Chem.*, 37: 365–371.

Furlani, R.P. and Godoy, H.T. (2008). Vitamins B1 and B2 contents in cultivated mushrooms. *Food Chem.*, 106: 816–819.

Gao, J.M., Xue, Y.C.Y., Wang, J. and Liu, K. (2001). Armillaramide a new sphingolipid from the fungus *Armillaria mellea. Fitoterapia*, 72: 858–864.

Gargano, M.L., Leo, J.L.D., Van, G., Omoanghe, S.I., Ulrike, L., Giuseppe, V., Solomon, P.W. and Zervakis, G.I. (2017). Medicinal mushrooms: Valuable biological resources of high exploitation potential. *Plant Biosystems—An International Journal Dealing with all Aspects of Plant Biology*, 151: 3, 548–565, Doi: 10.1080/11263504.2017.1301590.

Guillaumin, J.J., Legrand, P. and Gonthier, P.G. (2013). Armillaria root rots. *Infectious Forest Diseases*, CAB International Wallingford, UK, 4: 159–177.

Guo, W.J., Guo, S.X., Yang, J.S., Chen, X.M. and Xiao, P.G. (2007). Triterpenes and steroids from *Armillaria mellea. Vahlex Fr. Biochem. Systemat. Ecol.*, 35: 790–793.

Hannson, O. and Seifert, G. (1987). Effects of cultivation techniques and media on yields and morphology of the Basidiomycete *Armillaria mellea. Appl. Microbiol. Biotechnol.*, 26: 468–473.

Healy, V., O'Connell, T., McCarthy, V. and Doonan, S. (1999). The lysine specific proteinase from *Armillaria mellea* is a member of a novel class of metallo endopeptidases located in Basidiomycetes. *Biochem. Biophys. Res. Commun.*, 262: 60–63.

Hood, I.A., Redfern, D.B. and Kile, G.A. (1991). *Armillaria* in planted hosts. pp. 122–149. *In*: Shaw II, C.G. and Kile, G.A. (eds.). *Armillaria Root Disease*, USDA Forest Service, Washington DC.

Isbister, G.K., Bowe, S.J., Dawson, A. and Whyte, I.M. (2004). Relative toxicity of selective serotonin reuptake inhibitors (SSRIs) in overdose. *J. Toxicol. Clin. Toxicol.*, 42: 277–285.

Kalyoncu, F., Oskay, M., Sağlam, H., Erdoğan, T.F. and Tamer, A.U. (2010). Antimicrobial and antioxidant activities of mycelia of 10 wild mushroom species. *J. Med. Food.*, 13(2): 415–419.

Karkocha, I. (1964). Badania wartości odżywczej gąski zielonej (Tricholoma equestre) i opieńki miodowej (*Armillaria mellea*). *Rocz. PZH*, 15: 311–315.

Kemp, K. (2002). Heavy metals in edible mushrooms. *Int. J. PIXE*, 12: 117–124.

Kile, G.A., McDonald, G.I. and Byler, J.W. (1991). Ecology and disease in natural forests. pp. 102–121. *In*: Shaw III, C.G. and Kile, G.A. (eds.). *Armillaria Root Disease*, USDA for Serv. Agric. Hand. No. 691, USDA Forest Service, Washington DC.

Kong, X.W. and Jiang, L. (2007). Effects on the immunological function of exopolysaccharide from *Armillaria mellea* in mice. *J. Anhui. Univ.*, 31: 87–90.

Kong, X.W., Shen, Y.S., Wang, M.C. and Hong, D.H. (2003). Study on some properties and structure of Am-a polysaccharide from *Armillaria mellea. Food Sci.*, 24: 23–26.

Kuo, M. (2007). *100 Edible Mushrooms*. Ann Arbor Michigan: The University of Michigan Press, pp. 244-6. ISBN 978-0-472-03126-9.

Lee, S.Y., Kim, J.S., Kim, J.E., Sapkota, K., Shen, M.H., Kim, S., Chun, H.S., Yoo, H.S.J.C., Choi, H.S., Kim, M.K. and Kim, S.J. (2005). Purification and characterisation of fibrinolytic enzyme from the cultured mycelia of *Armillaria mellea. Protein Express. Purif.*, 43: 10–17.

Leu, Y.S. (2019). Induction of autophagic death of human hepatocellular carcinoma cells by armillaridin from *Armillaria mellea. Am. J. Chin. Med.*, 47(6): 1365–1380.

Li, L.H., Wang, Z. and Xie, D.P. (2003). Study of polysaccharide production from *Armillaria mellea's* submerged fermentation. *J. Hunan Agr. Univ.*, 29: 428–430.

Li, Y.P., Wu, K.F. and Liu, Y. (2005). Protective effect of *Armillaria mellea* polysaccharide on mice bone marrow cell damage caused by cyclophosphamide, China. *J. Chin. Mater. Med.*, 30: 283–286.

Lin, Z.B., Zhang, D.M. and Xia, D. (1988). Effects of *Armillaria mellea* polysaccharide on immunological function in mice. *Chin. Pharmacol. Bull.*, 4: 93–96.

Liu, J., Yuan, S. and Tian, Z.H. (2003). Study on the active constituents and application in functional food of *Armillaria mellea. Food Sci.*, 24: 165–168.

Lobo, V., Patail, A., Phatak, A. and Chandra, N. (2010). Free radicals antioxidants and functional foods impact on human health. *Pharmacogn. Rev.*, 4: 118–126.

Lu, G.P.C.Q., Wang, Z. and Cai, Q. (2002). Pharmacological and clinical researches of Gastrodin injection. *Chin. Tradit. Herb. Drugs*, 33: 449–450.

Majdańska, M. (2007). Badanie właściwości redukcyjnych i antyoksydacyjnych wyciągów wodnych i metanolowych otrzymanych z grzybni Armillaria mellea nieselenowanej i selenowanej. Warszawski Uniwersytet Medyczny. Praca dyplomowa.

Mattila, P., Könkö, K., Eurola, M., Pihlava, J.M., Astola, J., Vahteristo, L., Hietaniemi, V., Kumpulainen, J., Valtonen, M. and Piironen, V. (2001). Contents of vitamins, mineral elements, and some phenolic compounds in cultivated mushrooms. *J. Agric. Food Chem.*, 49: 2343–2348.

Midland, S.L., Izac, R.R., Wing, R.M., Zaki, A.I. and Munnecke, D.E. (1982). Melleolide, a new antibiotic from *Armillaria mellea. Tetrahedron Lett.*, 23: 2515.

Muszyńska, B., Maślanka, A., Ekiert, H. and Sułkowska-Ziaja, K. (2001). Analysis of indole compounds in *Armillaria mellea* fruiting bodies. *Acta Pol. Pharm.*, 68: 93–97.

New Drug Group. (1977). Department of Pharmacology, Institute of Materia Medica of Chinese Academy of Medical Science, Pharmacological actions of Gastrodia watery preparation and fermentation liquid of *Armellaria mellea* on nervous system. *Chin. J. Med.*, 8: 470–472.

Obuchi, T.H.N., Kondoh, M., Watanabe, S., Tamai, J.S., Omura, X.T. and Yan, L. (1990). Armillaric acid a new antibiotic produced by *Armillaria mellea. Planta Med.*, 56: 198–201.

Ojemann, L.M., Nelson, W.L., Shin, D.S., Rowe, A.O. and Buchanan, R.A. (2006). Tian-Ma, an ancient Chinese herb, offers new options for the treatment of epilepsy and other conditions. *Epilepsy Behav.*, 8(2): 376–383.

Pegler, D.N. (2000). Taxonomy, nomenclature and description of *Armillaria*. pp. 81–93. *In*: Fox, R.T.V. (ed.). *Armillaria Root Rot: Biology and Control of Honey Fungus*, Intercept Ltd. ISBN 1-898298-64-5.

Phillips, R. (2010). *Mushrooms and Other Fungi of North America*. Buffalo, NY: Firefly Books P: 42. ISBN 978-1-55407-651-2.

Podlewska, A. (2006). Porównanie właściwości biobójczych wyciągów alkoholowych i wodnych otrzymanych z grzybni Armillaria mellea. Uniwesytet Medyczny w Warszawie. Praca dyplomowa.

Przybyłowicz, P. and Donoghue, J. (1988). Nutritional and health aspects of shiitake. *In*: Shiitake Growers Handbook. Dubuque, IA: Kendall Hunt Publishing Co.

Roberts, P.S. and Evans, L. (2011). *The Book of Fungi: A Life-size Guide to Six Hundred Species from around the World*. University of Chicago Press, Chicago IL USA, p. 63.

Roupas, P., Keogh, J., Noakes, M., Margetts, C. and Taylor, P. (2012). The role of edible mushrooms in health: Evaluation of the evidence. *Funct. J. Foods*, 4: 687–709.

Shen, Y.S. and Hong, Y. (1998). Isolation, purification and some properties of polysaccharides from the rhizomorph of *Armillariella mellea*. *Edible Fungi China*, 18: 38–40.

Shi, L., Cao, R.M., Lu, S.X. and Wu, G.X. (1998). Isolation purification and identification a new compound from the fruiting body of *Armillaria mellea*. *J.N. Bethune Univ. Med. Sci.*, 24: 343.

Siu, K.C., Xu, L., Chen, X. and Wu, J.Y. (2016). Molecular properties and antioxidant activities of polysaccharides isolated from alkaline extract of wild *Armillaria* ostoyae mushrooms. *Carbohydrate Polymers*, 137: 739–746.

Soković, M., Ćirić, A., Glamočlija, J. and Stojković, D. (2017). The bioactive properties of mushrooms. pp. 83–122. *In*: Ferreira, I.C.F.R., Morales, P. and Barros, L. (eds.). *Wild Plants Mushrooms and Nuts Functional Food Properties and Applications*. Wiley-Blackwell, UK, 1st ed., vol. 4.

Souilem, F., Fernandes, Â., Calhelha, R.C., Barreira, J.C.M., Barros, L., Barros Martins, A. and Ferreira, I.C.F.R. (2017). Wild mushrooms and their mycelia as sources of bioactive compounds, antioxidant anti-inflammatory and cytotoxic properties. *Food Chem.*, 230: 40–48.

Stasiak, A.A. (2008). Badanie właściwości antyoksydacyjnych i stopnia zmiatania wolnych rodników wyciągów metanolowych i wodnych otrzymanych z grzybni nieselenowanej i selenowanej Armillaria mellea. Warszawski Uniwersytet Medyczny. Praca dyplomowa.

Stone, T.W., Mackay, G.M., Forrest, C.M., Clark, C.J. and Darlington, L.G. (2003). Tryptophan metabolites and brain disorders. *Clin. Chem. Lab. Med.*, 41: 852–859.

Sun, Y.Y. (2004). Preliminary study on the antibacterial action of the *Armillaria mellea*. *Food Sci. Technol.*, 8: 51–52.

Sun, Y.X., Liang, H.T., Zhang, X.T., Tong, H.B. and Liu, J.C. (2009). Structural elucidation and immunological activity of a polysaccharide from the fruiting body of *Armillaria mellea*. *Bioresour. Technol.*, 100: 1860–1863.

Svoboda, L., Havlíčková, B. and Kalač, P. (2006). Contents of cadmium, mercury and lead in edible mushrooms growing in a historical silver-mining area. *Food Chem.*, 96: 580–585.

Tan, Z.J., Xie, D.P., Wang, Z. and Li, L.H. (2002). Study on the production conditions of exopolysaccharide from *Armillaria mellea*. *Microbiology*, 29: 33–37.

Tang, W.G. (1992). *Gastrodia elata B₁*. Chinese Drugs of Plant Origin, Berlin/Heidelberg, Springer-Verlag, pp. 545–548.

Toledo, C.V., Barroetaveña, C.A., Fernandes Barros, L. and Ferreira, I.C.F.R. (2016). Chemical and antioxidant properties of wild edible mushrooms from native Nothofagus spp. forest argentina. *Molecules*, 21: 1201. Doi: 10.3390/ molecules2109120.

Wang, H.G., Guan, H.Q., Zhao, Y.N. and Li, X.H. (2007). Effects of *Armillaria mellea* polysaccharide on IL-2 and TGF-1 in mice. *Prog. Modern Biomed.*, 7: 1306–1307.

Wang, X.Q., Yin, J.J., Meng, F.W., Peng, Y.H. and Lin, F. (2007). 65 cases of clinical observation of applying compound *Armillaria mellea* tablets in vertebro-basilar. *Chin. J. Clin. Healthcare*, 10: 60–62.

Watanabe, N., Obuchi, T., Tamai, M., Araki, H., Omura, S., Yang, J.S., Yu, D.Q., Liang, X.T. and Huan, J.H. (1990). A novel N 6-substituted adenosine isolated from *mihuan jun* (*Armillaria mellea*) as a cerebral-protecting compound. *Planta Med.*, 56: 48–52.

Weinhold, A.R. (1963). Rhizomorph production by *Armillaria mellea* induced by ethanol and related compounds. *Science*, 142: 1065–1066.

Weitz, H.J., Ballard, A.L., Campbell, C.D. and Killham, K. (2001). The effect of culture conditions on the mycelial growth and luminescence of naturally bioluminescent fungi. *FEMS Microbiol. Let.*, 202: 165–170.

Wu, J.Y., Xiao, Z., Wang, P. and Li, Q.B. (2001). Spectrophotometric determination of extracellular laccase activity of *Armillaria mellea. J. Xiamen Univ.*, 40: 893–898.

Wu, J., Zhou, J. and Lang, Y. (2012). A polysaccharide from *Armillaria mellea* exhibits strong *in vitro* anticancer activity via apoptosis-involved mechanisms. *International Journal of Biological Macromolecules*, 51(4): 663–667.

Wu, J.J., Zhou, Y., Lang, L., Yao, H., Xu, H., Shi, H. and Xu, S. (2012). A polysaccharide from *Armillaria mellea* exhibits strong *in vitro* anticancer activity via apoptosis-involved mechanisms. *Int. J. Biol. Macromol.*, 51(4): 663–667.

Xiao, Y.Z.J., Wang, Y.P., Wang, C.L., Pu, Y. and Shi, Y. (2002). Studies on production, purification and partial characteristics of the extracellular laccase from *Armillaria mellea. Chin. J. Biotechnol.*, 18: 457–462.

Xiong, J.J. and Huang, H. (1998). The A1- and non-A1-effects of N6-(5-hydroxy 2-pyridyl)-methyl-adenosine on rat vas deferens. *Acta Pharmaceut. Sin.*, 33: 175–179.

Xu, J. and Guo, S. (2000). Retrospect on the research of the cultivation of *Gastrodia elata B₁*, a rare traditional Chinese medicine. *Chin. Med. J.*, 113: 686–692.

Xue, M., Wang, Q.Y. and Fan, J.Y. (2004). Primary study on fruiting condition of fruit body of *Armillaria mellea. Sci. Technol. Food Ind.*, 25: 58–60.

Yamac, M. and Bilgili, F. (2006). Antimicrobial activities of fruit bodies and/or mycelial cultures of some mushroom isolates. *Pharmaceut. Biol.*, 44: 660–667.

Yang, J.S., Chen, Y.W., Feng, X.Z., Yu, D.Q. and Liang, X.T. (1984). Chemical constituents of *Armillaria mellea* mycelium. I. Isolation and characteriSation of *Armillarin* and *Armillaridin. Planta Med.*, 50: 288–290.

Yang, J.S., Chen, Y.W., Feng, X.Z., Yu, D.Q., He, C.H., Zheng, Q.T. and Yang, X.T. (1989). Isolation and structure elucidation of *Armillaricin. Planta Med.*, 55: 564–565.

Yang, J.S., Su, Y.L., Wang, Y.L., Feng, X.Z., Yu, D.Q., Cong, P.Z., Tamai, M., Obuchi, T., Kondoh, H. and Liang, X.T. (1989). Isolation and structures of two new sesquiterpenoids aromatic esters: *Armillarigin* and *Armillarikin. Planta Med.*, 55: 479–481.

Yang, J.S., Su, Y.L., Wang, Y.L., Feng, X.Z., Yu, D.Q., Liang, X.T., He, C.H., Zheng, Q.T., Yang, J.J. and Yang, J. (1990). Chemical constituents of *Armillaria mellea* mycelium. VI. Isolation and structure of *Armillaripin. Acta Pharmaceut. Sin.*, 25: 353–356.

Yang, J.S., Su, Y.L., Wang, Y.L., Feng, X.Z., Yu, D.Q. and Liang, X.T. (1990). Studies on the chemical constituents of *Armillaria mellea* mycelium. V. Isolation and characterisation of *Armillarilin* and *Armillaria. Acta Pharmaceut. Sin.*, 25: 24–28.

Yang, J.S., Su, Y.L., Wang, Y.L., Feng, X.Z., Yu, Q.D. and Liang, X.T. (1991). Two novel protoilludane non-sesquiterpenoids esters: *Armillasin* and *Armillatin* from *Armillaria mellea. Planta Med.*, 57: 478–480.

Yang, J.S., Su, Y.L., Wang, Y.L., Feng, X.Z., Yu, Q.D. and Liang, X.T. (1991). Chemical constituents of *Armillaria mellea* mycelium. VII. Isolation and characterisation of chemical constituents of the acetone extract. *Acta Pharmaceut. Sin.*, 26: 117–22.

Yang, L.H., Huang, Q.R., Feng, P.Y., Jiang, H., Cai, D.H. and Liu, J. (2007). Isolation and identification of polysaccharides from *Armillaria mellea* and their effects on scavenging oxygen free radicals. *Food Sci.*, 28: 30.

Yang, S. (2018). Polysaccharide-enriched fraction from *Amillariella mellea* fruiting body improves insulin resistance. *Molecules*, 24(1).

Yu, L., Shen, Y.S. and Miao, H.C. (2006). Study on the anti-vertigo function of polysaccharides of *Gastrodia elata* and polysaccharides of *Armillaria mellea. Chin. J. Information, TCM*, 13(8): 29–36.

Yu, L., Shen, Y.S., Wu, J.M. and Xie, Q.Y. (2006). Anti-aging effects of polysaccharides from rhizomorph of *Armillaria mellea. Chin. Tradit. Patent Med.*, 28: 994–996.

Yu, M. and Shen, Y.S. (2002). The effect of polysaccharide from the rhizomorph of *Armillaria mellea* (AMP) on the blood glucose and acutely toxic in mice. *Edible Fungi China*, 21: 35–37.

Yu, M.Y., Shen, M., Mei, Y.D. and Wang, F. (2001). Study on effects of polysaccharide from the rhizomorph of *Armillaria mellea* on immunity. *J. Biol.*, 18: 6–18.

Zavastin, D., Elena, B., Gabriela, D., Lisa, G., Gherman, S., Petronela, B., Gabriel, M. and Anca, S.C. (2018). Metal content and crude polysaccharide characterization of selected mushrooms growing in Romania. *Subtropical Plant Science*, 67: 149–158.

Zavastin, D.E., Mircea, C.A., Aprotosoaie, S., Gherman, M., Hancianuand, A. and Miron, L. (2015). *Armillaria mellea* phenolic content *in vitro* antioxidant and antihyperglycemic effects. *Rev. Med. Chir. Soc. Med. Nat. Iasi.*, 119: 273–280.

Zhang, L.M., Zhang, T.Z. and Yao, G.B. (1983). Preliminary study on the effect of *Armillaria mellea* tablets on high lipemia. *Chin. Drug Res.*, 11: 19–20.

Zhang, Q.L., Zhang, H.J. and Wang, B.J. (1995). Isolation purification and compositional analysis of polysaccharide of *Armillaira mellea. Chin. Pharmaceut. J.*, 30: 401–403.

Zhang, S., Liu, X., Yan, L., Zhang, Q., Zhu, J., Huang, N. and Wang, Z. (2015). Chemical compositions and antioxidant activities of polysaccharides from the sporophores and cultured products of *Armillaria mellea. Molecules*, 20: 5680–5697. Doi: 10.3390/molecules20045680. PMID: 25838171; PMCID: PMC6272248.

Zhang, T. (2019). Study on antidepressant-like effect of protoilludane sesquiterpenoids aromatic esters from *Armillaria mellea. Nat Prod Res.*, pp. 1–4.

Zhang, Y.Y. and Shen, S. (2001). Optimise the conditions of submerged fermentation and isolate the polysaccharide from *Armillaria mellea. Edible Fungi*, 2: 5–7.

Zhang, Y.Y., Shen, S., Ge, J.Z., Yi, R.C. and Kong, X.W. (2002). Reverse effects of polysaccharide from *Armillaria mellea* mutated by N + ion beam implantation. *Acta Laser Biol. Sin.*, 11: 272–275.

Zhang, Y.Y., Shen, S., Zhang, X.Z. and Cheng, T. (2001). Studies on extracting conditions and bio-activities of polysaccharide from the rhizomorph of *Armillaria mellea. Edible Fungi China*, 20: 38–40.

Źródłowski, Z. (1995). The influence of washing and peeling of mushrooms *Agaricus bisporus* on the level of heavy metal contaminations. *Pol. J. Food Nutr. Sci.*, 4/45: 26–33.

Zsigmondy, R., Andreea, V., Krisztina, H. Sándor, B. and Edina, I.U. (2015). Elemental profile of edible mushrooms from a forest near a major Romanian city. *Acta Universities Sapientiae, Agriculture and Environment*, 7(1): 98–107.

CHAPTER 5

Jelly Ear
(Auricularia auricula-judae)

Mona Kejariwal

1. Introduction

The wood ear fungus, i.e., *Auricularia auricula judae* is mainly found in winter and spring seasons in all parts of the world. Like other wood-rot fungi, it also prefers to grow on bark, dead logs as well as hard wood. A few so-called jelly fungi are as soft as jelly, occasionally eaten with pudding and belong to Basidiomycetes. Many people can reconstruct and continue to produce the fungi with the help of spores and providing optimum and suitable conditions for growth. It grows on the branches of dying old trees and as weak parasites on the barks of growing tree-trunks. This fungus is rarely found in other types of deciduous trees while it prefers to grow on trees with moist bark. In Australia, the fungus prefers to grow on fallen branches of long trees of eucalyptus and similar species. Wood ear mushrooms are found all the year round, but are most common in late summer and autumn. In India, it dominates from August to November and on high-altitude forests, especially the Western Ghats.

In 1789, this jelly wood ear was first described by a French theologian and archaeologist Jean Baptiste François as *Tremella auricula judae*. After him, the fungus was redefined and given many names before getting its current genus in 1897, by Austrian plant lover and mycologist Richard Wettstein. The gelatinous fungus, thallus, has distinctive aroma and taste due to its smooth and crunchy texture. Since nineteenth century, *Auricularia* genus gained popularity and high consumption in China and Southeast Asia as shown in many reports (Beall, 1932; Lamb and Wellington, 1975; Liu et al., 2021).

The word *Auricularia*, which has Latin origin, means 'ear lobe'; judae means 'Judas', the Jew who it is said deceived Jesus. As per the older field guides, the species was commonly known as Jew's ear fungus (Liu et al., 2021). There are various names as synonyms have been given to this omnipresent mushroom's genus, *Auricularia*.

RD and SH National College, Bandra (west), Mumbai-50.
Email: mona.kejriwal@gmail.com

Such as Sir François called it *Tremella auricula* L. and Bolton has described it as *Peziza auricula-judae* (Bull.). Similarly other scientist in nineteenth and twentieth centuries have named this fungus with different genus, as *Tremella* and *Exidia* by Bull; *Hirneola* by Berk, *Auricularia* by Quel and Underw. The type species of the genus *Auricularia* is *Auricularia mesenterica* which is also known as 'tripe' or 'grey brain fungus' due to its smaller size, greyish-brown appearance with brain-like infolding in the form of bracket-like tiers on various timber trees, like *sal* on the Western Ghats and it has broad leaves. The species has more hair and is gelatinous in nature in comparison to *Auricularia auricula judae*. The myth, which dates back more than 400 years, suggests that the mushroom name emerged from the elderwood trees, which are still believed to be connected with visible residues of Judas' tormented spirit. For identification, one considers the fruiting bodies and spores of the fungus. The outer surface of the lobed fruit is light brown with a hint of purple and covered with fine light grey hair with smooth inner surface. The individual lobes of *Auricularia auricular-judae* range between 3–10 cms in diameter. The spores are allantoid, 1618×68 μm long. Ear jelly is used in cooking, where it imparts not only scintillating texture but also provides extraordinary flavour and fragrance. Young, fresh fruit bodies are generally considered the best. A similar and closely related species is an edible mushroom very popular in the Far East, especially in China, where it is also used medicinally (Yao et al., 2015).

As per the reports provided by Society of Mycologist, the genus *Auricularia* has 17% of total global mushroom production which is considered as the third most cultivated mushroom genus after two other economically important mushrooms—*Lentinula* and *Pleurotus* with 22% and 19% respectively of global mushroom (Royse et al., 2017). *Auricularia* is important as a food source, but it also has other uses. The species can be used to produce new therapeutic medications by being used as medicinal resources for a variety of human ailments, including the possibility to treat cancer and related disorders (Wu et al., 2014). The mycological substances, like polysaccharide-based compounds from *Auricularia* species are well known for their antimicrobial properties. They have shown characteristics, like antitumour, antioxidant, anti-coagulant and immune-modulating (Ma et al., 2010; 2018) activities.

The bioactive components of *Auricularia* must be understood in order to advance our understanding. It could lead to the creation of new therapeutic medications as a result of their pharmacology. This chapter will provide a comprehensive overview of *Auricularia* application in the context of scientific and neutraceutical aspects. The chapter will give a detailed review on the fungal cultivation and applications.

2. *Auricularia auricula-judae*: Nutritional Profile

Depending on the species, the nutritional value of edible *Auricularia* varies. Dried *Auricularia* has a carbohydrate content of 79.9–93.2% and a crude protein content of 6.5–13% on an average.

Carbohydrates make up the majority of food we eat. In case of *Auricularia auricula-judae*, carbohydrate is represented in the proximate composition, for example: 81.0% carbohydrates, 8.1% crude protein, 1.5% crude fat, 6.9% crude fibre and 9.4% ash (Chang and Hayes, 2013).

The bioavailable, easily digestible polysaccharides are formed of water-soluble mannan and glucan, pectin, chitin and cellulose. The proportion of sugars in the polysaccharide defines its digestibility. The degree of wood formation also determines the digestibility, whereas non-digestible components, like non-starch polysaccharides and fibres, are important sources of prebiotics. In addition to these major nutrients, edible mushrooms have various micronutrients that most human foods lack (Carrasco et al., 2018). Islam et al. (2021) gave a comprehensive view of nutritional values of mushroom as shown in Fig. 1.

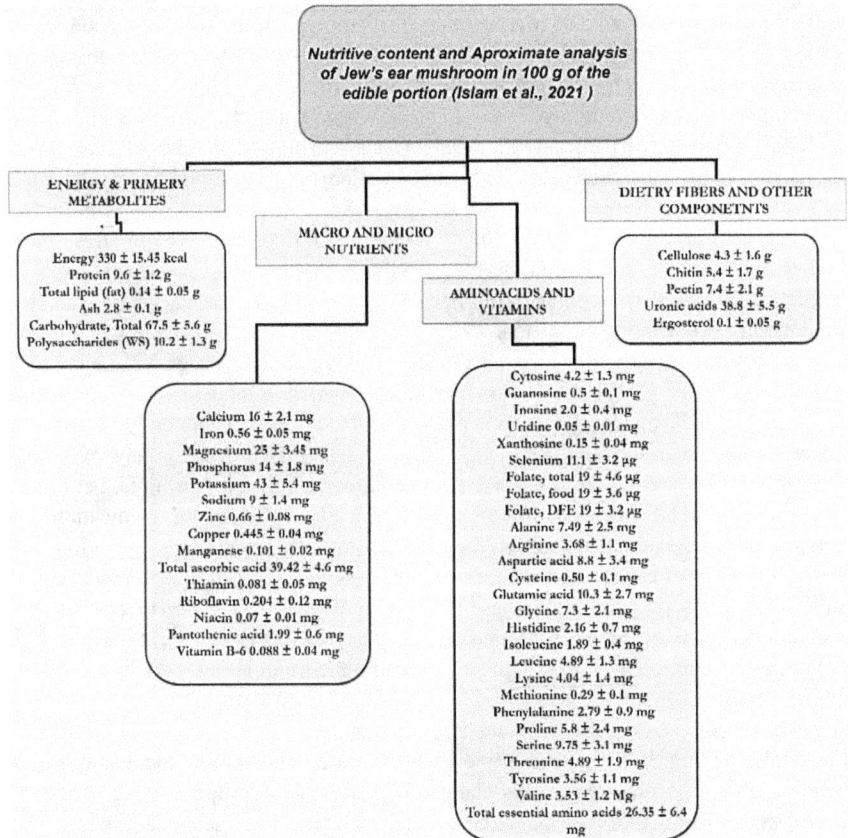

Fig. 1. Nutritional profile of *A. auricula-judae* extracts (Islam et al., 2021).

3. Production and Cultivation of *Auricularia auricula judae*

China has been cultivating *Auricularia auricula-judae*, also known as wood ear mushroom for over 2,100 years (Yao et al., 2018). The commercial cultivation of *A. cornea* and *A. heimuer* is currently under way in South Asian countries, like Indonesia, Malaysia, the Philippines, Thailand and Vietnam besides China. *Auricularia* cultivars are mostly produced in China, and in 2017, over 6.3 billion kg were produced, making it the second most extensively produced food, particularly in countries which are flag bearers in mushroom cultivation.

The fungus has well-developed hyphae, primary, secondary and tertiary mycelium, dolipore septa and pinched joints. Cultivation of this mushroom involves two important methods – spawning and fruiting body production. Grains are laid, using mycelium derived from tissue culture. The stem method, artificial stem method and plastic bag method are generally used for fruit production.

Several kinds of culture media are now used to produce *Auricularia*, each with its own nutritional profile, as well as appropriate temperatures and pH ranges; for example Yeast Mannitol Agar (YMA), Yeast Mannitol Broth (YEB) and Leonian medium, as described by Yu et al., and Jo et al., in 2014. The cultivation and production of edible *Auricularia* necessitates ideal mycelia growing conditions. For example, *A. auricula-judae* was found to grow best on potato dextrose agar/broth and similar medium at 25–30°C or room temperature or on Czapek dox or Leonian, but not on pH 6 or 9 (Jo et al., 2014). Mycelium of *Auricularia villosula* grows significantly with potato juice, sugar, soybean powder and 0.5% phosphate. The ideal condition for cultivation of fungi is pH 8 and temperature of 30°C.

Spawn production is done through tissue culture by generation of mycelium. This is followed by fruiting body either by tree log or artificial tree trunk, or polybag method which are the main farming methods accomplished till date (Priya et al., 2016). Cheap production of *Auricularia* species occurs in many Southeast Asian countries through use of farm manure, organic waste and agricultural waste; for example, compost consisting mainly of corn cobs and rice harvesting waste. Straw, hardwood sawdust, gypsum stone, cotton bran, wheat bran, rice bran, quicklime are used as supplements. In our laboratory, we have succeeded in cultivating *A. auricula-judea* on sugarcane molasses. Similarly, in Taiwan, as low-cost agricultural waste, for example cotton used scraps, rice straw, wood shavings and sawdust are often used to make wood ear mushrooms (Shrikhandia and Sumbali, 2022).

New techniques are employed to address the interspecies and intra-species genetic variability, i.e., Random Amplified Polymorphic DNA (RAPD) fingerprinting analysis. This gives an opportunity to breeders to experiment with fungal quality and improved neutraceutical profile of the fungus (Yan et al., 2004).

4. Polysaccharide Extraction and Purification of *Auricularia auricula-judae*

The neutraceutical, pharmaceutical and medicinal value of this mushroom is mostly accounted on the various polysaccharides obtained from it. Therefore it is important to understand the process of extraction and purification of these polysaccharides. As explained by Chen et al. (2020) in their review, the process includes the following steps:

1. Pre-treatment of fruiting bodies of mushroom is carried through washing, drying and pulverisation, followed by defatting of material. Pure fermented cultures are directly subjected to extraction and purification methods.

2. Solid residues obtained through this process are subjected to various kinds of extraction methods, like hot-water extraction, alkali or acid-based extraction, ultrasound/microwave/enzyme-assisted extraction or polyethylene glycol-based ultrasound extraction. These extractions are monitored through various optical extraction parameters for maximum and optimum output.

3. Supernatant collected through these extractions are then completed through ethanol precipitation or freeze drying or dialysis, resulting in crude polysaccharides.

4. These crude polysaccharides are used for various experiments after purification through various kinds of chromatography methods which lead to separation of neutral and acidic polysaccharides resulting in purified polysaccharide fractions.

5. Medicinal Efficacy and Pharmacological Use of Wood Ear Mushrooms

5.1 Ethnobotanical Uses

Since ages, Asians and Europeans have been known to use *Auricularia* species in their traditional medicinal systems as an alternative source of medicine for various ailments. For example *A. auricula-judea* has been conventionally used for healing poisonous bites of snakes and plants as mentioned by Fuller and other scientists in 2005. It has also been used to treat fear as a part of traditional treatment by the Tzeltal community, as described by Guzman in 2008. Besides it is used in alternative medicinal systems due to its high medicinal properties, as described by various scientists across the globe. Now *A. auricula-judea* has been adopted and integrated into Western medicine.

5.2 Medicinal Uses

Due to the high concentration of water-soluble carbohydrates, such as glucans in the fungus, it is designed specifically for its therapeutic potential. Considerable research has been conducted on the healing properties and pharmacological impact of various tested isolates from *Auricularia* (Misaki et al., 1981; Sekara et al., 2015). It exhibits action against clotting, allergies, tumours, pathogens and reactive species (Yoon et al., 2003; Ukai et al., 1983; Kiho et al., 1985; Damte et al., 2011; Gbolagade and Fasidi, 2005; Sun et al., 2010).

Additionally, it has an antitumour, anticancer and antiviral action (Nguyen et al., 2012). According to Sheu et al. (2004), *Auricularia* possesses anti-inflammatory and immunomodulatory potential. *Auricularia* can indeed treat hypolipidemia, high cholesterol and both hypoglycemia and diabetes (Yuan et al., 1998; Reza et al., 2015).

5.2.1 Antidiabetic Activity

A. auricula-judae polysaccharides are considered as a very good antidiabetic agent as the hot water extract increase muscle glucose uptake through reduced levels of glucose production in the pancreas and the liver. It also inhibits α-amylase activity in several tests performed on various experimental setups (Wu et al., 2014). The dietary fibres or polysaccharides obtained from the mushroom specifically target the intestinal tract to absorb low levels of glucose, helping the diabetic patient to avoid hyperglycaemia. These fibres also work as prebiotics for gut microflora, inducing the growth of good bacteria in the intestinal tract.

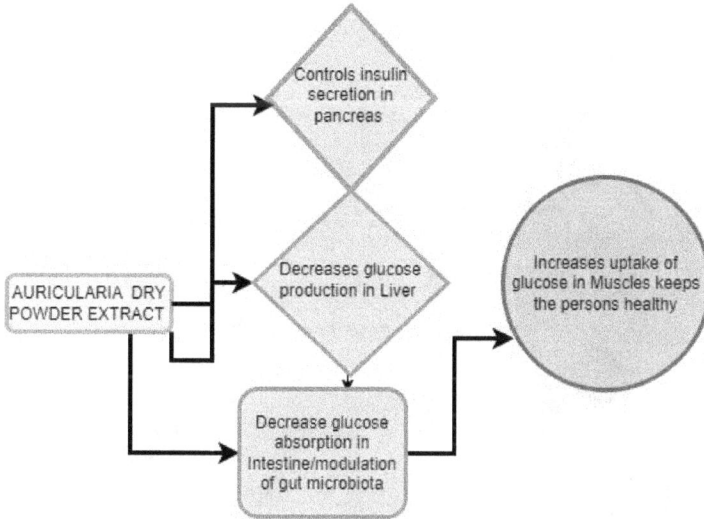

Fig. 2. Diabetes management through *A. auricula-judae* extract.

5.2.2 Antioxidant Activity

Polysaccharides found in a crude extract of *A. auricula judae* have been proven to possess substantial powerful antioxidants, including radical-scavenging activity, like chelation of Fe2+, reducing power assays (Cai et al., 2015). A prior study found that the large amounts of polysaccharides present in bread formed from flour produced from Jew's ear as evidenced by its DPPH (Fan et al., 2007), ABTS and antioxidant capabilities, and free radical-scavenging properties of hydroxyl (Khaskheli et al., 2015). Similarly, Ng and See (2016) and Islam et al. (2015) showed that the fungal phenolic chemicals produced significant results in TPC, TFC and FRAP tests, showing substantial antioxidant capabilities (Fig. 3).

5.2.3 Cholesterol-lowering Effect

Presence of various bioactive compounds in *Auricularia* leads to decrease of LDL levels in blood. Higher cholesterol levels result in thickening of artery walls and clotting of blood vessels. High lipid diets, like cheese and red meat, increase the risk of hyper-cholesterol in individuals and heart ailments and related complications. The biopolymers isolated from *A. auricula-judae* fruiting body also have hypolipidemic effects, lower plasma triglycerides, total cholesterol, especially LDL cholesterol and atherogenic index, leading to increase in HDL cholesterol levels (Jeong et al., 2007). This data suggests that *A. auricula-judae* polysaccharides have a significant potential to be used as anti-hyperlipidaemia drugs (Fig. 4).

5.2.4 Anti-obesity Activity

Black ear fungi are an excellent source of new chemical compounds with a wide range of pharmaceutical properties. Anti-obesity activity has also been recognised in the hot-water extract fungi. Three types of *A. auricula-judae* suppressed adipocyte development in preadipocyte 3T3-L1 cells significantly (4.43%). The basic molecular

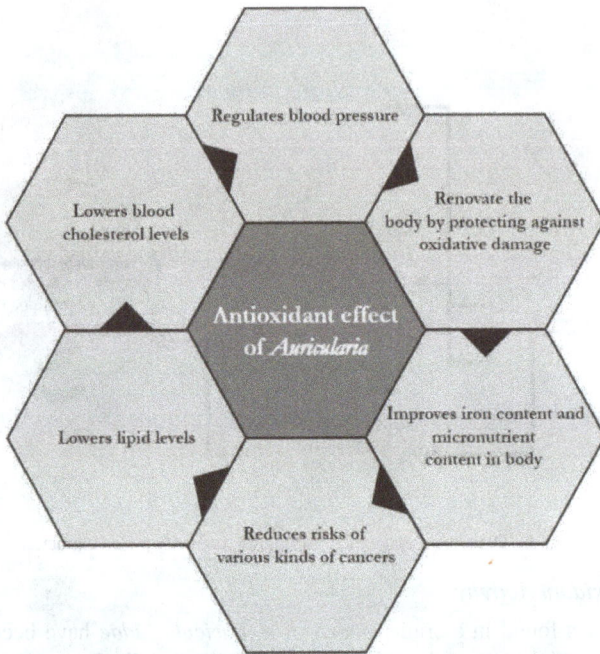

Fig. 3. Antioxidant effect of *A. auricula-judae* extract.

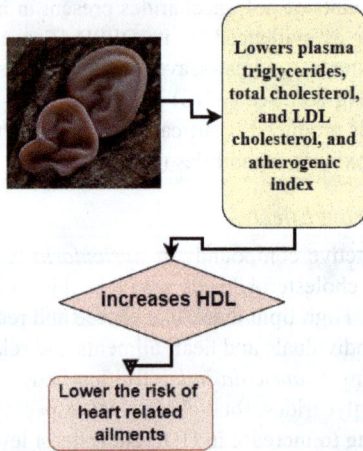

Fig. 4. Anti-hyperlipidaemic activity of *A. auricula-judae.*

mechanism appears to depend on the ability of polysaccharides to activate the AMPK signalling pathway and increase adiponectin and leptin gene expression in adipose tissue while suppressing inflammatory signals (Park et al., 2017).

5.2.5 Anti-inflammatory and Anticancerous Activity

Wang et al., 2019 revealed that *A. auricula-judae* has potent anti-inflammatory action and reduces the production of cytokines that cause inflammation during *in vitro* and

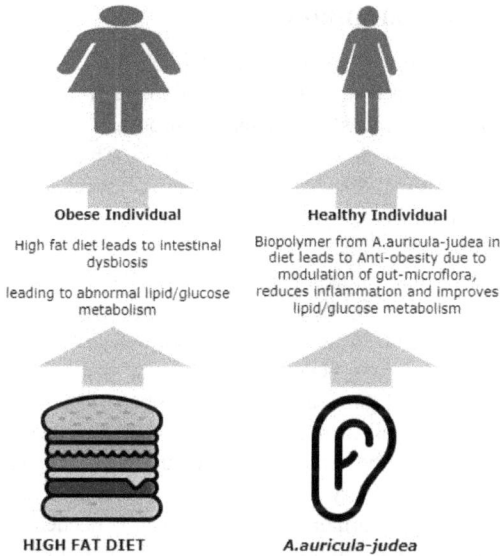

Fig. 5. Anti-obesity activity of *A. auricula-judae.*

in vivo experiments. *A. Auricula-judae's* extracted molecules show the mechanism to activate cytokines that fight inflammation and inhibit the immunological response and formation of cancer cells. The polysaccharides contained in *A. auricular-judae* have also been proven to reduce LPS-induced acute lung injury by reducing oxidative stress and blocking TNF- and IL-6 blood production. Beta-glucans are abundant in *A. auricula-judae*, and they protect against inflammatory diseases, strengthen the immune system and may have anticancer properties. The anti-proliferative efficacy of *A. auricula-judae* aqueous and ethanolic extracts on breast cancer cells has been demonstrated with significant IC value in various studies done in the last two decades (Wang and Guo, 2020).

5.2.6 Anticoagulation Activity

Various extracts from *A. auricula-judae* are responsible for inhibiting thrombin, a potent-coagulating protein in blood. The water, alkaline and acidic extracts prepared by various methods were responsible for giving positive results in assays, like partial thromboplastin time and prothromoboplastin time. The findings are very promising in establishing polysaccharides of black ear mushroom as a neutraceutical source for mitigating anti-cancerous activity in healthy individuals, who are exposed to various kinds of pollutions (Yoon et al., 2003).

5.2.7 Anti-microbial Activity and Antiviral Activity

Auricularia auricula-judea polysaccharides have been proven as antimicrobial agents through various studies conducted on bacteria and fungi. Polysaccharide extractions though ear fungi have antibacterial potency against *Cryptococcus* species (Cai et al., 2015). Sulphated polysaccharide fractions of Jew's ear fungi show antiviral effect in Newcastle Disease Virus (NDV) as per the findings of MTT assay (Nguyen et al., 2012).

6. Conclusion and Summary

A. auricula-judea contains a number of phytochemicals that have antioxidant properties that help in reducing the amount of oxygen-mediated radicals in our bodies and maintaining homeostasis. This mushroom is a popular fungus with a high concentration of active chemicals in South Asia. China is a major producer of these mushrooms. Various studies have found that polysaccharide fractions obtained from these fungi play an important bioactive role in *A. auricula-judea*. They have many health benefits by keeping a person away from all sorts of lifestyle diseases, such as diabetes, abnormal cholesterol metabolism, obesity, inflammation, cancer and other diseases. In addition, these polysaccharides have a positive effect on the health of the gut microbiota. In this regard, the current investigation is not enough. Animal research can only be beneficial. Although many chemical compounds, derivatives, and effects of *A. auricular-judae* have been identified, various complex structures and their unexplained mechanisms need to be investigated. In addition, a detailed investigation of the structure and operation of the various components is still ongoing. It is necessary to make and establish *A. auricula-judae* in a systematic way. As a result, further medical research is needed to better understand the cellular mechanisms, bioactive chemicals found internally and their molecular mechanisms in *A. auricula-judae*, as it can help prevent diseases in humans and create pharmacological treatments and effective food products. This fungus can be tested on the Indian subcontinent due to its cooking and medicinal properties like that of its other counterparts.

References

Beall, G. (1932). The life history and behaviour of the European earwig, *Forficula auricularia* L., in British Columbia. *In: Proceedings of the Entomological Society of British Columbia*, vol. 29, pp. 28–43.

Cai, M., Lin, Y., Luo, Y.L., Liang, H.H. and Sun, P. (2015). Extraction, antimicrobial, and antioxidant activities of crude polysaccharides from the wood ear medicinal mushroom *Auricularia auricula-judae* (higher Basidiomycetes). *International Journal of Medicinal Mushrooms*, 17(6).

Carrasco, J., Zied, D.C., Pardo, J.E., Preston, G.M. and Pardo-Giménez, A. (2018). Supplementation in mushroom crops and its impact on yield and quality. *AMB Express*, 8(1): 1–9.

Chen, N., Zhang, H., Zong, X., Li, S., Wang, J., Wang, Y. and Jin, M. (2020). Polysaccharides from *Auricularia auricula*: Preparation, structural features and biological activities. *Carbohydrate Polymers*, 247: 116750.

Chang, S.T. and Hayes, W.A. (eds.). (2013). *The Biology and Cultivation of Edible Mushrooms*. Academic Press.

Damte, D., Reza, M., Lee, S.J., Jo, W.S. and Park, S.C. (2011). Anti-inflammatory activity of dichloromethane extract of *Auricularia auricula-judae* in RAW264, 7 cells. *Toxicological Research*, 27(1): 11–14.

Fan, L., Zhang, S., Yu, L. and Ma, L. (2007). Evaluation of antioxidant property and quality of breads containing *Auricularia auricula* polysaccharide flour. *Food Chemistry*, 101(3): 1158–1163.

Fuller, R., Buchanan, P. and Roberts, M. (2005). Medicinal uses of fungi by New Zealand Maori people. *International Journal of Medicinal Mushrooms*, 7: 402.

Gbolagade, J.S. and Fasidi, I.O. (2005). Antimicrobial activities of some selected Nigerian mushrooms. *African Journal of Biomedical Research*, 8(2): 83–87.

Islam, K.M., Islam, M.D., Rauf, S.M.A., Khan, A., Hossain, M.K., Sarkar, S. and Rahman, M. (2015). Effects of climatic factors on prevalence of developmental stages of *Fasciola gigantica* infection in Lymnaea snails (*Lymnaea auricularia var rufescens*) in Bangladesh. *Archives of Razi Institute*, 70(3): 187–194.

Islam, T., Ganesan, K. and Xu, B. (2021). Insights into health-promoting effects of Jew's ear (*Auricularia auricula-judae*). *Trends in Food Science & Technology*, 114: 552–569.

Jeong, H., Yang, B.K., Jeong, Y.T., Kim, G.N., Jeong, Y.S., Kim, S.M. and Song, C.H. (2007). Hypolipidemic effects of biopolymers extracted from culture broth, mycelia, and fruiting bodies of *Auricularia auricula-judae* in dietary-induced hyperlipidemic rats. *Mycobiology*, 35(1): 16–20.

Jo, W.S., Kim, D.G., Seok, S.J., Jung, H.Y. and Park, S.C. (2014). The culture conditions for the mycelial growth of *Auricularia auricula-judae*. *Journal of Mushroom*, 12(2): 88–95.

Khaskheli, S.G., Zheng, W., Sheikh, S.A., Khaskheli, A.A., Liu, Y., Soomro, A.H. and Huang, W. (2015). Characterisation of *Auricularia auricula* polysaccharides and its antioxidant properties in fresh and pickled product. *International Journal of Biological Macromolecules*, 81: 387–395.

Kiho, T., Sakai, M., Ukai, S., Hara, C. and Tanaka, Y. (1985). Anti-inflammatory effect of the polysaccharide from the fruit bodies of *Auricularia* species. *Carbohydrate Research*, 142(2): 344–351.

Lamb, R.J. and Wellington, W.G. (1975). Life history and population characteristics of the European Earwig, *Forficula Auricularia* (Dermaptera: Forficulidae), at Vancouver, British Columbia. *The Canadian Entomologist*, 107(8): 819–824.

Liu, E., Ji, Y., Zhang, F., Liu, B. and Meng, X. (2021). Review on *Auricularia auricula-judae* as a functional food: Growth, chemical composition, and biological activities. *Journal of Agricultural and Food Chemistry*, 69(6): 1739–1750.

Ma, Z., Wang, J., Zhang, L., Zhang, Y. and Ding, K. (2010). Evaluation of water soluble β-D-glucan from *Auricularia auricular-judae* as potential antitumour agent. *Carbohydrate Polymers*, 80(3): 977–983.

Ma, F., Wu, J., Li, P., Tao, D., Zhao, H., Zhang, B. and Li, B. (2018). Effect of solution plasma process with hydrogen peroxide on the degradation of water-soluble polysaccharide from *Auricularia auricula*. II: Solution conformation and antioxidant activities *in vitro*. *Carbohydrate Polymers*, 198: 575–580.

Misaki, A., Kakuta, M., Sasaki, T., Tanaka, M. and Miyaji, H. (1981). Studies on interrelation of structure and antitumor effects of polysaccharides: Antitumor action of periodate-modified, branched (1→ 3)-β-D-glucan of *Auricularia auricula-judae*, and other polysaccharides containing (1→ 3)-glycosidic linkages. *Carbohydrate Research*, 92(1): 115–129.

Ng, Z.X. and See, A.N. (2019). Effect of *in vitro* digestion on the total polyphenol and flavonoid, antioxidant activity and carbohydrate hydrolyzing enzymes inhibitory potential of selected functional plant-based foods. *Journal of Food Processing and Preservation*, 43(4): e13903.

Nguyen, T.L., Chen, J., Hu, Y., Wang, D., Fan, Y., Wang, J. and Dang, B.K. (2012). *In vitro* antiviral activity of sulfated *Auricularia auricula* polysaccharides. *Carbohydrate Polymers*, 90(3): 1254–1258.

Priya, R.U., Geetha, D. and Darshan, S. (2016). Biology and cultivation of black ear mushroom— *Auricularia* spp. *Advances in Life Sciences*, 5(22): 10252–10254.

Reza, M.A., Hossain, M.A., Damte, D., Jo, W.S., Hsu, W.H. and Park, S.C. (2015). Hypolipidemic and hepatic steatosis preventing activities of the wood ear medicinal mushroom *Auricularia auricula-judae* (higher Basidiomycetes) ethanol extract *in vivo* and *in vitro*. *International Journal of Medicinal Mushrooms*, 17(8).

Royse, D.J. and Tan, J. (2017). Current overview of mushroom production in the world. pp. 5–13. *In*: Zied, D.C. and Gimenez A.P. (eds.). *Edible and Medicinal Mushrooms: Technology and Applications*. John Wiley & Sons Ltd, West Sussex, UK.

Sekara, A., Kalisz, A., Grabowska, A. and Siwulski, M. (2015). *Auricularia* spp.—mushrooms as novel food and therapeutic agents—A review. *Sydowia*, 67: 1–10.

Sheu, F., Chien, P.J., Chien, A.L., Chen, Y.F. and Chin, K.L. (2004). Isolation and characterisation of an immunomodulatory protein (APP) from the Jew's Ear mushroom *Auricularia polytricha*. *Food Chemistry*, 87(4): 593–600.

Shrikhandia, S.P., Devi, S. and Sumbali, G. (2022). Lignocellulosic waste management through cultivation of certain commercially useful and medicinal mushrooms: Recent scenario. In *Biology, Cultivation and Applications of Mushrooms*. Springer, Singapore, pp. 497–534.

Sun, Y.X., Liu, J.C. and Kennedy, J.F. (2010). Purification, composition analysis and antioxidant activity of different polysaccharide conjugates (APPs) from the fruiting bodies of *Auricularia polytricha*. *Carbohydrate Polymers*, 82(2): 299–304.

Ukai, S., Hara, C., Kuruma, I. and Tanaka, Y. (1983). Polysaccharides in fungi, XIV. Anti-inflammatory effect of the polysaccharides from the fruit bodies of several fungi. *Journal of Pharmacobiodynamics*, 6(12): 983–990.

Wang, J., Zhang, T., Liu, X., Fan, H. and Wei, C. (2019). Aqueous extracts of se-enriched *Auricularia auricular* attenuates D-galactose-induced cognitive deficits, oxidative stress and neuro-inflammation via suppressing RAGE/MAPK/NF-κB pathway. *Neuroscience Letters*, 704: 106–111.

Wang, Y. and Guo, M. (2020). Purification and structural characterisation of polysaccharides isolated from *Auricularia cornea* var. Li. *Carbohydrate Polymers*, 230: 115680.

Wu, N.J., Chiou, F.J., Weng, Y.M., Yu, Z.R. and Wang, B.J. (2014). *In vitro* hypoglycemic effects of hot water extract from *Auricularia polytricha* (wood ear mushroom). *International Journal of Food Sciences and Nutrition*, 65(4): 502–506.

Yan, P.S., Luo, X.C. and Zhou, Q. (2004). RAPD molecular differentiation of the cultivated strains of the jelly mushrooms, *Auricularia auricula* and *A. polytricha*. *World Journal of Microbiology and Biotechnology*, 20(8): 795–799.

Yao, F., Zhang, Y., Lu, L. and Fang, M. (2015). Research progress on genetics and breeding of *Auricularia auricula-judae*. *Journal of Fungal Research*, 13(3): 125–128.

Yoon, S.J., Yu, M.A., Pyun, Y.R., Hwang, J.K., Chu, D.C., Juneja, L.R. and Mourão, P.A. (2003). The nontoxic mushroom *Auricularia auricula* contains a polysaccharide with anticoagulant activity mediated by antithrombin. *Thrombosis Research*, 112(3): 151–158.

Yu, J., Sun, R., Zhao, Z. and Wang, Y. (2014). *Auricularia polytricha* polysaccharides induce cell cycle arrest and apoptosis in human lung cancer A549 cells. *International Journal of Biological Macromolecules*, 68: 67–71.

Yuan, Z., He, P., Cui, J. and Takeuchi, H. (1998). Hypoglycemic effect of water-soluble polysaccharide from *Auricularia auricula-judae* Quel. on genetically diabetic KK-Ay mice. *Bioscience, Biotechnology and Biochemistry*, 62(10): 1898–1903.

CHAPTER 6

Cloud Ear Fungus
(Auricularia polytricha)

Siriporn Chuchawankul,* Sharika Rajasekharan,
Nichaporn Wongsirojkul and *Sunita Nilkhet*

1. Introduction

Wild and cultivated edible mushrooms are rich sources of moisture, fibre, amino acids, carbohydrates, fats, essential vitamins and minerals that fulfil human dietary and energy requirements. *Auricularia* spp., have been reported to be one among the most cultivated mushrooms in the world and widely cultivated in China (Cheung, 2008) for culinary and medicinal uses. The preference to natural product-based therapeutic approach has opened the possibility of using edible food as preventive and protective agents against a vast number of ailments. The present chapter focuses on the recent findings, highlighting the important nutraceutical properties of *Auricularia polytricha*.

2. Ecological Importance, Ethno Biology and Cultivation

Auricularia polytricha (AP) belongs to the *Auriculariaceae* family of class *Agaricomycetes* and order *Basidomycota*, which includes the saprophytic fungi distinguishable by their gelatinous ear-like shaped fruiting body. Currently around 37 species have been identified across the world (Sekara et al., 2015; Wu et al., 2021). *Auricularia* spp., are widely found in tropical forests inhabiting live tree trunks, fallen branches as well as dead or degrading wood, contributing to the ecosystem (Priya et al., 2016; Wu et al., 2021). The fruiting body is distinctly shaped, resembling a floppy ear or cup-shaped structure with size ranging between 3–12 cms across which

Immunomodulation of Natural Products Research Unit, Department of Transfusion Medicine and Clinical Microbiology, Faculty of Allied Health Sciences, Chulalongkorn University, Bangkok, Thailand.
* Corresponding author: Siriporn.ch@chula.ac.th

is it attached to the substrate with the stalk (Priya et al., 2016). Owing to its shape, AP was popularly known as Jew's ear. An epithet derived from Judas's ear as the species was found to grow commonly, though not exclusively, on elder tree, associated with the death of Judas Iscariot. Presently other names, such as jelly ear, wood ear or cloud ear are also used. Earlier records reveal that it was commonly used in European folk-medicine till the 18th and 19th centuries to treat throat and eye ailments. Since ancient times *Auricularia auricula-judae* (balck ear, wood ear) and *Auricularia polytricha* (red ear, cloud ear) have an irreplaceable association with Asian food and traditional medicine. It has been popularly used to strengthen the body as well as to treat various ailments, such as fever, haemorrhoids, haemoptysis, angina and the prevention and treatment of thrombosis (Sekara et al., 2015).

The fruiting body has a gelatinous and elastic texture with smooth and bright reddish-brown colour with a purple tint that tends to grow darker as it ages, tiny, downy grey hair along with smooth or undulating folds and wrinkles. The inner surface has a greyish brown appearance with similar folds and wrinkles, giving the fruiting body a resemblance to the ear. Dried fruiting body becomes hard and brittle compared to the fresh ones. The spores are white to cream and yellowish in colour with the appearance resembling a sausage with a size range between 16–18 µm and 6–8 µm thickness. Environmental factors along with exposure to light, age of the specimen as well as availability of moisture play a major role in the morphology of *Auricularia* spp. Occurrence of morphological plasticity, resulting in difficulty in distinguishing various species, has led to opting of molecular marker-based characterisation approaches over the conventional Lowy's method based on the hyphal structure of internal stratification of Basidiome (Bandara et al., 2015; Du et al., 2011; Onyango et al., 2016; Priya et al., 2016). The development of EST and SSR marker-based identification techniques has eased the analysis and identification of genetic variations while helping in accelerating the breeding methods in the future (Yu et al., 2008; Zhou et al., 2014).

Over the past few decades, domestication and cultivation of *Auricularia* spp., has increased due to its extensive demand as a culinary ingredient and also its medicinal properties. New research strategies are being applied for the economical cultivation and extraction of bioactive compounds. Previous records from the South Asian countries reveal the cultivation of *Auricularia* spp., as a traditional ingredient in many medicinal concoctions. Cultivation of *Auricularia* spp., as early as Tang dynasty has been recorded in the Li Shih-Chen's *Pen Tsao Kang Mu* in Ming dynasty stating: "Put the steamed bran on logs, then cover with straw; wood ear will grow"; though conclusive records are not available (Chang, 1977). Available knowledge on origin and cultivation of mushrooms is scanty. Present-day cultivation process progresses through spawn preparation to fruiting body production which involve various steps. Commercial production of mother spawn usually starts from tissue-cultured mycelia. Bottles or polypropylene bags filled with softened grains along with bran as substrates are sterilised through autoclaving at 121°C for one hour. Transfer of inoculum as well as storage requires a contamination-free environment. The bags or bottles are stored for about seven to 20 days in dark, provided with ambient temperature. The supplementation of mixture with protein-rich media also increases the mycelia growth. Furthermore, the fully grown mycelia are used for growing the fruiting body through various methods, such as the log method. The log pieces are inoculated with the *Auricularia* spawn by drilling a hole of 1 cm depth and these are piled up together. The production of fruiting body takes a lot of time by this method. The artificial log method comprises of two variations. One method is cultivation in sawdust, wherein sawdust is soaked in water, sun-dried and mixed with rice bran, white sugar and calcium carbonate; the mixture is then filled in polypropylene bags and sterilised. Once the inoculum is introduced, the bags are incubated in sterile conditions for mycelial growth and later on, cuts are drawn on bags for facilitating growth of the fruiting body. Compost cultivation method also utilises sawdust mixed with rice bran, white sugar and calcium carbonate, except that it has a moisture content of about 65–70%. The replacement of sawdust with rubber-tree sawdust was found to improve the overall yield of *Auricularia* spp., as well as the supplementation of wheat bran also increased the yield during harvest (Onyango et al., 2013; Priya et al., 2016). The polybag method of cultivation, using wheat straw supplemented with wheat and rice bran, is a high-yielding method of cultivation for *Auricularia polytricha* (Priya et al., 2016).

Currently edible mushroom-based products and their derivatives are of high popularity and high demand commercially owing to their health-promoting trait. Recent years have seen a growth of cutting-edge research, including muti-omics and sequencing studies that aim to understand the potential bio-actives present (Miao et al., 2020) as well as the regulation of gene expression to complement our knowledge on the regulation of development of molecular, metabolic and physiological attributes (Wang et al., 2021). Nutritional values show 100 g of dried *Auricularia polytricha* provides about 293 kcal consisting of 81 g of carbohydrates, 8.1 g of protein, 1.5 g of fats, 9.4 g of ash and 6.9 g of crude alkaloids, thiamine, riboflavin, ascorbic acid, vitamin D_2 and other minerals (Sekara et al., 2015). The polysaccharides from the fruiting body isolated through various extraction methods also showed high antioxidant capacity. Large-scale production of polysaccharide through submerged culture was proposed by (Wu et al., 2006). Apart from health benefits, *Auricularia polytricha* is well known for its ability to serve as an absorbent for removing emulsified oil from water (Yang et al., 2014). The larger biomass, strong texture and high bio-absorption capacity has gained its acceptance as a medium to absorb oil and heavy metal absorption (Zheng et al., 2014). Other innovative uses include hydrogel preparation from *A. polytricha*-derived β-glucan, for intestinal bio-release of Vitamin B (Zhu et al., 2018).

The present chapter focuses on the nutraceutical benefits imparted by *Auricularia polytricha* and its derived compounds through induction of regulation of host-regulatory players. The mushroom reported to display antioxidant activity (Chen and Xue, 2018) and various protective effects, including anticancerous (Song and Du, 2010; Yu et al., 2014), anti-inflammatory and immunomodulatory (Sangphech et al., 2021; Sillapachaiyaporn et al., 2019), neuroprotective (Bennett et al., 2013; Sillapachaiyaporn et al., 2021), hepatoprotective (Chellappan et al., 2016), antihypercholestric (Zhao et al., 2015), antidiabetic activity (Wu et al., 2014), will be discussed in detail in the chapter (Fig. 1).

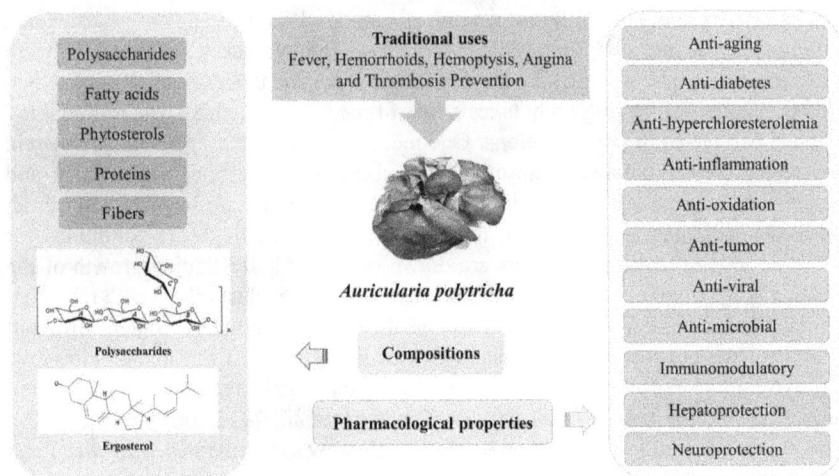

Fig. 1. The graphical overview of *Auricularia polytricha.*

3. Chemical Composition of *Auricularia polytricha*

Auricularia polytricha has been investigated for its composition through many methodologies, reporting carbohydrates (Chen and Xue, 2018), phytosterols, proteins, fatty acids and vitamins (Table 1). The most well-known mushroom component is polysaccharides, especially from water extraction (Yu et al., 2014). Polysaccharides are composed of different linked molecules and backbones, resulting in various molecular weights. The crude water extract of *Auricularia polytricha* has been reported to consist of $(1\rightarrow3)$-linked-β-D-glucopyranosyl polysaccharides as the backbone (Song and Du, 2010; Yu et al., 2014). The low molecular weight of polysaccharides was also purified, which carried out subfraction of polysaccharides (Sun et al., 2010). The polysaccharides were separated for their characteristics by size and structural characteristics. Methodologies have been applied to determine the attributes of monosaccharides, e.g., HPLC, GC, NMR, FT-IR, GPC, GC-MS and LC-MS/MS. Monosaccharides are composed of the smallest unit of carbohydrates and can be distinguished in substance form (Sun et al., 2010; Xiang et al., 2021). The previous study had reported that enzymatic-hydrolysed polysaccharides and purified polysaccharides shared most of the monosaccharide patterns; still, the mass percentage of sugar ratio differed, leading to differences in sizes and biological activities (Song et al., 2021).

Apart from polysaccharides, fatty acids and phytosterols were also discovered in mushrooms. *Auricularia polytricha* itself showed abundant sterols in a crude hexane extract reported in the study, considered a lipophilic fraction (Sillapachaiyaporn et al., 2019). In AP studies, linoleic acid and oleic acid were the primary fatty acids (Saini et al., 2021). As the vitamin D_2 precursor, the critical compound during UV radiation exposure, the ergosterol, was transformed into vitamin D_2. Ergosterol transformation was examined by HPLC with UV-B radiation exposure increasing the level of vitamin D in mushrooms (Huang et al., 2015). This phenomenon has been clarified in *Agaricus bisporus* and *Cordyceps militaris* mushrooms and provides that edible mushrooms

Table 1. The chemical composition of *Auricularia polytricha*.

Substances		Composition	Contents	Analytical Techniques	References
Polysaccharides					
	Crude water-soluble polysaccharides	Carbohydrates	80.69 ± 1.25% of total carbohydrates	HPLC[a], GC[b]	Chen and Xue, 2018
	Crude polysaccharide APPS	(1→3)-linked-β-D-glucopyranosyl and (1→3, 6)-linked-β-d-glucopyranosyl residues in a 2:1 ratio, (1→)-β-d-glucopyranosyl at the O-6 position of (1→3, 6)-linked-β-d-glucopyranosyl	The crude sample (2.0 g) was contained of AAPS-1 (192 mg), AAPS-2 (137 mg), and AAPS-3 (98 mg)	HSCCC[c], NMR[d] spectroscopy (^{1}H, ^{13}C)	Song and Du, 2010
	Low molecular-weight polysaccharide (APP)	(1→3)-linked-b-D-mannopyranosyl, (1→3)-linked-b-D-galactopyranosyl and (1→3, 6)-linked-b-D-mannopyranosyl with one single terminal (1→3)-b-D-glucopyranosyl at the O-6 position of (1→3,6)-linked-b-D-mannopyranosyl	CAPP (36.2 g) from 0.5 kg of AP dry weight	NMR[d] spectroscopy (^{1}H, ^{13}C), DEAE–cellulose anion exchange Chromatography	Sun et al., 2010
	Purified polysaccharides	APPsA-1, APPsB-1, APPsB-2 and APPsC-1		Size-exclusion chromatography and ion-exchange chromatography	Sun et al., 2010
		AAP, AAP-10, APP, and APP-10		HPLC[a]	Xiang et al., 2021
		Enzymatic-hydrolysed polysaccharides		HPLC[a], FT-IR[e], GPC[f]	Song et al., 2021

Table 1 contd. ...

...Table 1 contd.

Substances	Composition	Contents	Analytical Techniques	References
Fatty acids and sterols				
Ergosterol and fatty acids	Ergosterol, Ergosta-5,7-dien-3b-ol, Ergosta-7-3b-ol and Ergosta-7,22-dien-3b-ol, linoleic, palmitic, oleic and stearic acid	Total sterols (100.4 mg/100 g DW), total fatty acids (0.53% DW)	GC-MS[g]	Saini et al., 2021
Crude hexane extract	Palmitic acid, linoleic acid, oleic acid, steric acid, anthraergostapentene, and ergosterol		GC-MS[g]	Sillapachaiyaporn et al., 2019
UV-B radiation on fresh mushroom powder	Vitamin D_2	60.29 ± 3.59 mg/g dry matter	HPLC[a]	Huang et al., 2015
Others				
Fibrinolytic enzymes		Fibrinolytic activity at 10.83U mg of freeze-dry *A. polytricha*	SDS-PAGE[h]	Mohamed Ali et al., 2014
Carbohydrate-active enzymes	5 glyoxal oxidase, AA5 galactose oxidase, glycoside hydrolase (GH) family 20 hexosaminidase, and GH47 alpha-mannosidase		LC-MS/MS[i]	Jia et al., 2017

[a]HPLC = High performance liquid chromatography, [b]GC = Gas chromatography, [c]HSCCC = High-speed counter current chromatography, [d]NMR = Nuclear magnetic resonance, [e]FT-IR = Fourier-transform infrared spectroscopy, [f]GPC = Gel permeation chromatography, [g]Gas chromatography-mass spectrometry, [h]SDS-PAGE = Sodium dodecyl sulphate polyacrylamide gel electrophoresis, [i]LC-MS/MS = Liquid chromatography-tandem mass spectrometry.

potentially are vitamin D resources (Hu et al., 2021). Moreover, other active enzymatic activities have been investigated in the proteomic and metabolic analysis. For example, fibrinolytic enzymes, derived from *Auricularia polytricha*, were reported with fibrinolytic activity and purified by the SDS-PAGE technique (Mohamed Ali et al., 2014). Furthermore, the proteomic analysis of *Auricularia polytricha* was firstly done by LC-MS/MS to identify the proteins. In addition, carbohydrate-active enzymes were diagnosed (Jia et al., 2017). According to literature, *Auricularia polytricha* was rich with polysaccharides, sterols and proteins and the diversity of patterns depends on the methodologies of extraction process and chemical analysis.

4. Pharmacological Properties of *Auricularia polytricha*

Auricularia Polytricha (AP), an edible cloud ear-like mushroom, is recognised in Chinese traditional medicine and is widely used as medicinal, food and antibiotic material. It contains plenty of nutraceutical and pharmaceutical compounds and exists as a rich source of dietary fibre. Interestingly, the active components exhibit diverse biological functions, including anti-tumour, anti-oxidation, anti-inflammation, attenuating lipid accumulation, hypoglycemia, anti-HIV, relieving constipation, and neuroprotective (Table 2). The mechanisms behind this are very attractive to investigate. Herein, the recent discovery of some mechanisms is summarised and discussed in this chapter.

4.1 Antitumour Activity

Cancer results from uncontrolled cell division and spread to surrounding tissues. Several cancerous cells grow and develop into tumour cells. Polysaccharides become interesting candidates in the fields of medicine and pharmacy due to their high therapeutic efficacy and low toxicity. They are short-chain carbohydrates (glucan) containing at least 10 monosaccharides via glycosidic bonds. The composition, lengthening and molecular weight are the key factors for their structures and biological functions (Li et al., 2021). Crude polysaccharides of AP are recognised as the potent active ingredients for immunomodulatory and anti-tumour effects by suppressing the proliferation and DNA synthesis of human lung cancer A549 and enhancing apoptosis (Yu et al., 2014). The isolated polysaccharide fractions of AP (APIIA, 10 kDa) which is β configuration polysaccharide containing mannose and xylose, showed significant *in vivo* anti-tumour test by enhancing the phagocytosis of macrophages and promoting the release of nitric oxide (NO) and cytokines (Yu et al., 2009). β-D-glucan from AP, separated by high-speed counter-current chromatography (AAPS-2), showed the highest anti-tumour effects in mice (Song and Du, 2010). Another study in the S180 tumour cell showed the suppression effects on tumour growth of the novel α β-glucan polysaccharide with an average molecular weight of 1.62×10^5 Da (Song and Du, 2012). The higher molecular weight of glucan is reported to have a higher efficacy against tumour proliferation (El Enshasy and Hatti-Kaul, 2013). Recently, polysaccharides containing β-glucan, a linear polysaccharide of D-glucose monomers linked via β-glycosidic acid, are the beneficial agents for use against tumours since they can bind to the receptors on the surface of immune cells, which then activate the pro-inflammatory response of innate immune cells and anti-tumour defence (Wu et al., 2021).

Table 2. Pharmacological findings of *Auricalaria polytricha*.

Sources/Bioactive	Extraction Methods	Cell Types/Model Use	Description of Findings	References
Protein	5% cold acetic acid in the presence of 0.1% 2-mercaptoethanol, followed by ammonium sulphate fractionation, DE-52 and Mono Q anion-exchange chromatography.	Murine macrophage cell line RAW 264.7.	Promoting cell proliferation and IFN-γ secretion by lymphocytes, increases NO and TNF-α production by LPS-induced macrophages and strengthens the immune response.	Sheu et al., 2004
Polysaccharides	Water extract (100°C, 2 hr), centrifugation, evaporation, precipitated using ethanol, removed protein using dialysed 3 kDa MW cut-off against water, precipitated using 95% ethanol for 24 hrs 4°C	Male-specific pathogen-free BALB/c albino mice; RAW 264.7 cells.	A fraction of polysaccharides (namely APPIIA) showed a significant anti-tumour effect, enhanced the neutral red phagocytosis of macrophages.	Yu et al., 2009
Crude polysaccharides	95% ethanol extract at 75°C under reflux to remove lipid, followed by water extract 75°C, and precipitated by 95% ethanol.	*In vitro*.	Antioxidative modulator (uronic acid dose-dependent).	Sun et al., 2010
Crude polysaccharides	Water extract 90°C, precipitated by ethanol 4°C overnight, deprotonated and dialyzed against water 48 hr.	Swiss albino female mice four to six weeks old.	Three polysaccharides (AAPS-1, AAPS-2 and, AAPS-3) were separated from a crude extract of *A. polytricha* by HSCCC. AAPS-2 showed excellent inhibition against the growth of S180 sarcoma in mice.	Song and Du, 2010
Glucan	Protease-assisted aqueous extraction followed by ethanol precipitation and purified by gel filtration chromatography.	S180-bearing mouse model.	Anti-tumour immunomodulator in both *in vivo* and *in vitro* experiments with non-toxic effects.	Song and Du, 2010
Whole components	Commercial cultivated strain of AP (spray-dry).	Male Sprague-Dawley rats.	1) Delay the progression of non-alcoholic fatty liver disease (NAFLD). 2) Lessened lipid accumulation. 3) Increase in antioxidant levels.	Chiu et al., 2014

Total polysaccharide	Water extract at 95°C; 3 hrs, freeze-dried, stored at −20°C.	*In vitro.*	Contribute hypoglycemia state by suppressing the activity of alpha-amylase (digestion of polysaccharides enzyme).	Wu et al., 2014
Crude polysaccharides	95% ethanol extract at 75°C under reflux to remove lipid, followed by water extract 75°C and precipitated by ethanol.	A549 cells.	Promote apoptosis by suppressed proliferation and DNA synthesis of human lung cancer A549 cell is dose-dependent.	Yu et al., 2014
Polysaccharides	Rehydrated fruiting bodies in reverse osmosis (RO) water (five times) at 126°C, high pressure (1.2 kg/cm^2) for 30 min, ultrasonicated for 1 hr, filtration (130–140 mesh), spray-dried and ground to fine powder.	Forty-six-week old male Sprague-Dawley rats.	Rats receiving either 0.75% or 1.5% AP exhibited effective interruption of NAFLD progression, by decreased lipid accumulation and elevated antioxidative status.	Chiu et al., 2014
Crude polysaccharides	Water extract at 95°C, centrifuge 7000 rpm 20 min, evaporation, precipitated by ethanol.	Male Sprague-Dawley rats.	Suppress serum concentration of blood lipid to the normal level, especially the cholesterol level is significantly reduced.	Zhao et al., 2015
Polyphenolic and flavonoid content	Soxhlet extraction using ethanol and water 12 hrs, rotary vacuum evaporator, filter.	*In vitro.*	Both water and ethanol extracts have antioxidant activity. In addition, ethanol extracts had antimicrobial activity on tested organisms, but antimicrobial properties were not determined in distilled water extracts.	Emre et al., 2016
Polysaccharides	Mascerated water extract (75–80°C) 48 hrs, freeze-dried.	Sprague–Dawley rats.	Extract treatment significantly attenuated the paracetamol-induced increase in AST, ALT, ALP, LDH, TB, TG and cholesterol and increased the diminished TP in dose-dependent manner.	Chellappan et al., 2016
Crude polysaccharides denoted as APP3a	Water extract at 80°C, centrifuged 4500 rpm, 20 min, precipitated with 95% ethanol 4°C for 48 hrs.	*In vitro.*	APP3a had strong reducing power and strong scavenging activity against hydroxyl radicals, superoxide radicals and DPPH radicals in a dose-dependent manner.	Chen and Xue, 2018

Table 2 contd. ...

...*Table 2 contd.*

Sources/ Bioactive	Extraction Methods	Cell Types/Model Use	Description of Findings	References
Crude hexane extract	Macerated in hexane 72 hrs, filtered and evaporated.	*In vitro.*	Inhibited HIV-1 PR activity	Sillapachaiyaporn et al., 2019
Ultrasonic-assisted melanin extract	Ultrasonication.	*In vitro.*	Providing good scavenging activity on 2,2'-azinobis-(3-ethylbenzthiazoline-6-sulphonate) (ABTS); 1,1-diphenyl-2-picrylhydrazyl (DPPH) and hydroxyl radical, as well as had an obvious protective effect on H_2O 2-induced L02 cell damage.	Xin et al., 2019
Auricularia polytricha-derived dietary fiber	Dried at 60°C and milled by grinder into smaller size and filter through 100-mesh sieve.	Male Sprague-Dawley rats.	High constipation-relieving activity.	Jia et al., 2020
Polyphenol and flavonoid	Macerated in ethanol 72 hrs, filtered and evaporated.	STZ-induced diabetic rats.	AP can prevent and attenuate diabetes-induced testicular dysfunction by alleviating DNA damage and levels of serum oxidative stress markers in treating diabetes-induced testicular dysfunction.	Abang and Anyanwu, 2020
Heteropolysaccharides	Water extract at 105°C, precipitated by ethanol and deproteinisation using ethanol.	Male Sprague-Dawley rats.	Relieve oxidative stress and diabetic pathogenesis in streptozotocin-induced diabetic mice.	Xiang et al., 2021
Ethanol extract	Sequential maceration for 72 hrs, filtered and evaporated.	RAW 264.7 cells.	APE crude extract promotes lipid accumulation and anti-inflammatory activity by suppressing nitric oxide production.	Sangphech et al., 2021

Auricularia polytricha noodles	Wheat flour mixed with *A. polytricha* granules (particle size of 200 mesh).	Seven-week-old C57BL/6 N male mice.	A. polytricha consumption could decrease the serum concentrations of blood lipid induced by high-fat diet feeding by enriching the diversity of gut microbiota and improve dysbiosis.	Fang et al., 2021
Ethanol extract	Macerated in ethanol 72 hrs, filtered and evaporated.	HT-22 mouse hippocampal cell line, caenorhabditis elegans.	Ethanolic extract exhibited the most potent effects on the inhibition of intracellular ROS accumulation and protection against neuronal death in glutamate-induced HT-22 hippocampal cells. Moreover, APE could alleviate the expressions of antioxidant-related genes, including Sod2 and Gpx. *In vivo* investigations showed that APE promoted longevity and improved the health span of the wild-type C elegans model.	Sillapachaiyaporn et al., 2021

4.2 Immunomodulation

Immune systems generate several types of immunomodulatory, both immunostimulatory and immunosuppressive, for protection and maintaining homeostasis. Various bioactive substances of AP provide immunomodulatory functions including polysaccharides, proteins, glycopeptides and terpenes (Wang et al., 2012). Even though carbohydrate isolated from AP provides anti-tumour activity as mentioned above, it was found that AP protein which is free of carbohydrate isolated enhances the proliferation of murine splenocyte via the increased level of NO and pro-inflammatory cytokines (Sheu et al., 2004). AP protein could be an immune stimulant boosting the immune response. A study of the purified AP protein in macrophages is in agreement with the earlier-mentioned results that boost the immunity system by enhancing the production of TNF-α and NO (Chang et al., 2007). Moreover, some specific crude polysaccharides with 23.51 kDa can display immunomodulatory activities by enhancing the cell proliferation of RAW264.7 macrophage, NO production and phagocytosis which is further related to the anticancer activity (Bao et al., 2020). Glycoproteins and terpenes (the five-carbon isoprene) isolated from mushrooms can also provide specific immunomodulatory activities. Instead of enhancing cytotoxic activity, they tend to proliferate and activate the immune system substances, including natural killer cells, neutrophils and macrophages (Hassan et al., 2015).

4.3 Anti-inflammation Activity

Inflammation is a natural response to injury or damaged conditions from external stimuli (Medzhitov, 2008). Inflammation is exactly related to cancer. In agreement with anticancer properties, AP could play an anti-inflammatory role. Inflammation is the multiplex network of several cytokines and chemokine cascades. The pathogenesis of inflammation is the result of the elevated level of NO and some pro-inflammatory cytokines, including tumour necrosis factor-α (TNF-α), interferon-γ (IFN-γ), interleukin-6 (IL-6) and interleukin-1β (IL-1β). Several studies showed that AP can attenuate the production of NO and pro-inflammatory factors (Sangphech et al., 2021; Xiang et al., 2021). AP consists of saccharides, proteins, fatty acids, phenolic compounds, indole compounds, vitamins, terpenoids and micro-elements which could play a specific activity in promoting anti-inflammation. Polysaccharides play the role in influencing the immunity and alteration of defence responses. It is described that disaccharides, such as trehalose which can inhibit the cyclooxygenase-2 (proinflammatory proteins, COX-2), and activate nitric oxide synthase and interleukin-1 (IL-1), serve as anti-inflammation mediators (Muszyńska et al., 2018). Furthermore, polysaccharides, such as β-glucan, alter both pro-inflammatory and anti-inflammatory cytokines to inhibit the inflammation by suppressing COX-2 proteins and IL-1 (Muszyńska et al., 2018; Sangphech et al., 2021). In addition, the particular constituents of amino acids, such as leucine, isoleucine, tyrosine and phenylalanine from mushrooms influence the anti-inflammation activity (Jedinak et al., 2011). L-ergothioneine is another example which is also found to provide both anti-inflammation and anti-oxidation activities by suppressing both H_2O_2 and TNF-α as well as scavenging free radicals (Gunawardena et al., 2014). Moreover, the ingredients of mushroom are rich in unsaturated fatty acids (PUFAs) which are the precursors of eicosanoids. Eicosanoids are closely related to inflammation activity. Fatty acid contained in AP is mainly composed of linoleic (C18:2 n-6), linolenic (C18:3 n-3) and oleic (C18:1 n-9), which exhibit anti-inflammation activity. Ergosterol, the particular mushroom sterol and the precursor of vitamin D, also acts as an anti-inflammatory bioactive agent. The phenolic compounds display a major role in anti-oxidation and can also inhibit inflammatory enzymes, including cyclooxygenase, lipoxygenase, microsomal monooxygenase, and NADH oxygenase (Muszyńska et al., 2018).

4.4 Antioxidation Activity

Oxidative stress is the state of inequality between oxidants and antioxidants. Oxidants or reactive oxygen species (ROS) are free radical compounds, such as superoxide, peroxynitrite and hydroxyl (Siti et al., 2015). The increase of oxidative stress leads to cell injury. Therefore, oxidative stress and inflammation are closely related. Oxidative stress can generate inflammation. As mentioned above, AP can mediate anti-inflammation activity as well as anti-oxidative stress which is one response against anti-inflammation. Several AP polysaccharides with various contents of monosaccharides were reported to repress the hydroxyl radicals and decompose peroxides. One of them consisting of arabinose: mannose: glucose: galactose in the ratio of 1:1.33:1.06:1.23 exhibits antioxidant activity by scavenging hydroxyl radicals and providing hydrogen for stable radical formation (Chen and Xue, 2018). Interestingly, the uronic acid content can alter the anti-oxidation level. The higher the uronic acid concentration the stronger the anti-oxidation displayed (Sun et al., 2010). One possible reason is the higher capability of uronic acid to donate the hydrogen atom and produce a harmless product. In addition, phenolic compounds are considered to be the key anti-oxidative mediators. They can easily donate electrons and neutralise ROS. The anti-oxidative compounds found in mushroom are gallic, protocatechuic, p-hydroxybenzoic, p-coumaric, cinnamic, and caffeic acids (Ferreira et al., 2009). The higher the total phenolic content the higher the level of anti-oxidation.

4.5 Antidiabetic Activity

Diabetes mellitus (DM) is a chronic metabolic disease. There are several treatments for hypoglycemic conditions, including insulin injection and synthetic pharmaceutical medicine. Interestingly, AP constituents are related to hypoglycemic-inducing activity (Chen and Xue, 2018). AP has an abundance of soluble viscous fibres, such as cellulose. It was found that fibres are prone to bind glucose, keeping the intestinal lumen free of glucose (Wu et al., 2014). Also, fibres act as a barrier for disrupting the α-amylase activity. Moreover, it is considered that polysaccharides from AP containing particular monosaccharides could serve as mediators in NF-κB signalling pathways (Xiang et al., 2021). The glucan-rich polysaccharides increase body weight and insulin content but reduce blood glucose levels comparable to metformin (Kanagasabapathy et al., 2012). It is very interesting to further investigate the mechanisms and the development of this novel hypoglycemic compound.

4.6 Anti-hypercholesterolemic Activity

Hypercholesterolemia is the key factor in several diseases, including atherosclerosis, hypertension and cardiovascular diseases. The level of blood triglyceride, cholesterol and lipoprotein (LDL-C and VLDL-C) are commonly co-elevated. However, recent medicines to attenuate the lipid, i.e., fibrates, statins and niacin have low efficacy and cause several side effects (Lee et al., 2013). Therefore, the development of high-efficacy anti-hypercholesterolemia substances is still challenged. Numerous polysaccharides exhibit anti-hypercholesterolemia and anti-hyperlipidemia in clinical trials (Miao et al., 2020). Oral administration of soluble polysaccharides from AP showed a lower level of total cholesterol, triglycerides and LDL-cholesterol close to normal levels in hypercholesterolemic rats in 28 days (Zhao et al., 2015). It was mentioned that the molecular weight and viscosity of polysaccharides are the vital factors in cholesterol bio-accessibility and bioavailability (Silva et al., 2021). Also, the lower molecular weight and viscosity of AP polysaccharides by gamma irradiation showed high potential in anti-hypercholesterolemia in C57 mice (Li et al., 2022). Also, it was reported that AP also inhibits the absorption of lipids by competitive binding to lipid molecules in a lumen. Other factors to enhance the efficacy are still required.

4.7 Antiviral and Antimicrobial Activity

A great variety of micro-organisms are harmful to other living organisms. Antimicrobial agents are assisted to protect against undesirable effects, including odours, colouring, taste and contacting diseases. The current synthetic antimicrobial agents may provide undesirable side effects and may be toxic. The safe antimicrobial and antiviral agents are very attractive to identify. Interestingly, AP extract from ethanol and hexane inhibited the HIV-1 PR activity. Lipid contents, including two triacylglycerols, linoleic acid and ergosterol are responsible for inhibitory activity (Sillapachaiyaporn et al., 2019). In addition, polysaccharides and terpenes from mushrooms play a role in antiviral activity (Dasgupta and Acharya, 2019; Faccin et al., 2007). Moreover, anti-microbial activity of AP polysaccharide, chitosan, was found to inhibit the growth of both Gram-positive and negative (*Staphylococcus aureus* and *Escherichia coli*, respectively) compared to the commercial chitosan (Chang et al., 2019). Additionally, the β-(1,3)-glucan-exopolysaccharides are reported to be another anti-microbial agent, protecting the host by activating the phagocytes (Perera et al., 2020). The informative mechanisms of both antiviral and antimicrobial activity are in line for in-depth investigation.

4.8 Neuroprotective Activity

Alzheimer's disease is a neurodegenerative disorder due to the impairment of amyloid plaques and the phosphorylated tau protein related to the inflammation cascade. Mushrooms are an abundant source of anti-inflammation and anti-oxidation potency. Moreover, the ethanolic extract of AP could protect against neural death in glutamate-induced HT-22 hippocampal cells (Sillapachaiyaporn et al., 2021). In addition, the activity of beta-secretase (BACE1), the repressed target of Alzheimer's disease, can be suppressed by AP polysaccharides (Bennett et al., 2013). AP aqueous extracts acted against the seizure which is the initial risk in Alzheimer's disease (Gupta et al., 2019). Although recent reports showed the potency of AP on neurodegenerative diseases, *in vivo* model is opened for elucidation.

5. Conclusion

Auricularia polytricha is a popular edible mushroom which is highly regarded for it therapeutic value in oriental medicine as well as a culinary delight. Recent research strategies have brought out the various pharmacological properties of the mushroom proving it to be an excellent functional dietary food, which provides energy as well as improves the overall health of an individual. *In vitro* and *in vivo* studies show immunomodulatory effects via the regulation of innate, complement mediated and adaptive immunity by enhancing the active mechanism of immune system, such as the macrophages, IL, TNF-α, IFN-γ, NO and the complement system as well as ROS scavenging and antioxidant activity along with desirable qualities, such as antidiabetic and anti-hyperchloresterolemic effect and protection against neurodegenerative disorders. Apart from health benefits, the hydrogel formulation produced from the fruiting body serves to be an excellent drug delivery system which needs to be studied in depth. In the light of the above-mentioned pharmaceutical and chemical properties discussed in the chapter, *Auricularia polytricha* can be considered beneficial to human diet and health.

References

Abang, A.C. and Anyanwu, G. (2020). *Auricularia polytricha* (mushroom) regulates testicular DNA expression and oxidative stress markers of Streptozotocin-induced diabetic male Wistar rat. *Int. J. Nutr.*, 5(3): 7–15.

Bandara, A.R., Chen, J., Karunarathna, S., Hyde, K.D. and Kakumyan, P. (2015). A*uricularia thailandica* sp. nov. (*Auriculariaceae, Auriculariales*) a widely distributed species from southeastern Asia. *Phytotaxa*, 208(2): 147–156.

Bao, Z., Yao, L., Zhang, X. and Lin, S. (2020). Isolation, purification, characterisation and immunomodulatory effects of polysaccharide from *Auricularia auricula* on RAW264, 7 macrophages. *J. Food Biochem.*, 44(12): e13516.

Bennett, L., Sheean, P., Zabaras, D. and Head, R. (2013). Heat-stable components of wood ear mushroom, *Auricularia polytricha* (higher Basidiomycetes) inhibit *in vitro* activity of beta secretase (BACE1). *Int. J. Med. Mushrooms*, 15(3).

Chang, A.K.T., Frias, Jr. R.R., Alvarez, L.V., Bigol, U.G. and Guzman, J.P.M.D. (2019). Comparative antibacterial activity of commercial chitosan and chitosan extracted from *Auricularia* sp. *Biocatalysis and Agricultural Biotechnology*, 17: 189–195.

Chang, H.-H., Chien, P.-J., Tong, M.-H. and Sheu, F. (2007). Mushroom immunomodulatory proteins possess potential thermal/freezing resistance, acid/alkali tolerance and dehydration stability. *Food Chem.*, 105(2): 597–605.

Chang, S.-T. (1977). The origin and early development of straw mushroom cultivation. *Econ. Bot.*, 31(3): 374–376.

Chellappan, D.K., Ganasen, S., Batumalai, S., Candasamy, M., Krishnappa, P., Dua, K., Chellian, J. and Gupta, G. (2016). The protective action of the aqueous extract of *Auricularia polytricha* in paracetamol induced hepatotoxicity in rats. *Recent Patents on Drug Delivery & Formulation*, 10(1): 72–76.

Chen, Y. and Xue, Y. (2018). Purification, chemical characterisation and antioxidant activities of a novel polysaccharide from *Auricularia polytricha. Int. J. Biol. Macromol.*, 120: 1087–1092.

Cheung, P.C. (2008). Nutritional value and health benefits of mushrooms. *Mushrooms as Functional Foods*, 2: 71–109.

Chiu, W.-C., Yang, H.-H., Chiang, S.-C., Chou, Y.-X. and Yang, H.-T. (2014). *Auricularia polytricha* aqueous extract supplementation decreases hepatic lipid accumulation and improves antioxidative status in animal model of nonalcoholic fatty liver. *Biomedicine*, 4(2): 1–10.

Dasgupta, A. and Acharya, K. (2019). Mushrooms: An emerging resource for therapeutic terpenoids. *3 Biotech.*, 9(10): 1–14.

Du, P., Cui, B. and Dai, Y. (2011). Genetic diversity of wild *Auricularia polytricha* in Yunnan province of south-western China revealed by sequence-related amplified polymorphism (SRAP) analysis. *Journal of Medicinal Plants Research*, 5(8): 1374–1381.

El Enshasy, H.A. and Hatti-Kaul, R. (2013). Mushroom immunomodulators: Unique molecules with unlimited applications. *Trends in Biotechnol.*, 31(12): 668–677.

Emre, A., Cagatay, G., Avci, G.A., Suiçmez, M. and Cevher, S.C. (2016). An edible mushroom with medicinal significance; *Auricularia polytricha. Hittite Journal of Science and Engineering*, 3(2): 111–116.

Faccin, L., Benati, F., Rincão, V., Mantovani, M., Soares, S., Gonzaga, M., Nozawa, C. and Carvalho Linhares, R. (2007). Antiviral activity of aqueous and ethanol extracts and of an isolated polysaccharide from *Agaricus brasiliensis* against poliovirus type 1. *Lett. Appl. Microbiol.*, 45(1): 24–28.

Fang, D., Wang, D., Ma, G., Ji, Y., Zheng, H., Chen, H., Zhao, M., Hu, Q. and Zhao, L. (2021). *Auricularia polytricha* noodles prevent hyperlipemia and modulate gut microbiota in high-fat diet fed mice. *Food Science and Human Wellness*, 10(4): 431–441.

Ferreira, I.C., Barros, L. and Abreu, R. (2009). Antioxidants in wild mushrooms. *Curr. Med. Chem.*, 16(12): 1543–1560.

Gong, P., Wang, S., Liu, M., Chen, F., Yang, W., Chang, X., Liu, N., Zhao, Y., Wang, J. and Chen, X. 2020. Extraction methods, chemical characterizations and biological activities of mushroom polysaccharides: A mini-review. *Carbohydr. Res.*, 494: 108037.

Gunawardena, D., Bennett, L., Shanmugam, K., King, K., Williams, R., Zabaras, D., Head, R., Ooi, L., Gyengesi, E. and Münch, G. (2014). Anti-inflammatory effects of five commercially available mushroom species determined in lipopolysaccharide and interferon-γ activated murine macrophages. *Food Chem.*, 148: 92–96.

Gupta, G., Pathak, S., Dahiya, R., Awasthi, R., Mishra, A., Sharma, R.K., Agrawal, M. and Dua, K. (2019). Aqueous extract of wood ear mushroom, *Auricularia polytricha* (Agaricomycetes), demonstrated antiepileptic activity against seizure induced by maximal electroshock and isoniazid in experimental animals. *Int. J. Med. Mushrooms*, 21(1).

Hassan, M.A.A., Rouf, R., Tiralongo, E., May, T.W. and Tiralongo, J. (2015). Mushroom lectins: Specificity, structure and bioactivity relevant to human disease. *Int. J. Mol. Sci.*, 16(4): 7802–7838.

Hu, D., Yang, X., Hu, C., Feng, Z., Chen, W. and Shi, H. (2021). Comparison of ergosterol and vitamin D_2 in Mushrooms *Agaricus bisporus* and *Cordyceps militaris* using ultraviolet irradiation directly on dry powder or in ethanol suspension. *ACS Omega*, 6(44): 29506–29515.

Huang, S.J., Lin, C.P. and Tsai, S.Y. (2015). Vitamin D_2 content and antioxidant properties of fruit body and mycelia of edible mushrooms by UV-B irradiation. *J. Food Compost. Anal.*, 42: 38–45.

Jedinak, A., Dudhgaonkar, S., Wu, Q.-l., Simon, J. and Sliva, D. (2011). Anti-inflammatory activity of edible oyster mushroom is mediated through the inhibition of NF-κB and AP-1 signalling. *Nutr. J.*, 10(1): 1–10.

Jia, D., Wang, B., Li, X., Peng, W., Zhou, J., Tan, H., Tang, J., Huang, Z., Tan, W. and Gan, B. (2017). Proteomic analysis revealed the fruiting-body protein profile of *Auricularia polytricha. Curr. Microbiol.*, 74(8): 943–951.

Jia, F., Yang, S., Ma, Y., Gong, Z., Cui, W., Wang, Y. and Wang, W. (2020). Extraction optimization and constipation-relieving activity of dietary fibre from *Auricularia polytricha. Food Bioscience*, 33: 100506.

Kanagasabapathy, G., Kuppusamy, U.R., Abd Malek, S.N., Abdulla, M.A., Chua, K.-H. and Sabaratnam, V. (2012). Glucan-rich polysaccharides from *Pleurotus sajor-caju* (Fr.) Singer prevents glucose intolerance, insulin resistance and inflammation in C57BL/6J mice fed a high-fat diet. *BMC Complement. Altern. Med.*, 12(1): 1–9.

Lee, H., Choi, J., Kim, Y., Kim, I., Kim, B. and Lee, C. (2013). Effects of the *Cynanchum wilfordii* ethanol extract on the serum lipid profile in hypercholesterolemic rats. *Preventive Nutrition and Food Science*, 18(3): 157.

Li, N., Wang, C., Georgiev, M.I., Bajpai, V.K., Tundis, R., Simal-Gandara, J., Lu, X., Xiao, J., Tang, X. and Qiao, X. (2021). Advances in dietary polysaccharides as anticancer agents: Structure-activity relationship. *Trends Food Sci. Technol.*, 111: 360–377.

Li, P., Xiong, C. and Huang, W. (2022). Gamma-irradiation-induced degradation of the water-soluble polysaccharide from *Auricularia polytricha* and its Anti-hypercholesterolemic activity. *Molecules*, 27(3): 1110.

Medzhitov, R. (2008). Origin and physiological roles of inflammation. *Nature*, 454(7203): 428–435.

Miao, J., Regenstein, J.M., Qiu, J., Zhang, J., Zhang, X., Li, H., Zhang, H. and Wang, Z. (2020). Isolation, structural characterisation and bioactivities of polysaccharides and its derivatives from Auricularia—A review. *Int. J. Biol. Macromol.*, 150: 102–113.

Mohamed Ali, S., Ling, T.C., Muniandy, S., Tan, Y.S., Raman, J. and Sabaratnam, V. (2014). Recovery and partial purification of fibrinolytic enzymes of *Auricularia polytricha* (Mont.) Sacc by an aqueous two-phase system. *Sep. Purif. Technol.*, 122: 359–366. Doi:https://doi.org/10.1016/j.seppur.2013.11.016.

Muszyńska, B., Grzywacz-Kisielewska, A., Kała, K. and Gdula-Argasińska, J. (2018). Anti-inflammatory properties of edible mushrooms: A review. *Food Chem.*, 243: 373–381.

Onyango, B., Mbaluto, C., Mutuku, C. and Otieno, D. (2013). Effect of wheat bran supplementation with fresh and composted agricultural wastes on the growth of Kenyan native wood ear mushrooms [*Auricularia auricula* (L. ex Hook.) Underw.]. *African Journal of Biotechnology*, 12(19).

Onyango, B., Otieno, C. and Palapala, V.A. (2016). Molecular characterisation of wood ear mushrooms [*Auricularia* sp.] from Kakamega Forest in western Kenya. *Current Research in Environmental & Applied Mycology*, 6(1): 51–60.

Perera, N., Yang, F.-L., Chiu, H.-W., Hsieh, C.-Y., Li, L.-H., Zhang, Y.-L., Hua, K.-F. and Wu, S.-H. (2020). Phagocytosis enhancement, endotoxin tolerance and signal mechanisms of immunologically active glucuronoxylomannan from *Auricularia auricula-judae*. *Int. J. Biol. Macromol.*, 165: 495–505.

Priya, R., Geetha, D. and Darshan, S. (2016). Biology and cultivation of black ear mushroom—*Auricularia* spp. *Advances in Life Sciences*, 5(22): 10252–10254.

Saini, R.K., Rauf, A., Khalil, A.A., Ko, E.-Y., Keum, Y.-S., Anwar, S., Alamri, A. and Rengasamy, K.R. (2021). Edible mushrooms show significant differences in sterols and fatty acid compositions. *S. Afr. J. Bot.*, 141: 344–356.

Sangphech, N., Sillapachaiyaporn, C., Nilkhet, S. and Chuchawankul, S. (2021). *Auricularia polytricha* ethanol crude extract from sequential maceration induces lipid accumulation and inflammatory suppression in RAW264, 7 macrophages. *Food Funct.*, 12(21): 10563–10570.

Sekara, A., Kalisz, A., Grabowska, A. and Siwulski, M. (2015). *Auricularia* spp.-mushrooms as novel food and therapeutic agents—A review. *Sydowia*, 67: 1–10.

Sheu, F., Chien, P.-J., Chien, A.-L., Chen, Y.-F. and Chin, K.-L. (2004). Isolation and characterisation of an immunomodulatory protein (APP) from the Jew's Ear mushroom *Auricularia polytricha*. *Food Chem.*, 87(4): 593–600.

Sillapachaiyaporn, C., Nilkhet, S., Ung, A.T. and Chuchawankul, S. (2019). Anti-HIV-1 protease activity of the crude extracts and isolated compounds from *Auricularia polytricha*. *BMC Complement. Altern. Med.*, 19(1): 1–10.

Sillapachaiyaporn, C., Rangsinth, P., Nilkhet, S., Ung, A.T., Chuchawankul, S. and Tencomnao, T. (2021). Neuroprotective effects against Glutamate-induced HT-22 hippocampal cell damage and *Caenorhabditis elegans* lifespan/healthspan enhancing activity of *Auricularia polytricha* mushroom extracts. *Pharmaceuticals*, 14(10): 1001.

Silva, I., Machado, F., Moreno, M.J., Nunes, C., Coimbra, M.A. and Coreta-Gomes, F. (2021). Polysaccharide structures and their hypocholesterolemic potential. *Molecules*, 26(15): 4559.

Siti, H.N., Kamisah, Y. and Kamsiah, J. (2015). The role of oxidative stress, antioxidants and vascular inflammation in cardiovascular disease (A review). *Vascul. Pharmacol.*, 71: 40–56.

Song, G. and Du, Q. (2010). Isolation of a polysaccharide with anticancer activity from *Auricularia polytricha* using high-speed countercurrent chromatography with an aqueous two-phase system. *J. Chromatogr.*, 1217(38): 5930–5934.

Song, G. and Du, Q. (2012). Structure characterisation of a α β-glucan polysaccharide from *Auricularia polytricha*. *Nat. Prod. Res.*, 26(21): 1963–1970.

Song, X., Pang, H., Cui, W., Zhang, J., Li, J. and Jia, L. (2021). Renoprotective effects of enzyme-hydrolysed polysaccharides from *Auricularia polytricha* on adenine-induced chronic kidney diseases in mice. *Biomed. Pharmacother.*, 135: 111004.

Sun, Y., Li, T. and Liu, J. (2010). Structural characterisation and hydroxyl radicals scavenging capacity of a polysaccharide from the fruiting bodies of *Auricularia polytricha*. *Carbohydr. Polym.*, 80(2): 377–380.

Sun, Y.X., Liu, J.C. and Kennedy, J.F. (2010). Purification, composition analysis and antioxidant activity of different polysaccharide conjugates (APPs) from the fruiting bodies of *Auricularia polytricha*. *Carbohydr. Polym.*, 82(2): 299–304.

Wang, W., Wang, Y., Gong, Z., Yang, S. and Jia, F. (2021). Comparison of the nutritional properties and transcriptome profiling between the two different harvesting periods of *Auricularia polytricha*. *Frontiers in Nutrition*, 8.

Wang, X.F., Su, K.Q., Bao, T.W., Cong, W.R., Chen, Y.F., Li, Q.Z. and Zhou, X.W. (2012). Immunomodulatory effects of fungal proteins. *Current Topics in Nutraceutical Research*, 10(1).

Wu, F., Tohtirjap, A., Fan, L.-F., Zhou, L.-W., Alvarenga, R.L., Gibertoni, T.B. and Dai, Y.-C. (2021). Global diversity and updated phylogeny of Auricularia (*Auriculariales, Basidiomycota*). *Journal of Fungi*, 7(11): 933.

Wu, J., Ding, Z.-Y. and Zhang, K.-C. (2006). Improvement of exopolysaccharide production by macro-fungus *Auricularia auricula* in submerged culture. *Enzyme Microb. Technol.*, 39(4): 743–749.

Wu, L., Zhao, J., Zhang, X., Liu, S. and Zhao, C. (2021). Antitumour effect of soluble β-glucan as an immune stimulant. *Int. J. Biol. Macromol.*, 179: 116–124.

Wu, N.J., Chiou, F.J., Weng, Y.M., Yu, Z.R. and Wang, B.J. (2014). *In vitro* hypoglycemic effects of hot water extract from *Auricularia polytricha* (wood ear mushroom). *Int. J. Food Sci. Nutr.*, 65(4): 502–506.

Xiang, H., Sun-Waterhouse, D. and Cui, C. (2021). Hypoglycemic polysaccharides from *Auricularia auricula* and *Auricularia polytricha* inhibit oxidative stress, NF-κB signaling and proinflammatory cytokine production in streptozotocin-induced diabetic mice. *Food Science and Human Wellness*, 10(1): 87–93.

Xin, L., Yuan, Y., Ruo-lin, H., Liang-tao, C., Xiao-ping, W., Ming-feng, Z. and Jun-sheng, F. (2019). Optimisation of extraction conditions of melanin from *Auricularia polytricha* and its antioxidant activities *in vitro*. *Natural Product Research and Development*, 31(10): 1688.

Yang, X., Guo, M., Wu, Y., Wu, Q. and Zhang, R. (2014). Removal of emulsified oil from water by fruiting bodies of macro-fungus (*Auricularia polytricha*). *PLoS ONE*, 9(4): e95162.

Yu, J., Sun, R., Zhao, Z. and Wang, Y. (2014). *Auricularia polytricha* polysaccharides induce cell cycle arrest and apoptosis in human lung cancer A549 cells. *Int. J. Biol. Macromol.*, 68: 67–71.

Yu, M., Ma, B., Luo, X., Zheng, L., Xu, X. and Yang, Z. (2008). Molecular diversity of *Auricularia polytricha* revealed by inter-simple sequence repeat and sequence-related amplified polymorphism markers. *Curr. Microbiol.*, 56(3): 240–245.

Yu, M., Xu, X., Qing, Y., Luo, X., Yang, Z. and Zheng, L. (2009). Isolation of an anti-tumour polysaccharide from *Auricularia polytricha* (Jew's ear) and its effects on macrophage activation. *Eur. Food Res. Technol.*, 228(3): 477–485.

Zhao, S., Rong, C., Liu, Y., Xu, F., Wang, S., Duan, C., Chen, J. and Wu, X. (2015). Extraction of a soluble polysaccharide from *Auricularia polytricha* and evaluation of its anti-hypercholesterolemic effect in rats. *Carbohydr. Polym.*, 122: 39–45.

Zheng, S., Huang, H., Zhang, R. and Cao, L. (2014). Removal of Cr (VI) from aqueous solutions by fruiting bodies of the jelly fungus (*Auricularia polytricha*). *Appl. Microbiol. Biotechnol.*, 98(20): 8729–8736.

Zhou, Y., Chen, L., Fan, X. and Bian, Y. (2014). *De novo* assembly of *Auricularia polytricha* transcriptome using Illumina sequencing for gene discovery and SSR marker identification. *PLoS ONE*, 9(3): e91740.

Zhu, K., Chen, X., Yu, D., He, Y. and Song, G. (2018). Preparation and characterisation of a novel hydrogel based on *Auricularia polytricha* β-glucan and its bio-release property for vitamin B_{12} delivery. *J. Sci. Food Agric.*, 98(7): 2617–2623.

CHAPTER 7

Giant Puffball
(Calvatia gigantea)

Rohit Shukla,[1] Mohee Shukla,[2] Saket Jha,[2,] Ravikant Singh,[3]*
Ashutosh Pathak[4] and Anupam Dikshit[2]

1. Introduction

Mushrooms are widely used in all possible fields, such as food, cosmetics, medicines and many more (Valverde et al., 2015; Zhang et al., 2016). Edible mushrooms are highly nutritive and can be consumed raw as well as boiled or fried (Wang et al., 1990). They are rich in fibres, vitamins and digestible proteins (Ouzouni et al., 2009; Reis et al., 2012). There are about 3,000 approximately edible mushrooms present on earth and among which approximately 200 species are consumed by humans (Kalač, 2013) and still need more to be highlighted and cultivated. Apart from being edible, mushrooms are widely used in pharmaceutical and nutraceutical industries as these are considered to be medicinal herbs. Globally mushrooms contribute approximately 25% of the molecules used by pharma industries to make medicines (Newman and Cragg, 2012). Mushrooms generally have high nutritious and medicinal value and have bioactive components, like 150–300 g of protein, 15–50 g of lipid and 75–150 g of dry weight of it (per kg). The nutritional energy is found to be least in their dry matter with only 300–500 Kcal per kg (Kalač, 2013).

Recently, researchers have shown a keen interest in other wild mushrooms which have nutraceutical capabilities. These are found to be very rich in proteins, amino contents and dietary fibres and least fat content (Barros et al., 2007). They are rich in several essential vitamins, like vitamin A, vitamin C, D and K, and also vitamin B complex with several rare earth elements (Heleno et al., 2010; Kalač, 2010) and other bioactive compounds, like carotenes, unsaturated fatty acids, phenols, etc.

[1] Centre of Science and Society (under IIDS), University of Allahabad, Prayagraj-211002.
[2] Biological Product Laboratory, Department of Botany, University of Allahabad, Prayagraj-211002.
[3] Department of Biotechnology, Swami Vivekanand University, Sagar-470228.
[4] Wood Anatomy Discipline, Forest Research Institute, Dehradun, Uttarakhand-248006.
* Corresponding author: jhasaket90@gmail.com

The present chapter throws light on the wild mushroom *Calvatia gigantea* and on the nutraceutical properties as well as pharmaceutical value. This species belongs to the family *Agaricaceae* and is normally called 'puffballs' (Fig. 1). Earlier it was placed in *Lycoperdaceae* family, but on the basis of recent molecular work which supports the expansion of family *Agaricaceae* to include some taxa with 'puffballs' Basidiome form; member of *Lycoperdaceae* hich shows close affinity to *Lapiota* genus (Krüger et al., 2001). There are 140 species of *Calvatia* recorded globally, in which only 58 are validated and accepted among taxonomists (Table 1, Verma et al., 2018). Genus *Calvatia* was first described as Bovista craniformis by Lewis David de Schweinitz in 1832 but Elias Fries (1849), transferred it to the newly circumscribed genus *Calvatia craniiformis* as the type and only species. Most of the species are edible only at the young stage and eaten by almost every country where it found in a wild state. Another species of *Calvatia* is *C. fumosa* which has a very unpleasant smell and taste and due to its pungent odour, it is not consumable. *Calvatia gigantea* (Batsch) Lloyd and *Calvatia utriformis* were reported as edible (Rubel and Arora, 2008).

CLASSIFICATION

- KINGDOM • Fungi
- PHYLUM • Basidiomycota
- CLASS • Agaricomycetes
- ORDER • Agaricales
- FAMILY • Agaricaceae
- GENUS • Calvatia
- SPECIES • *C. gigantea* (Batsch) Lloyd

Fig. 1. Classification of *C. gigantea.*

2. Diversity and Distribution

C. gigantea is globally considered as a giant puffball, found widespread in the temperate zone (Kreisel, 1994) and known to be amongst the largest edible mushrooms and usually found in summer and autumn season. It looks like a white ball measuring up to 30–150 cm in diameter (Leffingwell and Alford, 2011). It is consumed when it is young and has white colour. As the fruiting body grows wider, it becomes more unpleasant in smell and taste. The white coloured fruiting body grows faster and changes to brown colour, releasing millions of spores. Generally, it is consumed by

Table 1. Major *Calvatia* species with their bioactivity.

Name of Species	Bioactivity	References
Calvatia bovista (L.) Pers.	Gastro-intestinal irritation or mild toxicity.	Sharma et al., 2021; Verma et al., 2018
Calvatia caelata (Bull.) Morgan	Anti-proliferative and anti-mitogenic.	Lam et al., 2001; Verma et al., 2018
Calvatia candida (Rostk.) Hollós	Anti-bacterial, contain alkaloids, steroids, terpenoids and have potent anti-oxidant activities.	Veremii et al., 2021
Calvatia craniiformis (Schwein.) Fr. ex De Toni	Anti-fungal, anti-carcinogenic, anti-inflammatory and anti-tumour effect.	Jameel and Al-Ezzy, 2017; Mohammed et al., 2017; Takaishi et al., 1997
Calvatia cyathiformis (Bosc) Morgan	Cure for leucorrhoea or pneumaturia, and tumour inhibitors.	Læssøe and Spooner, 1994; Lucas et al., 1958, Verma et al., 2018
Calvatia excipuliformis (Scop.) Perdeck	Most active bio-accumulator of ^{210}Po.	Coetze and van Wyk, 2009; Verma et al., 2018
Calvatia fenzlii (Reichardt) Kawam.	Check the growth pf K562 leukemia cells in human via enhancing the ceasing of cell cycle through PPAR-γ.	Eroğlu et al., 2016; Verma et al., 2018
Calvatia fragilis (Quél.) Morgan	–	Verma et al., 2018
Calvatia gardneri (Berk.) Lloyd	–	Verma et al., 2018
Calvatia gigantea (Batsch) Lloyd	Used in beekeeping, as tumour inhibitors, protection against infection by the poliomyelitis virus, ECHO enteroviruses; activity against the A2/Japan 305 virus, and A/PR8 influenza virus; anesthetic, anti-cancerous, antidiabetic, antioxidant.	Læssøe and Spooner, 1994; Lucas et al., 1958; Verma et al., 2018; Cochran et al., 1966; Kivrak et al., 2016; Roland et al., 1960; Valverde et al., 2015; Zhang et al., 2016; Coetze and van Wyk, 2009
Calvatia lilacina (Mont. & Berk.) Henn.	Anti-tumour activity.	Kim et al., 1992; Verma et al., 2018
Calvatia longicauda (Henn.) Lloyd	–	Verma et al., 2018
Calvatia lycoperdoides A.H. Sm.	–	Verma et al., 2018
Calvatia pachyderma (Peck) Morgan	–	Verma et al., 2018
Calvatia pyriformis (Lév.) Kreisel	–	Verma et al., 2018
Calvatia rubroflava (Cragin) Lloyd	Rubroflavin and leucorubroflavin rich	Le Goff and Ouazzani, 2014; Verma et al., 2018

frying its slice in butter, garlic and spices and herbs. It basidiocarp and basidium are among the largest to arise from single primordium (Clémençon, 1990) and very rarely found in fairy rings as it usually grows singly (Gube, 2007). The fairy rings are reported to be associated with *Agrostis capillaris* L., *Festuca nigrescens* L., or *Lolium perenne* L., on the fields (Li, 2011).

C. *gigantea* is among the well-known members of Basidiomycetes and generally causes type I allergies (Simon-Nobbe et al., 2008). By the inward movement of gleba, it creates more pressure to eject the spores. This is very similar to ignition created in jet planes. This type of spore dispersal is very common in Basidiomycetes but these puffballs do not possess bellow mechanism for release of basidiospores. The liberation of spores is done by irregular piercing of peridium which exposes the huge mass of basidiospores into the sky (Fig. 2). Their basidiospores independently get blown away by the wind and the surrounding sky is full of huge mass of its spores (Ingold, 1965). Buller (1909) first reported the amount of spores released by the mushroom which is calculated as seven trillion from a single gleba.

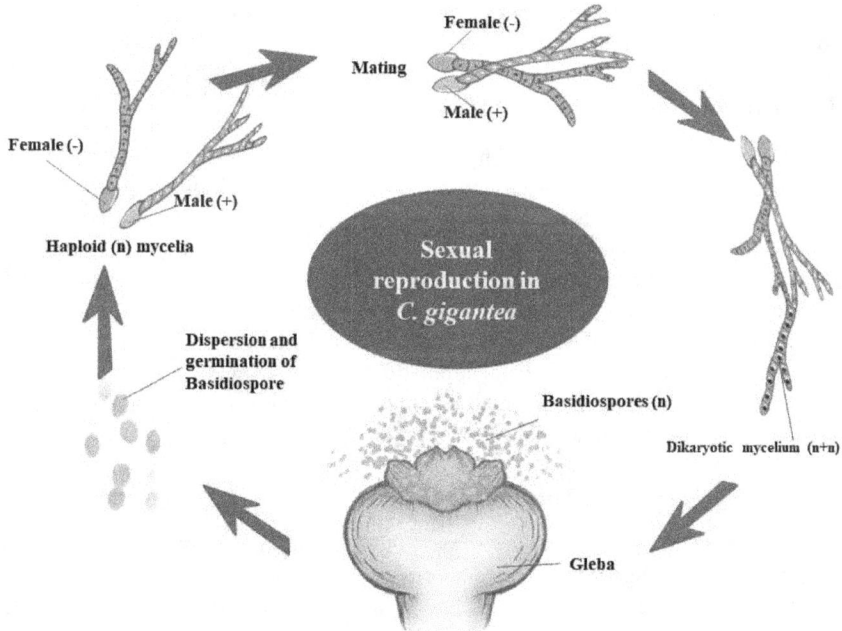

Fig. 2. Life cycle (sexual reproduction) *C. gigantea.*

2.1 Fruiting Body of C. gigantea

Generally, the fruiting body weighs more than 100 kg containing 5–160 trillion basidiospores (Murat et al., 2008). The outer layer of peridium, i.e., exoperidium is without pseudo-parencymatous endostratum tissue and is only single-layered thick (Gube, 2007) because of which the peridia of *C. gigantea* are very light weight and constitute at least 10% of the gross weight of the fruit. Basidiocarp of *C. gigantea* lacks a sterile base. The gleba has uniformity in both colour and texture (Li, 2011). The palisade structures which form the hymenium stand on fan-shaped branches of hyphae and differ from species of *Lycoperdaceae* along with different species of *Calvatia*. According to Bates et al. (2009) ultra-structural study of *Calvatia* species with the help of scanning electron microscopy analysed that the ornamentation characteristic feature of basidiospores is significant in the point of taxonomy. The pedicels and sterigmal of basidiospore fragments are important in view of taxonomy for differentiating *Lycoperdon* species (Bates, 2004).

2.2 *Life Cycle of* C. **gigantea**

The life cycle of *Calvatia gigantea* is very similar to the *Agaricus* species and till date, only sexual mode of reproduction is reported (Walton, 2018). In *Calvatia gigantea* the life cycle starts with ripening of gleba, as shown in Fig. 2. The basidiospores emerge out and disperse with the help of air when the ripened gleba busts. After the dispersion of basidiospores, it starts the process of germination and haploid mycelia grow with each basidiospore. Mating of haploid male (+) and haploid female (–) mycelia takes place and dikaryotic mycelium (n + n) develops. With the help of dikaryotic mycilium (n + n), premordia formation takes place and finally the fruiting body of *Calvatia gigantea* develops again.

3. Nutraceutical Values of *C. gigantea*

The macronutrients, proteins, carbohydrates and moisture contents are reported to be high in *C. gigantea*. Edible mushrooms are well known for their highly nutritive, less calories and low fat constituents with high water content, making them suitable for all types of persons to eat. The need of nutritive constituents creates the curiosity to study *C. gigantea* as their exploration reveals many surprises and several nutraceutical constituents. Apart from having vitamins, rare earth elements, proteins and vitamins are also present. These mushrooms are now found to be have unsaturated fatty acids, digestive sugars as well as several aromatic bioactive compounds (Kivrak et al., 2016). These are well known for their essential amino acids, which make them more significant by way of nutraceutical aspects (Coetze and van Wyk, 2009).

3.1 *Carbohydrate Content*

Many carbohydrates are derived from this mushroom, mostly belonging to disaccharides and oligosaccharides. Majority of the species belong to genus Calvatia and are rich in disaccharides and constitute the major source of soluble sugars. Soluble sugars are found to impart sweetness to basidiocarp at the younger stage. Due to this, the mushroom is found to be sweeter in taste in its young fruiting stage as compared to its older counterpart mainly due to the degradation of these soluble sugars which may be replaced by other carbohydrate derivatives (Kivrak et al., 2016).

3.2 *Protein and Amino Acid Contents*

The mushroom has essential monomers of proteins which are vital for the growth of the human body. Due to these vital amino constituents this mushroom is more significant for eating for the proper body growth among children. All types of amino acids are found in the mushroom and these are essential as well as non-essential for the human body (Coetze and van Wyk, 2009). *C. gigantea* has majority of amino compounds out of which arginine, proline, 4-hydroxy proline and tyrosine are non-essential types of amino acids, whereas leucine, phenylalanine, tryptophan and valine are essential types of amino acids that are found in significant amounts (Ibrahim et al., 1987). Substrate-dependent protein contents are found in *C. gigantea*. Shannon and Stevenson (1975) reported a good amount of protein as well as sodium content (Vetter, 2003; Agrahar-Murugkar and Subbulakshmi, 2005). Several trace elements (such as Fe, Co, Zn, Se, etc.) were also reported by various researchers in this mushroom (Borovička and Řanda, 2007).

3.3 Fats and Oil Contents

The extraction and analysis of fatty acids and their derivatives as well as saturated and unsaturated oils has always been a major challenge in the past five to six decades and researchers and industrialists are focusing on it so as to develop a simple and potential method or technique to isolate and analyse the oils and fats in these mushrooms. After knowing the proper analysis, majority of the saturated and unsaturated oils are reported from these mushrooms and are found to be a good source of oils and fats to extract. These mushrooms are rich in linoleic acids, oleic acid and palmitic acids (Borovička and Řanda, 2007). Trans-fatty acids are found to be more significant for the body growth and development, but these oils and fats also cause severe cardiovascular problems in humans due to their richness in cholesterol constituents. Apart from the above-mentioned fatty acids, *C. gigantea* are found to be rich in lauric acid, decanoic acid, octanoic acid, etc. This mushroom also has small amounts of phosphatidylcholine and phosphatidylethanolamine, which are derivatives of glycerol-phospholipids (Proštenik et al., 1983).

3.4 Other Compounds found in C. gigantea

Several other compounds, like aromatic constituents, lipase enzyme, etc., were also reported in *C. gigantea* and these are discussed below with their subheadings:

3.4.1 Aroma Compound

Several researchers found that apart from amino acids, proteins, fats and oils, these puffballs are also rich in various aromatic compounds which mostly belong to the class of esters, alcohols, etc. The major compounds which have been identified till date are hexane-aldehyde, benzaldehyde, 2-pentyl-furan, etc. These are generally responsible for the taste and odour in mushroom but also differ due to the difference in their growing area, place of origin and several other external factors. These external environmental stresses lead to the production of these aromatic constituents, in which the most common fatty acid is linoleic acids. This was found to be present in different ratios in every fruiting body, depending on their size and age. This fatty acid is a major compound which is the main precursor to several aromatic compounds in puffballs. Most common are octane-alcohol constituents which further help in production of several enzymes, like fatty acid oxygenase and hydro-peroxidase lyase, etc. A mushroom alcohol, namely octane-alcohol, is also found in *C. gigantea* (Dijkstra, 1976) and is believed to be responsible for the odour, colour and taste. It was also found that the amino acid (glutamic acid) and 5-guanosine monophosphate contributed to the flavour in several mushrooms including puffballs.

3.4.2 Phenolic Contents

The mushroom is also rich in maximum phenolic constituents which help in production of protocatechuic and *p*-hydroxybenzoic acids. Apart from these phenlic constituents, several other compounds are also present, such as *para*-coumaric acid, *para*-hydroxybenzoic acid, gentisic acid and protocatechuic acid. These facilitate the several metabolic activities in mushrooms and also act as major components in its defence metabolism. The mushroom is found to be a great source of phenolics and other secondary metabolites which can be potentially used to reduce the oxidative stress. The phenolic constituents found in *C. gigantea* are able to deduce the various health diseases because of their high antioxidant activity.

3.4.3 Lipase Enzymes Contents

The mushroom is rich in several enzymes, such as lipase hydrolyses which are derived from other fatty acids and glycerol which are major constituents of several microorganisms (Ibrahim et al., 1987). From all the microbes, moulds are found to be very good source of lipase and its derivatives. These puffball mushrooms are rich in lipase and thus are involved in several applications related to lipase activity (Chander and Klostermeyer, 1983). In recent years, several researchers have reported the potential activity of lipase enzyme as it is able to catalyse the hydrolysis of fats and oils and helps in synthesis of several esters. The esters have industrial applications and provide various types of flavouring in foods and preservatives (Langrand et al., 1990). Lipase synthesis in mushroom is enhanced by growing these mushrooms in a high nutrient-growth medium and enzymes similar to lipase produces high lipase activities. It was reported that up to 90% enhancement of lipase activity occurs under high carbon and nitrogen sources with optimum pH and temperature (Langrand et al., 1990).

4. Pharmaceutical Values of *C. gigantea*

Several experimental investigations provide direct evidence of the potentially healthy nutritive value of *C. gigantea* and thus, from ancient times, tribals and rural people used to consume puffballs. The puffballs are also consumed mainly by diabetic people as it has low sugar concentration and also helps in reducing the glucose level in the bloodstream (Zhang et al., 2016; Vamanu, 2018). Many mushroom species are well known for their medicinal properties and are thus widely used in pharmaceuticals (Table 1; Fig. 3; Li et al., 2018). *Calvatia* species are widely known for their medicinal properties and are mainly used in folk-medicine for treatment of many diseases and as wound dressing (Læssøe and Spooner, 1994). *C. gigantea* and its other species are reported to have inhibitory effect on tumour-inducing cells (Lucas et al., 1958). The chemical, craniformin, isolated from the *Calvatia* species by Gill and Steglich (1987) and Lam et al. (2001) reported the efficacy of an ubiquitin-like peptide which acts as anti-proliferative agent against breast cancerous cells. The mushroom also has anti-mitogenic activity and it was found that the *Calvatia* has oncostatic properties which further help in antiviral activity (Cochran, 1978). The compound calvatic acid has successively been isolated in *C. gigantea* (Okuda and Fujiwara, 1982). Calvatic acid was found antimicrobial by Viterbo et al., (1975) but many other researchers do

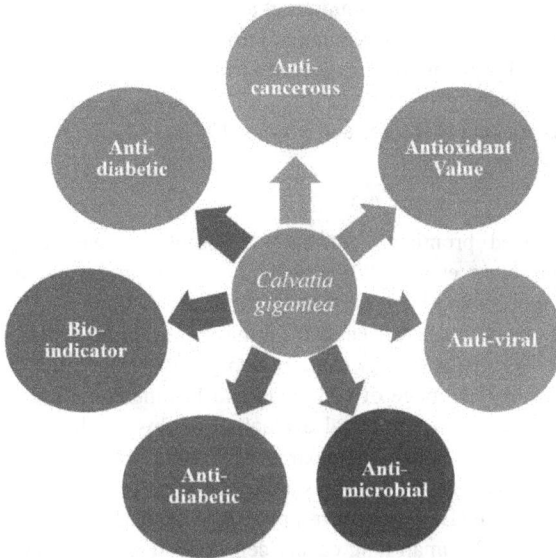

Fig. 3. Pharmaceutical values of *Calvatia gigantea.*

not support this study. Recently, calvatic acid was found lethal against the growth of *Helicobacter pylori* (Sorba et al., 2001).

Calvatic acid is also known to have anti-tumour effect and is used for inhibition of sarcoma cells in cell culture. Besides this, it also increases the survival time duration in mice suffering from leukemia (Umezawa et al., 1975). Calvatic acid has direct evidence to produce the particular inhibitor of glutathione tansferease-P1-1 and due to this, it is effective as anti-cancerous material (Antonini et al., 1997). A novel protein, called calcaelin, was also isolated from puffball mushroom having ribosomal inactivation expression which was found to be more effective in reducing the breast cancerous cells but not efficient against fungal and bacterial pathogens (Ng et al., 2003). Recently, methanolic extract from this mushroom showed antibacterial but less efficacy against fungal pathogens (Suay et al., 2000; Dulger, 2005). But several other researchers found it to be used in Japanese and Chinese traditional medicine and thus is contradictory to its previous report. This motivates the researchers to continue with pharmaceutical investigations. Ergosterol was extracted from *Calvatia* species and it is a poly-oxygenated ergosterol derivative that has potential against several pathogens (Kwon et al., 1980).

The *Calvatia* species has been used since ancient times by Japanese and Chinese traditional medical practitioners. It was also used by tribals of South America and Asia. Widely used as a wound-healing agent (Rai et al., 1993), it is also used against leucorrhoea, pneumaturia, inflammation, diarrhoea, etc. (Læssøe and Spooner, 1994). *C. gigantea* was found effective against several viruses, such as the influenza virus studied on the mice model (Piearce, 1981; Kekos and Macris, 1983; 1987). This mushroom is able to produce various phenolic compounds which are mostly tannins as these compounds are a direct source of carbon for these giant puffballs (Galiotou-Panayotou and Macris, 1986). These puffball mushroom release various enzymes which have the ability to digest catechin, which is the core unit of all major tannins

(Alexopoulos and Mims, 1979; Komninos et al., 1988) and thus these mushrooms are capable of breakdown of complex tannins into simpler forms. This mushroom can release some fungal proteins when exposed to a higher carbon diet and rich in tannins. Due to this, enhancement in eminent levels of alpha-amylase enzymes in plenty of tannin-rich medium occurs (Kekos and Macris, 1987).

4.1 In Cancer

C. gigantea showed promising anti-tumour activity as well as promise as an anesthetic agent (Nam et al., 2001). Several proteins and peptides are extracted from *Calvatia* species and have anti-cancerous activity. Lam et al. (2001) reported about its effectiveness on breast cancer. Furthermore, it was described that several bio-chemicals or agents in *C. gigantea* are capable of altering tumour proliferation while stimulating the stoppage of cell cycle and leading to cell death (Hseu et al., 2014). One anti-tumour experimental evaluation of this mushroom was performed and the researchers evaluated that the extract of this mushroom showed enhancement in expression of several genes, enzymes and cells, viz., down regulation of CCND1, CCND2, CDK4, Akt and Bcl-2 as well as up regulate expression of Bax, p53, caspase-3 and caspase-9 in A549 human lung cancer cell line, suggesting that the *C. gigantea* extract can induce cell cycle arrest and apoptosis in A549 cell (Ren et al., 2014). Various researchers reported the biological efficacy of *C. gigantea* by way of being antimicrobial, antioxidant and anti-cancerous activity. Extracts from its fruiting body have several bioactive compounds, such as calvacin which is found more potential as compared to standard drugs (Ren et al., 2014) and has anti-cancerous activity (Roland et al., 1960).

4.2 In Diabetes

These giant puffball mushrooms are capable of inhibiting more than 50% of enzymatic activity which is responsible for the increase in glucose content in blood. These are also rich in several proteins, vitamins and minerals and are found to have lesser fats (Wang et al., 2014). Various research investigations directly or indirectly showed the nutritious potential of this mushroom and its suitability for controlling diabetic and heart disease because it has low fat and cholesterol contents. *C. gigantea* showed various biological metabolisms by which it helps to control blood sugar level (Ye et al., 2017; Lee et al., 2018). These are reported to control the blood sugar and lower the glucose level in blood (De Silva et al., 2012; Wasser, 2014; Zhang et al., 2016). This is due to the presence of several biologically active constituents which help in smooth functioning of several vital organs and glands in the human body. These bioactive constituents also help in synthesis and regulation of insulin (Ogbole et al., 2019) and regulate several hormones and enzymes, thus playing a crucial role in proper and smooth metabolism in the human body (Calvo et al., 2016). *Calvatia* species also have *beta*-glucans which restore the pancreatic cells and stimulate the insulin produced by beta-cells. This increased insulin helps in removable of glucose from blood and leads to improving the sensitivity to the neighbouring tissues as they utilise additional glucose from the blood (Chen et al., 2013; Sari et al., 2017).

4.3 Antioxidant Value

These mushrooms are highly rich in compounds responsible for antioxidant activities and are thus a potential source of natural antioxidants. Various methods are introduced to identify and analyse the antioxidants found in these mushrooms. Different solvents give different types of extractants which have various types of compounds (Kivrak et al., 2016). These compounds show different antioxidant potentiality and provide vast variations in producing the different natural antioxidants. This variable in producing different antioxidant compounds makes them more crucial for consumption as well for herbal drug production. Due to their high ability to produce compounds that lead to antioxidant efficacy, they are interesting natural microbes in the nutraceutical industry (Kivrak et al., 2016). They are potential in antioxidant activity but less useful in producing standard drugs and yet are much higher in comparison to other mushrooms. This antioxidant activity is conducted by inhibiting lipid peroxidation, using beta carotene-linoleic acid method (Aissous et al., 2021). This method includes the reduction of several free radicals which are found to be harmful in the biological system. These free radicals are the main cause of several diseases in human beings but can be controlled by antioxidants because they inhibit the process of oxidation in an oxidising substrate when used in low amounts in comparison to the substrate (Aissous et al., 2021). In this method, the linoleic acid oxidises to release peroxyl-free radicals which further degrade to release the hydrogen atom from diallylic methylene groups of linoleic acid. This results in participation of free radicals in attacking the chromophore beta carotene, which reduces or breaks lipid molecules. The presence of mushroom antioxidants leads to utilisation of free radicals to reduce or degrade the lipid compounds (Aissous et al., 2021). The methanolic and hexane extracts from the basidiocarp were found significant in degrading the lipids, using the β-carotene–linoleic acid method. The free radical-scavenging activity from this method depends on the reaction between the free radicals released from several carotenoids or other compounds with linoleic acid and beta carotene. Ethyl acetate extract of *C. gigantea* showed antioxidant activity against bleaching by β-carotene and it was similar to *Agaricus bitorquis* (Kivrak et al., 2016).

4.4 As Bioindicators

All *Calvatia* species are terrestrial saprophytes and no single species has been found till today as parasitic. *C. gigantea* was found to have mycorrhizal association with *Pinus* species and *Picea* species (Trappe, 1962). According to Rimóczi (1987), *C. gigantea* powerfully degrades the nitrogen compounds and favours the nitrogen-rich soil habitat. It was also found that these puffballs love the acidic or partial acidic adaphic habitat and generally avoid soils of basic nature. Regarding temperature along with humidity, *Calvatia* is mesophilic and needs shade in the early stage of development. These mushrooms are strictly found in semi-arid to high arctic deserts (Lange, 1990) which act as geo-bioindicators.

4.5 Supplementary Uses of C. gigantean

From the brewery wastes, this mushroom produces fungal proteins which are used as digestive agents in several biochemical oxygen demands. It will also release some hydrolytic enzymes which can be used in various bio-engineering applications

(Shannon and Stevenson, 1975). It was also seen that the *C. gigantea* are active in producing alpha amylase when grown in a starch medium (Kekos and Macris, 1983; 1987). These alpha amylases are found to be tannin resistant and inhibit several enzymatic activities in organisms (Komninos et al., 1988). *C. gigantea* is capable of consuming toxic components, such as phenolic, polyphenolic constituents and condensed form of tannins. Galiotou-Panayotou et al. (1988) extracted the several enzymes which are able to degrade catechin. They have affinity for the amylase and biomass production from the carbon compound sources containing toxic substances (Kekos and Macris, 1987). The fruiting body of *Calvatia* is a potential source of chitinase and chitobiase enzymes (Zikakis and Castle, 1988). According to Tracey (1955) chitinolytic activity is seen in aqueous extract of ripened basidiocarp, but lesser activity was found in its dried and young fruit or other parts of the puffball. The extracts of *C. gigantea* are capable of degrading chitin and perform the chitinase as well as chitobiase activities which were also confirmed by Zikakis and Castle (1988). This mushroom is considered a good source of lipase production (Christakopoulos et al., 1992) and produces the valuable source for lipase enzymes used in nutraceutical factories in preparation of several diet supplements (Goud et al., 2009).

Clavatia species are also used in chasing bees as these were used in apiculture for driving bees out of their hives and the smoke which smolders from fruiting bodies aids to calm the bees (Coetze and van Wyk, 2009; Wood, 1983). The exact chemical is still unknown but Wood (1983) reported this activity due to H_2S and HCN as well as various unknown compounds showing high anesthetic (sleeping) behaviour. This sleeping effect in bees due to these chemicals makes these giant puffballs excellent bee-repellent. Duke (1926) reported hut-fumigation in Kenya region by smoke of fruiting bodies of *C. gigentea* but it also showed an adverse effect on human eyes (Walleyn and Rammeloo, 1994). *C. utriformis* is another species of *Calvatia* which is an excellent source of amadou (Læssøe and Spooner, 1994) and *C. gigantea* can be used for the preservation of *C. utriformis* due to the presence of saltpeter. Because of its igneous nature, it can also be used as an igneous substance (Coetze and van Wyk, 2009).

4.6 Disease caused by C. gigantean

Although *C. gigantea* is chiefly used as major eatable mushroom, but it is safe to consume only during its young stage with the white gleba. When the gleba of giant puffballs ripens and becomes mature, it causes several gastrointestinal infections/ diseases if consumed by someone (Læssøe and Spooner, 1994). Not only gleba, spores of *Calvatia* species (earlier this genus was considered under genera *Lycoperdon*) can cause acute respiratory disease, known as lycoperdonosis (Henriksen, 1976) when consumed in a large quantity during breathing and symptoms of disease easily can be seen in sensitive persons. The basidiospores can cause severe skin irritation in sensitive individuals and are mainly responsible for type I allergies. These skin infections are diagnosed by radio-allergo-sorbent test reactivity (Levetin et al., 1992).

5. Conclusion

Calvatia gigantea is also called a giant puffball and belongs to family Agaricaceae. The mushroom is known to have highly nutraceutical properties and is eaten only when its fruiting body is young and whitish in colour. Apart from this, it is also known to have pharmaceutical properties and is widely used in Japanese and Chinese traditional medicine. It is consumed by tribals and rural people. It is also found more effective on several pathogens and shows the potential for antioxidant activity. Also, one of its compounds, called calvicin extracted from the fruiting body, has anti-cancerous and antidiabetic properties. More emphasis should be lain on use and cultivation by farmers to increase the yield of this mushroom. This can be beneficial in rural welfare and is applicable in meeting the nutritional requirements and malnutrition.

Acknowledgements

All contributing authors are thankful to Head, Department of Botany for allowing the laboratory use and the UGC for providing the fellowship.

References

Agrahar-Murugkar, D. and Subbulakshmi, G.J.F.C. (2005). Nutritional value of edible wild mushrooms collected from the Khasi Hills of Meghalaya. *Food Chemistry*, 89(4): 599–603. https://doi.org/10.1016/j.foodchem.2004.03.042.

Aissous, I., Benrebai, M., Cacan, E., Caglar, B., Erenler, R., Ameddah, S. and Bensouici, C. (2021). Antioxidant and anti-proliferative activities of the butanol extract of Ball aerial parts. *Current Issues in Pharmacy and Medical Sciences*, 34(1): 5–11. https://doi.org/10.2478/cipms-2021-0002.

Alexopoulos, C.J. and Mims, C.W. (1979). *Introductory Mycology*. John Wiley and Sons.

Antonini, G., Pitari, G., Caccuri, A.M., Ricci, G., Boschi, D., Fruttero, R. and Ascenzi, P. (1997). Inhibition of human Placenta Glutathione transferase Pl-1 by the antibiotic calvatic acid and its diazocyanide analogue: Evidence for multiple catalytic intermediates. *European Journal of Biochemistry*, 245(3): 663–667. https://doi.org/10.1111/j.1432-1033.1997.00663.x.

Barros, L., Baptista, P., Correia, D.M., Casal, S., Oliveira, B. and Ferreira, I.C. (2007). Fatty acid and sugar compositions and nutritional value of five wild edible mushrooms from northeast Portugal. *Food Chemistry*, 105(1): 140–145. https://doi.org/10.1016/j.foodchem.2007.03.052.

Bates, S.T. (2004). Arizona Members of the Geastraceae and Lycoperdaceae (Basidiomycota, Fungi). Doctoral Dissertation, Arizona State University, Arizona.

Bates, S.T., Roberson, R.W. and Desjardin, D.E. (2009). Arizona gasteroid fungi I: *Lycoperdaceae* (Agaricales, Basidiomycota). *Fungal Diversity*, 37: 153.

Borovička, J. and Řanda, Z. (2007). Distribution of iron, cobalt, zinc and selenium in macrofungi. *Mycological Progress*, 6(4): 249–259. https://doi.org/10.1007/s11557-007-0544-y.

Buller, A.H.R. (1909). *Researches on Fungi*. vol. I, Longmans, Green and Co., London, pp. 287.

Calvo, M.S., Mehrotra, A., Beelman, R.B., Nadkarni, G., Wang, L., Cai, W. and Uribarri, J. (2016). A retrospective study in adults with metabolic syndrome: Diabetic risk factor in response to daily consumption of *Agaricus bisporus* (white button mushrooms). *Plant Foods for Human Nutrition*, 71(3): 245–251. https://doi.org/10.1007/s11130-016-0552-7.

Chander, H. and Klostermeyer, H. (1983). Production of lipase by *Geotrichum candidum* under various growth conditions. *Milchwissenschaft*, 38(7): 410–412.

Chen, J., Zhang, X.D. and Jiang, Z. (2013). The application of fungal beta-glucans for the treatment of colon cancer. *Anti-cancer Agents in Medicinal Chemistry*, 13(5): 725–730.

Christakopoulos, P., Tzia, C., Kekos, D. and Macris, B.J. (1992). Production and characterisation of extracellular lipase from *Calvatia gigantea*. *Applied Microbiology and Biotechnology*, 38(2): 194–197. https://doi.org/10.1007/BF00174467.

Clémençon, H. (1990). Fixierung, Einbettung und Schnittfärbungen für die plectologische Untersuchung von Hymenomyceten mit dem Lichtmikroskop. *Mycol. Helvet.*, 3: 451–466.

Cochran, K.W. (1978). Medical effects. pp. 169–187. *In*: Chang, S.T. and W.A. Hayes (eds.). *The Biology and Cultivation of Edible Mushrooms*, Academic Press, New York, USA.

Cochran, K.W., Nishikawa, T. and Beneke, E.S. (1966). Botanical sources of influenza inhibitors. *Antimicrobial Agents and Chemotherapy*, 6: 515–520.

Coetze, J.C. and van Wyk, A.E. (2009). The genus *Calvatia* ('Gasteromycetes', *Lycoperdaceae*): A review of its ethnomycology and biotechnological potential. *African Journal of Biotechnology*, 8(22): 6007–6015. https://doi.org/10.5897/AJB09.360.

De Silva, D.D., Rapior, S., Hyde, K.D. and Bahkali, A.H. (2012). Medicinal mushrooms in prevention and control of diabetes mellitus. *Fungal Diversity*, 56(1): 1–29. https://doi.org/10.1007/s13225-012-0187-4.

Dijkstra, F.Y. (1976). Studies on mushroom flavours 3. Some flavour compounds in fresh, canned and dried edible mushrooms. *Zeitschrift für Lebensmittel-Untersuchung und Forschung*, 160(4): 401–405. https://doi.org/10.1007/BF01106331.

Duke, M.M. (1926). Fungi from Kenya colony. *Bulletin of Miscellaneous Information* (Royal Botanical Gardens, Kew), 305–320. https://doi.org/10.2307/4118196.

Dulger, B. (2005). Antimicrobial activity of ten Lycoperdaceae. *Fitoterapia*, 76(3-4): 352–354. https://doi.org/10.1016/j.fitote.2005.02.004.

Eroğlu, C., Seçme, M., Atmaca, P., Kaygusuz, O., Gezer, K., Bağcı, G. and Dodurga, Y. (2016). Extract of *Calvatia gigantea* inhibits proliferation of A549 human lung cancer cells. *Cytotechnology*, 68(5): 2075–2081. https://doi.org/10.1007/s10616-016-9947-4.

Fries, E.M. (1849). *Summa vegetabilium Scandinaviae* (in Latin). 2. *Uppsala*, Sweden, *Typographia Academica*, 442.

Galiotou-Panayotou, M. and Macris, B.J. (1986). Degradation of condensed tannins by *Calvatia gigantea*. *Applied Microbiology and Biotechnology*, 23(6): 502–506. https://doi.org/10.1007/BF02346069.

Galiotou-Panayotou, M., Rodis, P., Macris, B.J. and Stathakos, D. (1988). Purification of a novel enzyme involved in catechin degradation by *Calvatia gigantea*. *Applied Microbiology and Biotechnology*, 28(6): 543–545. https://doi.org/10.1007/BF00250409.

Gill, M. and Steglich, W. (1987). Pigments of fungi (Macromycetes). *Fortschritte der Chemie organischer Naturstoffe/Progress in the Chemistry of Organic Natural Products*, 1–297. https://doi.org/10.1007/978-3-7091-6971-1_1.

Goud, M.J.P., Suryam, A., Lakshmipathi, V. and Charya, M.S. (2009). Extracellular hydrolytic enzyme profiles of certain South Indian Basidiomycetes. *African Journal of Biotechnology*, 8(3): 354–360.

Gube, M. (2007). The gleba development of *Langermannia gigantea* (Batsch: Pers.) Rostk. (Basidiomycetes) compared to other Lycoperdaceae, and some systematic implications. *Mycologia*, 99(3): 396–405. https://doi.org/10.1080/15572536.2007.11832564.

Heleno, S.A., Barros, L., Sousa, M.J., Martins, A. and Ferreira, I.C. (2010). Tocopherols composition of Portuguese wild mushrooms with antioxidant capacity. *Food Chemistry*, 119(4): 1443–1450. https://doi.org/10.1016/j.foodchem.2009.09.025.

Henriksen, N.T. (1976). Lycoperdonosis. *Acta Pædiatrica*, 65(4): 643–645. https://doi.org/10.1111/j.1651-2227.1976.tb04945.x.

Hseu, Y.C., Lee, C.C., Chen, Y.C., Kumar, K.S., Chen, C.S., Huang, Y.C. and Yang, H.L. (2014). The anti-tumour activity of *Antrodia salmonea* in human promyelocytic leukemia (HL-60) cells is mediated via the induction of G1 cell-cycle arrest and apoptosis *in vitro* or *in vivo*. *Journal of Ethnopharmacology*, 153(2): 499–510. https://doi.org/10.1016/j.jep.2014.03.012.

Ibrahim, C.O., Hayashi, M. and Nagai, S. (1987). Purification and some properties of a thermostable lipase from *Humicola lanuginosa* No. 3. *Agricultural and Biological Chemistry*, 51(1): 37–45. https://doi.org/10.1080/00021369.1987.10867997.

Ingold, C.T. (1965). Spore liberation. *Spore Liberation*, pp. 210.

Jameel, G.H. and Al-Ezzy, A.I.A. (2017). Evaluation of antifungal activity of *Calvatia craniiformis* and Ivermectin as novel alternative therapies for *Aspergillus niger*-associated acute otitis media with special refer to socio demographic factors among rural children of Diyala province, Iraq. *International Journal of Pharmaceutical and Clinical Research*, 9(8): 581–589. https://doi.org/10.25258/ijpcr.v9i08.9583.

Kalač, P. (2010). Trace element contents in European species of wild growing edible mushrooms: A review for the period 2000–2009. *Food Chemistry*, 122(1): 2–15. https://doi.org/10.1016/j. foodchem.2010.02.045.

Kalač, P. (2013). A review of chemical composition and nutritional value of wild-growing and cultivated mushrooms. *Journal of the Science of Food and Agriculture*, 93(2): 209–218. https:// doi.org/10.1002/jsfa.5960.

Kekos, D. and Macris, B.J. (1983). Production and characterisation of amylase from *Calvatia gigantea*. *Applied and Environmental Microbiology*, 45(3): 935–941. https://doi.org/10.1128/ aem.45.3.935-941.1983.

Kekos, D. and Macris, B.J. (1987). Effect of tannins on growth and amylase production by *Calvatia gigantea*. *Enzyme and Microbial Technology*, 9(2): 94–96. https://doi.org/10.1016/0141-0229(87)90149-9.

Kim, B.K., Kwun, J.Y., Park, Y.I., Bok, J.W. and Choi, E.C. (1992). Antitumour components of the cultured mycelia of *Calvatia craniformis*. *Journal of the Korean Cancer Association*, 24(1): 1–18.

Kivrak, I., Kivrak, S. and Harmandar, M. (2016). Bioactive compounds, chemical composition, and medicinal value of the giant puffball, *Calvatia gigantea* (higher Basidiomycetes), from Turkey. *International Journal of Medicinal Mushrooms*, 18(2): https://doi.org/10.1615/ IntJMedMushrooms.v18.i2.10.

Komninos, J., Kekos, D., Macris, B.J. and Galiotou-Panayotou, M. (1988). Tannin-resistant α-amylase from *Calvatia gigantea*. *Biotechnology and Bioengineering*, 32(7): 939–941. https:// doi.org/10.1002/bit.260320717.

Kreisel, H. (1994). Studies in the *Calvatia* complex (Basidiomycetes) 2. *Feddes Repertorium*, 105(5-6): 369–376. https://doi.org/10.1002/fedr.19941050516.

Krüger, D., Binder, M., Fischer, M. and Kreisel, H. (2001). The Lycoperdales. A molecular approach to the systematics of some gasteroid mushrooms. *Mycologia*, 93(5): 947–957. https://doi.org/10 .1080/00275514.2001.12063228.

Kwon, T.J., Lee, C.O., Kang, C.Y., Kim, B.K. and Park, D.W. (1980). Studies on the constituents of higher fungi of Korea (XXI)—A sterol from *Calvatia saccatum* (Vahl.) Fr., *The Korean Journal of Mycology*, 8(1): 25–28.

Læssøe, T. and Spooner, B. (1994). The uses of 'Gasteromycetes'. *Mycologist*, 8(4): 154–159.

Lam, Y.W., Ng, T.B. and Wang, H.X. (2001). Antiproliferative and antimitogenic activities in a peptide from puffball mushroom *Calvatia caelata*. *Biochemical and Biophysical Research Communications*, 289(3): 744–749. https://doi.org/10.1006/bbrc.2001.6036.

Lange, M. (1990). Arctic gasteromycetes II. *Calvatia* in greenland. *Svalbard and Iceland*. *Nordic Journal of Botany*, 9(5): 525–546. https://doi.org/10.1111/j.1756-1051.1990.tb00545.x.

Langrand, G., Rondot, N., Triantaphylides, C. and Baratti, J. (1990). Short chain flavour esters synthesis by microbial lipases. *Biotechnology Letters*, 12(8): 581–586. https://doi.org/10.1007/ BF01030756.

Le Goff, G. and Ouazzani, J. (2014). Natural hydrazine-containing compounds: Biosynthesis, isolation, biological activities and synthesis. *Bioorganic and Medicinal Chemistry*, 22(23): 6529–6544. https://doi.org/10.1016/j.bmc.2014.10.011.

Lee, S., Lee, D., Lee, J.C., Kang, K.S., Ryoo, R., Park, H.J. and Kim, K.H. (2018). Bioactivity-guided isolation of anti-inflammatory constituents of the rare mushroom *Calvatia nipponica* in LPS-stimulated RAW 264.7 macrophages. *Chemistry and Biodiversity*, 15(9): e1800203. https://doi. org/10.1002/cbdv.201800203.

Leffingwell, J.C. and Alford, E.D. (2011). Volatile constituents of the giant puffball mushroom (*Calvatia gigantea*). *Leffingwell Reports*, 4: 1–17.

Levetin, E., Horner, W.E. and Lehrer, S.B. (1992). Morphology and allergenic properties of Basidiospores from four *Calvatia* species. *Mycologia*, 84(5): 759–767. https://doi.org/10.1080/ 00275514.1992.12026202.

Li, D.W. (2011). Five trillion basidiospores in a fruiting body of *Calvatia gigantea*. *Mycosphere*, 2(4): 457–462. https://www.mycosphere.org/pdfs/MC2_4_No7.pdf.

Li, H., Wu, S., Ma, X., Chen, W., Zhang, J., Duan, S. and Dong, Y. (2018). The genome sequences of 90 mushrooms. *Scientific Reports*, 8(1): 1–5. https://doi.org/10.1038/s41598-018-28303-2.

Lucas, E.H., Byerrum, R.U., Clarke, D.A., Reilly, H.C., Stevens, J.A. and Stock, C.C. (1958). Production of oncostatic principles *in vivo* and *in vitro* by species of the genus *Calvatia*. pp. 493–496. *In*: Welch, H. and Marti-Ibañez, F. (eds.). *Antibiot. Annu*, 1958–1959, Medical Encyclopedia Inc., New York. https://europepmc.org/article/med/13637789.

Mohammed, I.H., Jameel, G.H. and Al-Ezzy, A.I.A. (2017). Immuno-histochemical detection of *caspase8* expression and apoptotic index activities of *Calvatia craniiformis* crude extract in balb/c mice inoculated with h22 cells. *International Journal of Pharmaceutical Sciences and Research*, 8(6): 2504–2515. http://dx.doi.org/10.13040/IJPSR.0975-8232.8.

Murat, C., Mello, A., Abbà, S., Vizzini, A. and Bonfante, P. (2008). Edible mycorrhizal fungi: Identification, life cycle and morphogenesis. pp. 707–732. *In*: *Mycorrhiza*. Springer, Berlin, Heidelberg. https://doi.org/10.1007/978-3-540-78826-3_33.

Nam, K.S., Jo, Y.S., Kim, Y.H., Hyun, J.W. and Kim, H.W. (2001). Cytotoxic activities of acetoxyscirpenediol and ergosterol peroxide from *Paecilomyces tenuipes*. *Life Sciences*, 69(2): 229–237. https://doi.org/10.1016/S0024-3205(01)01125-0.

Newman, D.J. and Cragg, G.M. (2012). Natural products as sources of new drugs over the 30 years from 1981 to 2010. *Journal of Natural Products*, 75(3): 311–335. https://doi.org/10.1021/np200906s.

Ng, T.B., Lam, Y.W. and Wang, H. (2003). Calcaelin, a new protein with translation-inhibiting, antiproliferative and antimitogenic activities from the mosaic puffball mushroom *Calvatia caelata*. *Planta Medica*, 69(3): 212–217. https://doi.org/10.1055/s-2003-38492.

Ogbole, O.O., Nkumah, A.O., Linus, A.U. and Falade, M.O. (2019). Molecular identification, *in vivo* and *in vitro* activities of *Calvatia gigantea* (macro-fungus) as an antidiabetic agent. *Mycology*, 10(3): 166–173. https://doi.org/10.1080/21501203.2019.1595204.

Okuda, T. and Fujiwara, A. (1982). Calvatic acid production by the Lycoperdaceae, *2: Distribution among the Gasteromycetes*. Transactions of the Mycological Society of Japan (Japan).

Ouzouni, P.K., Petridis, D., Koller, W.D. and Riganakos, K.A. (2009). Nutritional value and metal content of wild edible mushrooms collected from west Macedonia and Epirus, Greece. *Food Chemistry*, 115(4): 1575–1580. https://doi.org/10.1016/j.foodchem.2009.02.014.

Piearce, G.D. (1981). Zambian mushrooms—Customs and folklore. *Bulletin of the British Mycological Society*, 15(2): 139–142. https://doi.org/10.1016/S0007-1528(81)80013-2.

Proštenik, M., Častek, A., Ćosović, Č., Gospočić, L., Jandrić, Z., Kljaić, K. and Ondrušek, V. (1983). The patterns of the constituent fatty acids of glycerophospholipids in mushrooms: Lipids of higher fungi. *Experimental Mycology*, 7(1): 74–81. https://doi.org/10.1016/0147-5975(83)90077-4.

Rai, B.K., Ayachi, S.S. and Rai, A. (1993). A note on ethno-myco-medicines from Central India. *Mycologist* (United Kingdom).

Reis, F.S., Barros, L., Martins, A. and Ferreira, I.C. (2012). Chemical composition and nutritional value of the most widely appreciated cultivated mushrooms: An inter-species comparative study. *Food and Chemical Toxicology*, 50(2): 191–197. https://doi.org/10.1016/j.fct.2011.10.056.

Ren, L., Hemar, Y., Perera, C.O., Lewis, G., Krissansen, G.W. and Buchanan, P.K. (2014). Antibacterial and antioxidant activities of aqueous extracts of eight edible mushrooms. *Bioactive Carbohydrates and Dietary Fibre*, 3(2): 41–51. https://doi.org/10.1016/j.bcdf.2014.01.003.

Rimóczi, I. (1987). Ecology, cenology and distribution of the giant puff-ball [*Langermannia gigantea* (Batsch ex Pers.) Rostk.] in Hungary. *Acta Botanica Hungarica*, 33(3): 279–294.

Roland, J.F., Chmielewicz, Z.F., Weiner, B.A., Gross, A.M., Boening, O.P., Luck, J.V. and Stevens, J.A. (1960). Calvacin: A new antitumour agent. *Science*, 132(3443): 1897–1897. https://doi.org/10.1126/science.132.3443.1897.

Rubel, W. and Arora, D. (2008). A study of cultural bias in field guide determinations of mushroom edibility using the iconic mushroom, *Amanita muscaria*, as an example. *Economic Botany*, 62(3): 223–243. https://doi.org/10.1007/s12231-008-9040-9.

Sari, M., Prange, A., Lelley, J.I. and Hambitzer, R. (2017). Screening of beta-glucan contents in commercially cultivated and wild growing mushrooms. *Food Chemistry*, 216: 45–51. https://doi.org/10.1016/j.foodchem.2016.08.010.

Shannon, L.J. and Stevenson, K.E. (1975). Growth of *Calvatia gigantean* and *Candida steatolytica* in brewery wastes for microbial protein production and BOD reduction. *Journal of Food Science*, 40(4): 830–832. https://doi.org/10.1111/j.1365-2621.1975.tb00568.x.

Sharma, R., Sharma, Y.P., Hashmi, S.A.J., Kumar, S. and Manhas, R.K. (2021). *Ethnomycological Study of Wild Edible and Medicinal Mushrooms in District Jammu, J&K*, India. https://doi.org/10.21203/rs.3.rs-1054070/v1.

Simon-Nobbe, B., Denk, U., Pöll, V., Rid, R. and Breitenbach, M. (2008). The spectrum of fungal allergy. *International Archives of Allergy and Immunology*, 145(1): 58–86. https://doi.org/10.1159/000107578.

Sorba, G., Bertinaria, M., Di Stilo, A., Gasco, A., Scaltrito, M.M., Brenciaglia, M.I. and Dubini, F. (2001). Anti-helicobacter pylori agents endowed with H_2-antagonist properties. *Bioorganic & Medicinal Chemistry Letters*, 11(3): 403–406. https://doi.org/10.1016/S0960-894X(00)00671-5.

Suay, I., Arenal, F., Asensio, F.J., Basilio, A., Angeles Cabello, M., Teresa Díez, M. and Francisca Vicente, M. (2000). Screening of basidiomycetes for antimicrobial activities. *Antonie van Leeuwenhoek*, 78(2): 129–140. https://doi.org/10.1023/A:1026552024021.

Takaishi, Y., Murakami, Y., Uda, M., Ohashi, T., Hamamura, N., Kidota, M. and Kadota, S. (1997). Hydroxyphenylazoformamide derivatives from *Calvatia craniformis*. *Phytochemistry*, 45(5): 997–1001. https://doi.org/10.1016/S0031-9422(97)00066-6.

Tracey, M.V. (1955). Chitin. pp. 264–274. *In*: Peach, K. and M.V. Tracey. *Modern Methods of Plant Analysis*. vol. II. Springer-Verlag, Berlin.

Trappe, J.M. (1962). Fungus associates of Ectotrophic mycorrhizae. *The Botanical Review*, 28(4): 538–606. https://doi.org/10.1007/BF02868758.

Umezawa, H., Takeuchi, T., Iinuma, H., Ito, M., Ishizuka, M., Kurakata, Y. and Tanabe, O. (1975). A new antibiotic, calvatic acid. *The Journal of Antibiotics*, 28(1): 87–90. https://doi.org/10.7164/antibiotics.28.87.

Valverde, M.E., Hernández-Pérez, T. and Paredes-López, O. (2015). Edible mushrooms: Improving human health and promoting quality life. *International Journal of Microbiology*. https://doi.org/10.1155/2015/376387.

Vamanu, E. (2018). Complementary functional strategy for modulation of human gut microbiota. *Current Pharmaceutical Design*, 24(35): 4144–4149. https://doi.org/10.2174/13816128246661 81001154242.

Veremii, Y., Sokolovskyi, B., Nikulin, D., Tsvyd, N., Iukhymenko, V., Chernyak, V. and Martysh, E. (2021). Influence of plasma micro-and coronary discharge on the development of higher fungi myceles, *Медична фізика–сучасний стан, проблеми, шляхи розвитку. Новітні технології*. 117. https://vomfi.univ.kiev.ua/wp-content/uploads/2021/09/zbirnik-MF-20200110.pdf#page=117.

Verma, R.K., Mishra, S.N., Pandro, V. and Thakur, A.K. (2018). Diversity and distribution of *Calvatia* species in India: A new record from central India. *International Journal of Current Microbiology and Applied Science*, 7(9): 2540–2551. https://doi.org/10.20546/ijcmas.2018.709.316.

Vetter, J. (2003). Data on sodium content of common edible mushrooms. *Food Chemistry*, 81(4): 589–593. https://doi.org/10.1016/S0308-8146(02)00501-0.

Viterbo, D., Gasco, A., Serafino, A. and Mortarini, V. (1975). p-Carboxyphenylazoxycyanide–dimethyl sulphoxide: An antibacterial and antifungal compound from *Calvatia lilacina*. *Acta Crystallographica Section B: Structural Crystallography and Crystal Chemistry*, 31(8): 2151–2153. https://doi.org/10.1107/S0567740875007091.

Walleyn, R. and Rammeloo, J. (1994). *The Poisonous and Useful Fungi of Africa, South of the Sahara: A Literature Survey*. National Botanic Garden of Belgium.

Walton, J. (2018). *The Cyclic Peptide Toxins of Amanita and Other Poisonous Mushrooms*. https://doi.org/10.1007/978-3-319-76822-9.

Wang, L., Yugue, L. and Yan, X. (1990). Analysis of amino acid content of 30 varieties of edible fungi. *Mushroom Journal of the Tropics*, 10: 74–78.

Wang, X.M., Zhang, J., Wu, L.H., Zhao, Y.L., Li, T., Li, J.Q. and Liu, H.G. (2014). A mini-review of chemical composition and nutritional value of edible wild-grown mushroom from China. *Food Chemistry*, 151: 279–285. https://doi.org/10.1016/j.foodchem.2013.11.062.

Wasser, S. (2014). Medicinal mushroom science: Current perspectives, advances, evidences, and challenges. *Biomedical Journal*, 37(6): 345–356. https://doi.org/10.4103/2319-4170.138318.

Wood, W.F. (1983). Anaesthesia of honeybees by smoke from the pyrolysis of puffballs and keratin. *Journal of Agricultural Research*, 22(2): 107–110. https://doi.org/10.1080/00218839.1983.111 00569.

Ye, Y., Liu, K., Zeng, Q. and Zeng, Q. (2017). Antimicrobial activity of puffball (*Bovistella radicata*) and separation of bioactive compounds. *Amb Express*, 7(1): 1–10. https://doi.org/10.1186/ s13568-017-0402-5.

Zhang, J.J., Li, Y., Zhou, T., Xu, D.P., Zhang, P., Li, S. and Li, H.B. (2016). Bioactivities and health benefits of mushrooms mainly from China. *Molecules*, 21(7): 938. https://doi.org/10.3390/ molecules21070938.

Zikakis, J.P. and Castle, J.E. (1988). Chitinase-chitobiase from soybean seeds and puffballs. *Methods in Enzymology*, 161: 490–497, Academic Press. https://doi.org/10.1016/0076-6879(88)61064-0.

CHAPTER 8

Girolle (*Cantharellus cibarius*)

Manjula Rai,[1] Shuvadip Mondal,[2] Rupa Sanyal,[3]
Abhijit Dey[4,*] and Surjit Sen[2,*]

1. Introduction

Mushrooms are a well-known food material for mankind from centuries past and in present times, these are becoming a boosting dietary supplement because of their various nutritional characteristics and healing capabilities (Vlasenko et al., 2019; Daba et al., 2020; El-Hagrassi et al., 2020; Elkhateeb, 2020; Elkhateeb and Daba, 2021). Some wild-grown edible mushrooms, which are appreciated due to their unique chemical constituents, serve as a highly nutritive and tasty food in multiple corners of the world (Gry et al., 2012; Falandysz and Drewnowska, 2015; Kozarski et al., 2015; Deshaware et al., 2021). As mushrooms contain various drug materials, they have a potent medicinal impact as well as a culinary attribute because of possessing high amounts of protein, low total fat level, high ratio of polyunsaturated fatty acids (PUFA), thus making them an effective low-calorie diet (Chang and Buswell, 2003; Grangeia et al., 2011; Kumari et al., 2011; Teplyakova et al., 2012; Kozarski et al., 2015; Vlasenko et al., 2019). The use of edible mushroom as a medicine has long been accepted for its antimicrobial, cholesterol-lowering, anticancer and antioxidant properties (Ribeiro et al., 2006; Barros et al., 2007; Aryantha et al., 2010; Kuka et al., 2014; Lemieszek et al., 2018). Reports indicate that mushrooms possess antitumour, anti-infection, immunomodulation and anti-hypertensive activities (Kozarski et al., 2015; Chen et al., 2017). Diseases like hypertension, stroke and some degenerative neurological disorders are proved to be prevented or treated with this mushroom due to its beneficial role (Nyman et al., 2016; Elkhateeb and Daba, 2021).

[1] Department of Botany, St. Joseph's College, Darjeeling, West Bengal – 734104, India.
[2] Department of Botany, Fakir Chand College, Diamond Harbour, West Bengal – 743331, India.
[3] Department of Botany, Bhairab Ganguly College, Belgharia, Kolkata, West Bengal – 700056, India.
[4] Department of Life Sciences, Presidency University, 86/1 College Street, Kolkata: 700073.
* Corresponding authors: surjitsen09@gmail.com, abhijit.dbs@presiuniv.ac.in

One of the best known wild edible macrofungi is yellow-silk fungus or golden Chanterelle *Cantharellus cibarius* Fr. (Chen et al., 2017; Falandysz and Drewnowska, 2017). A mixture of pigment composed of β-carotene, lycopene, α-carotene, xanthophyll and canthaxanthin gives the characteristic golden-yellow colour to this fungus (Deshaware et al., 2021). It has been used as medicine in different parts of Asian countries for over 2000 years (Nyman et al., 2016). The fungus has been extensively studied and investigated for its chemical composition. *C. cibarius* belongs to Basidiomycota; Class: Agariomycetes; Order: Cantharellales and Family: *Cantharellaceae* (Kozarski et al., 2015; Elkhateeb and Daba, 2021). It is popular, not only for its taste and aroma, but also for its cosmopolitan distribution (Kozarski et al., 2015; Chen et al., 2017).

Mycorrhizal fungi are known to play a mutually beneficial role by making an association with the roots of many green plants. In a wild condition, *C. cibarius* makes its habitat in deciduous and coniferous forests, thus making a symbiotic ectomycorrhizal association with the roots of pine, fir, oak, beech and hornbeam (Pilz et al., 2003; Muszyńska et al., 2013; Bulam et al., 2021). It shows its growing capabilities in a variety of soils, but the best growth of mycelium can be observed in a well-drained forest soil having pH range of 4–5.5 and low nitrogen content (Deshaware et al., 2021). The wild-grown Chanterelles are a very common occurrence, but the cultivation practice has been a difficult job for a long time (Deshaware et al., 2021). The reason behind the difficulty to cultivate *C. cibarius* is the presence of bacteria and other foreign microorganisms with the sporocarp tissue of the fungus (Kozarski et al., 2015). Being a mycorrhizal fungus, this species is unable to grow in cultivated environments (Kozarski et al., 2015). In spite of this difficulty, researchers are trying to cultivate it, but the process is still under development (Kozarski et al., 2015).

Like other culinary medicinal mushrooms, *C. cibarius*, is beneficial for health, both as food and medicine. It is notably the best known species of the genus *Cantharellus* that is used as a healthy delicious food, as it is rich in carbohydrates, proteins, vitamins, dietary fibres, minerals and is low in fats (Kumari et al., 2011; Kozarski et al., 2015; Deshaware et al., 2021). Furthermore, it is rich in minerals like selenium, calcium, phosphorus, potassium and various vitamins (Kozarski et al., 2015; Deshaware et al., 2021; Elkhateeb and Daba, 2021). Chemical analysis of this species shows the presence of different fatty acids, organic acids and amino acids (Kuka et al., 2014). Oxidation of linoleic acid in the fruit bodies of *C. cibarius* yield 1-octen-3-ol, which impart the characteristic apricot-like flavour to it (Muszyńska et al., 2013). The rich chemical composition not only makes *C. cibarius* a nutritive mushroom but also acts as a good source of antioxidant. This species contains catechin, pyrogallol, myricetin, and different phenolic acids, thus making this species high in antioxidant activities than any other known mushrooms (Kuka et al., 2014). This rich source of phytochemicals and antioxidants serves a potent medicine for different diseases. As in medicinal purposes, this species is reported to be beneficial for the stomach, liver, lung, lowering blood-sugar level, improving eyesight and as a supplementary medicine in treatment of cancer (Chen et al., 2017; Deshaware et al., 2021). Decoction of this species was recently applied as a food ingredient to produce functional frankfurters that can act as low sensorial modifications and antioxidant and antimicrobial agents (Bulam et al., 2021). Resultantly, these bioactive properties make this mushroom a promoter of human health as a common dietary material. The present chapter is aimed at discussing the nutritional and medicinal aspects of *C. cibarius* and its application as a functional food or as a basis of nutraceuticals for maintenance and promotion of human health and life quality.

2. Distribution

Increasing uses of *C. cibarius* as a food and medicine has been observed in different corners of the world. The species has been known by many local/vernacular names in different countries – Tavuk Mantari, Sari Mantar in Turkey (Bulam et al., 2021), Tavuk Kirmati, Horoz Mantari, Yumurta Mantari in eastern Black Sea regions (Peksen et al., 2016), Girolle in France, Capo gallo in Italy, Chanterelle or Golden Chanterelle in Britain (Kozarski et al., 2015), Amarillo in Spain, Anzutake in Japan, Csirke gomba in Hungary, Dotterpilz in Germany and Huangzhi-gu in China (Pilz et al., 2003).

The Golden Chanterelle has a world-wide distribution, ranging from Scandinavia to Mediterranean region in Europe (Kozarski et al., 2015) and North America to the northern regions of Asia, including the Himalayas and the Yunnan province of China (Muszyńska et al., 2013; Drewnowska and Falandysz, 2014; Chen et al., 2017) and in Africa (Bulam et al., 2019). In Asia, it is reported from Pakistan, India, China, Thailand, Malaysia, Japan and Philippines; in Australia, New Guinea, New Caledonia this species (var. *australiensis*) grows in the eucalyptus forest (Pilz et al., 2003).

3. Production and Commerce

Amongst the European countries, Poland is best known for the production and export of *C. cibarius* (Falandysz et al., 2015). The total collection of this species in Poland is nearly 4,000 tonnes and was exported to different markets of Germany, France and United States in 2012 (Drewnowska and Falandysz, 2014; Falandysz and Drewnowska, 2015). In Europe, it is harvested as a wild edible mushroom and is ranked in the second position with an estimated economic value of nearly 1 billion Euro per year (Lovric et al., 2020; 2021). In Asia, it has been reported that the consumption of *C. cibarius* exceeds 20 kg of fresh product per capita per year in Yunnan province of China, and in some remote parts of the United Kingdom (Barret et al., 1999; Zhang et al., 2010). Japan also imports *C. cibarius* from France and is sold for over $100 per pound at the Nishiki market in Kyoto (Pilz et al., 2003).

4. General Description

Morphologically *C. cibarius* bears a largely curved, flat or superficially depressed cap, a fleshy stipe and false gills (Elkhateeb and Daba, 2021). It is also characterised by its fruity, apricot-like odour (Elkhateeb and Daba, 2021). Carpophores are scattered on the forest floor covered with leaf litters (Fig. 1). The fruit body of the species consists of the following structures:

Pileus: 3–5 cm broad, convex when young, then flattened and finally broadly funnel-shaped, involute margin at first becomes deeply depressed in the centre and with age becomes expanded and wavy, fleshy, surface dry, smooth, whitish yellow to orange yellow. *Pileus hyphae*: cylindric, thin-walled septate, 3–12.5 μm in diameter, clamp connection present in all the hyphae (Rana, 2016). *Context*: thick, firm, orange-yellow to yellow; red spots seen on the damaged cap. *Lamellae*: decurrent, forked,

Fig. 1. Fruiting bodies of *Cantharellus cibarius* in natural habitat.

extends down the stalk, loosely arranged, well-separated, narrow and blunt, colour orange-yellow to yellow. *Stipe*: usually thick and short 2–6 cm long, 5–25 mm thick, central cylindrical, equal in thickness throughout or generally tapering at the base, solid, dry, glabrous, orange-yellow to yellow. *Basidia*: 60–100 × 6–9 μm cylindric, hyaline, thin-walled apiculate, non-amyloid. *Basidiospores*: yellowish, elliptical, wrinkled with furrows and folds. *Cystidia*: absent. *Spore print*: ochraceous, white, creamy or yellow (Elkhateeb and Daba, 2021).

5. Nutritional Composition of *Cantharellus cibarius*

Due to food shortages during World War II, importance was given to the nutritional benefits of mushrooms. Since then, scientists have been researching on the dietary elements of edible mushrooms in the hope of discovering an alternative food source, particularly for people who are protein deficient. Even in developed countries, the consumption of edible macrofungi is increasing due to a high content of proteins that contain all the required amino acids, as well as a large amounts of carbohydrate, dietary fibre, fat, vitamins and minerals (Thimmel and Kluthe, 1998). Studies show that the chemical constituents of *C. cibarius* vary according to its origin, conservation procedure, ecology, environmental conditions, etc., the fruit body of *C. cibarius* had been analysed and reveals different levels of carbohydrates, proteins, vitamins and minerals (Kozarski et al., 2015).

5.1 Carbohydrates

Carbohydrates are the primary source of energy and account for the majority of total daily intake due to their ease of digestion. Carbohydrates account for 50–65% of the dry weight of mushroom fruiting bodies (Wani et al., 2010). The carbohydrate content of *C. cibarius* varies substantially, ranging between 31.9–47 g/100 g dry weight (dw), indicating that the mushrooms could be a source of considerable carbohydrates. Digestible and non-digestible carbohydrates make for the total carbohydrate content of mushrooms. Mannitol is one of the dominant digestible carbohydrates which constitute about 8.56 g/100 g. Moreover, glucose and trehalose are also abundant sugars, constituting about 7.98 g/100 g and 6.68 g/100 g, respectively (Kumar et al., 2013).

5.2 Proteins

Protein is the most important component that contributes the proximate value of food. Protein makes up more than half of total nitrogen in mushrooms and their content varies, depending on the substratum, harvest period and species. Primary role of protein is to provide enough amounts of amino acids to the body. One of the key components of Chanterelle is crude protein. The molecules make up 47–53.7% of the total, although free amino acids make up only around 1% of the total dw (Kuka et al., 2014). The fruit body of *C. cibarius* contains almost all the free amino acids but the concentration of glutamic acid (29.99 mg/g dw) is highest followed by lysine (5.74 mg/g dw) and threonine (8.98 mg/g dw). It has also been reported that Chanterelle contains proline, isoleucine and valine (bitter amino acids) in higher concentrations. The presence of 5'-GMP (guanosine monophosphate) is responsible for the meaty flavour of mushrooms. It is reported that the savoury flavour of *C. cibarius* may be augmented due to the synergistic activity of the taste of 5'-GMP and amino acids, like monosodium glutamate (Beluhan and Ranogajec, 2011).

5.3 Fat

Mushrooms have a very low fat content (approximately 2–8% dw basis), which is predominantly by unsaturated fatty acids (Hong et al., 2012). Chantcrelle contains different types of fatty acids, such as linoleic, oleic, arachidonic, pentadecanoic, stearic, lauric, myristic, palmitic, palmitoleic, heptadecanoic, cis-11,14-eicosadienoic, cis-8,11,14-eicosatrienoic, behenic, tricosanoic, and lignoceric acids. The major fatty acids are linoleic (654.706 mg/kg dw) and oleic acid (148.168 mg/kg dw). Saturated fatty acids are more (926.953 mg/kg dw) than polyunsaturated (655.176 mg/kg dw) or monounsaturated fatty acids (148.493 mg/kg dw) in Chanterelle (Ribeiro et al., 2009).

5.4 Vitamins

Chanterelle is rich in natural vitamin C and vitamin D (Kozarski et al., 2015). Vitamin D concentration varies greatly but remains high even while the mushroom is dried and kept for up to six years (Rangel-Castro et al., 2002). Chanterelle is an excellent choice for vegetarians since it is one of the most abundant natural dietary sources of vitamin D. The fresh fruiting body contains ergosterol, ergosta-7, 22-dienol, ergosta-5, 7-dienol and ergosta-7-enolod liver oil. Canned Chanterelle contains less ergosterol, but more of other sterols: ergosta-7, 22-dienol, ergosta-5, 7-dienol, ergosta-7-enol and more vitamin D_2. It has been reported that consumption of about 30-50g of fresh or canned *C. cibarius* is sufficient to fulfil the daily vitamin D requirements (Phillips et al., 2011). Chanterelle fresh fruiting bodies are the main source of vitamin D_2 (ergocalciferol) (14.2 µg/100 g fresh weight) while dried fruiting bodies contain 0.12–6.3 µg/g dw of ergocalciferol. The level of ergocalciferol in *C. cibarius* fruit-body is affected by UV exposure (Mui et al., 1998; Barros et al., 2008).

5.5 Fibre

Dietary fibre was recognised as a nutrient by the Nutrition Labeling and Education Act of 1993 and is composed of dietary residues that are difficult to digest and absorb in the small intestine of humans. Hemicellulose, cellulose, lignin, oligosaccharides, pectins, gums and waxes are examples of non-digestible carbohydrate polymers. Fibre-rich diets have physiological benefits including laxation, controlled blood glucose level, reduced intestinal travel times, toxic agent entrapment, stimulation of the growth of gastrointestinal microbiota and protection from large bowel cancer. As a consequence, there has been a recent tendency in the food industry to seek out new sources of dietary fibre. The fresh fruiting bodies of *C. cibarius* contain dietary fibre ranges between 1.40–11.2 g/100 g of dw (Crisan and Sands, 1978; Kumar et al., 2013).

5.6 Moisture and Ash

Moisture or total water content of a food product is commonly given as a percentage by weight on a wet basis. It influences food qualities, like physical appearance, flavour, texture, weight, shelf-life, purity and resistance to microbial contamination. High moisture content in food products can accelerate growth of microorganisms, reducing the quality and shelf-life. Moisture content has become an important part of the food industry (Zambrano et al., 2019). The moisture content of *C. cibarius* ranges between 5.44–39.88 g/100 g dw and ash between 7.76–13.2 g/100 g dw (Agraharmurugkar and Subbulakshmi, 2005; Kumar et al., 2013).

5.7 Minerals

Studies and analysis have shown that the various essential elements are found in fruit bodies of *C. cibarius* (Drewnowska et al., 2015) (Table 1). The mineral composition

Table 1. Mineral profile of *Cantharellus cibarius*.

Types	Elements	Symbol	Value Range (gm/kg Dry Matter)	References
Macro	Potassium	K	40 ± 8 to 60 ± 4	Kumar et al., 2013; Drewnowska et al., 2015; Falandysz et al., 2015
	Phosphorus	P	4.4 ± 0.7 to 6.9 ± 0.5	
	Magnesium	Mg	0.97 ± 0.08 to 1.2 ± 0.1	
	Rubidium	Rb	0.62 ± 0.17 to 1.5 ± 0.24	
	Sodium	Na	0.61 ± 0.012 to 0.14 ± 0.061	
	Calcium	Ca	0.21 ± 0.059 to 0.68 ± 0.062	
Trace	Iron	Fe	0.068 ± 0.009 to 0.12 ± 0.076	
	Manganese	Mn	0.027 ± 0.008 to 0.081 ± 0.062	
	Copper	Cu	0.036 ± 0.011 to 0.056 ± 0.009	
	Zinc	Zn	0.074 ± 0.018 to 0.11 ± 0.009	
	Cobalt	Co	0.0002 ± 0.00012 to 0.001 ± 0.0002	
	Chromium	Cr	0.00014 ± 0.00002 to 0.00016 ± 0.00004	
	Aluminum	Al	0.094 ± 0.017 to 0.28 ± 0.19	
	Nickel	Ni	0.00082 ± 0.00025 to 0.0022 ± 0.0004	

of Chanterelle was dominated by K, P and Mg (Falandysz et al., 2015). It has been reported that the concentration of essential macro elements varies with different geographical distribution. The fruit bodies of Chanterelle were low in macromineral Na and Ca. The mineral Rb is considered as the macromineral in *C. cibarius* due to its abundance in the fruit body (Falandysz et al., 2015). Many trace elements are found in *C. cibarius* and among them Fe, Mn, Cu, Co, Cr, Al and Ni are important (Drewnowska et al., 2015).

6. Medicinal Value and Bioactive Compounds

Mushrooms with enhanced nutraceutical qualities are well known as a healthy food and medicine source. Mushroom extracts are utilised as dietary supplements to treat patients (Wasser, 2010). Medicinal mushrooms include bioactive substances that are used to treat a variety of ailments and infections. Chanterelle contains a range of metabolites which are essential because of their nutritional and medicinal properties. As a conventional remedy, *C. cibarius* is considered to benefit the stomach, eliminate toxins from the liver, cleanse the lungs and enhance vision (Chen et al., 2017). The main physiologically bioactive primary and secondary metabolites present in *C. cibarius* are polysaccharides, phenolic compounds, indole, flavonoids, amino acids, vitamins, organic acids, carotenoids, sterols and enzymes (Daniewski et al., 2012; Kuka et al., 2014; Jabłónska-Rýs et al., 2016; Muszyńska et al., 2016; Nyman et al., 2016; Bulam et al., 2019; Panchak et al., 2020; Thu et al., 2020) (Table 2). The crude extract has many therapeutic potentials like being antioxidant, antimicrobial, cardioprotective, anti-inflammatory, antiangiogenic, antigenotoxic, lipoxygenase (LOX) inhibition and wound-healing (Kozarski et al., 2015; Muszyńska et al., 2016; Nasiry et al., 2017; Nowacka-Jechalke et al., 2018; Turfan et al., 2019; Azeem et al., 2020; Badalyan and Rapior, 2020; Thu et al., 2020; Marathe et al., 2021; Uthan et al., 2021). One of the most important bioactive ingredients of Chanterelle is polysaccharides (Jechalke et al., 2018). They mainly belong to chitin, chitosans and most importantly β-glucan possessing multidirectional activities, including antioxidant, anticancer, anti-hyperglycemia, anti-hypercholesterolemia and anti-immunostimulating properties (Kalaĉ, 2009). They reduce the accumulation of carcinogens, their activation as well as the proliferation of neoplastic cells. They are also utilised to treat infectious diseases, increased proliferation, differentiation and development of immune cells, and stem cell activation due to their immunomodulating properties. Mannitol is a naturally-occurring polyol and Chanterelle contains significant amounts of mannitol. It is used in medicine as a potent osmotic diuretic to reduce intracranial pressure and to facilitate the flow of medicines through the blood-brain barrier in the treatment of life-threatening disorders (Saha and Racine, 2011). Chanterelles are distinct from other mushrooms due to their enhanced levels of vitamins including B complex (B_1, B_2, B_6, H), vitamins A, D_2, E and C. *C. cibarius* is rich in vitamin A and is believed to protect against nyctalopia, skin dryness and respiratory infections (Dulger et al., 2004). It has been reported that *C. cibarius* contains α, β, γ-tocopherol and carotenoids (Barros et al., 2008). The species was found to contain a variety of phenolic compounds, like pyrogallol, catechin, myricetin and phenolic acids, such as gallic, p-hydroxybenzoic, protocatechuic, gentisic, ferulic homogentisic and caffeic acids (Valentao et al., 2005). Cinnamic acid, a non-phenolic compound, is a precursor of many phenolic acids and alkaloids in Chanterelle (Barros et al., 2009). These substances have received much

Table 2. Bioactive compounds of *Cantharellus cibarius* and its bioactivity.

Bioactive Group	Compound	Bioactivity	References
Polysaccharide	*β-glucans*	Antioxidant, anticancer, immunomodulating activity, hyperglycemia and Hypercholesterolemia.	Kalač, 2009; Moro et al., 2012; Méndez-Espinoza et al., 2013; Elkhateeb and Daba, 2021
	Chitin		
	Chitosans		
Phenolic	*Cinnamic acids*	Antioxidant, anti-inflammatory, antimicrobial, hypoglycemic, immunostimulating, cardioprotective, anticancer.	Scalbert et al., 2005; Silva et al., 2005; Muszynska et al., 2013: 2016
	Ferulic acid		
	Gallic acid		
	Homogentisic acid		
	p-Hydroxybenzoic acid		
	Protocatechuic acid		
	Sinapic acid		
	Vanillic acid		
	Catechin		
	Myricetin		
	Pyrogallol		
Indole	*Indole*	Antioxidant, reduce lipid peroxidation, act as endogenous neurotransmitter.	Sewerynek et al., 2005; Pytka et al., 2016; Azouzi et al., 2017
	Melatonin		
	Serotonin		
	Tryptamine		
	L-Tryptophan		
	Indoleacetonitryle		
	Kynurenine sulfate		
Enzymes	*Homodimeric lactase*	Immunomodulating, antiproliferative.	Ng and Wang, 2004; Nakamura and Go, 2005; de Pinho et al., 2008; Ramĺrez-Anguiano et al., 2007
	Mn-dependent peroxidase		
	Tyrosinase		
Carotenoids	*α-carotene*	Antioxidant, reduce oxidative stress, nervous disorder, anticancer.	Barros et al., 2008; Fiedor and Burda, 2014; Kozarski et al., 2015
	β-carotene		
	γ-carotene		
	Lycopene		
Sterols	*Ergocalciferol*	Prevents colon cancer, anti-diabetes, glucose intolerance, anti-hypertension, osteoporosis and multiple sclerosis.	Rangel-Castro et al, 2002; Phillips et al., 2011
	Ergosterol		
	Ergosta-7,22-dienol		
	Ergosta-7-enol		

Table 2 contd. ...

...Table 2 contd.

Bioactive Group	Compound	Bioactivity	References
Tocopherols	*α-tocopherol*	Anti-inflammatory, antioxidant and anticancer activities; prevents DNA damage and decreases the risk of cardiovascular disease.	Barros et al., 2008; Phillips et al., 2011; Jiang, 2014
	β-tocopherol		
	γ-tocopherol		
Fatty Acid	*Acetylenic*	Antibacterial, antiviral, antifungal, antiparasitic, pesticidal, antitumour, anti-inflammatory.	Li et al., 2008; Ribeiro et al., 2009; Hong et al., 2012
	Arachidonic		
	Behenic		
	cis-11,14-eicosadienoic		
	Heptadecanoic		
	Lauric acid		
	Lignoceric		
	Linoleic		
	Myristic		
	Oleic		
	Palmitic		
	Palmitoleic		
	Pentadecanoic		
	Tricosanoic		
	Stearic		

interest because of their antioxidant properties; so they are preventive measures against a variety of ailments (da Silva and Jorge, 2012). Fatty acids have been shown to possess strong antifungal and antibacterial activities. Chanterelle contains many fatty acids but predominant are linoleic acid and oleic acid. Acetylenic acids possess strong antimicrobial, pesticidal and antitumour qualities (Li et al., 2008). Acetylenic acid (NPC235242), isolated from methanolic extracts of Chanterelle, showed significant transcriptional activity towards peroxisome proliferator-activated receptor γ which regulates cell proliferation, glucose and fat metabolism and inflammatory processes (Hong et al., 2012).

6.1 Polysaccharides

Polysaccharides are one of the essential bioactive components of Chanterelle. The composition, linking pattern and sequence of the monosaccharide are the key factors in determining the bioactivities of polysaccharides. Chanterelle polysaccharides, mainly the β-glucans, are used in ointments to protect the skin against UV radiation and in the treatment of various skin diseases (Muszynska et al., 2016). It has been found that protein-bound polysaccharide fractions possess immunostimulating properties (Kalaĉ, 2009; Moro et al., 2012; M´endez-Espinoza et al., 2013). It is well documented that crude polysaccharide extracts of Chanterelle inhibit the proliferation of colon cancer cells without showing toxicity to the normal cells (Kolundzic et al., 2017; Vlasenko et al., 2019). Moreover, polysaccharides extracted from Chanterelle have been studied and are found to exhibit antioxidation and hypoglycemic properties (Jin, 2011). Macrophages are a form of phagocytes that play a key part in innate immune responses and are used as model cells to study immunomodulatory properties. Mushroom polysaccharides have been reported to boost macrophage viability by binding to certain membrane receptors and activating the immune response (Park et al., 2012; Liao et al., 2016). A water-soluble polysaccharide (JP1) isolated from *C. cibarius* has been found to be non-toxic to a mouse macrophage cell line (RAW264.7). Moreover JP1 can help the mouse peritoneal macrophages to secrete NO and boost the secretion of cytokines IL-6 in RAW264.7 cells, indicating higher immunostimulating activity (Chen et al., 2017). COX-1 and COX-2 activities are suppressed by the polysaccharide fraction of Chanterelle, indicating reduced inflammation, pain and fever (Elkhateeb and Daba, 2021).

6.2 Phenolic Compounds

Phenolic compounds usually have a protective role against different diseases due to their antioxidant properties (Silva et al., 2004). Besides, they have anti-inflammatory, antimicrobial, hypoglycemic and immunostimulating activities and minimise the risk of cancer and heart ailments (Muszyńska et al., 2016). Free radical inhibitors, oxygen scavengers, peroxide decomposers or metal inactivators are considered as the main group of phenolic compounds (Yagi, 1970; Dziezak, 1986). *Cantharellus cibarius* harvested in Poland was found to contain phenolic acids, such as cinnamic, *p*-hydroxybenzoic, protocatechuic, sinapic and vanillic acids and their amounts range between 1.29–3.32 mg/kg dry weight. The highest concentration is noted for vanillic acid (3.32 mg/kg dry weight) (Muszyńska et al., 2013). *C. cibarius* from Spain contains phenolic compounds, such as pyrogallol, myricetin, catechin, caffeic acid, ferulic acid, gallic acid, *p*-hydroxybenzoic acid, homogentisic acid and protocatechuic acid (Barros et al., 2009). Studies indicate that Chanterelle contains higher caffeic acid and catechin than other examined species (*A. mellea, B. badius, B. edulis, L. delicious* and *P. ostreatus*) and perhaps these substances have greater antioxidative power than the other examined polyphenols (Palacios et al., 2011; Muszyńska et al., 2013). Methanolic extract of Chanterelle showed higher antioxidant activity in suppression of lipid peroxidation and iron-chelating activity compared to hydrogen atom and/or electron transfer (Kozarski et al., 2015).

6.3 Indole Compound

The culinary–medicinal mushrooms contain almost 140 different types of indole compounds (Homer and Sperry, 2017). Indole compounds are grouped into hallucinogenic and non-hallucinogenic types and mushrooms could have both hallucinogenic indoles (psilocybin, psilocin and baeocystin) and non-hallucinogenic, includes tryptophan derivatives (tryptamine, serotonin and melatonin). The indole compound, melatonin, showed a significant antioxidant impact. Inhibition of lipid peroxidation in a dose-dependent fashion is exhibited by melatonin and some other indole derivatives, like N-acetylserotonin, 6-methoxytryptamine (Sewerynek et al., 2005). Serotonin is one of the potent endogenous neurotransmitters as well as induce antioxidant activity (Pytka et al., 2016; Azouzi et al., 2017). Methanolic extracts of fruit bodies and *in vitro* mycelial culture of *C. cibarius* were investigated for similar indole compounds. The extracts were found to contain eight indole compounds—melatonin, serotonin, indole, methyltryptophan, kynurenine sulphate, L-tryptophan, 5-hydroxytryptophan and 5-indole-3-acetonitryle. Furthermore, fruit bodies also contain tryptamine, unlike mycelial cultures (Muszyńska et al., 2013). The quantitatively dominating compounds in either forms (fruiting bodies or mycelial culture) were recorded for kynurenine sulphate and serotonin and no significant difference was found. However, mycelial culture accumulates more serotonin (20.49 mg/100 g dw) and kynurenine sulphate (35.34 mg/100 g dw). However, a significant difference was recorded in the level of 5-hydroxytryptophan (0.01 mg/100 g dw in fruit body and 12.52 mg/100 g dw in mycelium) and kynurenine sulphate (3.62 mg/100 g dw in fruit body and 35.34 mg/100 g dw in mycelium). Muszyńska et al. (2013) observed that the melatonin concentration in the fruit body was much higher (0.11 mg/100 g dw) compared to mycelium (0.01 mg/100 g dw).

6.4 Carotenoids

Plants, fungi and some other microorganisms contain carotenoids, which are complex organic molecules having a tetraterpene structure (Fiedor and Burda, 2014). All the carotenoids act as antioxidants. Studies indicate that carotenoids protect our body from oxidative stress, neurological and circulatory disorders as well as different forms of cancers (Fiedor and Burda, 2014). Chanterelle contains α-carotene, β-carotene, γ-carotene and lycopene but β-carotene (5.77 µg/g) and lycopene (1.95 µg/g) are more (Barros et al., 2008). Ribeiro et al. (2011) identified the presence of β-carotene only in *C. cibarius* among the 17 species by using the HPLC–DAD (high-performance liquid chromatography linked to a diode array detector) technique. Carotenoids from Chanterelle are used in the food industry as a natural source of pigments due to their high concentration (Barros et al., 2008). The presence of β-carotene is considered to contribute to the orange colour of *C. cibarius*. The methanolic extract of *C. cibarius* was measured for antioxidant compounds and the results indicated that the levels of β-carotene and lycopene were equal (112.2 µg/g dw) (Kozarski et al., 2015).

6.5 Enzymes

Cantharellus cibarius contains active enzymes like tyrosinase and Mn-dependent peroxidase. The mushroom contains homodimeric lactase which is a lignolytic enzyme used in biotechnology (Ng and Wang, 2004; Nakamura and Go, 2005; Ramírez-Anguiano et al., 2007). This enzyme has immunomodulating, antiproliferative and HIV-1 reverse transcriptase-suppressing properties (Ng and Wang, 2004). Oxidation of free linoleic acid by the enzyme oxidoreductases yields 1-octen-3-ol, which is responsible for the typical flavour of Chanterelle (de Pinho et al., 2008).

6.6 Biological Activity of Cantharellus cibarius Fruit Body Extract

Chanterelle exhibits an array of biological activities, such as antioxidant, antimicrobial, cytotoxic, antihypoxic, antihyperglycemic, anti-inflammatory, and wound healing activity. The antimicrobial activities of methanolic extract of *C. cibarius* were studied using a disk diffusion and broth microdilution method against Gram-positive (*Staphylococcus aureus, Enterococcus faecalis, Bacillus cereus* and *Listeria monocytogenes*) and Gram-negative (*Escherichia coli, Salmonella enteritidis, Shigella sonnei,* and *Yersinia enterocolitica*) bacteria and the results indicated that the extract has a potential bactericidal activity (Kozarski et al., 2015). Broth microdilution assay was performed to investigate the antimicrobial activity of various extracts (cyclohexane, dichloromethane, methanol and aqueous) of Chanterelle with special importance on *Helicobacter pylori.* All extracts were active against tested organisms but methanolic extract showed highest activity against *Helicobacter pylori* with minimal inhibitory concentration values. Another study performed by Dulger et al. (2004) showed acetone, chloroform, ethanol and ethyl acetate extracts of Chanterelle as being positive for antimicrobial activity against Gram (+) and Gram (−) bacteria, yeasts, filamentous fungi and actinomycetes.

The antioxidant and antihyperglycemic effects of both ethanolic and hydromethanolic extracts of *C. cibarius* were assessed. Ethanol extracts exhibited high ferrous ion chelating, lipoxygenase and α-glucosidase inhibitory properties (Zavastin et al., 2016).

The antioxidant activity of different fractions of the methanolic extracts, including n-hexane, chloroform, ethyl acetate, n-butanol and water was tested. It was found that nitric oxide scavenging activity, total phenols and flavonoids were greatest in the n-hexane fraction while the aqueous fraction had higher reducing activity and the ethyl acetate fraction demonstrated strongest DPPH scavenging activity (Ebrahimzadeh et al., 2015).

The methanolic extract of mushroom was studied *in vitro* for cytotoxic activities. Compared to normal control human fetal lung fibroblasts MRC-5 and human lung bronchial epithelial cells BEAS-2B, the extract demonstrated significant selectivity in cytotoxicity on human cervix adenocarcinoma HeLa, breast cancer MDAMB-453 and human myelogenous leukemia K562 (Kozarski et al., 2015).

The antigenotoxic activity of *C. cibarius* aqueous extracts was assessed by comet assay. Human mononuclear cells exposed to methyl methanesulphonate (MMS) were used as a test system to evaluate the inhibitory efficiency. *In vitro* 3 protocols (previous, simultaneous and posterior) included five extracts in different concentrations (0.0125, 0.025, 0.05, 0.1 and 0.2%) and found aqueous extracts in simultaneous and previous treatment with efficiently reduced DNA damage. The maximum reduction of DNA damage was showed by 0.0125% extract from the previous protocol and the 0.2% extract from the simultaneous one. No dose dependent activity was recorded (Mendez-Espinoza et al., 2013).

When compared to the non-treated and vehicle-treated groups, Chanterelle methanolic crude extract demonstrated noteworthy wound-healing and anti-inflammatory activity. Histological changes exhibit complete repair of the epidermal layer, enhanced production of collagen and a significant degree of neovascularisation and epithelisation in methanolic extract which could be the scientific basis for the therapeutic use of *C. cibarius* as a wound healer (Luo et al., 2004; Nasiry et al., 2017).

7. Conclusion

Cantharellus cibarius holds a long history of being used as an edible foodstuff and a well-known natural source of medicine. Extensive research findings indicate that this mushroom can be used as a low calorie diet due to its high proximate composition.

Thus, incorporation of this important macrofungi in our daily life may be beneficial for us. The rich sources of carbohydrates, amino acids, vitamins, minerals are definitely a good adjunct to its unique pharmaceutical composition by having a number of bioactive compounds with exceptional potential for treating a variety of diseases. Antimicrobial, anti-oxidative, anticancer, immunomodulatory, anti-inflammatory, anti-genotoxic, and anti–diabetic activities are the significant aspects of *C. cibarius*. This chapter summarises all the possible nutritive and medicinal hallmarks of this species. However, most of the experimentations are based on *in vitro* trials and a very limited *in vivo* study has been made. So in future investigations, clinical trials are essential to approve the safety and efficacy of these mushroom-derived bioactive compounds. Moreover, detailed studies should be performed to understand the mechanism of action of various bioactive compounds as we have very limited information in this regard. Finally, collaborative research work among the mycologist, nutritionist, chemist and pharmaceutics are essential to enhance the application of *C. cibarius* as a functional food or as a source of nutraceuticals for better health and life quality.

References

Agrahar-Murugkar, D. and Subbulakshmi, G. (2005). Nutritional value of edible wild mushrooms collected from the Khasi Hills, Meghalaya. *Food Chem.*, 89: 599–603.

Aryantha, I.N.P., Kusmaningati, A.B., Sutjiatmo, A.B., Sumartini, Y., Nursidah, A. and Narvikasari, S. (2010). The effect of *Laetiporus* sp. (Bull. ex. Fr.) *bond. et sing.* (Polyporaceae) extract on total blood cholesterol. *Biotechnology*, 9: 312–318.

Azeem, U., Hakeem, K.R. and Ali, M. (2020). Commercialisation and conservation. pp. 97–106. *In: Fungi for Human Health.* Switzerland: Springer. https://Doi.org/10.1007/978-3-030-58756-7_8.

Azouzi, S., Santuz, H., Morandat, S., Pereira, C., Côté, F., Hermine, O., El Kirat, K., Colin, Y., Etchebest, C. and Amireault, P. (2017). Antioxidant and membrane binding properties of serotonin protect lipids from oxidation, *Biophys. J.*, 112: 1863–1873. https ://Doi.org/10.1016/j.bpj.2017.03.037.

Badalyan, S.M. and Rapior, S. (2020). Perspectives of biomedical application of macrofungi. *Curr. Trends Biomed. Eng. Biosci.*, 19: 556024. http://Doi.org/10.19080/CTBEB.2020.19.556024.

Barret, C.L., Beresford, N.A., Self, P.L., Howard, B.J., Frankland, J.C., Fulker, M.J., Dodd, B.A. and Marriott, I.V.R. (1999). Radiocaesium activity concentrations in the fruit-bodies of macrofungi in Great Britain and an assessment of dietary intake habits. *Sci. Tot. Environ.*, 231: 67–83.

Barros, L., Cruz, T., Baptista, P., Estevinho, L.M. and Ferreira, I.C. (2008). Wild and commercial mushrooms as source of nutrients and nutraceuticals. *Food Chem. Toxicol.*, 46(8): 2742–2747.

Barros, L., Duenas, M., Ferreira, I.C., Baptista, P. and Santos-Buelga, C. (2009). Phenolic acids determination by HPLC-DAD-ESI/MS in 16 different Portuguese wild mushrooms species. *Food Chem. Toxicol.*, 47(6): 1076–1079.

Barros, L., Ferreira, M., Ferreira, I., Queiros, B. and Baptista, P. (2007). Total phenols, ascorbic acid, β-carotene, and lycopene in Portuguese wild edible mushrooms and their antioxidant activities. *Food Chem.*, 103: 413–419.

Beluhan, S. and Ranogajec, A. (2011). Chemical composition and non-volatile components of croatian wild edible mushrooms. *Food Chem.*, 124: 1076–1082. https://Doi.org/10.1016/j.foodchem.2010.07.081.

Bulam, S., Pekşen, A. and Üstün, N.Ş. (2019). Yenebilir ve tıbbi mantarların gıda ürünlerinde kullanım potansiyeli. *Mantar Dergisi*, 10: 137–151. https://Doi.org/10.30708.mantar.644367.

Bulam, S., Şule, N., Üstün, N.S. and Pekşen, A. (2021). Effects of different processing methods on nutrients, bioactive compounds, and biological activities of Chanterelle mushroom (*Cantharellus cibarius*): A review. *Eur. Food Sci. Eng.*, 2(2): 52–58.

Chang, S.T. and Buswell, J.A. (2003). Medicinal mushrooms—A prominent source of nutriceuticals for the 21st century. *Curr. Top. Nutraceutical Res.*, 1(3): 257–283.

Chen, L., Peng, X., Lv, J., Liao, S., Ou, S. and Shen, Y. (2017). Purification and structural characterisation of a novel water-soluble neutral polysaccharide from *Cantharellus cibarius* and its immunostimulating activity in RAW264.7 cells. *Int. J. Poly. Sci.*, 3074915. https://Doi. org/10.1155/2017/3074915.

Crisan, E.V. and Sands, A. (1978). Nutritional value. pp. 137–168. *In*: Chang, S.T. and W.A. Hayer (eds.). *The Biology and Cultivation of Edible Mushrooms.* Academic Press, New York, USA.

da Silva, A.C. and Jorge, N. (2012). Influence of *Lentinus edodes* and *Agaricus blazei* extracts on the prevention of oxidation and retention of tocopherols in soybean oil in an accelerated storage test. *J. Food Sci. Technol.* Doi:10.1007/s13197-13012-10623-13191.

Daba, G.M., Elkhateeb, W., El-Dein, A.N., Fadl E Ahmed, E. and El-Hagrasi, A. (2020). Therapeutic potentials of n-hexane extracts of the three medicinal mushrooms regarding their anticolon cancer, antioxidant, and hypocholesterolemic capabilities. *Biodiversitas J. Biol. Diversity*, 21(6): 2437–2445.

Daniewski, W.M., Danikiewicz, W., Gołębiewski, W.M. et al. (2012). Search for bioactive compounds from *Cantharellus cibarius. Nat. Prod. Comm.*, 7(7): 917–918.

de Pinho, P.G., Ribeiro, B., Goncalves, R.F., Baptista, P., Valentao, P., Seabra, R.M. and Andrade, P.B. (2008). Correlation between the pattern volatiles and the overall aroma of wild edible mushrooms. *J. Agric. Food Chem.*, 56(5): 1704–1712.

Deshaware, S., Marathe, S.J., Bedade, D., Deska J. and Shamekh, S. (2021). Investigation on mycelial growth requirements of *Cantharellus cibarius* under laboratory conditions. *Arch. Microbiol.* Doi: 10.1007/s00203-020-02142-0.

Drewnowska, M. and Falandysz, J. (2014). Investigation on mineral composition and accumulation by popular edible mushroom common Chanterelle (*Cantharellus cibarius*). *Ecotoxicol. Environ. Saf.*, 113: 9–17. Doi.org/10.1016/j.ecoenv.2014.11.028.

Dulger, B., Gonuz, A. and Gucin, F. (2004). Antimicrobial activity of the macro fungus *Cantharellus cibarius. Pak. J. Biol. Sci.*, 7(9): 1535–1539.

Dziezak, J.D. (1986). Antioxidants—The ultimate answer to oxidation. *Food Technol.*, 40: 94.

Ebrahimzadeh, M.A., Safdari, Y. and Khalili, M. (2015). Antioxidant activity of different fractions of methanolic extract of the golden Chanterelle mushroom *Cantharellus cibarius* (higher basidiomycetes) from Iran. *Int. J. Med. Mushrooms*, 17(6): 557–65.

El-Hagrassi, A., Daba, G.M., Elkhateeb, W., Ahmed, E. and El-Dein, A.N. (2020). *In vitro* bioactive potential and chemical analysis of the n-hexane extract of the medicinal mushroom, *Cordyceps militaris*, Malays. *J. Microbiol.*, 16(1): 40–48.

Elkhateeb, W.A. (2020). What medicinal mushroom can do? *Chem. Res. J.*, 5(1): 106–118.

Elkhateeb, W.A. and Daba, G.M. (2021). Highlights on the golden Mushroom *Cantharellus cibarius* and unique shaggy ink cap Mushroom *Coprinus comatus* and smoky bracket Mushroom *Bjerkandera adusta* ecology and biological activities. *J. Mycol. Mycological. Sci.*, 4(2): 000143. Doi: 10.23880/oajmms-16000143.

Falandysz, J. and Drewnowska, M. (2015). Macro and trace elements in common Chanterelle (*Cantharellus cibarius*) mushroom from the European background areas in Poland: Composition, accumulation, dietary exposure and data review for species. *J. Environ. Sci. Health*, 50: 374–387. Doi: 10.1080/03601234.2015.1000190.

Falandysz, J. and Drewnowska, M. (2017). Cooking can decrease mercury contamination of a mushroom meal: *Cantharellus cibarius* and Amanita fulva. *Environ. Sci. Pollut. Res. Int.* Doi 10.1007/s11356-017-8933-5.

Fiedor, J. and Burda, K. (2014). Potential role of carotenoids as antioxidants in human health and disease. *Nutrients*, 6: 466–488. https ://Doi.org/10.3390/nu602 0466.

Grangeia, C., Heleno, S.A., Barros, L., Martins, A. and Ferreira, I.C.F.R. (2011). Effects of trophism on nutritional and nutraceutical potential of wild edible mushrooms. *Food Res. Int.*, 44: 1029–1035.

Gry, J., Andersson, C., Kruger, L., Lyran, B., Jensvoll, L., Matilainen, N., Nurttila, A., Olafsson, G. and Fabech B. (2012). *TemaNord.* Nor. C. Min. Copenhagen, Denmark, p. 81.

Homer, J.A. and Sperry, J. (2017). Mushroom-derived indole alkaloids. *J. Nat. Prod.*, 80: 2178–2187. https ://Doi.org/10.1021/acs.jnatp rod.7b003 90.

Hong, S.S., Lee, J.H., Jeong, W., Kim, N., Jin, H.Z., Hwang, B.Y., Lee, H., Sung-Joon Lee, Sik Jang, D. and Lee, D. (2012). Acetylenic acid analogues from the edible mushroom Chanterelle (*Cantharellus cibarius*) and their effects on the gene expression of peroxisome proliferator-activated receptor-gamma target genes. *Bioorg. Med. Chem. Lett.*, 22: 2347–349. https://Doi.org/10.1016/j.bmcl.2012.01.070.

Jabłónska-Rýs, E., Sławińska, A. and Szwajgier, D. (2016). Effect of lactic acid fermentation on antioxidant properties and phenolic acid contents of oyster (*Pleurotus ostreatus*) and Chanterelle (*Cantharellus cibarius*) mushrooms. *Food Sci. Biotechnol.*, 25(2): 439–444.

Jechalke, N.N., Nowak, R., Juda, M., Malm, A., Lemieszek, M., Rzeski, W. and Kaczyński, Z. (2018). New biological activity of the polysaccharide fraction from *Cantharellus cibarius* and its structural characterisation. *Food Chem.*, 268: 355–361.

Jiang, Q. (2014). Natural forms of vitamin E: Metabolism, antioxidant, and anti-inflammatory activities and their role in disease prevention and therapy. *Free Radic. Biol. Med.*, 72: 76–90. Doi: 10.1016/j.freeradbiomed.2014.03.035.

Jin, W.J. (2011). Study on extraction of *C. cibarius* polysaccharide and its antioxidant activity. *J. Anhui. Agric. Sci.*, 2: 641–657.

Kalaĉ, P. (2009). Chemical composition and nutritional value of European species of wild growing mushrooms: A review. *Food Chem.*, 113: 9–16.

Kolundzic, M., Stanojkovic, T., Radovic, J., Tacic, A., Dodevska, M., Milenkovic, M., Sisto, F., Masia, C., Farronato, G., Nikolic, V. and Kundakovic, T. (2017). Cytotoxic and antimicrobial activities of *Cantharellus cibarius* Fr. (Cantarellaceae). *J. Med. Food.*, 20(8): 790–796.

Kozarski, M., Anita Klaus, A., Jovana Vunduk, J., Zizak, Z., Niksic, M., Jakovljevic, D., Miroslav M., Vrvic, M.M. and Van Griensven, L.J.L.D. (2015). Nutraceutical properties of the methanolic extract of edible mushroom *Cantharellus cibarius* (Fries): Primary mechanisms. *Food Funct.*, 6(6): 1875–86. Doi: 10.1039/c5fo00312a. PMID: 25943486.

Kuka, M., Cakste, I. and Galoburda, R. (2014). Chemical composition of Latvian wild edible mushroom *Cantharellus cibarius*. In: *Proceedings of the in Baltic Conference on Food Science and Technology food for Consumer Well-Being*. Foodbalt, Jelgava, Latvia.

Kumar, R., Tapwal, A., Pandey, S., Borah, R.K., Borah, D. and Borgohain, J. (2013). Macro-fungal diversity and nutrient content of some edible mushrooms of Nagaland, India. *Nusantara Biosci.*, 5(1): 1–7.

Kumari, D., Reddy, M.S. and Upadhyay, R.C. (2011). Antioxidant activity of three species of wild mushroom genus *Cantharellus* collected from North-Western Himalaya, India. *Int. J. Agric. Biol.*, 13: 415–418.

Lemieszek, M.K., Nunes, F.M., Cardoso, C., Marques, G. and Rzeski, W. (2018). Neuroprotective properties of *Cantharellus cibarius* polysaccharide fractions in different *in vitro* models of neurodegeneration. *Carbo. Polym.*, 197: 598–607. Doi.org/10.1016/j.carbp ol.2018.06.038.

Li, X.C., Jacob, M.R., Khan, S.I., Ashfaq, M.K., Babu, K.S., Agarwal, A.K., Elsohly, H.N., Manly, S.P. and Clark, A.M. (2008). Potent *in vitro* antifungal activities of naturally occurring acetylenic acids. *Antimicrob. Agents Chemother.*, 52(7): 2442–2448. https://Doi.org/10.1128/AAC.01297-07.

Liao, W., Lai, T., Chen, L., Fu, J., Sreenivasan, S.T. and Ren, J. (2016). Synthesis and characterisation of a walnut peptides-zinc complex and its antiproliferative activity against human breast carcinoma cells through the induction of apoptosis. *J. Agric. Food Chem.*, 64(7): 1509–1519.

Lovrić, M., Da Re, R., Vidale, E., Prokofieva, I., Wong, J., Pettenella, D., Verkerk, P.J. and Mavsar, R. (2020). Nonwood forest products in Europe—A quantitative overview. *For. Policy Econ.*, 116: 102175. https://Doi.org/10.1016/j.forpol.2020.102175.

Lovrić, M., Da Re, R., Vidale, E., Prokofieva, I., Wong, J., Pettenella, D., Verkerk, P.J. and Mavsar, R. (2021). Collection and consumption of non-wood forest products in Europe. *Forestry: Int. J. For. Res.*, 94: 757–770. https://doi.org/10.1093/forestry/cpab018.

Luo, H., Mo, M., Huang, X., Li, X. and Zhang, K. (2004). Coprinus comatus: A basidiomycete fungus forms novel spiny structures and infects nematode. *Mycologia.*, 96(6): 1218–1224.

Marathe, S.J., Hamzi, W., Bashein, A.M., Deska, J., Seppänen-Laakso, T., Singhal R.S. and Shamekh, S. (2021). Anti-angiogenic effect of *Cantharellus cibarius* extracts, its correlation with lipoxygenase inhibition, and role of the bioactives therein. *Nutr. Cancer*, 1–11. https://Doi.org/1 0.1080/01635581.2021.1909739.

Mendez-Espinoza, C., Garcia-Nieto, E., Esquivel, A.M., Gonzalez, M.M., Bautista, E.V., Ezquerro, C.C. and Santacruz, L.J. (2013). Antigenotoxic potential of aqueous extracts from the Chanterelle mushroom, *Cantharellus cibarius* (higher Basidiomycetes) on human mononuclear cell cultures. *Int. J. Med. Mushrooms*, 15(3): 325–332.

Méndez-Espinoza, C., García-Nieto, E. and Montoya Esquivel, A. (2013). Antigenotoxic potential of aqueous extracts from the Chanterelle mushroom, *Cantharellus cibarius* (higher Basidiomycetes), on human mononuclear cell cultures. *Int. J. Med. Mushrooms*, 15(3): 325–332.

Moro, C., Palacios, I., Lozano, M., d'Arrigo, M., Guillamon, E. Garicochea, A.V., Martinez, J.A. and García-Lafuente, A. (2012). Anti-inflammatory activity of methanolic extracts from edible mushrooms in LPS activated RAW 264.7 macrophages. *Food Chem.*, 130(2): 350–355.

Mui, D., Feibelman, T. and Bennett, J.W. (1998). A preliminary study of the carotenoids of some North American species of Cantharellus. *Int. J. Plant Sci.* 159(2): 244–247. https://Doi.org/10.1086/297545.

Muszyńska, B., Kała, K., Firlej, A. and Sułkowska-Ziaja, K. (2016). *Cantharellus cibarius*— Culinary-medicinal mushroom content and biological activity. *Acta Poloniae Pharm. Drug Res.*, 73: 589–598.

Muszyńska, B., Sułkowska-Ziaja, K. and Ekiert, H. (2013). Analysis of indole compounds in methanolic extracts from the fruiting bodies of *Cantharellus cibarius* (the Chanterelle) and from the mycelium of this species cultured *in vitro*. *J. Food Sci. Technol.*, 50(6): 1233–1237. Doi: 10.1007/s13197-013-1009-8.

Nakamura, K. and Go, N. (2005). Function and molecular evolution of multicopper blue proteins. *Cell Mol. Life Sci.*, 62: 2050–2066.

Nasiry, D., Khalatbary, A.R. and Ebrahimzadeh, M.A. (2017). Anti-inflammatory and wound-healing potential of Golden Chanterelle mushroom, *Cantharellus cibarius* (Agaricomycetes). *Int. J. Med. Mushrooms*, 19: 893–903. https://Doi.org/10.1615/IntJMedMushrooms.2017024674.

Ng, T.B. and Wang, H.X. (2004). A *homodimeric laccase* with unique characteristics from the yellow mushroom *Cantharellus cibarius*. *Biochem. Biophys. Res. Commun.*, 313(1): 37–41.

Nowacka-Jechalke, N., Nowak, R., Juda, M., Malm, A., Lemieszek, M., Rzeski, W. and Kaczyński, Z. (2018). New biological activity of the polysaccharide fraction from *Cantharellus cibarius* and its structural characterisation. *Food Chem.*, 268: 355–361. https://Doi.org/10.1016/j.foodchem.2018.06.106.

Nyman, A.A.T., Aachmann, F.L., Rise, F., Ballance, S. and Samuelsen, A.B.C. (2016). Structural characterisation of a branched $(1 \rightarrow 6)$-α-mannan and β-glucans isolated from the fruiting bodies of *Cantharellus cibarius*. *Carbohyd. Polym.*, 146: 197–207. http://dx.Doi.org/10.1016/j.carbpol.2016.03.052.

Palacios, I., Lozano, M., Moro, C., Arrigo, M., Rostagno, M.A., Martinez, J.A., Garcia-Lafuente, A., Guillamon, E. and Villares, A. (2011). Antioxidant properties of phenolic compounds occurring in edible mushrooms. *Food Chem.*, 128: 674–678. http://dx.Doi.org/10.1016/j.foodchem.2011.03.085.

Panchak, L.V., Ya Antonyuk, L., Zyn, A.R. and Antonyuk, V. (2020). Extractive substances of fruit body Golden Chanterelle (*Cantharellus cibarius* Fr.) and Hedgehog mushroom (*Hydnum repandum* Fr.). *Em. J. Food Agric.*, 32: 826–834. https://Doi.org/10.9755/ejfa.2020.v32.i11.2195.

Park, D.K., Hayashi, T. and Park, H.J. (2012). Arabinogalactan-type polysaccharides (APS) from *Cordyceps militaris* grown on germinated soybeans (GSC) induces innate immune activity of THP-1 monocytes through promoting their macrophage differentiation and macrophage activity. *Food Sci. Biotechnol.*, 21(5): 1501–1506.

Peksen, A., Bulam, S. and Üstün, N.Ş. (2016). Edible wild mushrooms sold in Giresun local markets. *1st International Mediterranean Science and Engineering Congress* (IMSEC 2016), Adana, Turkey.

Phillips, K.M., Ruggio, D.M., Horst, R.L., Minor B. and Simon R.R. (2011). Vitamin D and sterol composition of 10 types of mushrooms from retail suppliers in the United States. *J. Agric. Food Chem.*, 59: 7841–53. Doi: 10.1021/jf104246z.

Pilz, D., Norvell, L., Danell, E. and Molina, R. (2003). Ecology and management of commercially harvested Chanterelle mushrooms. *Gen. Tech. Rep.*, Department of Agriculture, Forest Service, Portland. Doi.org/10.2737/PNW-GTR-57613.

Pytka, K., Podkowa, K. and Rapacz, A. (2016). The role of serotonergic, adrenergic and dopaminergic receptors in antidepressant-like effect. *Pharmacol. Reports*, 68: 263–274. https ://Doi. org/10.1016/j.phare p.2015.08.007.

Ramírez-Anguiano, A., Santoyo, S., Reglero, G. and Soler-Rivas, C. (2007). Radical scavenging activities, endogenous oxidative enzymes and total phenols in edible mushrooms commonly consumed in Europe. *J. Agric. Food Chem.*, 12: 2272. Doi: 10.1002/jsfa.2983.

Rana, R. (2016). Systematic studies on wild edible Mushroom *Cantharellus cibarius* fries collected from Shimla hills of Himachal Pradesh, India. *IJIRSET*, 5(12): 20651–20654. Doi:10.15680/ IJIRSET.2016.0512093.

Rangel-Castro, J.I., Danell, E. and Staffas, A. (2002). The ergocalciferol content of dried pigmented and albino *Cantharellus cibarius* fruit bodies. *Mycol. Res.*, 106: 70–73. Doi: 10.1017} S0953756201005299.

Ribeiro, B., Andrade, P., Silva, B., Baptista, P., Seabra, R. and Valentao, P. (2006). Contents of carboxylic acids and two phenolics and antioxidant activity of dried Portuguese wild edible mushrooms. *J. Agri. Food Chem.*, 54: 8530–8537.

Ribeiro, B., de Pinho, P., Andrade, P.B., Baptista, P. and Valento, P. (2009). Fatty acid composition of wild edible mushrooms species: A comparative study. *Microchem. J.*, 93: 29–35.

Ribeiro, B., de Pinho, P.G., Andrade, P.B. and Patrícia, V. (2011). Do bioactive carotenoids contribute to the color of edible mushrooms? *Open Chem. Biomed. Methods. J.*, 4: 14–18. https://Doi. org/10.2174/1875038901 10401 0014.

Saha, B.C. and Racine, F.M. (2011). Biotechnological production of mannitol and its applications. *Appl. Microbiol. Biotechnol.*, 89(4): 879–91. Doi: 10.1007/s00253-010-2979-3.

Scalbert, A., Manach, C., Morand, C., Remesy, C. and Jimenez J. (2005). Dietary polyphenols and the prevention of diseases. *Crit. Rev. Food Sci. Nutr.*, 45(4): 287–306. Doi: 10.1080/1040869059096.

Sewerynek, E.M., Stuss, K., Oszczygiel, J. and Kulak, A. (2005). Protective effect of indole compounds on lipopolysaccharideinduced lipid peroxidation in *in vitro* conditions. *Endokrynol. Pol.*, 56(4): 508.

Silva, B.M., Andrade, P.B., Valentão, P., Ferreres, F., Seabra, R.M. and Ferreira, M.A. (2004). Quince (*Cydonia oblonga* Miller) fruit (pulp, peel, and seed) and jam: Antioxidant activity. *J. Agric. Food Chem.*, 52: 4705–4712.

Teplyakova, T.V., Psurtseva, N.V., Kosogova, T.A., Mazurkova, N.A., Khanin, V.A. and Vlasenko, V.A. (2012). Antiviral activity of polyporoid mushrooms (higher Basidiomycetes) from Altai Mountains (Russia). *Int. J. Med. Mushrooms*, 14(1): 37–45. Doi: 10.1615/intjmedmushr.v14. i1.40.

Thimmel, R. and Kluthe, R. (1998). The nutritional database for edible mushrooms. *Ernahrung.*, 22(2): 63–65.

Thu, Z.M., Myo, K.K., Aung, H.T., Clericuzio, M., Armijos, C. and Vidari, G. (2020). Bioactive phytochemical constituents of wild edible mushrooms from Southeast Asia. *Molecules*, 25: https://Doi.org/10.3390/molecules25081972.

Turfan, N., Ayan, S., Akın, Ş.S. and Akın, E. (2019). Nutritional and antioxidant variability of some wild and cultivated edible mushrooms from Kastamonu rural areas. *TURJAF*, 7: 11–16. https:// Doi.org/10.24925/turjaf.v7isp3.11-16.3094.

Uthan, E.T., Senturk, H., Uyanoglu, M. and Yamaç, M. (2021). First report on the *in vivo* prebiotic, biochemical, and histological effects of crude polysaccharide fraction of Golden Chantharelle mushroom, *Cantharellus cibarius* (Agaricomycetes). *Int. J. Med. Mushrooms*, 23: 67–77. https:// Doi.org/10.1615/IntJMedMushrooms.2021038233.

Valentao, P., Andrade, P.B., Rangel, J., Ribeiro, B., Silva, B.M., Baptista, P. and Seabra, R.M. (2005). Effect of the conservation procedure on the contents of phenolic compounds and organic acids in Chanterelle (*Cantharellus cibarius*) mushroom. *J. Agric. Food Chem.*, 53(12): 4925–4931.

Vlasenko, V., Turmunkh, D., Ochirbat, E., Budsuren, D., Nyamsuren, K., Samiya, J., Ganbaatar, B. and Vlasenko A. (2019). Medicinal potential of extracts from the Chanterelle mushroom,

Cantharellus cibarius (Review) and prospects for studying its strains from differs plant communities of ultra-continental regions of the Asia. *BIO Web of Conferences*, 16: 00039.

Wani, B.A., Bodha, R.H. and Wani, A.H. (2010). Nutritional and medicinal importance of mushrooms. *J. Med. Plants Res.*, 4(24): 259–604.

Wasser, S.P. (2010). Medicinal mushroom science: History, current status, future trends, and unsolved problems. *Int. J. Med. Mushrooms*, 12(1): 1–16.

Yagi, K. (1970). A rapid method for evaluation of autoxidation and antioxidants. *Agric. Biol. Chem.*, 34: 142.

Zambrano, M.V., Dutta, B., Mercer, D.G. and Touchie, M.F. (2019). Assessment of moisture content measurement methods of dried food products in small-scale operations in developing countries: A review. *Trends Food Sci. Technol.*, 88: 484–496. https://Doi.org/10.1016/j.tifs.2019.04.006.

Zavastin, D.E., Bujor, A., Tuchiluş, C., Mircea, C.G., Gherman, S.P., Aprotosoaie, A.C. and Miron, A. (2016). Studies on antioxidant, antihyperglycemic and antimicrobial effects of edible Mushrooms *Boletus edulis* and *Cantharellus Cibarius*. *J. Plant Develop.*, 23: 87–95.

Zhang, D., Frankowska, A., Jarzyńska, G., Kojta, A.K., Drewnowska, M., Wydmańska, D., Bielawski, L., Wang, J. and Falandysz, J. (2010). Metals of King Bolete (*Boletus edulis*) collected at the same site over two years. *African J. Agric. Res.*, 5: 3050–3055.

Chapter 9

Black Trumpet
(*Craterellus cornucopioides*)

Karla Hazel Ozuna-Valencia,[1] *Ana Laura Moreno-Robles,*[2]
Francisco Rodríguez-Félix,[1] *Maria Jesús Moreno-Vásquez,*[2]
Carlos Gregorio Barreras-Urbina,[1,3] *Tomás Jesús Madera-*
Santana,[3] *Saúl Ruíz-Cruz,*[1] *Ariadna Thalia Bernal-Mercado,*[1]
Lorena Armenta-Villegas[2] and *José Agustín Tapia-Hernández*[1,*]

1. Introduction

It is estimated the at least 12,000 species of fungus are present in the world, where only 2,000 of them are suitable for human consumption (Gîrd and Mocanu, 2020). Edible fungi are a resource that can contribute to the economy, ecology, health, the food industry and the diversification of rural environment (Jiménez et al., 2017). These mushrooms are a resource collected in Asian countries for the treatment and prevention of diseases; even in Mexico, it is quite common to obtain wild mushrooms, both for self-consumption and for trade due to their high nutritional value (Ruan, 2018; Yang et al., 2018).

Within the category of edible mushrooms is *Craterellus cornucopioides*, also known as Black trumpet; it is a fungus found mostly in China, Europe, and North America (Gîrd and Mocanu, 2020). Due to its high nutritional content, it is the basis of many Asian diets; it is high in protein, low in fat, in addition to containing vitamins and minerals, such as iron (Painuli et al., 2020; Sarikurkcu et al., 2020). From a nutritional point of view, its composition is so attractive that its contribution is even compared with eggs, milk, or meat (Sánchez, 2017).

[1] Departamento de Investigación y Posgrado en Alimentos (DIPA), Universidad de Sonora, Blvd. Luis Encinas y Rosales, S/N, Colonia Centro, 83000 Hermosillo, Sonora, Mexico.
[2] Departamento de Ciencias Químico Biológicas, Universidad de Sonora, Blvd. Luis Encinas y Rosales, S/N, Colonia Centro, 83000 Hermosillo, Sonora, Mexico.
[3] Centro de Investigación en Alimentación y Desarrollo, A. C., Coordinación de Tecnología de Alimentos de Origen Vegetal, Carretera Gustavo Enrique Astiazarán Rosas Núm. 46. La Victoria, C.P. 83304, Hermosillo, Sonora, Mexico.
* Corresponding author: joseagustin.tapia@unison.mx

It is important to mention that the interest in edible mushrooms lies in their functional properties (Sánchez, 2017). *C. cornucopioides* has been studied because its concentrates have a great antioxidant, antimicrobial, anticancer, antihypertensive, anti-inflammatory, antidiabetic capacity, to mention a few examples (Guo et al., 2017; Liu et al., 2020). Polysaccharides, polyphenolic compounds, terpenes, and proteins are responsible for the functional activities of this precious fungus (Guo et al., 2019). For this reason, recent studies have focused on using them in the pharmacology and food industries (Bains et al., 2021).

This chapter will review some of the more notable points about the fungus *C. cornucopioides*. Firstly, the subject of its structure, taxonomy and nutritional composition will be discussed; its value as a nutraceutical food; the role it plays as food with functional properties, especially in the antioxidant profile and the relevance of the active compounds present in the fungus. It will also talk a little about how it has been used for the treatment and control of conditions, such as obesity, cancer, diabetes, hypertension and cardiovascular problems. Finally, a little will be said about its use in the development of functional foods.

2. Structure and Taxonomy of *Craterellus cornucopioides* (Black Trumpet)

2.1 Structure

Craterellus cornucopioides is a fungus commonly known as Black trumpet, Black chanterelle, Horn of plenty, or Trumpet of the dead. It is an edible mushroom of ash grey colour that turns black with humidity and is in the shape of a trumpet that can measure between 5–15 cm; it has an irregular shape (Pilz et al., 2003). However, the characteristics that make up the fruiting body of the Black trumpet are shown in detail in Table 1. Its growth occurs in groups or in a caespitose form, that is, it can form various stems (Nabors, 2006). Its spores have an ovoid or elliptical shape and can measure 12–15 × 7–8.5 μm, the spore print is white. Figure 1 shows the structure of the fungus.

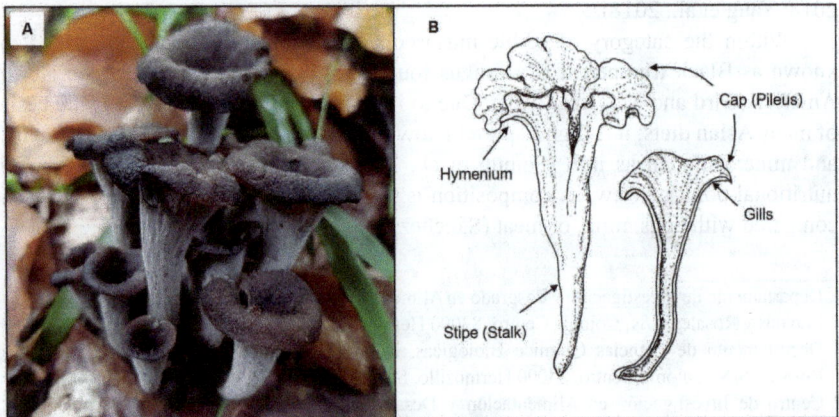

Fig. 1. A: *Craterellus cornucopioides* mushroom in its natural habitat (image: Creative Commons by Frank, licenced under BY CC) B. Description of the fruiting body of *Craterellus cornucopioides*.

Table 1. Description of the fruiting body of *Craterellus cornucopioides* (Black trumpet).

Fruiting Body	
Cap (Pileus)	Narrow, wavy and trumpet-shaped (3–8 cm)
Hymenium	Almost smooth and without folds
Gills	Veins, wrinkles or poorly defined folds
Stipe (Stalk)	Narrow at the base, tubular and pointed in shape
Surface	Scaly

Source: Adapted from Nabors (2006); Pilz et al. (2003).

2.2 Taxonomy

The basonym of *C. cornucopioides* is *Peziza cornucopioides L.* 1753. This term is commonly used in botany and is regulated by the International Code of Botanical Nomenclature (ICBN), which belongs to the originally registered scientific name (Sigwart, 2018). However, this species has various synonyms with which it can be identified – *Craterellus ochrosporus* Burt, *Merulius cornucopioides* (L.) With., *Merulius cornucopioides* (L.) Pers., *Merulius purpureus* With., *Pezicula cornucopioides* (L.) Paulet, *Pleurotus cornucopioides* (Pers.) Gillet, *Sterbeeckia cornucopioides* (L.) Dumor (Pilz et al., 2003). Table 2 shows the taxonomy and description of *C. cornucopioides*.

Table 2. *Craterellus cornucopioides* taxonomy.

Taxonomy		Description
Kingdom	Fungi	A group of eukaryotic organisms that include moulds, yeasts and mushroom growers.
Division	*Basidiomycota*	Within this division are fungi that produce basidia with basidiospores.
Class	*Agaricomycetes*	Includes all fungi that produce basidiocarps of varied sizes.
Order	*Cantharellales*	Fungi with coral morphology.
Family	*Cantharellaceae*	Funnel-shaped or trumpet-shaped cap.
Gender	*Craterellus*	Edible mushrooms.
Species	*Craterellus cornucopioides*	This type of mushroom is edible.

Source: Adapted from Nabors, 2006; Pilz et al., 2003.

3. Chemical Composition of *Craterellus cornucopioides* (Black Trumpet)

Due to their nutritional content, mushrooms are considered an important source of nutrients; therefore, they are excellent for use in low-calorie diets (Sarikurkcu et al., 2020). One of the great advantages of edible mushrooms is their high content of protein, vitamins and essential minerals, such as iron (Fe), potassium (K), phosphorus (P), magnesium (Mg), manganese (Mn), zinc (Zn) and calcium (Ca) (Sarikurkcu et al., 2020; Tel-çayan et al., 2018). It is worth mentioning that mushrooms also can accumulate substantial amounts of heavy metals, such as lead (Pl), cadmium (Cd), and mercury (Hg), so that, by regularly consuming these foods, there is a greater dietary exposure to cadmium compared to other products (Dospatliev and Ivanova, 2017; Sarikurkcu et al., 2020).

Yabani et al. (2009), conducted a study to find out the nutritional content of *C. cornucopioides* in which it was shown that the caloric energy they possess is 388.29 ± 15.70 kcal/100 g, while $50.10 \pm 2.90\%$ protein, $34 \pm 7\%$ carbohydrates, $5.89 \pm 0.01\%$ fat, and $10.26 \pm 0.80\%$ ash. While (Tel-çayan et al., 2018), determined the content of metals present in *C. cornucopioides*, for essential minerals he obtained 619 ± 6 mg/kg of Na, 498 ± 4 mg/kg of Mg, 41.4 ± 1 mg/kg of Ca, 76.4 ± 2.84 mg/kg of Mn, 77.1 ± 3 mg/kg of Fe and 16.7 ± 0.8 mg/kg of Zn. Based on the results of both studies, it can be confirmed that mushrooms are rich in protein and minerals.

Among the essential minerals is iron, which corresponds to one of the main components in plants (Dospatliev and Ivanova, 2021). Humans can obtain iron through the diet, which is necessary because its adequate consumption prevents anemia (Tel-çayan et al., 2018). For this reason, (Dospatliev and Ivanova, 2021) conducted a study on the Fe content; the result obtained was an average of 82.69 ± 1.78 in mg/kg, which complies with the provisions of the FAO and the WHO. Finally (Dospatliev and Ivanova 2017), determined the content of cadmium and lead, which gave 0.087 mg/kg and 0.413 mg/kg, respectively.

Also, it should be mentioned that mushrooms have a high content of vitamins, such as vitamin A, B complex vitamins (Thiamine B_1, Riboflavin B_2, Pantothenic Acid B_3, Nicotinic Acid B_5, etc.), vitamin C and vitamin D (Kaliyaperumal et al., 2018). In a study conducted by (Watanabe et al., 2012), they characterised the content of vitamin B_{12} where they concluded that it is enough to consume approximately 100 g of Black trumpet to satisfy the content of vitamin B_{12} that is required daily. Compared to mushrooms, about 1kg is required to equal the amount of vitamins. Finally, it should be noted that the chemical and nutritional content of *C. cornucopioides* is highly beneficial for health.

4. Nutraceutical Value of *Craterellus cornucopioides* (Black Trumpet)

It has been reported that edible mushrooms have a high nutritional content; however, they manage to provide favourable health characteristics (Kaliyaperumal et al., 2018). Thus, mushrooms have been considered an important source of nutraceutical foods (Alkan et al., 2020; Flores, 2020; Singh et al., 2021; Turfan et al., 2018; Upadhyaya, 2018). Likewise, the term 'nutraceutical' comes from the combination between nutrition and pharmaceutical (García, 2018). In this sense, a nutraceutical is one that, despite being considered as a food or as part of it, manages to provide various health benefits, which allows a disease to be prevented, or treated (Barros et al., 2008; Painuli et al., 2020).

The nutraceutical power present in mushrooms corresponds to the content of polysaccharides, proteins, glycoproteins, unsaturated fatty acids, phenolic compounds, tocopherols, ergosterols, and lectins (Ma et al., 2018; Upadhyaya, 2018). Based on the above, researchers have developed studies focused on the therapeutic potential of nutraceutical compounds present in mushrooms (Painuli et al., 2020). Regarding the bioactive properties of *C. cornucopioides*, Remya et al. (2019) mention that myricetin is one of the active constituents of Black trumpet, which is responsible for providing antioxidant activity. This activity is attributed to the ability to donate the hydrogens of its hydroxyl groups (Chowdhury et al., 2020).

Myricetin (3,5,7-trihydroxy-2-(3,4,5-trihydroxyphenyl)-4-chromenone), shown in Fig. 2, is a flavonoid found in vegetables and fruits (Stoll et al., 2019). It is interesting how (Chowdhury et al., 2020) refer to flavonoids as Nature's doctors; it is also mentioned that these correspond to phytohormones, which are precisely the secondary metabolites of polyphenols. In this way, it has been determined that myricetin has various health benefits, such as inhibition of hyperglycemia, decrease in triglycerides, reduction of oxidative stress and cholesterol reduction (Imran et al., 2021). However, the bioactive properties in which it stands out the most are antioxidant, anti-inflammatory, and antitumour (Stoll et al., 2019).

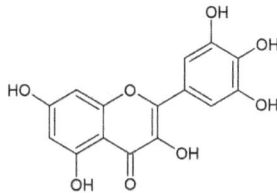

Myricetin

Fig. 2. Chemical structure of myricetin, a bioactive compound present in *C. cornucopioides* (adapted from Stoll et al., 2019).

In the paper of Guo et al. (2019), they isolate two new illudane sesquiterpenoids which are Craterelins D and E and a new menthane monoterpene, called 4-Hydroxy-4-isopropenyl-cyclohexane methanol acetate as bioactive compounds from *C. cornucopioides* cultures. Right here, the researchers performed a cytotoxicity assay on five tumour cell lines. Figure 3 shows the structures belonging to the new compounds isolated from cultures of *C. cornucopioides*.

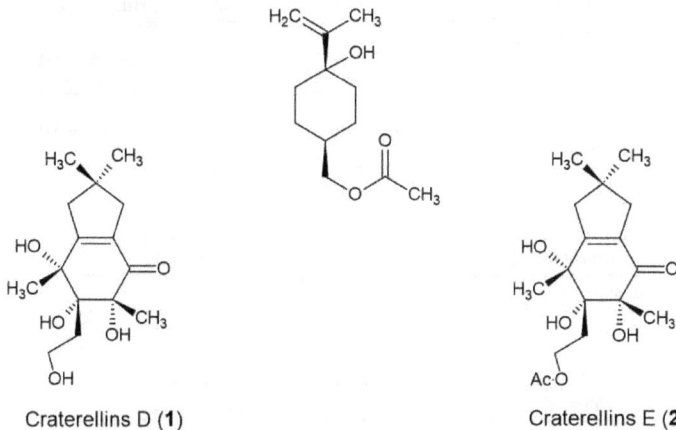

4-Hydroxy-4-isopropenyl-cyclohexanemethanol acetate

Craterellins D (**1**)

Craterellins E (**2**)

Fig. 3. Structures isolated from cultures of *C. cornucopioides* menthane monoterpene 4-Hydroxy-4-isopropenyl-cyclohexane methanol acetate and illudane sesquiterpenoids Craterelins D and E (adapted from Guo et al., 2019).

5. Bioactive Properties of *Craterellus cornucopioides* (Black Trumpet)

Edible mushrooms have gained importance due to their high nutritional value and functional attributes (Yadav and Negi, 2021). They can present antioxidant, antimicrobial, anti-inflammatory, anticancer, antidiabetic, immunoprotective properties among others through various metabolic pathways (Azeem et al., 2020). This wide range of benefits that mushroom extracts possess is attributed to the secondary metabolites found in them, such as polysaccharides, terpenes, proteins, amino acids, alkaloids, sterols, polyunsaturated fatty acids and phenolic compounds (Abdelshafy et al., 2021; Azeem et al., 2020).

Most of the biological activities of fungi are due to the content of phenolic acids, which in turn is considered a group of phenolic compounds with the highest abundance in edible mushrooms (Abdelshafy et al., 2021). Table 3 shows the concentration obtained by employing HPLC of some of these compounds present in the extract of *C. cornucopioides*. In this table, the variety of phenolic acids present can be highlighted, in addition to the high content of quercetin (22.8 µg/g), a polyphenol widely used in the development of active packaging (De Araújo et al., 2021). In the study by (Costea et al., 2020) they focus on the concentration of gallic acid (2.75 g/100 g dry extract), due to its importance in the food area.

The polysaccharides obtained from *C. cornucopioides* are ingredients with great medicinal potential as anticoagulants or in lowering blood lipids (Guo et al., 2019). In the paper from Liu et al. (2021) the polysaccharides of *C. cornucopioides* were characterised and it was observed that they mostly contained α and β glucoside bonds formed mostly of mannose and xylose. The structure of edible mushroom polysaccharides, such as β-glucan, presents a triple helix that activates RAW 264.7 macrophages that regulate immunosuppressive and immunomodulatory activity in rodent studies (Guo et al., 2019; Maity et al., 2021; Thu et al., 2020). In the following

Table 3. Phenolic compounds obtained from the acetone extract of *C. cornucopioides*.

Phenolic Compounds	µg Phenolic Compound/g Dry Mushroom
Phenolic acids	
Gallic acid	3.5 ± 0.1
p-cumaric acid	3.5 ± 0.1
Chlorogenic acid	2.7 ± 0.1
Cafeic acid	–
Syringic acid	0.8 ± 0.1
Feluric acid	4.7 ± 0.9
Flavonols	
Rutine	7.3 ± 0.2
Quercetin	22.8 ± 0.5
Flavan-3-ol	
Catechin	2.1 ± 0.1

Source: Adapted from Kosanić et al. (2019).

sections, some of the bioactive properties of edible mushrooms will be discussed in more detail.

5.1 *Antioxidant*

In physiological matters, an antioxidant protects living beings from free radical damage and, even at low concentrations, significantly inhibit the oxidation of a substrate (Abo et al., 2021). Oxidative damage is strongly associated with many chronic diseases in humans (Yang et al., 2018). This imbalance results from a series of cellular functions where reactive oxygen species (ROS) exceed the antioxidant content of the organism (Gulcin, 2020). An overview of the reactions that contribute to ROS formation is shown in Fig. 4.

Antioxidants can work in two ways: first, a primary antioxidant acts as a hydrogen donor to react with radicals and terminate the radical chain process; on the other hand, the secondary antioxidant traps oxygen to reduce compounds, chelate metals, absorb UV rays, act against singlet oxygen and participate in synergy with other compounds (Kumar et al., 2020). Most of the antioxidants present in edible mushrooms are of the secondary type, which is why they are of great interest both in health and in food (Alkan et al., 2020).

There are several works where the antioxidant properties of *C. cornucopioides* are evaluated (Kol et al., 2018a; Kosanić et al., 2019; Liu et al., 2020; Yang et al., 2018). Antioxidant activity is mainly evaluated by DPPH methods, a stable radical that is used to determine the reducing capacity of a compound; ABTS, where the radical $ABTS^{\cdot+}$ is reduced; FRAP, which determines the ability to reduce Fe^{3+} to Fe^{2+}, and finally the metal chelation activity that focuses on Fe^{2+} ions (Xiao et al., 2020).

In the study by Yang et al. (2018) the antioxidant activity was evaluated through the DPPH assay where it was found that the fractions at a concentration of polysaccharides of 400 µg/mL from *C. cornucopioides* managed to reduce the DPPH radical at 81.2%. In the same way, the ABTS method was carried out and exhibited an antioxidant power of 99.4% with the same concentration of black fungus polysaccharides. On the other hand, in the study by Kosanić et al. (2019), the antioxidant capacity of *C. cornucopioides* extracts was compared employing DPPH where it was observed that compared to ascorbic acid (6.4 µg/mg IC_{50}) the mushroom extracts have the highest antioxidant activity (19.7 µg/mg IC_{50}).

Something interesting that is presented in the article by Liu et al. (2020) is that the metal chelation activity of the fungus is high (28%) even if the sample is fried and with a 1:50 dilution and the same occurs in the DPPH trial (45%). In the same study, it is mentioned that this can be attributed to the stability of the compounds that provide antioxidant properties to *C. cornucopioides*, such as sugars, amino acids, and polyphenols. Even so, it is better to use extraction techniques with solvents, such as methanol where, at concentrations of 10 mg/mL, up to 75% reduction capacity of the DPPH reagent is obtained (Kol et al., 2018a).

As mentioned above, polyphenols have a lot to do with the active properties of *C. cornucopioides*, since they are one of the compounds found in the greatest quantity in this type of fungi (Kol et al., 2018a). The chemical structure of the polyphenolic compounds isolated from the black fungus (Fig. 5) have shown to be excellent for inactivating metals, trapping oxygen, inhibiting the formation of free radicals and

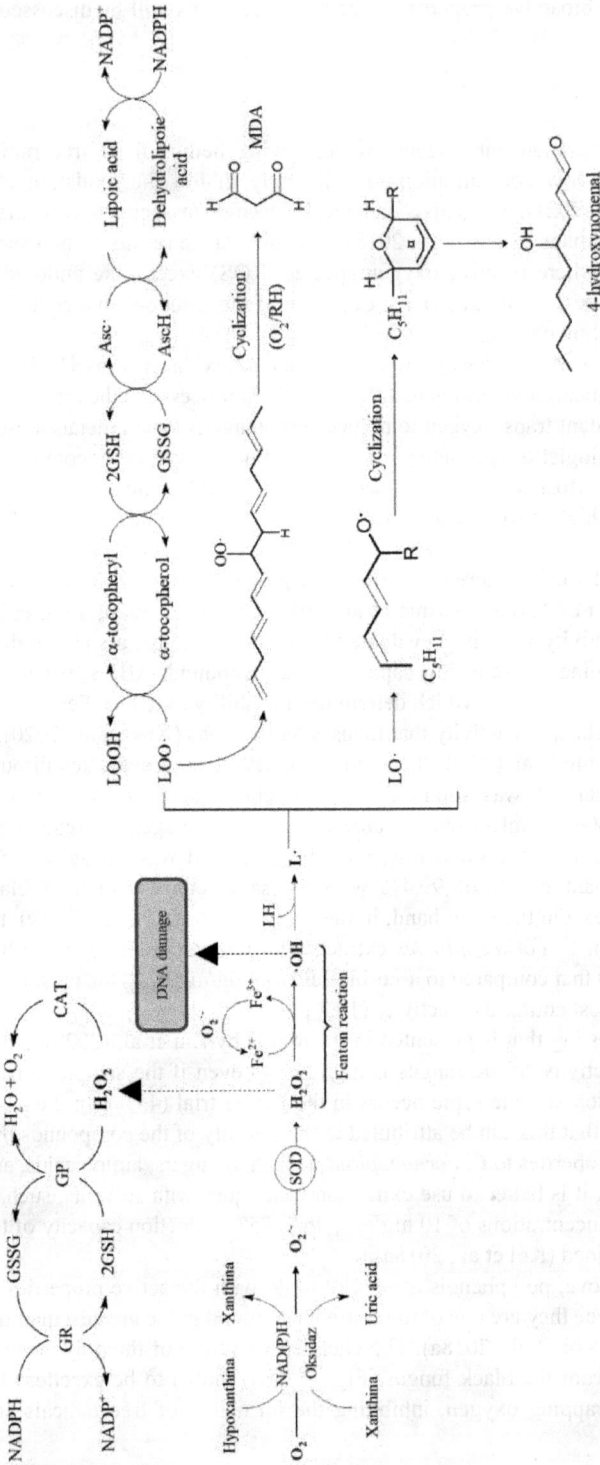

Fig. 4. General reactions of ROS formation and their effects. *GSH* glutathione, *GSSG* glutathione disulfide, *GR* glutathione reductase, *GP* glutathione peroxidase, *CAT* catalase, *SOD* superoxide dismutase (adapted from Gulçin, 2020).

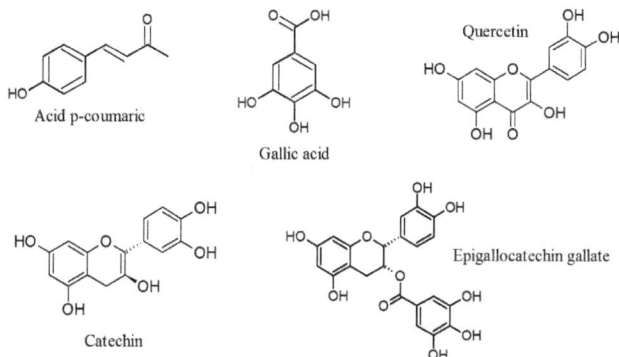

Fig. 5: Chemical structure of the main phenolic compounds isolated from extracts and concentrates of *C. cornucopioides.*

breaking down peroxidase—an enzyme responsible for darkening in plant structures (Elkhateeb et al., 2019).

5.2 Antimicrobial

Currently one of the biggest problems in the health area is microbial resistance to antibiotics since millions of deaths each year are attributed to this situation (Khameneh et al., 2019). Similarly, the attack of yeasts, moulds, bacteria, parasites and their metabolic products also compromise food safety, which can cause a loss of up to 70% in food products (Guzman, 2017). Due to this, the antimicrobial properties of various wild mushrooms have recently been studied and are extremely attractive to consumers seeking natural solutions (Shen et al., 2017).

The mechanism carried out by an antimicrobial compound is based on stopping protein biosynthesis, destroying the cell wall and membrane, preventing DNA replication and repair, and inhibiting the metabolic pathways of the microorganism (Khameneh et al., 2019). The effectiveness of the antibacterial compound depends on the concentration and the type of bacteria to be attacked, for example, Gram-negative bacteria may become less resistant because its cell wall is more porous and hydrophilic (Bains et al., 2021; Liu et al., 2017).

Figure 6 shows the general mechanism of action of an active antimicrobial compound. The active compound, depending on its structure, will attack Gram-positive and Gram-negative bacteria differently, especially considering the double cell membrane that a Gram-negative bacterium possesses. This attack can lead to DNA damage and loss of cytoplasm; In addition, this type of active compound can act by chelating metals, which prevents the bacteria from obtaining the micronutrients to conduct basic functions (Liu et al., 2017).

In the work of Özcan and Ertan (2018), the antimicrobial activity of the methanolic and acetone extracts of *C. cornucopioides* was evaluated. With a concentration of 200 mg/mL of both fungus extracts, the zone of inhibition (ZI) for *Klebsiella pneumoniae* was higher compared to the control (15 mm). This is attributed to the β-glucan present in *C. cornucopioides*. On the other hand, the water and methanol extracts of the same are more effective (6–8 mm of ZI) than gentamicin

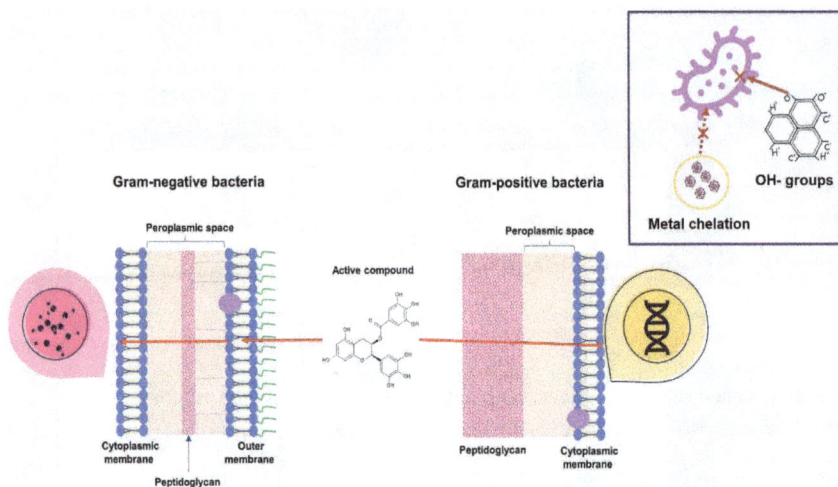

Fig. 6. General scheme of the attack of an antimicrobial active compound against Gram-negative and Gram-positive bacteria.

(2.3–2.7 mm of IZ), a broad-spectrum antibiotic against *Pseudomonas aureaginosa* (Kol et al., 2018a).

The microorganisms these studies focus on include Gram-negative bacteria, such as *Escherichia coli* and Gram-positive such as S*taphylococcus aureus*, although some efforts have also focused on pathogens, such as *Mycobacterium tuberculosis* (Kol et al., 2018b; Novakovic, 2021; Yadav and Negi, 2021). Due to this, the use of *C. cornucopioides* can be an alternative for the development of both food preservatives and in health to be used in the synthesis of natural antibiotics.

5.3 *Chronic Degenerative Diseases*

Chronic-degenerative diseases are irreversible and incurable, so the affected person is forced to change his daily life to do everything within his reach and thus avoid complications that prevent him from leading a normal life (Pinedo et al., 2017). These diseases are one of the biggest problems in the health area. Since treatment is usually expensive, there is a high incidence and mortality is high (Franco et al., 2018). These slowly progressive conditions are responsible for approximately 63% of deaths worldwide (Villa et al., 2017).

Some studies ensure that due to COVID-19 in some countries, the percentage of the incidence of these diseases increased, which in the long term can have deadly consequences (Ayala et al., 2021). These diseases include diabetes, obesity, hypertension, cancer and other cardiovascular conditions (Franco et al., 2018). The countries that are most affected by this type of illness are those of low and medium development, where despite having a percentage of more than 80% of people with some of these illnesses, less than 10% end up being treated for economic reasons (Pinedo et al., 2017).

The use of plants as a treatment against diseases goes back a long time, although before it was only based on their consumption, today their active compounds can be isolated to increase the effectiveness of the treatment (Martel et al., 2017). Now, mushrooms are constituents that are widely used in the development of products with active ingredients focused on treating some chronic-degenerative diseases (Abdelshafy et al., 2021). For this reason, this section will review some of the applications of *C. cornucopioides* in the treatment of this type of disease (Turfan et al., 2018).

5.4 Anticancerogenic

Fungi with biological metabolites have been recorded that can develop anticancer activity (Alkan et al., 2020; Habtemariam, 2020; Ishara et al., 2022). It is important to mention that cancer is one of the leading causes of death in the world (Zhang et al., 2016). According to the WHO, in 2018, the figures of mortality from cancer in the world are 9.6 million, which places this condition as the second cause of death (Habtemariam, 2020). While Nowakowski et al. (2021) report that 19.3 million people were diagnosed with cancer in 2020, of which almost 10 million (approximately 52%) died.

Therefore, several researchers carry out studies that allow the development of new drugs to combat this disease more efficiently. To date, chemotherapy is the main control measure for cancer (Habtemariam, 2020). On the one hand, the fight against cancer cells within the human body involves the defence system that is controlled by immune responses mediated by immune cells and their secretory substances. (Wong et al., 2011). On the other hand, cancer has often been associated with chronic inflammation, which is responsible for maintaining the immunosuppressive network (Del Cornò et al., 2020). Therefore, various efforts have been made to develop new drugs and methods of cancer detection (Nowakowski et al., 2021).

It should be noted that the types of cancer with the highest mortality rate are breast, lung, stomach, liver and colorectal, while the most common but treatable types of cancer are prostate and skin (Habtemariam, 2020). For example, Zhang et al. (2016) report that there are numerous studies in which it is proven that fungi have significant inhibitory activity against cancer of the breast, cervix, pancreas, gastric, hepatocellular carcinoma and leukemia. Therefore, studies have been conducted to know the chemical nature and mechanisms of the action of fungi.

The medicinal use of mushrooms has been part of traditional Chinese medicine for 2,000 years (Habtemariam, 2020; Kaliyaperumal et al., 2018). In ancient Chinese pharmacopeia, herbs and fungi were included as the most effective natural remedies to combat various types of tumours (Blagodatski et al., 2018). Commonly in China and Japan, tea preparations were made with mushroom powder; the purpose was that the tea served traditional medicine (Habtemariam, 2020). In ancient times, mushrooms have been called by various names: the Romans referred to mushrooms as 'the food of the gods', the first Egyptians called them 'gifts of the God of Osiris', while the Chinese called them 'the elixir of life' (Kaliyaperumal et al., 2018).

It is believed that the first fungus cultured was *Auricularia atria* in the year AD 600; subsequently the record of the culture of *Flammulina velutipes* was in AD 800 (Kaliyaperumal et al., 2018). At present, the cultivation of mushrooms has increased greatly, in particular in Europe, America and Asia (Upadhyaya, 2018). However, there is a great variety of these, since they can be classified into two, such as: cultivated mushrooms and wild mushrooms (Díaz and Téllez, 2021). In addition to the scientific research carried out on fungi, the first occurred in the mid-twentieth century where the anti-tumour activity of *Boletus edulis* in edible and medicinal mushrooms was studied (Blagodatski et al., 2018).

As for the bioactive metabolites present in fungi, they inhibit cancer development by activating the host's natural immune system, which is responsible for providing immunomodulatory activity (Ishara et al., 2022). Likewise, Guo et al. (2020) report that polysaccharides that have been isolated from the fruiting bodies and mycelia of fungi exert immunomodulatory functions and antitumour activities. Thus, Fig. 7 is shown, which represents the anticancer potential of bioactive metabolites in distinct types of cancer.

Elkhateeb et al. (2019) comment that the anticancer potential is complex since it is not only given by the inhibition of the specific processes of cancer or by the directed activation of the apoptosis of the tumour but also conducted by means of indirect actions, such as immunomodulation. Another case to combat this disease is the combination of common therapies against cancer and extracts of fungi. Nowakowski et al. (2021) report that the mixture between several fungi rich in β-1,3-glucan, chitosan or vitamin C causes an increase in the anti-tumour activity of common therapies; in addition, when mixed with chemotherapies which benefit the treatment.

As already mentioned, in *C. cornucopioides* or Black trumpet there is myricetin, which corresponds to the active compound of this species. It is important to mention that myricetin counts as an anticancer potential due to the ability to induce apoptosis and the regulation of the mutation (Imran et al., 2021). Therefore, it allows the use of *C. cornucopioides* to be useful to counteract the diverse types of cancer, either by its content of flavonoids, such as myricetin or by the content of polysaccharides extracted from it.

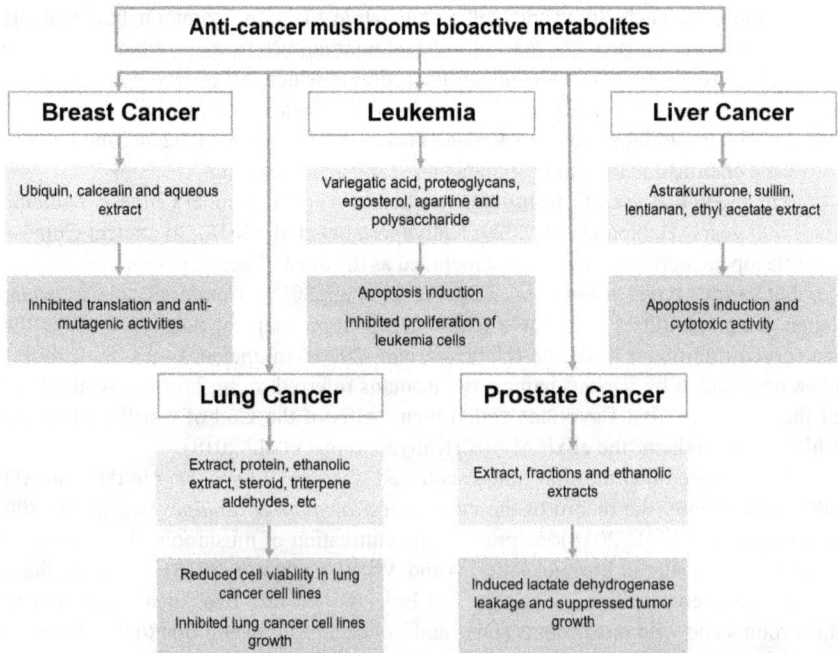

Fig. 7. Anticancer potential of bioactive metabolites presents in fungi against distinct types of cancer (adapted from Ishara et al., 2022).

5.5 Antidiabetic

Diabetes mellitus is an autoimmune disorder that can affect the β-cells of the pancreas (type 1) or insulin resistance by receptors in different cells may occur (Dubey et al., 2019). Type 2 diabetes is the most common; it is estimated that 463 million adults suffer from this type of diabetes, and it is expected that by 2030, the figure will increase to more than 500 million (Khursheed et al., 2020). Because of this, natural alternatives are sought for the treatment of those affected and fungi are an excellent option, thanks to their low glucose level and active properties (Gargano et al., 2017).

The antidiabetic mechanism of fungi is based on distinct aspects, such as the increase in insulin and the reduction of pancreatic glucagon; improvement in insulin sensitivity by activation of adenosine monophosphate-activated kinase; reduction of insulin resistance; increased hepatic glycogen, ROS chelation, inhibition of the enzyme α-glucosidase, among others (Khursheed et al., 2020). On the other hand, edible mushrooms contain a lot of indigestible carbohydrates that are ideal for a high-fibre diet, which is recommended for patients with diabetes (Dubey et al., 2019; Gargano et al., 2017).

Consumption of *C. cornucopioides* and other edible mushrooms is recommended because it improves nutrient absorption, adipogenesis, energy performance, insulin sensitivity and gut microbiota composition (Martel et al., 2017). These functions are attributed to the type of phenolic compounds present in these species of fungi and especially in the content of β-glucans (Abdelshafy et al., 2021). *C. cornucopioides* is a fungus with a high concentration of β-D-glucans, responsible for reducing the activity of α-amylase, α-glucosidase and facilitating pathways related to glucose homeostasis (Das et al., 2022).

As mentioned above, myricetin is a polyphenol present in *C. cornucopioides* and exhibits distinct bioactive properties, including its function against diabetes (Gupta et al., 2020). In a study that was conducted in China, the effect of myricetin on the diet was evaluated, since it is one of the countries that consumes the most foods that include this flavonoid (Yao et al., 2019). These authors concluded that myricetin consumption and the prevalence of type 2 diabetes mellitus are inversely proportional, so a diet that includes *C. cornucopioides* could prevent this disease. In addition, it was also found that this polyphenol can regulate blood glucose levels.

For his part, he evaluated the effect of myricetin consumption in rats with diabetes. Among its results, it was observed that complications caused by diabetes, such as osteoporosis, were reduced with the consumption of 100 mg of myricetin per kilogram of the body per day for 12 weeks. This is attributed to the osteoblastic effect that prevents apoptosis of dead cells induced by high glucose levels. They also mention that consumption for six months of myricetin (200 mg/kg/day) can significantly reduce diabetes (Ying et al., 2020).

5.6 Antihypertensive

Arterial hypertension is a multifunctional and chronic pathophysiological condition characterised by elevated elevation of blood pressure (Mohamed et al., 2014). Hypertension in adults is generally defined as systolic blood pressure (SBP) as ≥ 140 mm Hg and diastolic blood pressure (SBP) as ≥ 90 mm Hg (Gitu and Hahn,

2021). Unlike normal blood pressure, it maintains systolic blood pressure with an average of 120 mm Hg and diastolic blood pressure with an average of 80 mm Hg (Mohamed et al., 2014). Hypertension is the leading cause of death, this disease reduces life years based on disability worldwide—10.4 million deaths have occurred annually (Gitu and Hahn, 2021).

This disease has become a major global challenge due to the frequency of cases, the factors involved are: change in lifestyle, urbanisation, malnutrition and unbalanced food intake (Mohamed et al., 2014). At the time of suffering from hypertension, there is a risk of increasing the muscle mass index (BMI); it is known that obesity increases the probability of suffering from high blood pressure (Ganesan and Xu, 2018). The exact mechanism of the development of hypertension is unclear, but in most cases, in adults, the causes correspond to the following factors: genetic, environmental, neutral, hormonal and anatomical (Gitu and Hahn, 2021).

It is important to mention that mushrooms have always been appreciated for having a vital role in the cure or prevention of hypertension (Ma et al., 2018). As mentioned by Ishara et al. (2022) fungi feature D-mannitol, D-glucose, D-galactose, D-mannose and tripeptides that serve as angiotensin-converting enzyme (ACE) inhibitors that help lower blood pressure. Finally, the antihypertensive effects are conducted by inhibiting the angiotensin-converting enzyme that converts the inactive Ang I protein into active Ang II, which increases the vasoconstrictor action.

5.7 Anti-obesogenic

Obesity has been classified as a global epidemic that, in addition to being characterised by excessive fat deposition, is attributed to significant complications in health (Iñiguez et al., 2018). This disease is defined as the body mass index (BMI) greater than 30 Kg/m^2, where the improvement of body fat storage is deposited in adipose tissue (Ganesan and Xu, 2018). Obesity and overweight are some of the factors for several diseases, such as diabetes, various types of cancer, cardiovascular diseases, hypertension, lung diseases, obstructive sleep apnea and osteoarthritis (Ganesan and Xu, 2018; Okati et al., 2022).

In 2016, the WHO reported that more than 1,900 million adults are overweight, while more than 650 million are obese. However, this data has been increasing due to the global COVID-19 pandemic, where the most affected countries are those with middle and low incomes (Salsinha et al., 2021). This is due to the confinement and change in the routine of people, forcing them to follow a more sedentary lifestyle. It should be noted that this disease not only affects the health of those who suffer from it, it also affects their economy since it is presented as a financial burden for people and society in general (Okati et al., 2022).

The etiology that includes obesity is due to the excess in the consumption of nutrients and lack of physical activity; it can also be generated by hereditary diseases, medications, or disorders, whether mental or endocrine (Ganesan and Xu, 2018). However, in the market, there are therapeutic drugs for obesity, such as orlistat (Xenical) and sibutramine (Reductil), which cause severe complications in the consumer, such as emesis, insomnia, headache, myocardial infarction, stomach pain, and constipation (Ganesan and Xu, 2018).

As mentioned throughout the chapter, studies have been conducted showing that mushrooms have an anti-obesogenic effect (Ma et al., 2018; Thu et al., 2020). In a study conducted by Ganesan and Xu (2018), they showed that *Agaricus bisporus* counts as a protector against increased body weight due to poor diet and associated hepatic steatosis, while Ishara et al. (2022), comments that fatty acids and dietary fibre extracted from mushrooms can reduce cholesterol (serum, total, and LDL) as well as redisplay serum lipid levels and plasma levels of triglycerol.

In the review conducted by Ganesan and Xu (2018), they talk about the effects that edible and medicinal mushrooms have against obesity. Here they mention that the bioactive compounds of the fungus function as cardiac markers and are hypercholesteremic—the latter are related to lipid metabolism. Fungi has shown a reduction of very low-density lipoprotein cholesterol (VLDL), the catalytic functions of HMG-CoA (or 3-hydroxy-3-methylglutaryl-coenzyme A or β-hydroxy-β-methylglutaryl-coenzyme A) and amplification of the catabolic rate of cholesterol.

As for the effects of *C. cornucopioides*, focus is on the effect of myricetin. In a study mentioned in Imran et al. (2021), they found that the use of myricetin as supplementation significantly improved hepatic steatosis and systemic insulin resistance along with a reduction in body weight. The main mechanism by which the anti-obesity effect is conducted is given by the ability of myricetin to modulate thermogenic regulatory proteins. These proteins are considered fat burners due to the direct effect they have on the hormones that regulate metabolism.

5.8 Cardiovascular Diseases

Cardiovascular diseases are the main cause of death worldwide; the prediction of this condition is a great challenge for the medical sector (Nadakinamani et al., 2022). This disease has caused 17.3 million deaths annually, yet the death rate continues to rise day by day (Khodadi, 2020). It is important to note that cardiovascular diseases are difficult to detect because they have several risk factors, such as high blood pressure, cholesterol and abnormal pulse (Nadakinamani et al., 2022). Also when suffering from obesity or hypertension, the risk of presenting cardiovascular diseases increases (Mohamed et al., 2014).

Because it is a delicate disease, it must be managed with care as it can cause damage to the heart or even death; so it is necessary to protect patients or perform early detection (Nadakinamani et al., 2022). In addition, platelets must be taken into account, since they have a very important role in thrombosis and atherosclerosis – both are diseases that lead to death (Khodadi, 2020). Due to this serious disease, it is important to look for alternatives that allow prevention or cure of cardiovascular problems.

In Khodadi (2020), the importance of platelets is explained since they influence lipid metabolism in cardiovascular diseases. So, lipids can activate platelets and they have receptors for peptide hormones that could activate and cause thrombosis. In addition, platelets are activated by oxidative stress and the production of reactive oxygen species (ROS), which, in turn, interact with other cells, causing thrombosis. Also, the activation of platelets is regulated by platelet signalling, secreted proteins, and molecular pathways in thrombosis. Therefore, it is important to take these into account to prevent cardiovascular diseases.

As for medicinal and edible mushrooms, the bioactive compounds they possess help cause cardiovascular disease. In the review conducted by Ishara et al. (2022), they mention that in some countries, they prescribe some mushrooms to prevent or reduce the risks of this disease since their consumption can alter the metabolic markers of cholesterol (total, LDL and HDL) along with triglycerides. In this same review, they mention that they attribute this reduction to the action of eritadenin, which inhibits the enzyme S-adenosyl homocyteine hydrolase (SAHH), which is key to phospholipid metabolism.

6. Uses of *Craterellus cornucopioides* (Black Trumpet) in Functional Foods

A functional food can be described as one that provides basic nutrients and in turn, has one or more important components that can improve the functions of the body that consumes it (Arias et al., 2018). The development of this type of food has been growing since the interest of the consumer is no longer focused only on low-cost products, but seeks to obtain some benefit in their health (Cortés et al., 2016). As reviewed throughout this chapter, edible mushrooms possess a broad branch of functional properties that are mostly used in Asian countries (Yang et al., 2018).

The collection of this type of fungi in the wild can be a risk due to the accumulation of toxic elements or heavy metals; so controlled cultivation is an excellent option to avoid this problem and thus take advantage of its excellent benefits (Turfan et al., 2018). Once the fungus is previously washed and disinfected, it can be used in diverse ways; cooking it or preparing it for direct consumption; with infusions; using it as a dietary supplement, or through direct consumption (Gargano et al., 2017; Liu et al., 2020; Ying et al., 2020).

Liu et al. (2020) evaluated the study on the effects of different cooking methods of *C. cornucopioides*. The techniques used were boiling, steaming, frying and microwave irradiation. According to research, the chemical, physical and enzymatic modifications that occur during treatments can become hugely different from each other. On the one hand, the use of microwaves caused the food in question to release nutrients, such as proteins, fatty acids and minerals more easily into the stomach. On the other hand, with the frying, many sugars and antioxidant compounds were lost since these are usually extremely sensitive to elevated temperatures.

Another critical issue of fungi, such as *C. cornucopioides*, is its functionality as food. It is relevant to review these properties during the development of a new product, since it is based on issues like the capacity of retention of water, emulsification, molding, gelatinization, its density, among other issues (Kumar and Sánchez, 2020). Because of this, mushrooms can be amazing ingredients not only to increase the nutritional value or obtain a bioactive benefit, but also that their use can have a positive impact on the development of another food product as long as you take into account what characteristics you want to improve (Ishara et al., 2022).

6.1 *Antiviral*

It is well known that more than 60% of infectious disease epidemics are caused by viruses (Guo et al., 2021). As mentioned above, fungi can be used to fight microorganisms, and this includes viruses as well (Yildiz et al., 2017). The polysaccharides present in fungi are responsible for antiviral activity by blocking the pathways the virus follows or simply eliminating them altogether (He et al., 2020). The use of *C. cornucopioides* can help treat or prevent viral infections. In fact, there are some other fungi that are able to prevent a complete invasion of hepatitis in the initial stages.

6.2 Anti-inflammatory

Inflammation is a natural response of the immune system to alert us of some damage and although it can be treated quickly, this condition can become chronic and cause a lot of discomfort to the user (Muszyńska et al., 2018). They evaluated the effectiveness of polysaccharides, β-glucans present in (Li et al., 2021) *C. cornucopioides* and concluded that this fungus can regulate enzymes, such as caspase-1, which are the precursors of inflammatory reactions. Therefore, these authors say that this product is an excellent functional food to combat developing inflammation or to prevent it if it is a chronic problem.

6.3 Immunomodulatory

Acquired immunodeficiencies are a major current concern even in highly developing countries and are characterised by deterioration of one to several cellular components of the immune system, causing the individual to be more susceptible to infectious diseases (Morris et al., 2016). The function of an immunomodulator is to improve and increase the functions of immune cells to promote their proliferation, in addition to the production of antibodies and cytokines (Medina et al., 2019). Self-modulating substances have been isolated from the mycelium and fruiting bodies of different fungi, such as β-glucans and lectins, which are also found in *C. cornucopioides* (Fernández et al., 2020).

7. Conclusion

C. cornucopioides is a black fungus that is physically distinguished by its trumpet shape, capable of forming various stems thanks to its ceptitious growth. Its genus *Cratellus* places it within the edible fungi. The black mushroom can have a protein content of 50%, which indicates that it is a mushroom with a high protein level. In addition, it is also distinguished by its high content of minerals, such as magnesium, iron, calcium and especially vitamin B_{12}. The compound responsible for the nutraceutical value of *C. cornucopioides* is myricetin, a flavonoid used as antioxidant, anti-inflammatory and antitumour. Similarly, the B-glucans present in this species of fungi are effective in inhibiting different bacteria. Similarly, the B-glucans present in this species of fungi are effective in inhibiting different bacteria. It has been used to treat chronic-degenerative diseases, such as cancer since it accelerates apoptosis; diabetes, especially type 2; hypertension, because it increases vasoconstrictor action and antiobesogenic, due to its high fibre content. *C. cornucopioides* mushrooms can be consumed directly or cooked by different methods to take advantage of their its, in addition to their functional properties to improve emulsification and gelatinisation. Black trumpet mushrooms are an excellent nutritional, nutraceutical and functional option – qualities that can be further explored around food and health.

References

Abdelshafy, A.M., Belwal, T., Liang, Z., Wang, L., Li, D., Luo, Z. and Li, L. (2021). A comprehensive review on phenolic compounds from edible mushrooms: Occurrence, biological activity, application and future prospective. *Critical Reviews in Food Science and Nutrition*. https://Doi.org/10.1080/10408398.2021.1898335.

Abo, H.H., Darwish, A.M.G., Abo, Y.H., Elsayed, A., Abdel, M.A. and Abdel, A.M. (2021). Fungi as a gold mine of antioxidants. *In*: *Industrially Important Fungi for Sustainable Development, Fungal Biology*, pp. 73–113, Springer Nature, Switzerland AG. https://Doi.org/10.1007/978-3-030-85603-8_2.

Alkan, S., Uysal, A., Kasik, G., Vlaisavljevic, S., Berežni, S. and Zengin, G. (2020). Chemical characterisation, antioxidant, enzyme inhibition and antimutagenic properties of eight mushroom species: A comparative study. *Journal of Fungi*, 6(3): 1–19. https://Doi.org/10.3390/jof6030166.

Arias, D., Montaño, L.N., Velasco, M.A. and Martínez, J. (2018). Functional foods: Application advances in agribusiness. *Tecnura*, 22(57): 55–68. https://Doi.org/10.14483/22487638.12178.

Ayala, M.R., Arévalo, J.C., Keita, H., Meneses, D.M., Azures, T.H., Castañeda, C.A. and Vergara, A. (2021). Implications of obesity and chronic-degenerative diseases in complications from COVID-19: Systematic review. *Revista Del Centro de Investigación de La Universidad La Salle*, 14(55): 11–24. https://Doi.org/10.26457/recein.v14i55.2689.

Azeem, U., Hakeem, K.R. and Ali, M. (2020). Bioactive constituents and pharmacological activities. *In*: *Fungi for Human Health*. https://Doi.org/10.1007/978-3-030-58756-7_7.

Bains, A., Chawla, P., Kaur, S., Najda, A., Fogarasi, M. and Fogarasi, S. (2021). Bioactives from mushroom: Health attributes and food industry applications. *Materials*, 14(24). https://Doi.org/10.3390/ma14247640.

Barros, L., Cruz, T., Baptista, P., Estevinho, L.M. and Ferreira, I.C.F.R. (2008). Wild and commercial mushrooms as source of nutrients and nutraceuticals. *Food and Chemical Toxicology*, 46(8): 2742–2747. https://Doi.org/10.1016/j.fct.2008.04.030.

Blagodatski, A., Yatsunskaya, M., Mikhailova, V., Tiasto, V., Kagansky, A. and Katanaev, V.L. (2018). Medicinal mushrooms as an attractive new source of natural compounds for future cancer therapy. *Oncotarget*, 9(49): 29259–29274.

Chowdhury, S., Bhuiya, S., Haque, L. and Das, S. (2020). A spectroscopic approach towards the comparative binding studies of the antioxidising flavonol myricetin with various single-stranded RNA. *Chemistry Select*, 5(45): 14430–14437. https://Doi.org/10.1002/slct.202003601.

Cortés, A.D.J., León, J.R., Jiménez, F.J., Díaz, M. and Villanueva, A. (2016). Alimentos funcionales, alfalfa y fitoestrógenos. *Revista Mutis*, 6(1): 28. https://Doi.org/10.21789/22561498.1110.

Costea, T., Hudiță, A., Olaru, O.T., Gălățeanu, B., Gîrd, C.E. and Mocanu, M.M. (2020). Chemical composition, antioxidant activity and cytotoxic effects of Romanian *Craterellus cornucopioides* (L.) pers. Mushroom. *Farmacia*, 68(2): 3400347. https://Doi.org/10.31925/farmacia.2020.2.21.

Das, A., Chen, C.M., Mu, S.C., Yang, S.H., Ju, Y.M. and Li, S.C. (2022). Medicinal components in edible Mushrooms on diabetes mellitus treatment. *Pharmaceutics*, 14(2): 436. https://Doi.org/10.3390/pharmaceutics14020436.

De Araújo, F.F., De Paulo, D., Neri, I.A. and Pastore, G.M. (2021). Polyphenols and their applications: An approach in food chemistry and innovation potential. *Food Chemistry*, 338(March 2020): 127535. https://Doi.org/10.1016/j.foodchem.2020.127535.

Del Cornò, M., Gessani, S. and Conti, L. (2020). Shaping the innate immune response by dietary glucans: Any role in the control of cancer? *Cancers*, 12(1). https://doi.org/10.3390/cancers12010155.

Díaz-Godínez, G. and Téllez-Téllez, M. (2021). Mushrooms as Edible Foods. In *Fungi in Sustainable Food Production* (pp. 143–164). Springer Nature Switzerland AG. https://doi.org/10.1007/978-3-030-64406-2_9.

Dospatliev, L. and Ivanova, M. (2017). Concentrations and risk assessment of lead and cadmium in wild edible mushrooms from the Batak mountain, Bulgaria. *Oxidation Communications*, 40(2): 993–1001.

Dospatliev, L. and Ivanova, M. (2021). Relationship between cambisols soil characteristics and iron content in wild edible mushrooms (*Cantharellus cibarius, Craterellus cornucopioides, Trichotomy equestre*). *Oxidation Communications*, 416(2): 409–416.

Dubey, S.K., Chaturvedi, V.K., Mishra, D., Bajpeyee, A., Tiwari, A. and Singh, M.P. (2019). Role of edible mushroom as a potent therapeutics for the diabetes and obesity. *In: 3 Biotech.* (vol. 9, Issue 12), Springer Verlag. https://Doi.org/10.1007/s13205-019-1982-3.

Elkhateeb, W.A., Daba, G.M., Thomas, P.W. and Wen, T.-C. (2019). Medicinal mushrooms as a new source of natural therapeutic bioactive compounds. *Egyptian Pharmaceutical Journal,* 18(2): 88–101. https://doi.org/10.4103/epj.ep.

Fernández, P., Haza, A.I. and Morales, P. (2020). Functional properties of edible mushrooms. *Agro. Sur.,* 48(1): 11–24. https://Doi.org/10.4206/agrosur.2020.v48n1-02.

Flores, J. (2020). Bioactive constituents and pharmacological activities. *In: Fungi for Human Health,* pp. 50–95, Springer Nature, Singapore AG.

Franco, M.B.E., Soto, A.K., García, A.A. and Ontiveros, E. (2018). Lifestyle in patients with chronic degenerative diseases. *Coloquio Panamericano de Investigación in Nursing,* 1–7.

Ganesan, K. and Xu, B. (2018). Anti-obesity effects of medicinal and edible mushrooms. *Molecules,* 23(11). https://doi.org/10.3390/molecules23112880.

Gargano, M.L., Van Griensven, L.J.L.D., Isikhuemhen, O.S., Lindequist, U., Venturella, G., Wasser, S.P. and Zervakis, G.I. (2017). Medicinal mushrooms: Valuable biological resources of high exploitation potential. *In: Plant Biosystems,* 151(3): 548–565, Taylor and Francis Ltd. https://Doi.org/10.1080/11263504.2017.1301590.

Gîrd, C.E. and Mocanu, M. (2020). Chemical composition, antioxidant activity and cytotoxic effects of romanian *Craterellus cornucopioides* (L.) Pers. Mushroom. *Farmacia,* 68(340–347). https://Doi.org/doi.org/10.31925/farmacia.2020.2.21.

Gitu, A.C. and Hahn, H.s. (2021). Evaluation and management of hypertension in Adults. *Relias Media,* 61(9): 907–911. https://doi.org/10.7727/wimj.2012.063.

Gulcin, İ. (2020). Antioxidants and antioxidant methods: An updated overview. *Archives of Toxicology,* 94(3): 651–715. https://Doi.org/10.1007/s00204-020-02689-3.

Guo, H., Diao, Q.P., Hou, D.Y., Li, Z.H., Zhou, Z.Y., Feng, T. and Liu, J.K. (2017). Sesquiterpenoids from cultures of the edible mushroom *Craterellus cornucopioides.* *Phytochemistry Letters,* 21(March): 114–117. https://Doi.org/10.1016/j.phytol.2017.06.007.

Guo, M., Meng, M., Zhao, J., Wang, X. and Wang, C. (2020). Immunomodulatory effects of the polysaccharide from *Craterellus cornucopioides* via activating the TLR4-NF κB signaling pathway in peritoneal macrophages of BALB/c mice. *International Journal of Biological Macromolecules,* 160: 871–879. https://doi.org/10.1016/j.ijbiomac.2020.05.270.

Guo, M.Z., Meng, M., Duan, S.Q., Feng, C.C. and Wang, C.L. (2019). Structure characterisation, physicochemical property, and immunomodulatory activity on RAW264.7 cells of a novel triple-helix polysaccharide from *Craterellus cornucopioides.* *International Journal of Biological Macromolecules,* 126: 796–804. https://Doi.org/10.1016/j.ijbiomac.2018.12.246.

Guo, Y., Chen, X. and Gong, P. (2021). Classification, structure and mechanism of antiviral polysaccharides derived from edible and medicinal fungus. *In: International Journal of Biological Macromolecules,* 183: 1753–1773, Elsevier B.V. https://Doi.org/10.1016/j.ijbiomac.2021.05.139.

Gupta, G., Siddiqui, M.A., Khan, M.M., Ajmal, M., Ahsan, R., Rahaman, M.A., Ahmad, M.A., Arshad, M. and Khushtar, M. (2020). Current pharmacological trends on myricetin. *In: Drug Research,* 70(10): 448–454, Georg Thieme Verlag. https://Doi.org/10.1055/a-1224-3625.

Guzman, K. (2017). *Quality in Perishable Food Logistics* [Universidad Nacional Agraria La Molina]. http://repositorio.lamolina.edu.pe/bitstream/handle/UNALM/3100/guzman-huaman-kelly.pdf?sequence=3&isAllowed=y.

Habtemariam, S. (2020). *Trametes versicolor* (Synn. *Coriolus versicolor*) polysaccharides in cancer therapy: Targets and efficacy. *Biomedicines,* 8(135): 1–26.

He, X., Fang, J., Guo, Q., Wang, M., Li, Y., Meng, Y. and Huang, L. (2020). Advances in antiviral polysaccharides derived from edible and medicinal plants and mushrooms. *In: Carbohydrate Polymers,* vol. 229, Elsevier Ltd. https://Doi.org/10.1016/j.carbpol.2019.115548.

Imran, M., Saeed, F., Hussain, G., Imran, A., Mehmood, Z., Gondal, T.A., Ghorab, A., Ahmad, I., Pezzani, R., Arshad, M.U., Bacha, U., Shariarti, M.A., Rauf, A., Muhammad, N., Shah, Z.A., Zengin, G. and Islam, S. (2021). Myricetin: A comprehensive review on its biological potentials. *Food Science and Nutrition,* 9(10): 5854–5868. https://Doi.org/10.1002/fsn3.2513.

Iñiguez, M., Pérez-Matute, P., Villanueva-Millán, M.J., Recio-Fernández, E., Roncero-Ramos, I., Pérez-Clavijo, M. and Oteo, J.A. (2018). *Agaricus bisporus* supplementation reduces high-fat diet-induced body weight gain and fatty liver development. *Journal of Physiology and Biochemistry*, 74(4): 635–646. https://doi.org/10.1007/s13105-018-0649-6.

Ishara, J., Buzera, A., Mushagalusa, G.N., Hammam, A.R.A., Munga, J., Karanja, P. and Kinyuru, J. (2022). Nutraceutical potential of mushroom bioactive metabolites and their food functionality. *Journal of Food Biochemistry*, 46(1): e14025. https://doi.org/10.1111/jfbc.14025.

Jiménez, A., Thomé, H., Espinoza, A. and Bordi, I.V. (2017). Recreational use of wild edible mushrooms: Cases of mycotourism in the world with emphasis on Mexico. *Bosque*, 38(3): 447–456. https://Doi.org/10.4067/S0717-92002017000300002.

Kaliyaperumal, M., Kezo, K. and Gunaseelan, S. (2018). A global overview of edible Mushrooms. *In*: *Biology of Macrofungi*, pp. 15–56. https://Doi.org/10.1007/978-3-030-02622-6.

Khameneh, B., Iranshahy, M., Soheili, V., Sedigheh, B. and Bazzaz, F. (2019). Review on plant antimicrobials: A mechanistic viewpoint. *Antimicrobial Resistance and Infection Control*, 8: 1–28.

Khodadi, E. (2020). Platelet function in cardiovascular disease: Activation of molecules and activation by molecules. *Cardiovascular Toxicology*, 20(1). https://doi.org/10.1007/s12012-019-09555-4.

Khursheed, R., Singh, S.K., Wadhwa, S., Gulati, M. and Awasthi, A. (2020). Therapeutic potential of mushrooms in diabetes mellitus: Role of polysaccharides. *In*: *International Journal of Biological Macromolecules*, 164: 1194–1205. Elsevier B.V. https://Doi.org/10.1016/j.ijbiomac.2020.07.145.

Kol, S., Bostanci, A., Kocabas, A., Uzun, Y. and Sadi, G. (2018a). Cell growth inhibitory potential of *Craterellus cornucopioides* (L.) Pers., together with antioxidant and antimicrobial properties. *Anatolian Journal of Botany*, 2(2): 60–64. https://Doi.org/10.30616/ajb.413645.

Kol, S., Bostanci, A., Kocabas, A., Uzun, Y. and Sadi, G. (2018b). Cell growth inhibitory potential of *Craterellus cornucopioides* (L.) Pers., together with antioxidant and antimicrobial properties. *Anatolian Journal of Botany*, 2(2): 60–64. https://Doi.org/10.30616/ajb.413645.

Kosanić, M., Ranković, B., Stanojković, T., Radović, M., Ćirić, A., Grujičić, D. and Milošević, O. (2019). *Craterellus cornucopioides* edible Mushroom as source of biologically active compounds. *Natural Product Communications*, 14(5). https://Doi.org/10.1177/1934578X19843610.

Kumar, N., Singh, H. and Sharma, S.K. (2020). Antioxidants: Responses and importance in plant defense system. *In*: *Sustainable Agriculture in the Era of Climate Change*, pp. 251–264, Springer Nature, Switzerland. https://Doi.org/10.1007/978-3-030-45669-6_11.

Li, J.J., Yuan, L.G., Zhong, Y.L., Wei, B.Z. and Zhang, N. (2021). Anti-inflammatory effect of a polysaccharide fraction from *Craterellus cornucopioides* in LPS stimulated macrophages. *Food Biochemistry*, 45: 1–9. https://Doi.org/10.1111/jfbc.13842.

Liu, J., Pu, H., Liu, S., Kan, J. and Jin, C. (2017). Synthesis, characterisation, bioactivity and potential application of phenolic acid grafted chitosan: A review. *Carbohydrate Polymers*, 174: 999–1017. https://Doi.org/10.1016/j.carbpol.2017.07.014.

Liu, Y., Duan, X., Zhang, M., Li, C., Zhang, Z., Liu, A., Hu, B., He, J., Wu, D., Chen, H. and Wu, W. (2020). Cooking methods effect on the nutrients, bioaccessibility and antioxidant activity of *Craterellus cornucopioides*. *Food Science and Technology*, 131(June): 109768. https://Doi.org/10.1016/j.lwt.2020.109768.

Liu, Y., Duan, X., Zhang, M., Li, C., Zhang, Z., Hu, B., Liu, A., Li, Q., Chen, H., Tang, Z., Wu, W. and Chen, D. (2021). Extraction, structure characterisation, carboxymethylation and antioxidant activity of acidic polysaccharides from *Craterellus cornucopioides*. *Industrial Crops and Products*, 159(October 2020): 113079. https://Doi.org/10.1016/j.indcrop.2020.113079.

Ma, G., Yang, W., Zhao, L., Pei, F., Fang, D. and Hu, Q. (2018). A critical review on the health promoting effects of mushrooms nutraceuticals. *Food Science and Human Wellness*, 7(2): 125–133. https://Doi.org/10.1016/j.fshw.2018.05.002.

Maity, P., Sen, I.K., Chakraborty, I., Mondal, S., Bar, H., Bhanja, S.K., Mandal, S. and Maity, G.N. (2021). Biologically active polysaccharide from edible mushrooms: A review. *International Journal of Biological Macromolecules*, 172: 408–417. https://Doi.org/10.1016/j.ijbiomac.2021.01.081.

Martel, J., Ojcius, D.M., Chang, C.J., Lin, C.S., Lu, C.C., Ko, Y.F., Tseng, S.F., Lai, H.C. and Young, J.D. (2017). Anti-obesogenic and antidiabetic effects of plants and mushrooms. *Nature Reviews Endocrinology*, 13(3): 149–160. https://doi.org/10.1038/nrendo.2016.142.

Martínez-Medina, G.A., Prado-Barragán, A., Martínez-Hernández, J.L., Ruíz, H.A., Rodríguez, R.M., Contreras-Esquivel, J.C. and Aguilar, C.N. (2019). Bio-functional peptides: Bioactivity, production and applications. *Journal of BioProcess and Chemical Technology*, 11(22): 1–7.

Mohamed Yahaya, N.F., Rahman, M.A. and Abdullah, N. (2014). Therapeutic potential of mushrooms in preventing and ameliorating hypertension. *Trends in Food Science and Technology*, 39(2): 104–115. https://doi.org/10.1016/j.tifs.2014.06.002.

Morris, H., Llaurado, G., Bermúdez, R.C. and Cos, P. (2016). Evaluation of the immunomodulatory activity of bioproducts obtained from the edible-medicinal mushroom *Pleurotus ostreatus*. *Revista Anales de La Academia de Ciencias de Cuba*, 8(1).

Muszyńska, B., Grzywacz, A., Kała, K. and Gdula, J. (2018). Anti-inflammatory properties of edible mushrooms: A review. *Food Chemistry*, 243(September 2017): 373–381. https://Doi.org/10.1016/j.foodchem.2017.09.149.

Nabors, M.W., González-Barreda, P., García Antón, M. and Moreno Sáiz, J.C. (2006). Introducción a la botánica. Madrid: Pearson Educación.

Nadakinamani, R.G., Reyana, A., Kautish, S., Vibith, A.S., Gupta, Y., Abdelwahab, S.F. and Mohamed, A.W. (2022). Clinical data analysis for prediction of cardiovascular disease using machine learning techniques. *Computational Intelligence and Neuroscience*, 2022. https://doi.org/10.1155/2022/2973324.

Novakovic, S. (2021). The potential of the application of Boletus edulis, *Cantharellus cibarius* and *Craterellus cornucopioides* in frankfurters: A review. *IOP Conference Series: Earth and Environmental Science*, 854(1). https://Doi.org/10.1088/1755-1315/854/1/012068.

Nowakowski, P., Markiewicz-Żukowska, R., Bielecka, J., Mielcarek, K., Grabia, M. and Socha, K. (2021). Treasures from the forest: Evaluation of mushroom extracts as anti-cancer agents. *Biomedicine and Pharmacotherapy*, 143(August). https://doi.org/10.1016/j.biopha.2021.112106.

Okati-Aliabad, H., Ansari-Moghaddam, A., Kargar, S. and Jabbari, N. (2022). Prevalence of obesity and overweight among adults in the middle east countries from 2000 to 2020: A systematic review and meta-analysis. *Journal of Obesity*, 2022: 8074837. https://doi.org/10.1155/2022/8074837.

Özcan, Ö. and Ertan, F. (2018). Beta-glucan content, antioxidant and antimicrobial activities of some edible Mushroom species. *Food Science and Technology*, 6(2): 47–55. https://Doi.org/10.13189/fst.2018.060201.

Painuli, S., Semwal, P. and Egbuna, C. (2020). Mushroom: Nutraceutical, mineral, proximate constituents and bioactive component. *Functional Foods and Nutraceuticals*, 307–336. https://Doi.org/10.1007/978-3-030-42319-3_17.

Pliz, D., Norvell, L., Danell, E. and Molina, R. (2003). Ecology and management of commercially harvested chanterelle mushrooms. *In: USDA Forest Service—General Technical Report PNW*, Issue 576. https://Doi.org/10.2737/PNW-GTR-576.

Pinedo, A., García, C., Lugo, E. and Enciso, Y. (2017). Chronic degenerative diseases in patients attending the Medical Specialties Unit. *Revista Iberoamericana de Ciencias*, 4(5): 92–97. www.reibci.org.

Remya, V.R., Chandra, G. and Mohanakumar, K.P. (2019). Edible Mushrooms as neuro- nutraceuticals: Basis of therapeutics. *In: Medicinal Mushrooms*, pp. 71–101, Springer Nature, Singapore.

Ruan. (2018). Collection of wild edible mushrooms and strategies for the recognition of toxic species among the Tsotziles of Chamula, Chiapas, Mexico. *Scientia Fungorum*, 48: 1–13.

Salsinha, A.S., Rodríguez-Alcalá, L.M., Relvas, J.B. and Pintado, M.E. (2021). Fatty acids role on obesity induced hypothalamus inflammation: From problem to solution—A review. *Trends in Food Science and Technology*, 112(March): 592–607. https://doi.org/10.1016/j.tifs.2021.03.042.

Sánchez, C. (2017). Bioactives from Mushroom and their application. *In: Food Bioactives*, pp. 23–57. https://Doi.org/10.1007/978-3-319-51639-4.

Sarikurkcu, C., Akata, I., Guven, G. and Tepe, B. (2020). Metal concentration and health risk assessment of wild mushrooms collected from the Black Sea region of Turkey. *Environmental Science and Pollution Research*, 26419–26441. https://Doi.org/10.1007/s11356-020-09025-3.

Shen, H.S., Shao, S., Chen, J.C. and Zhou, T. (2017). Antimicrobials from Mushrooms for assuring food safety. *Comprehensive Reviews in Food Science and Food Safety*, 16(2): 316–329. https://Doi.org/10.1111/1541-4337.12255.

Sigwart, J.D. (2018). *What Species Mean: A User's Guide to the Units of Biodiversity*. CRC Press. https://doi.org/10.1201/9780429458972.

Singh, A., Misra, M., Mishra, S. and Sachan, S.G. (2021). Fungal production of food supplements. *In: Fungi in Sustainable Food Production*, pp. 129–142.

Stoll, S., Bitencourt, S., Laufer, S. and Inês, M. (2019). Myricetin inhibits panel of kinases implicated in tumorigenesis. *Basic and Clinical Pharmacology and Toxicology*, 125(1): 3–7. https://Doi.org/10.1111/bcpt.13201.

Tel-çayan, G., Ullah, Z., Öztürk, M. and Yabanlı, M. (2018). Heavy metals, trace and major elements in 16 wild Mushroom species determined by ICP-MS. *Atomic Spectroscopy*, 29–37(February). https://Doi.org/10.46770/AS.2018.01.004.

Thu, Z.M., Ko, K., Aung, H.T., Clericuzio, M., Armijos, C. and Vidari, G. (2020). Bioactive phytochemical constituents of wild edible mushrooms from Southeast Asia. *Molecules*, 25(8). https://Doi.org/10.3390/molecules25081972.

Turfan, N., Pekşen, A., Kibar, B. and Ünal, S. (2018). Determination of nutritional and bioactive properties in some selected wild growing and cultivated mushrooms from Turkey. *Acta Scientiarum Polonorum, Hortorum Cultus*, 17(3): 57–72. https://Doi.org/10.24326/asphc.2018.3.6.

Upadhyaya, J. (2018). Analysis of nutritional and nutraceutical properties of wild-grown Mushrooms of Nepal. *Innovation at the NHP/Food Interface*, 5(August 2019). https://Doi.org/10.1055/s-0038-1644943.

Villa, L.S., Flores, F.Y. and Santibañez, L.P. (2017). *Dental Pulp Stem Cells (DPSC): Prospective Therapeutics in Chronic Degenerative Diseases*.

Watanabe, F., Schwarz, J., Takenaka, S., Miyamoto, E., Ohishi, N., Nelle, E., Hochstrasser, R. and Yabuta, Y. (2012). Characterisation of Vitamin B_{12} compounds in the wild edible Mushrooms black trumpet (*Craterellus cornucopioides*) and golden chanterelle (*Cantharellus cirbarius*). *Nutr. Sci. Vitaminol.*, 58(10 mL): 438–441.

Wong, K.H., Lai, C.K.M. and Cheung, P.C.K. (2011). Immunomodulatory activities of mushroom sclerotial polysaccharides. *Food Hydrocolloids*, 25(2): 150–158. https://doi.org/10.1016/j.foodhyd.2010.04.008.

Xiao, F., Xu, T., Lu, B. and Liu, R. (2020). Guidelines for antioxidant assays for food components. *Food Frontiers*, 1(1): 60–69. https://Doi.org/10.1002/fft2.10.

Yabani, B., Mantarların, Y. and İçeriği, B. (2009). Nutritional composition of some wild edible Mushrooms. *Chemical Analysis*, 34(1): 25–31.

Yadav, D. and Negi, P.S. (2021). Bioactive components of mushrooms: Processing effects and health benefits. *Food Research International*, 148(July): 110599. https://Doi.org/10.1016/j.foodres.2021.110599.

Yang, W.W., Wang, L.M., Gong, L.L., Lu, Y.M., Pan, W.J., Wang, Y., Zhang, W.N. and Chen, Y. (2018). Structural characterisation and antioxidant activities of a novel polysaccharide fraction from the fruiting bodies of *Craterellus cornucopioides*. *International Journal of Biological Macromolecules*, 117: 473–482. https://Doi.org/10.1016/j.ijbiomac.2018.05.212.

Yao, Z., Li, C., Gu, Y., Zhang, Q., Liu, L., Meng, G., Wu, H., Bao, X., Zhang, S., Sun, S., Wang, X., Zhou, M., Jia, Q., Song, K., Li, Z., Gao, W., Niu, K. and Guo, C. (2019). Dietary myricetin intake is inversely associated with the prevalence of type 2 diabetes mellitus in a Chinese population. *Nutrition Research*, 68: 82–91. https://Doi.org/10.1016/j.nutres.2019.06.004.

Yildiz, S., Yilmaz, A. and Can, Z. (2017). *In vitro* bioactive properties of some wild Mushrooms collected from Kastamonu pProvince. *Kastamonu Üniversitesi Orman Fakültesi Dergisi*, 17(3): 523–530. https://Doi.org/10.17475/kastorman.263875.

Ying, X., Chen, X., Wang, T., Zheng, W., Chen, L. and Xu, Y. (2020). Possible osteoprotective effects of myricetin in STZ induced diabetic osteoporosis in rats. *European Journal of Pharmacology*, 866. https://Doi.org/10.1016/j.ejphar.2019.172805.

Zhang, J.J., Li, Y., Zhou, T., Xu, D.P., Zhang, P., Li, S. and Li, H. Bin. (2016). Bioactivities and health benefits of mushrooms mainly from China. *Molecules*, 21(7): 1–16. https://doi.org/10.3390/molecules21070938.

CHAPTER 10

Velvet Shank
(*Flammulina velutipes*)

Vijay Kumar Veena,[1,*] *Adhikesavan Harikrishnan*[2,*]
and *Purushothaman Maheswari*[3]

1. Introduction

Flammulina velutipes, Basidiomycete macro-fungi is an important edible mushroom (Table 1). *Flammulina velutipes* is commonly consumed worldwide but majorly in Japan, Korea and China. Various types of biomolecules that include polysaccharides, proteins, glycoproteins and secondary metabolites have been reported for several biological activities (Fig. 1). However, several reports suggest that the exopolysaccharides seem to be used for edible and possess medicinal value. These exopolysaccharides have been produced by submerged culture, minimal nutritional requirements and easy adaptable environmental conditions by various researchers (Yang et al., 2011; 2016; Tan et al., 2016; Ikekawa et al., 1985). Other names for *Flammulina velutipes*, a golden-needle mushroom, enokis mushroom, winter mushroom, lily mushroom and velvet-shank mushroom in different countries (Miles and Chang, 2004). *Flammulina velutipes* has attracted the attention of many people for its aroma, taste, texture, medicinal values and functional characteristics (Yang et al., 2011; 2016; Tan et al., 2016). Several reports on *Flammulina velutipes* suggest the biological and medicinal values that include antimicrobial, antiviral, antifungal, antioxidant, anti-inflammatory and anticancer potentials (Badalyan and Hambardzumyan, 2001; Smiderle et al., 2006; Wang et al., 2012; Wu et al., 2014).

---

[1] Department of Core Allied Healthcare, School of Allied Health Sciences, Next to KIAD Building, White field, Bengaluru-560066.
[2] Department of Chemistry, School of Arts and Sciences, Vinayaka Mission Research Foundation-Aarupadai Veedu (VMRF-AV) campus, Paiyanoor, Chennai-603104, Tamil Nadu, India.
[3] Department of Chemistry, SRM Valliammai Engineering College, Katankulathur, Chennai-603203, Tamil Nadu, India.
* Corresponding authors: vk.veena@jainuniversity.ac.in; harich4@gmail.com

Table 1. Taxonomy of *Flammulina velutipes*.

Kingdom	Fungi
Phylum	*Basidiomycota*
Class	*Agaricomycetes*
Order	*Agaricales*
Family	*Physalacriaceae*
Genus	*Flammulina*
Species	*Velutipes*

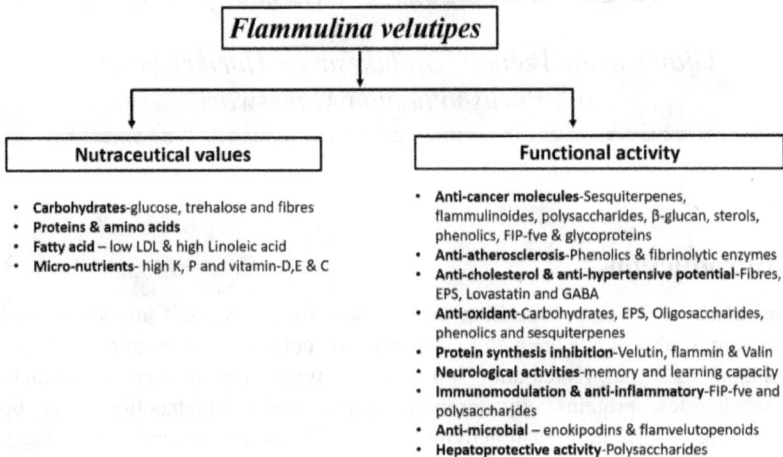

Flammulina velutipes

Nutraceutical values

- **Carbohydrates**-glucose, trehalose and fibres
- **Proteins & amino acids**
- **Fatty acid** – low LDL & high Linoleic acid
- **Micro-nutrients**- high K, P and vitamin-D,E & C

Functional activity

- **Anti-cancer molecules**-Sesquiterpenes, flammulinoides, polysaccharides, β-glucan, sterols, phenolics, FIP-fve & glycoproteins
- **Anti-atherosclerosis**-Phenolics & fibrinolytic enzymes
- **Anti-cholesterol & anti-hypertensive potential**-Fibres, EPS, Lovastatin and GABA
- **Anti-oxidant**-Carbohydrates, EPS, Oligosaccharides, phenolics and sesquiterpenes
- **Protein synthesis inhibition**-Velutin, flammin & Valin
- **Neurological activities**-memory and learning capacity
- **Immunomodulation & anti-inflammatory**-FIP-fve and polysaccharides
- **Anti-microbial** – enokipodins & flamvelutopenoids
- **Hepatoprotective activity**-Polysaccharides

Fig. 1. The nutritional and functional facts of *Flammulina velutipes*.

2. Nutritional Values Reported in *Flammulina velutipes*

Nutritional facts of *Flammulina velutipes* are reported to be naturally enriched with dietary fibres, proteinous part, essential-amino acids, vitamin(s) and essential mineral(s) (Table 2). Moreover, it has low amounts of sugars and no cholesterol (Karaman et al., 2010; Wu et al., 2010; Jing et al. 2014). Furthermore, it contains high amounts of calcium, phosphorous and vitamin-D (Na et al., 2005; Ko et al., 2007). Amino acids and protein seem to be the major nutritional components of *Flammulina velutipes*, which constitute about 64% of its dry weight in cultivated types (Dikeman et al., 2005; Ko et al., 2007; Beluhan et al., 2011; Akata et al., 2012; Pereira et al., 2012). However, the amount of protein seems to be influenced by growth substrate used, mushroom size, harvesting time and available source of nitrogen during the growth (Sing et al., 2001; Gupta et al., 2005). Various essential amino acids that include L-glutamic acid, L-alanine, glycine and L-lysine are reported in high concentrations

Table 2. Major nutritional facts of cultivated *Flammulina velutipes*.

Major Composition	(g/kg of Dry Weight)	References
Carbohydrates	580–871	Dikeman et al., 2005; Ko et al., 2007;
Proteins	178.9–279.5	Beluhan et al., 2011; Akata et al., 2012; Pereira et al., 2012
Fats	18.4–73.3	

(Smiderle et al., 2008; Kim et al., 2009). These amino acids contribute to the pleasant smell and sweet taste of food prepared with *Flammulina velutipes* (Smiderle et al., 2008; Beluhan et al., 2011).

Among the proteins, researchers have also reported the bioactive proteins that are known to possess the immunomodulatory, anticancer and anti-inflammatory properties (Chang et al., 2013). Flammin, velin, velutin and flammulin, the important proteins, are present in various parts of fruiting bodies that can be extracted and are known to be acting as ribosome inactivating proteins (RIP) (Wang and Ng, 2000; 2001; Ng, 2004). Few proteins have hemagglutinin properties and flammutoxin with cytolysis properties are reported (Tadjibaeva et al., 2000; Ng and Wang, 2006; Ng et al., 2000; Raymond et al., 2009). Asparginase and proflammin, anticancer enzymes, are reported in the mycelial extract of this mushroom (Maruyama et al., 2007; Eisele et al., 2011; Kotake et al., 2011; Kakkar and Bias, 2014).

As per the literature survey, the carbohydrates constitute half the dry weight of the mushroom (Dikeman et al., 2005; Ko et al., 2007; Beluhan et al., 2011; Akata et al., 2012; Pereira et al., 2012; Reis et al., 2012). The polysaccharides reported in this mushroom include β-glucan, chitin and starch; disaccharides like trehalose and sucrose and various monosachharides that includes fucose, mannose, glucose, ribose, xylose, fructose, mannitol and galactose (Dikeman et al., 2005; Kim et al., 2009; Beluhan et al., 2011). It is also reported that variations of carbohydrates are due to geographical conditions, cultural conditions and types of mushroom strains (Dikeman et al., 2005; Kim et al., 2009; Beluhan et al., 2011). The dietary fibre content in the mushroom have approximately 29.3% of its dry weight (Kalac et al., 2013; Tang et al., 2016). This high dietary fibre content of *Flammulina velutipes* is known to have the hypolipidemic effect in the animal model studies (Yeh et al., 2014).

Lipid is the macromolecule that is commonly observed in the range of 17.3–73.3 g/kg dry weight of the *Flammulina velutipes.* The fatty acids reported include sterols, sphingolipids and fats (Ko et al., 2007; Beluhan et al., 2011; Akata et al., 2012; Pereira et al., 2012; Reis et al., 2012). The total fatty acid content consists of a majority of monosaturated and polyunsaturated fatty acids (79.23%) and a little amount of unsaturated fatty acid (20.67%). Among the essential fatty acids, linoleic acid is known to be made up to 40.93–56.33% of total fatty acids, followed by oleic acid (15.08–16.43%); palmitic acid (10.31–14.56%) and stearic acid (1.38–3.56%) are reported in this mushroom (Pereira et al., 2012). Thus, *Flammulina velutipes* represents food items for human(s) and/or animal(s) to get enough quantity of essential fatty acids that cannot be synthesised in the body. Several studies have reported that mycosterols from *Flammulina velutipes* exhibit anti-proliferative activities in several cancer cells lines (Yi et al., 2012; 2013).

The micronutrients reported in the *F. velutipes* ranges between 5–12% of dry weight (Ko et al., 2007; Beluhan et al., 2011; Akata et al., 2012; Pereira et al., 2012). Among the minerals, potassium seems to be most abundantly seen (28–28.98 mg/g dry-weight) followed by phosphorous (8.8–9.4 mg/g of dry-weight). Hence, it contributes sufficient amounts of recommended level of intake per day as per FDA (Akhter et al., 2003). Higher potassium and lower sodium levels in mushroom is reported to be ideal for hypertension or people with heart disease (John et al., 2002; He et al., 2006). Many minerals, such as copper, magnesium, zinc, selenium, lithium, calcium, iron and sulphur are also reported (Smiderle et al., 2008; Zheng et al., 2012; Zeng et al., 2012).

Several vitamins have been detected that include vitamin D (toco-pherol; 0.6 µg/ml), vitamin C (ascorbic-acid; 238.0 µg/ml), β carotene (3.40 µg/ml) and lycopenes (0.2 µg/ml) (Pereira et al., 2012; Breene et al., 1990). These vitamins reported in *Flammulina velutipes* are known to act as antioxidants that play a very important role in maintenance of good health.

3. Functional Properties of *Flammulina velutipes*

The functional properties of *Flammulina velutipes* are well-documented by various researchers and also through traditional usages. However, the detailed anticancer, cholesterol-lowering properties, anti-inflammatory activities, antioxidant properties, antimicrobial and other important potentials have made the mushroom an attractive area for researchers (Fig. 1).

3.1 Anticancer Molecules in Flammulina velutipes

Cancer is the deadly neoplastic disease that is the second leading cause for death globally. The natural products seem to be important for drug development. Similarly, there are an array of bioactive molecules that are reported from the extracts and isolated molecules (Gu and Leonard, 2006; Smiderle et al., 2006; Ou et al., 2005; Wang et al., 2012; Yi et al., 2013). Moreover, the novel sesquiterpenes and nor-sequiterpenes are isolated from the *Flammulina velutipes* (Table 3). These molecules have reported anticancer, antibacterial and antioxidant potential (Wang et al., 2012) (Table 3). Several preclinical studies have reported that extracts of *Flammulina velutipes* exhibit anticancer properties. It is also being seen that the fruiting body extracts of *Flammulina velutipes* have anticancer activity against the breast cancer cell lines (Leonard et al., 2006). This study has shown that the extract of the mushroom is active in inhibiting the cancer cells estrogen positive (ER+; MCF-7) and triple-negative breast-cancer cells (MDA MB 231). Furthermore, they have shown apoptosis and inhibited the colony formation in MCF-7 cell lines (Gu and Leonard, 2006).

Several novel anticancer molecules were reported from the culture of *Flammulina velutipes* that includes flammulinol-A, sterpuric acid, isolactarane sesquiterpene and flammulinolides A-G (Wang et al., 2012). These are reported to inhibit the several cancer cells. Flammulinolides inhibit the human naso-pharyngeal cancers (KB cells) and liver cancer cells (Hep-G2) with a lower potency. Flammulinolide C has shown strong cell toxicity in cervical cancer cells (HeLa) (Wang et al., 2012). Sesquiterpenes has moderate toxicity against Hep G2, MCF-7, A549 and SGC7901cells (Wang et al., 2012). New non-sesquiterpene alkaloid isolated from *Flammulina velutipes* grown in fermented rice substrate is reported to inhibit the human KB cells (Xu et al., 2013). Other than sesquiterpenes, bio-active polysaccharide(s) from *Flammulina velutipes* have known to possess anticancer potential in the (SC180) mouse model(s) and (S180) mice tumour model(s) as well as hepatoma SMMMC 7721 cell line (Leung et al., 1997; Jiang et al., 1999). The crude extract that consists of the mycelium of *Flammulina velutipes* is found to inhibit the human lung cancer cells (BEL 7402) (Zhao et al., 2013). Similarly, triple-helix polysaccharide(s) isolated from ultrasonication have reported to inhibit the gastric cells (Yang et al., 2012). The β glucan is known for its anticancer potential from this mushroom (Zhang et al., 2007; Mantovani et al., 2008; Smiderle et al., 2006).

Table 3. Anticancer properties of the *Flammulina velutipes*.

Substance	Anticancer Experimental Details	References
Crude extracts	*In vitro* cell cytotoxicity against the cancer cell lines – BT20, MCF-7 and MDA MB-231 cells.	Gu and Leonard, 2006
Flammulinoide-A	*In vitro* cytotoxicity against cancer cell lines – HeLa, HepG2 and KB cells.	Wang et al., 2007
Flammulinoide-B		
Flammulinoide-C		
Flammulinoide-F		
Enokipodin-B	*In vitro* cytotoxic to cancer cell lines – HepG2, MCF 7, A549 and SGC 7901 cells.	Wang et al., 2012
Enokipodin-D		
Enokipodin-J		
2,5-cupradiene-1,4-dione	*In vitro* cytotoxic to cancer cell lines – HepG2, MCF 7, A549 and SGC 7901 cells.	Wang et al., 2012
Norsequiterpene alkaloids	*In vitro* cytotoxic in human-KB cancer cells.	Xu et al., 2013
Alkaline polysaccharides	*In vivo* anticancer activities in SC180 mouse models.	Leung et al., 1997
Hot water extract and polysaccharides	*In vivo* anticancer activity against SC180 mouse models and SMMC7721 liver cells.	Jiang et al., 1999
Polysaccharides	*In vitro* cytotoxicity against cancer cell lines – BEL 7402.	Zhao et al., 2013; Jeff et al., 2013
Triple-helices polysaccharides	*In vitro* cytotoxic in cancer cell lines – BCG823 and A549 cells.	Yang et al., 2012; 2013
Sterols	*In vitro* cytotoxic in cancer cell lines – SGC, Hep G2, A549 and U251 cells.	Yi et al., 2012; 2013; Chang et al., 2010; 2013
Fip-fve proteins	*In vitro* cytotoxic in cancer cell lines – A549, and BNL-liver mouse models.	Ikekawa et al., 1985

Sterol(s) reported in *Flammulina velutipes*, specially ergosterol(s) and dihydro-ergosterol are known to have the therapeutic molecules in stomach, lung, liver and neurological studies (Yi et al., 2012; 2013). Several studies report that sterol(s) can be encapsulated with liposomes and micelles-based nano-formulation and micro-emulsion to enhance the bioavailability, bio-distribution and solubility of the anticancer efficacy of sterols (Yi et al., 2013).

The phenolics that have reported anticancer properties have been isolated from *Flammulina velutipes* and include proto-catechuic acids, p coumaric and ellagic acid(s) (Rahman et al., 2015; Ferguson et al., 2005; Seeram et al., 2005). The p coumaric and ellagic acids isolated from this mushroom have the potential for anti-oxidative activities in colon cancer cells (HT 29 and HCT 16) respectively (Ferguson et al., 2005; Seeram et al., 2005). However, ellagic acid is reported to inhibit the P450 expression and its activity (Ahn et al., 1996). FIP fve, bio-active proteins obtained from *Flammulina velutipes* has been reported for multiple activities that include anticancer, immunomodulatory and anti-inflammatory (Chang et al., 2013; Chang et al., 2010). Further, FIP fve are shown to down-regulate the expressions in the RACGAP 1 gene in the lung cancer cells that are responsible for cancer cells survival and metastatic potential (Wang et al., 2011). Further research has shown that FIP-fve protein also increases the p53, a tumour-suppressor gene expression that is responsible for the proliferations in the lung cancer cells (A549) (Chang et al., 2010). Other studies have shown that oral intake of FIP-fve proteins inhibit the tumour growth in the BNL liver cancer-bearing mice suggesting that the anticancer activities are mediated through the inhibition of MAPK/ERK signalling pathways. This study has shown that FIP-fve protein also inhibits angiogenesis through inhibition of helper T cell-mediated cell-derived IFN-ɣ at the tumour microenvironment (Chang et al., 2010). Further, it is also shown to increase the expressions of MHC I, MHC II and costimulatory CD-80 molecules on blood mono-nuclear cells upon oral usage of FIP-fve with up-regulation of antigen presenting cells (APCs). These results co-relate with the maturation of spleen dendritic cells, APCs and stimulate the cytotoxic T cell response. Hence, FIP fve can be the adjuvant therapy along with the HPV 16E7 vaccines for tumour-containing mice that are known to show increase in the immune response (Ding et al., 2009). These research findings have strongly suggested the FIP fve has the potential for novel-type of adjuvants in cancer immunotherapy.

Glycoprotein(s) isolated from the fruiting bodies and mycelium of this mushroom have anticancer properties (Zhang et al., 2007). Proflamin, isolated from mycelium, is reported to increase the immune-suppressive process and inhibit the various cancer cells. Proflamin is shown to increase the survival of mice with 16-melanomas and 755-adenocarcinoma cell lines (Ikekawa et al., 1985; Ingold, 1980). When proflamin is combined with other therapies in vaccines or surgery, it has shown to inhibit the sarcoma S180, leukemia cells and metha-A fibrosarcoma. Similarly FIP fve are also shown to increase the innate and adaptive immunity (Maruyama et al., 2007). However, there are many bio-active molecules with effective anticancer and anti-tumour potential in *Flammulina velutipes*, suggesting as being an excellent source of drug discovery in various types of cancer.

3.2 Cholesterol-reducing Molecules in Flammulina velutipes

A great amount of attention in focused on *Flammulina velutipes* as it is an attractive source of several active compounds that include dietary fibre, polysaccharides and mycosterols which are reported to lower the cholesterol(s) and blood pressure (Table 4) (Yeh et al., 2014). It is known to have high dietary fibres than other edible fungi (Yang et al., 2001). The dry powdered extract of *Flammulina velutipes* is shown to have a higher amount of fibre. When given to male hamster(s) with high-fat diet(s), they show lowered lipid metabolism (Yen et al., 2001). This study suggests that 3% extract lowers the total cholesterol, triacylglycerol, LDL and HDL levels in serum and liver of hamster (Yeh et al., 2014). The exopolymer of *Flammulina velutipes* shows the hypo-lipidemic effects on high-fat diets in rats where it has reduced levels of total cholesterol(s), triacylglycerol, LDL and HDL in the serum and in the liver of the animal (Yang et al., 2002; Fukushima et al., 2001).

Lovastatin and ɣ-amino butyric acid (GABA) are reported in *Flammulina velutipes* in the fruiting body. Until now lovastatin, being clinical, used to inhibit the cholesterol production and thus reduce the risk of coronary heart diseases (Chen et al., 2012). GABA-containing foods and food products are reported to lower hypertension (Aoki et al., 2003; Inoue et al., 2003; Hayakawa et al., 2002). Research findings

Table 4. Anti-atherosclerosis activities of *Flammulina velutipes*.

Substance	Experimental Details	References
Methanolic extract reported protocatechuinic acid, p-couramic acid and ellagic acid	*In vitro* inhibition of LDL and TBARS.	Rahman et al., 2015
Thrombolytic enzymes isolated	*In vitro* coagulability and fibrinolytic assay.	Okamura et al., 2001; Park et al., 2007
FVP and water extract	Cholesterol lowering in male Syrian hamster.	Yan et al., 2014
Exopolymer and fibres	Cholesterol lowering in Sprague Dawley mice under hyperlipidemic male F344/DuCrj rat.	Yang et al., 2002; Fukushima et al., 2001
GABA isolated and mycelial extract	Reduced hypertension and inhibited the ACE.	Kim et al., 2002

show that addition of GABA-enriched powder of *Flammulina velutipes* at 0.9 mg/kg reduces blood pressure in hypertensive rats without any change in the normal mice blood pressure (Harada et al., 2011). GABA is known to inhibit the nor-adrenaline released from sympathetic-nervous system to increase the blood pressure (Hayakawa et al., 2002). Moreover, optimised culture medium of *Flammulina velutipes* is shown to inhibit the ACE-2 at the IC50 value of 22.6 mg/ml. This research finding suggests that ACE-2 inhibitors can be isolated from *Flammulina velutipes* to provide functional food-based anti-hypersensitive products which can be crucial in treating cardiovascular diseases (Kim et al., 2002).

3.3 Antioxidant Molecules in Flammulina velutipes

Although macro-nutrients and micronutrients are reported in *Flammulina velutipes*, the phenolic metabolites are known for their major antioxidant activity and are involved in various biological processes (Barros et al., 2007; Kim et al., 2008; Rahman et al., 2015). In general, phenolic metabolites are hetero-cyclic compounds well documented for their antioxidant properties associated with oxidative stress-related disorders that include cancer, cardiovascular diseases and diabetes. The various polyphenolic antioxidants reported in *Flammulina velutipes* include protocatechuic acid, p-coumaric acid, ellagic acid, gallic acid, pyrogallol, homo-gentisic, 5-sulfosalicylic acid, chlorogenic acid, caffeic acid, quercetin and ferulic acid (Table 5) in the range of 9.0–26 µg/g of dry weight of the mushroom (Kim et al., 2008). These phenolics are known to prevent atherosclerosis (Rahman et al., 2015) and have cancer-preventive activities (Kim et al., 2008). A study has also shown that the variations in phenolics of this mushroom are described under different geographical conditions and depend on the particular type of sub-species for synthesising phenolic molecules (Zeng et al., 2012). Several studies have reported that polysaccharides isolated from *Flammulina velutipes* have anti-oxidative potential (Ma et al., 2015; Xia et al., 2015). Rhamnose, a intracellular polysaccharide obtained from *Flammulina velutipes*, seems to be responsible for the anti-oxidative potential (Ma et al., 2015). Studies also suggest anti-aging properties of *Flammulina velutipes* through increase in the superoxide-dismutase (SOD) in the heart, kidney and heart with increase in the rhamnose dosage. Other studies show that

Table 5. Antioxidant properties of *Flammulina velutipes*.

Substance	Experimental Details	References
Intracellular polysaccharides isolated	Inhibition of ROS by DPPH assay; increase in SOD and *in vivo* studies to enhance SOD and GSH.	Ma et al., 2015
Exopolysaccharides isolated	Inhibition of ROS by DPPH assay; increase in SOD and *in vivo* studies to enhance SOD, catalase, TBARS and GSH.	Ma et al., 2015
Oligosaccharides	Inhibition of ROS by DPPH assay; increase in SOD and *in vivo* studies to enhance SOD, catalase, TBARS and GSH.	Xia et al., 2015
Water-soluble polysaccharides	*In vitro* DPPH assay.	Wu et al., 2014
Methanolic extract	*In vitro* inhibition of ROS and total phenol content.	Zhao et al., 2013
Water-soluble nucleotide(s) extract	*In vitro* inhibition of ROS and total phenol content.	Zeng et al., 2012; Cheng et al., 2012
Enokipodin-B	*In vitro* inhibition of ROS by ABTS and total phenol content.	Wang et al., 2012
Enokipodin-D	*In vitro* DPPH assay.	
Enokipodin-J	*In vitro* DPPH assay.	

exopolysaccharides (EPS) from *Flammulina velutipes*, that had rhamnose, have anti-oxidative potential (Ma et al., 2015). Oligosaccharides and FVP-1A polysaccharide isolated from *Flammulina velutipes* exhibit ant-oxidative potential (Xia et al., 2015; Wu et al., 2014; Zhao et al., 2013). However, crude polysaccharides are suggested to have the highest anti-oxidative potential than purified bioactive molecules, indicating the synergic antioxidant effects (Zhang et al., 2010; Zhang et al., 2013).

Phenolics that are major compounds are reported for the antioxidant properties in *Flammulina velutipes* (Barros et al., 2007; Karaman et al., 2010). It is also being identified that the phenolic content is found to be highest in *Flammulina velutipes* among the edible Australian mushrooms (Zeng et al., 2012; Bao et al., 2010).

The p-coumaric acid showed increased activities of SODs and inhibiting of the stress results in cardiac apoptosis in isoprenol-induced myocardial infarction in rat(s) (Stanely et al., 2013). Ellagic acid obtained in the fruiting bodies of *Flammulina velutipes* exhibited potent anti-oxidative potential (Kilic et al., 2014). Other than phenolics and carbohydrates, the other class of secondary metabolites have shown to be antioxidants. Nucleotides obtained from *Flammulina velutipes* are reported to possess mild anti-oxidation activity (Breene, 1990; Fu et al., 2002).

Sesquiterpene(s) isolated from *Flammulina velutipes* cultivated in cooked rice has shown to have anti-oxidative potential (Wang et al., 2012). Selenium present in *Flammulina velutipes* has anti-oxidant activity (Milovanovic et al., 2015). Based on the above research findings, it can be concluded that the anti-oxidant activity of *Flammulina velutipes* is attributed to its bioactive component(s) that include phenolics, rhamnose, polysaccharides, selenoproteins, vitamin C and nucleotide(s).

3.4 Anti-inflammatory Molecules in **Flammulina velutipes**

Flammulina velutipe has medicinal value attributed to the presence of immunomodulatory properties (Table 6). FIP is the major class of compounds that exhibit the immunomodulatory effects. FIP fvp is known to increase the expressions of IL 2, IL 4, IFN γ, TNF α, leukotriene(s) and IL 2R (Li et al., 2011). Further, it is also known to activate the lymphocyte proliferation by activation of MAPK pathway (Wang et al., 2004). Some other studies have reported that FIP fve inhibited the allergen-induced inflammation in mice (Chang et al., 2015; Chu et al., 2015). Chu and group have demonstrated that nasal application(s) of FIP fve reduced the dermatophagoide(s) microcera(s)-induced allergic response and decreased the inflammation by reducing the expression of cytokines in mice. Hence, FIP fve might be potentially utilised as a viral-based chemo-preventive. Hence, FIP fve can also be novel chemo-preventive drugs to counter the adverse effect of some drugs (Ou et al., 2015).

Other bioactive molecules isolated from *Flammulina velutipes* have shown to increase the nitric oxide production, TNF α and IL 1, in the macrophage(s) in a dose-dependent manner (Yin et al., 2010). Similarly, polysaccharides from *Flammulina velutipes* have known to increase the body weight, spleen and thymus growth in the mice model (Yan et al., 2014). The ethanolic extraction and hot water extracts of *Flammulina velutipes* have shown to display strong anti-complement activity.

Table 6. Anti-inflammatory and immunomodulatory properties of *Flammulina velutipes*.

Substance	Experimental Details	References
Fip-fve	*In vitro* cell proliferation, spleenocytes, cytokine signalling, *in vivo* food allergy in mouse model. Inhibited the cell cycle, calcium signalling and induced apoptosis in activated cells.	Hseih et al., 2003; Wang et al., 2004; Ou et al., 2009; Lee et al., 2013; Chu et al., 2015; Chang et al., 2014
Polysaccharides	Inhibited *in vitro* production of nitric oxide, TNF-α and IL-1 levels.	Yin et al., 2014
Water extracted polysaccharides	Inhibited *in vitro* production of nitric oxide, TNF-α and IL-1 levels.	Wu et al., 2014; Ryu et al., 2014
Hot water extracts of polysaccharides	Inhibited *in vitro* production of nitric oxide, TNF-α and IL-1 levels.	Gunawardera et al., 2014; Kang, 2012

3.5 Antimicrobial Activity in **Flammulina velutipes**

The antimicrobial properties of *Flammulina velutipes* have been observed since a long time. The chloroform extracts of fruiting bodies of *Flammulina velutipes* inhibit the pathogenic bacteria *Staphylococcus aureus* and *Bacillus subtilis* (Karaman et al., 2010). The methanolic extract of *Flammulina velutipes* is shown to have anti-bacterial effect against *B. subtilis*, *Bacillus pumilus*, *S. aureus* and *Pseudomonas aeruginosa* (Nedelkoska et al., 2013; Borhani et al., 2011). Some studies report that the extracts inhibit the pathogenic fungal adhesion, such as *Sporothrix schenckii* and *Candida albicans* towards the epithelial cells (Kashina et al., 2016).

Enokipodin(s), a α-cuparene type of sesquiterpenoid(s) isolated from *Flammulina velutipes* are reported to be major constituents for antimicrobial activities. Four types of enokipodins (A-D) with mild chemically different molecules are isolated from *Flammulina velutipes* (Ishikawa et al., 2000; 2001). The molecules showed anti-bacterial activities against the pathogenic bacteria (Ishikawa et al., 2000; Saito and Kuwahara, 2005). Simliarly, flamvelutpenoids A-D, isolated from *Flammulina velutipes* are shown to inhibit multi-drug-resistant pathogenic bacteria (Wang et al., 2012). Further, the same research group has isolated around six novel enokipodins E-J with anti-bacterial and antifungal properties.

3.6 Miscellaneous Activities

Other major biological activities reported from *Flammulina velutipes* include liver-damage protection and memory learning (Table 7) (Pang et al., 2007; Yang et al., 2011; 2015). The α-(1-4)-d glucan, polysaccharides (FVP-2) are reported to have hepato-protective properties in *Flammulina velutipes*. This study demonstrated that FVP-2 increased the growth of liver cells in mice in *in vitro* cultures (Pang et al., 2007). The mechanism of hepato-protection reported by this study was through inhibition of intracellular release of alanine amino-transferase (ALT) in the carbon tetrachloride-intoxicated hepatocytes. Further, the authors also showed reduction in lipid oxidation (Pang et al., 2007). The ergothioneine purified from *Flammulina velutipes* is known to reduce the activity of polyphenol-oxidase (PPO) that causes melanosis in shrimps and crabs by submerging them in the ergothioneine-rich mushroom extract (Encarnacion et al., 2010; 2011; 2012). This study demonstrated inhibition in the expression of PPO in hemocytes of crabs and shrimps with reduction in browning of crabs compared to untreated ones. Further studies showed that it can be the non-competitive inhibitors of PPO which interact with copper ions to reduce its activity. Hence, these extracts prevent the browning of fish meat myoglobin and oxy-myoglobin to oxidise to form met-myoglobin. The tyrosinase activity in melanin production was known to be inhibited by the ethanolic extracts of *Flammulina velutipes* grown in a glucose medium (Kim et al., 2011). These research findings demonstrate that *Flammulina velutipes* extracts can be effective a natural base for anti-melanosis, anti-decolourating and anti-browning agents that can be utilised in the food industry.

Ribosome-inactivating proteins (RIP) have known to possess various biological activities that include anticancer, immunomodulatory and antiviral activities (Ng et al., 1992). Velutin, a RIP identified from *Flammulina velutipes*, has been reported to have anti-HIV activities and inhibit glycol-hydrolase during HIV infections (Wang et al., 2001). Further, velutin is also reported to be non-teratogenic on the mouse embryo(s) (Ng et al., 2010). Flammin and velin, the other RIPs are also reported to be present in *Flammulina velutipes* but they do not exhibit any RNAs and proteolytic activities (Ng et al., 2004). Flammulina, similar to these RIPs, have also not shown any RNase activity (Wang et al., 2000).

Table 7. Influence on neurological behaviour by *Flammulina velutipes*.

Substance	Effects	Mode of Action	References
Polysaccharides	Memory and learning capacity	Restored neurotransmission in brain.	Yang et al., 2015
Water-extract of FVP	Memory and learning capacity	Enhanced the memory and learning in impaired rats.	Yang et al., 2011

4. Conclusion

Flammulina velutipes, an important edible mushroom is nutritionally rich in primary metabolites that include carbohydrates, proteins, essential amino-acids, unsaturated fatty acids, micronutrients, minerals and dietary fibre when compared to other plant and animal sources. Even though the nutritional and food values of *Flammulina velutipes* are well understood, it is the medicinal potential which has led to an increase in investigation in recent years. Several researchers have demonstrated that *Flammulina velutipes* can be used for medical disorders that include cancer, immune reaction and neurological defects. Among the macromolecules reported in *Flammulina velutipes*, polysaccharides seem to be more efficient. However, more research needs to be done for extraction and purification of these bioactive polysaccharides from *Flammulina velutipes* for commercial application. Deeper investigation on the FIP fve from *Flammulina velutipes* can be focused for fungal-based adjuvants and immunomodulatory activities. Other than these, *Flammulina velutipes* is documented for various medicinal properties that include anticancer, antioxidant, anti-atherosclerotic, anti-thrombotic, cholesterol reducing, anti-hypersensitivity, enhancing the memory and learning capacity, anti-complementation, anti-aging, antimicrobial, anti-inflammatory, inhibitor for melanosis, anti-browning and hepatoprotective capacity. These results suggest that structurally and functionally diverse micro- and micro-molecular structure can be identified for various treatment and functional food products from *Flammulina velutipes*. This part of the evidence-based discovery seems to be the next level of research required in *Flammulina velutipes*, especially for medicinal purposes. Hence, *Flammulina velutipes* seems to be a rich source of a natural product for development of various commercially important applications.

References

Ahn, D., Putt, D., Kresty, L., Stoner, G.D., Fromm, D. and Hollenberg, P.F. (1996). The effects of dietary ellagic acid on rat hepatic and esophageal mucosal cytochromes P450 and phase II enzymes. *Carcinogenesis*, 17: 821–828.

Akata, I., Ergonul, B. and Kalyoncu, F. (2012). Chemical compositions and antioxidant activities of 16 wild edible mushroom species grown in Anatolia. *Int. J. Pharm.*, 8: 134–138.

Aoki, H., Furuya, Y., Endo, Y. and Fujimoto, K. (2003). Effect of γ-aminobutyric acid-enriched tempeh-like fermented soybean (GABA-Tempeh) on the blood pressure of spontaneously hypertensive rats. *Biosci. Biotechnol. Biochem.*, 67: 1806–1808.

Badalyan, S.M. and Hambardzumyan, L.A. (2001). Investigation of immune–modulation activity of medicinal mushroom *Flammulina velutipes in vitro* cytokines induction by fruiting body extract. *Int. J. Med. Mush.*, 3(2-3): 110.

Bao, H.N.D., Shinomiya, Y., Ikeda, H. and Ohshima, T. (2009). Preventing discolouration and lipid oxidation in dark muscle of yellowtail by feeding an extract prepared from mushroom (*Flammulina velutipes*) cultured medium. *Aquaculture*, 295: 243–249.

Bao, H.N., Ochiai, Y. and Ohshima, T. (2010a). Antioxidative activities of hydrophilic extracts prepared from the fruiting body and spent culture medium of *Flammulina velutipes. Bioresour. Technol.*, 101: 6248–6255.

Bao, H.N., Osako, K. and Ohshima, T. (2010b). Value-added use of mushroom ergothioneine as a colour stabiliser in processed fish meats. *J. Sci. Food Agr.*, 90: 1634–1641.

Barros, L., Ferreira, M.-J., Queiros, B., Ferreira, I.C. and Baptista, P. (2007). Total phenols, ascorbic acid, -carotene and lycopene in Portuguese wild edible mushrooms and their antioxidant activities. *Food Chem.*, 103: 413–419.

Beluhan, S. and Ranogajec, A. (2011). Chemical composition and non-volatile components of Croatian wild edible mushrooms. *Food Chem.*, 124: 1076–1082.

Borhani, A., Badalyan, S.M., Garibyan, N.N., Mosazadeh, S.A. and Yasari, E. (2011). Flammulina velutipes (Curt.: Fr.) Singer: An edible mushroom in northern forest of Iran and its antagonistic activity against selected plant pathogenic fungi. *International Journal of Biology*, 3: 162–167.

Breene, W.M. (1990). Nutritional and medicinal value of specialty mushrooms. *J. Food Prot.*, 53: 883–899.

Chang, H.H., Hsieh, K.Y., Yeh, C.H., Tu, Y.P. and Sheu, F. (2010). Oral administration of an Enoki mushroom protein FVE activates innate and adaptive immunity and induces anti-tumor activity against murine hepatocellular carcinoma. *Int. Immunopharmacol.*, 10: 239–246.

Chang, Y.C., Chow, Y.H., Sun, H.L., Liu, Y.F., Lee, Y.T., Lue, K.H. et al. (2014). Alleviation of respiratory syncytial virus replication and inflammation by fungal immunomodulatory protein FIP-fve from *Flammulina velutipes. Antiviral Res.*, 110: 124–131.

Chang, Y.C., Hsiao, Y.M., Wu, M.F., Ou, C.C., Lin, Y.W., Lue, K.H. et al. (2013). Interruption of lung cancer cell migration and proliferation by fungal immunomodulatory protein FIP-fve from *Flammulina velutipes. J. Agric. Food Chem.*, 61: 12044–12052.

Chang, Y.C., Hsiao, Y.M., Hung, S.C., Chen, Y.W., Ou, C.C., Chang, W.T. et al. (2015). Alleviation of *Dermatophagoides microceras*-induced allergy by an immunomodulatory protein, FIP-fve, from *Flammulina velutipes* in mice. *Biosci. Biotechnol. Biochem.*, 79: 88–96.

Chen, S.-Y., Ho, K.-J., Hsieh, Y.-J., Wang, L.-T. and Mau, J.-L. (2012). Contents of lovastatin, γ-aminobutyric acid and ergothioneine in mushroom fruiting bodies and mycelia. *LWT Food Sci. Technol.*, 47: 274–278.

Chen, X.W.B., Sneed, K.B. and Zhou, S.-F. (2011). Pharmacokinetic profiles of anticancer herbal medicines in humans and the clinical implications. *Curr. Med. Chem.*, 18: 3190–3210.

Cheng, C.L., Wang, Z.Y., Cheng, L., Zhao, H.T., Yang, X., Liu, J.R. et al. (2012). *In vitro* antioxidant activities of water-soluble nucleotide-extract from edible fungi. *Food Sci. Technol. Res.*, 18: 405–412.

Chu, P.-Y., Sun, H.-L., Ko, J.-L., Ku, M.-S., Lin, L.-J., Lee, Y.-T. et al. (2015). Oral fungal immunomodulatory protein-Flammulina velutipes has influence on pulmonary inflammatory process and potential treatment for allergic airway disease: A mouse model. *J. Microbial. Immunol. Infect.* Doi: 10.1016/j.jmii.2015.07.013.

Dikeman, C.L., Bauer, L.L., Flickinger, E.A. and Fahey, G.C. Jr. (2005). Effects of stage of maturity and cooking on the chemical composition of select mushroom varieties. *J. Agric. Food Chem.*, 53: 1130–1138.

Ding, Y., Seow, S.V., Huang, C.H., Liew, L.M., Lim, Y.C., Kuo, I.C. et al. (2009). Coadministration of the fungal immunomodulatory protein FIP-Fve and a tumour-associated antigen enhanced anti-tumour immunity. *Immunology*, 128: e881–894.

Eisele, N., Linke, D., Bitzer, K., Na'amnieh, S., Nimtz, M. and Berger, R.G. (2011). The first characterised asparaginase from a Basidiomycete, *Flammulina velutipes. Bioresour. Technol.*, 102: 3316–3321.

Encarnacion, A.B., Fagutao, F., Hirono, I., Ushio, H. and Ohshima, T. (2010). Effects of ergothioneine from mushrooms (*Flammulina velutipes*) on melanosis and lipid oxidation of kuruma shrimp (*Marsupenaeus japonicus*). *J. Agric. Food Chem.*, 58: 2577–2585.

Encarnacion, A.B., Fagutao, F., Hirayama, J., Terayama, M., Hirono, I. and Ohshima, T. (2011a). Edible mushroom (*Flammulina velutipes*) extract inhibits melanosis in Kuruma shrimp (*Marsupenaeus japonicus*). *J. Food Sci.*, 76: C52–C58.

Encarnacion, A.B., Fagutao, F., Shozen, K.-I., Hirono, I. and Ohshima, T. (2011b). Biochemical intervention of ergothioneine-rich edible mushroom (*Flammulina velutipes*) extract inhibits melanosis in crab (*Chionoecetes japonicus*). *Food Chem.*, 127: 1594–599.

Encarnacion, A.B., Fagutao, F., Jintasataporn, O., Worawattanamateekul, W., Hirono, I. and Ohshima, T. (2012). Application of ergothioneine-rich extract from an edible mushroom *Flammulina velutipes* for melanosis prevention in shrimp, *Penaeus monodon* and *Litopenaeus vannamei*. *Food Res. Int.*, 45: 232–237.

Ferguson, L.R., Zhu, S.T. and Harris, P.J. (2005). Antioxidant and antigenotoxic effects of plant cell wall hydroxycinnamic acids in cultured HT-29 cells. *Mol. Nutr. Food Res.*, 49: 585–593.

Fu, H.Y., Shieh, D.E. and Ho, C.T. (2002). Antioxidant and free radical scavenging activities of edible mushrooms. *J. Food Lipis.* 9: 35–43.

Fukushima, M., Ohashi, T., Fujiwara, Y., Sonoyama, K. and Nakano, M. (2001). Cholesterol-lowering effects of maitake (*Grifola frondosa*) fibre, shiitake (*Lentinus edodes*) fibre and enokitake (*Flammulina velutipes*) fibre in rats. *Exp. Biol. Med.* (Maywood), 226: 758–765.

Gu, Y.H. and Leonard, J. (2006). *In vitro* effects on proliferation, apoptosis and colony inhibition in ER-dependent and ER-independent human breast cancer cells by selected mushroom species. *Oncol. Rep.*, 15: 417–423.

Gunawardena, D., Bennett, L., Shanmugam, K., King, K., Williams, R., Zabaras, D. et al. (2014). Anti-inflammatory effects of five commercially available mushroom species determined in lipopolysaccharide and interferon-gamma activated murine macrophages. *Food Chem.*, 148: 92–96.

Gupta, S., Lakshmi, A.J., Manjunath, M. and Prakash, J. (2005). Analysis of nutrient and antinutrient content of underutilised green leafy vegetables. *LWT Food Sci. Technol.*, 38: 339–345.

Harada, A., Nagai, T. and Yamamoto, M. (2011). Production of GABA-enriched powder by a brown variety of *Flammulina velutipes* (Enokitake) and its antihypertensive effects in spontaneously hypertensive rats. *J. Jpn. Soc. Food Sci.*, 58: 446–450.

Hayakawa, K., Kimura, M. and Kamata, K. (2002). Mechanism underlying γ-aminobutyric acid-induced antihypertensive effect in spontaneously hypertensive rats. *Eur. J. Pharmacol.*, 438: 107–113.

He, F.J., Nowson, C.A. and Macgregor, G.A. (2006). Fruit and vegetable consumption and stroke: Meta-analysis of cohort studies. *Lancet*, 367: 320–326.

Ikekawa, T., Maruyama, H., Miyano, T., Okura, A., Sawasaki, Y., Naito, K. et al. (1985). Proflamin, a new antitumor agent: Preparation, physicochemical properties and antitumour activity. *Jpn. J. Cancer Res.*, 76: 142–148.

Ingold, C. (1980). *Flammulina velutipes. Bull. Br. Mycol. Soc.*, 14: 112–118.

Inoue, K., Shirai, T., Ochiai, H., Kasao, M., Hayakawa, K., Kimura, M. et al. (2003). Blood-pressure-lowering effect of a novel fermented milk containing gamma-aminobutyric acid (GABA) in mild hypertensives. *Eur. J. Clin. Nutr.*, 57: 490–495.

Ishikawa, N.K., Fukushi, Y., Yamaji, K., Tahara, S. and Takahashi, K. (2001). Antimicrobial cuparene-type sesquiterpenes, Enokipodins C and D, from a mycelial culture of *Flammulina velutipes. J. Nat. Prod.*, 64: 932–934.

Ishikawa, N.K., Yamaji, K., Tahara, S., Fukushi, Y. and Takahashi, K. (2000). Highly oxidized cuparene-type sesquiterpenes from a mycelial culture of *Flammulina velutipes. Phytochemistry* 54: 777–782.

Ishikawa, N.K., Yamaji, K., Ishimoto, H., Miura, K., Fukushi, Y., Takahashi, K. et al. (2005). Production of enokipodins A, B, C, and D: A new group of antimicrobial metabolites from mycelial culture of *Flammulina velutipes. Mycoscience*, 46: 39–45.

Jeff, I.B., Yuan, X., Sun, L., Kassim, R.M., Foday, A.D. and Zhou, Y. (2013). Purification and *in vitro* anti-proliferative effect of novel neutral polysaccharides from *Lentinus edodes. Int. J. Biol. Macromol.*, 52: 99–106.

Jiang, S.M., Xiao, Z.M. and Xu, Z.H. (1999). Inhibitory activity of polysaccharide extracts from three kinds of edible fungi on proliferation of human hepatoma SMMC-7721 cell and mouse implanted S180 tumour. *World J. Gastroenterol.*, 5: 404–407.

Jing, P., Zhao, S.J., Lu, M.M., Cai, Z., Pang, J. and Song, L.H. (2014). Multiple-fingerprint analysis for investigating quality control of *Flammulina velutipes* fruiting body polysaccharides. *J. Agri. Food Chem.*, 62(50): 12128–12133.

John, J.H., Ziebland, S., Yudkin, P., Roe, L.S., Neil, H.A., Oxford, F. et al. (2002). Effects of fruit and vegetable consumption on plasma antioxidant concentrations and blood pressure: A randomised controlled trial. *Lancet*, 359: 1969–1974.

Kakkar, S. and Bais, S. (2014). A review on protocatechuic acid and its pharmacological potential. *ISRN Pharmacol.*, 2014: 952943.

Kalac, P. (2013). A review of chemical composition and nutritional value of wild-growing and cultivated mushrooms. *J. Sci. Food Agr.*, 93: 209–218.

Kang, H.-W. (2012). Antioxidant and anti-inflammatory effect of extracts from *Flammulina velutipes* (Curtis) Singer. *J. Korean Soc. Food Sci. Nutr.*, 41: 1072–1078.

Karaman, M., Jovin, E., Malbasa, R., Matavuly, M. and Popovic, M. (2010). Medicinal and edible lignicolous fungi as natural sources of anti-oxidative and antibacterial agents. *Phytotherapy Res.*, 24(10): 1473–1481.

Kashina, S., Flores Villavicencio, L.L., Balleza, M., Sabanero, G.B., Tsutsumi, V. and López, M.S. (2016). Extracts from *Flammulina velutipes* inhibit the adhesion of pathogenic fungi to epithelial cells. *Pharmacognosy Res.*, S56–60. Doi: 10.4103/0974-8490.178648.

Kilic, I., Yesiloglu, Y. and Bayrak, Y. (2014). Spectroscopic studies on the antioxidant activity of ellagic acid. *Spectrochim. Acta A. Mol. Biomol. Spectrosc.*, 130: 447–452.

Kim, J.M., Ra, K.S., Noh, D.O. and Suh, H.J. (2002). Optimisation of submerged culture conditions for the production of angiotensin converting enzyme inhibitor from *Flammulina velutipes*. *J. Ind. Microbiol. Biotechnol.*, 29: 292–295.

Kim, M.-Y., Chung, L.-M., Lee, S.-J., Ahn, J.-K., Kim, E.-H., Kim, M.-J., et al. (2009). Comparison of free amino acid, carbohydrates concentrations in Korean edible and medicinal mushrooms. *Food Chem.*, 113: 386–393.

Kim, M.Y., Seguin, P., Ahn, J.K., Kim, J.J., Chun, S.C., Kim, E.H., et al. (2008). Phenolic compound concentration and antioxidant activities of edible and medicinal mushrooms from Korea. *J. Agric. Food Chem.*, 56: 7265–7270.

Kim, S.Y., Kong, W.S. and Cho, J.Y. (2011). Identification of differentially expressed genes in *Flammulina velutipes* with anti-tyrosinase activity. *Curr. Microbiol.*, 62: 452–457.

Ko, W.C., Liu, W.C., Tsang, Y.T. and Hsieh, C.W. (2007). Kinetics of winter mushrooms (*Flammulina velutipes*) microstructure and quality changes during thermal processing. *J. Food Eng.*, 81(3): 587–598.

Kotake, T., Hirata, N., Degi, Y., Ishiguro, M., Kitazawa, K., Takata, R. et al. (2011). Endo-beta-1,3-galactanase from winter mushroom *Flammulina velutipes*. *J. Biol. Chem.*, 286: 27848–27854.

Leung, M.Y., Fung, K.P. and Choy, Y.M. (1997). The isolation and characterisation of an immunomodulatory and anti-tumour polysaccharide preparation from *Flammulina velutipes*. *Immunopharmacology*, 35: 255–263.

Li, Q., Huang, L., Wang, X., Li, X., Wu, S. and Zhou, X. (2011). Fungal immunomodulatory protein from *Flammulina velutipes* induces cytokine gene expression in mouse spleen cells. *Curr. Top. Nutroceutical Res.*, 9: 111–118.

Ma, Z., Cui, F., Gao, X., Zhang, J., Zheng, L. and Jia, L. (2015a). Purification, characterization, antioxidant activity and anti-aging of exopolysaccharides by *Flammulina velutipes* SF-06. *Antonie Van Leeuwenhoek*, 107: 73–82.

Ma, Z., Zhang, C., Gao, X., Cui, F., Zhang, J., Jia, M. et al. (2015b). Enzymatic and acidic degradation effect on intracellular polysaccharide of *Flammulina velutipes* SF-08. *Int. J. Biol. Macromol.*, 73: 236–244.

Mantovani, M.S., Bellini, M.F., Angeli, J.P.F. Oliveira, R.J., Silva, A.F. and Ribeiro, L.R. (2008). β-glucans in promoting health: Prevention against mutation and cancer. *Mutat. Res. Rev. Mutat.*, 658: 154–161.

Maruyama, H. and Ikekawa, T. (2007). Immunomodulation and anti-tumour activity of a mushroom product, Proflamin, isolated from *Flammulina velutipes* (W. Curt.: Fr.), Singer (Agaricomycetideae). *Int. J. Med. Mushrooms*, 9: 109–122.

Miles, P.G. and Chang, S.T. (2004). *Mushrooms: Cultivation, Nutritional Value, Medicinal Effect and Environmental Impact.* Second ed., CRC Press, London, UK, pp. 451.

Milovanovic, I., Stanojkovic, T., Stajic, M., Vukojevic, J. and Knezevic, A. (2015). Se effect on biological activity of Flammulina velutipes. *Ital. J. Food Sci.*, 27: 1–7.

Na, J.C., Jang, B.G., Kim, S.H., Kim, J.H., Kim, S.K., Kang, H.S., Lee, D.S., Lee, S.J., Cheong, J.C. and Lee, J.K. (2005). Influence of feeding *Flammuling veluipes* media on productivity and egg quality in laying hens. *Korean J. Poult. Sci.*, 32(2): 143–147.

Nedelkoska, D.N., Pancevska, N.A., Amedi, H., Veleska, D., Ivanova, E., Karadelev, M. et al. (2013). Screening of antibacterial and antifungal activities of selected Macedonian wild mushrooms. *Matica Srpska J. Nat. Sci.*, 124: 333–340.

Ng, T.B. (2000). Flammulin: A novel ribosomeinactivating protein from fruiting bodies of the winter mushroom *Flammulina velutipes. Biochem Cell Biol.*, 78: 699–702.

Ng, T.B. and Wang, H.X. (2004). Flammin and velin: New ribosome inactivating polypeptides from the mushroom *Flammulina velutipes. Peptides*, 25: 929–933.

Ng, T.B., Chan, W.Y. and Yeung, H.W. (1992). Proteins with abortifacient, ribosome inactivating, immunomodulatory, anti-tumour and anti-AIDS activities from *Cucurbitaceae* plants. *Gen. Pharmacol.*, 23: 579–590.

Ng, T.B., Lam, J.S., Wong, J.H., Lam, S.K., Ngai, P.H., Wang, H.X., et al. (2010). Differential abilities of the mushroom ribosome-inactivating proteins hypsin and velutin to perturb normal development of cultured mouse embryos. *Toxicol. in Vitro*, 24: 1250–1257.

Ng, T.B., Ngai, P.H. and Xia, L. (2006). An agglutinin with mitogenic and antiproliferative activities from the mushroom *Flammulina velutipes. Mycologia*, 98: 167–171.

Okamura, T., Ogata, T., Minamimoto, N., Takeno, T., Noda, H., Fukuda, S. et al. (2001). Characteristics of wine produces by mushroom fermentation. *Biosci. Biotechnol. Biochem.*, 65: 1596–1600.

Ou, C.C., Hsiao, Y.M., Wu, W.J., Tasy, G.J., Ko, J.L. and Lin, M.Y. (2009). FIP-fve stimulates interferon-gamma production via modulation of calcium release and PKC-alpha activation. *J. Agric. Food Chem.*, 57: 11008–11013.

Ou, C.-C., Hsiao, Y.-M., Hou, T.-Y., Wu, M.-F. and Ko, J.-L. (2015). Fungal immunomodulatory proteins alleviate docetaxel-induced adverse effects. *J. Funct. Foods*, 19: 451–463.

Ou, H.T., Shieh, C.J., Chen, J.Y. and Chang, H.M. (2005). The antiproliferative and differentiating effects of human leukemic U937 cells are mediated by cytokines from activated mononuclear cells by dietary mushrooms. *J. Agric. Food Chem.*, 53: 300–305.

Pang, X., Yao, W., Yang, X., Xie, C., Liu, D., Zhang, J. et al. (2007). Purification, characterization and biological activity on hepatocytes of a polysaccharide from *Flammulina velutipes* mycelium. *Carbohydr. Polym.*, 70: 291–297.

Park, S.E., Li, M.H., Kim, J.S., Sapkota, K., Kim, J.E., Choi, B.S. et al. (2007). Purification and characterization of a fibrinolytic protease from a culture supernatant of Flammulina velutipes mycelia. *Biosci. Biotechnol. Biochem.*, 71: 2214–2222.

Pereira, E., Barros, L., Martins, A. and Ferreira, I.C.F.R. (2012). Towards chemical and nutritional inventory of Portuguese wild edible mushrooms in different habitats. *Food Chem.*, 130: 394–403.

Rahman, M.A., Abdullah, N. and Aminudin, N. (2015). Antioxidative effects and inhibition of human low density lipoprotein oxidation *in vitro* of polyphenolic compounds in *Flammulina velutipes* (Golden Needle Mushroom). *Oxid. Med. Cell. Longev.*, 2015: 403023.

Raymond, J.A. and Janech, M.G. (2009). Ice-binding proteins from enoki and shiitake mushrooms. *Cryobiology*, 58: 151–156.

Reis, F.S., Barros, L., Martins, A. and Ferreira, I.C. (2012). Chemical composition and nutritional value of the most widely appreciated cultivated mushrooms: An inter-species comparative study. *Food Chem. Toxicol.*, 50: 191–197.

Ryu, H.-S., Kim, K.-O., Liu, Y., Yoon, L. and Kim, H.-S. (2014). Effects of edible mushrooms (Pleurotus ostreatus (Jacq.) P. Kumm., Pleurotus eryngil, Flammulina velutipes) extracts on immune cell activation in mice (830.17). *FASEB J.*, 28: 83.817.

Saito, M. and Kuwahara, S. (2005). Enantioselective total synthesis of enokipodins A-D, antimicrobial sesquiterpenes produced by the mushroom, *Flammulina velutipes. Biosci. Biotechnol. Biochem.*, 69: 374–381.

Seeram, N.P., Adams, L.S., Henning, S.M., Niu, Y., Zhang, Y., Nair, M.G. et al. (2005). *In vitro* antiproliferative, apoptotic and antioxidant activities of punicalagin, ellagic acid and a total pomegranate tannin extract are enhanced in combination with other polyphenols as found in pomegranate juice. *J. Nutr. Biochem.*, 16: 360–367.

Singh, G., Kawatra, A. and Sehgal, S. (2001). Nutritional composition of selected green leafy vegetables, herbs and carrots. *Plant Foods Hum. Nutr.*, 56: 359–364.

Smiderle, F.R., Carbonero, E.R., Mellinger, C.G., Sassaki, G.L., Gorin, P.A. and Iacomini, M. (2006). Structural characterisation of a polysaccharide and a beta-glucan isolated from the edible mushroom *Flammulina velutipes. Phytochemistry*, 67: 2189–2196.

Smiderle, F.R., Carbonero, E.R., Sassaki, G.L., Gorin, P.A.J. and Iacomini, M. (2008). Characterisation of a heterogalactan: Some nutritional values of the edible mushroom *Flammulina velutipes*. *Food Chem.*, 108: 329–333.

Stanely Mainzen Prince, P. and Roy, A.J. (2013). p-Coumaric acid attenuates apoptosis in isoproterenol-induced myocardial infarcted rats by inhibiting oxidative stress. *Int. J. Cardiol.*, 168: 3259–3266.

Tadjibaeva, G., Sabirov, R. and Tomita, T. (2000). Flammutoxin, a cytolysin from the edible mushroom *Flammulina velutipes*, forms two different types of voltage-gated channels in lipid bilayer membranes. *Biochem. Biophys. Acta*, 1467: 431–443.

Tan, H.-L., Chan, K.-G., Pusparajah, P., Saokaew, S., Duangjai, A., Lee, L.-H., et al. (2016b). Anticancer properties of the naturally occurring aphrodisiacs: Icariin and its derivatives. *Front. Pharmacol.*, 7: 191.

Tang, C., Hoo, P.C., Tan, L.T., Pushparajah, P., Khan, T.M., Lee, L., Goh, B. and Chan, K. (2016). Golden needle mushroom: A culinary medicine with evidence-based biological activities and health promoting properties. *Frontiers in Pharmacology*, 474(7): 1–27.

Wang, H. and Ng, T.B. (2001). Isolation and characterisation of velutin, a novel low-molecular-weight ribosome-inactivating protein from winter mushroom (*Flammulina velutipes*) fruiting bodies. *Life Sci.*, 68: 2151–2158.

Wang, H., Gao, J. and Ng, T.B. (2000). A new lectin with highly potent antihepatoma and antisarcoma activities from the oyster mushroom *Pleurotus ostreatus. Biochem. Biophys. Res. Commun.*, 275: 810–816.

Wang, H.M., Wang, L.W., Liu, X.M., Li, C.L., Xu, S.P. and Farooq, A.D. (2013). Neuroprotective effects of forsythiaside on learning and memory deficits in senescence-accelerated mouse prone (SAMP8) mice. *Pharmacol. Biochem. Behav.*, 105: 134–141.

Wang, H.X. and Ng, T.B. (2000). Flammulin: A novel ribosome-inactivating protein from fruiting bodies of the winter mushroom *Flammulina velutipes. Biochem. Cell Biol.*, 78: 699–702.

Wang, H., Khor, T.O., Shu, L., Su, Z., Fuentes, F., Lee, J.-H. et al. (2012). Plants against cancer: A review on natural phytochemicals in preventing and treating cancers and their drugability. *Anticancer Agents Med. Chem.*, 12: 1281.

Wang, P.-H., Hsu, C.-I., Tang, S.-C., Huang, Y.-L., Lin, J.-Y. and Ko, J.-L. (2004). Fungal immunomodulatory protein from *Flammulina velutipes* induces interferon-γ production through p38 mitogen-activated protein kinase signalling pathway. *J. Agric. Food Chem.*, 52: 2721–2725.

Wang, S.M., Ooi, L.L.P. and Hui, K.M. (2011). Upregulation of Rac GTPase-activating protein 1 is significantly associated with the early recurrence of human hepatocellular carcinoma. *Clin. Cancer Res.*, 17: 6040–6051.

Wang, Y., Bao, L., Liu, D., Yang, X., Li, S., Gao, H. et al. (2012). Two new sesquiterpenes and six norsesquiterpenes from the solid culture of the edible mushroom *Flammulina velutipes. Tetrahedron*, 68: 3012–3018.

Wang, Y., Bao, L., Yang, X., Li, L., Li, S., Gao, H. et al. (2012d). Bioactive sesquiterpenoids from the solid culture of the edible mushroom *Flammulina velutipes* growing on cooked rice. *Food Chem.*, 132: 1346–1353.

Wang, Y., Bao, L., Yang, X., Li, L., Li, S., Gao, H., Yao, X.S., Wen, H. and Liu, H.W. (2012). Bioactive sesquiterpenoids from the solid culture of the edible mushroom *Flammulina velutipes* growing on cooked rice. *Food Chem.*, 132(3): 1346–1353.

Wang, Y.-Q., Bao, L., Yang, X.-L., Dai, H.-Q., Guo, H., Yao, X.-S., et al. (2012b). Four new cuparene-type sesquiterpenes from *Flammulina velutipes. Helvetica Chimica Acta*, 95: 261–267.

Wu, D.M., Duan, W.Q., Liu, Y. and Cen, Y. (2010). Anti-inflammatory effect of the polysaccharides of golden needle mushroom in burned rats. *Int. J. Biol. Macromol.*, 46(1): 100–103.

Wu, M., Luo, X., Xu, X., Wei, W., Yu, M. and Jiang, N. (2014). Antioxidant and immune modulatory activities of a polysaccharide from *Flammulina velutipes. J. Tradit. Chin. Med.*, 34(6): 733–740.

Xia, Z. (2015). Preparation of the oligosaccharides derived from Flammulina velutipes and their antioxidant activities. Carbohydr. Polym. 118: 41–43.

Xu, Z.-Y., Wu, Z.-A. and Bi, K.-S. (2013). A novel norsesquiterpene alkaloid from the mushroom-forming fungus *Flammulina velutipes. Chin. Chem. Lett.*, 24: 57–58.

Yan, Z.F., Liu, N.X., Mao, X.X., Li, Y. and Li, C.T. (2014). Activation effects of polysaccharides of *Flammulina velutipes* mycorrhizae on the T lymphocyte immune function. *J. Immunol. Res.*, 2014: 285421.

Yang, B.K., Park, J.B. and Song, C.H. (2002). Hypolipidemic effect of exopolymer produced in submerged mycelial culture of five different mushrooms. *J. Microbiol. Biotechnol.*, 12: 957–961.

Yang, B., Zhao, M., Shi, J., Yang, N. and Jiang, Y. (2008). Effect of ultrasonic treatment on the recovery and DPPH radical scavenging activity of polysaccharides from longan fruit pericarp. *Food Chem.*, 106: 685–690.

Yang, J.-H., Lin, H.-C. and Mau, J.-L. (2001). Non-volatile taste components of several commercial mushrooms. *Food Chem.*, 72: 465–471.

Yang, W., Fang, Y., Liang, J. and Hu, Q. (2011). Optimisation of ultrasonic extraction of *Flammulina velutipes* polysaccharides and evaluation of its acetylcholinesterase inhibitory activity. *Food Res. Int.*, 44: 1269–1275.

Yang, W., Pei, F., Shi, Y., Zhao, L., Fang, Y. and Hu, Q. (2012). Purification, characterization and anti-proliferation activity of polysaccharides from *Flammulina velutipes. Carbohydr. Polym.*, 88: 474–480.

Yang, W., Yu, J., Pei, F., Mariga, A.M., Ma, N., Fang, Y. and Hu, Q. (2016). Effect of hot air drying on volatile compounds of *Flammulina velutipes* detected by HS-SPME–GC–MS and electronic nose. *Food Chem.*, 196(1): 860–866.

Yang, W., Yu, J., Zhao, L., Ma, N., Fang, Y., Pei, F. et al. (2015). Polysaccharides from *Flammulina velutipes* improve scopolamine-induced impairment of learning and memory of rats. *J. Funct. Foods*, 18: 411–422.

Yang, W.J., Fang, Y. and Liang, J. (2011). Optimisation of ultrasonic extraction of *Flammulina velutipes* polysaccharides and evaluation of its acetylcholinesterase inhibitory activity. *Food Res. Int.*, 44(5): 1269–1275.

Yeh, M.Y., Ko, W.C. and Lin, L.Y. (2014). Hypolipidemic and antioxidant activity of enoki mushrooms (*Flammulina velutipes*). *Biomed. Res. Int.*, 2014: 352385.

Yi, C., Fu, M., Cao, X., Tong, S., Zheng, Q., Firempong, C.K. et al. (2013a). Enhanced oral bioavailability and tissue distribution of a new potential anticancer agent, *Flammulina velutipes* sterols, through liposomal encapsulation. *J. Agric. Food Chem.*, 61: 5961–5971.

Yi, C., Sun, C., Tong, S., Cao, X., Feng, Y., Firempong, C.K. et al. (2013b). Cytotoxic effect of novel *Flammulina velutipes* sterols and its oral bioavailability via mixed micellar nanoformulation. *Int. J. Pharm.*, 448: 44–50.

Yi, C., Zhong, H., Tong, S., Cao, X., Firempong, C.K., Liu, H. et al. (2012). Enhanced oral bioavailability of a sterol-loaded microemulsion formulation of *Flammulina velutipes*, a potential antitumor drug. *Int. J. Nanomed.*, 7: 5067–5078.

Yin, H., Wang, Y., Wang, Y., Chen, T., Tang, H. and Wang, M. (2010). Purification, characterization and immuno-modulating properties of polysaccharides isolated from *Flammulina velutipes* mycelium. *Am. J. Chin. Med.*, 38: 191–204.

Zeng, X., Suwandi, J., Fuller, J., Doronila, A. and Ng, K. (2012). Antioxidant capacity and mineral contents of edible wild Australian mushrooms. *Food Sci. Technol. Int.*, 18: 367–379.

Zhang, M., Cui, S., Cheung, P. and Wang, Q. (2007). Antitumour polysaccharides from mushrooms: A review on their isolation process, structural characteristics and antitumour activity. *Trends Food Sci. Tech.*, 18: 4–19.

Zhang, R., Hu, D., Zhang, Z., Zuo, X., Jiang, R., Wang, H. et al. (2010). Development and characterisation of simple sequence repeat (SSR) markers for the mushroom *Flammulina velutipes. J. Biosci. Bioeng.*, 110: 273–275.

Zhang, Z., Jin, Q., Lv, G., Fan, L., Pan, H. and Fan, L. (2013). Comparative study on antioxidant activity of four varieties of *Flammulina velutipes* with different colour. *Int. J. Food Sci. Technol.*, 48: 1057–1064.

Zhao, C., Zhao, K., Liu, X.Y., Huang, Y.F. and Liu, B. (2013). *In vitro* antioxidant and antitumor activities of polysaccharides extracted from the mycelia of liquid-cultured *Flammulina velutipes. Food Sci. Technol. Res.*, 19: 661–667.

CHAPTER 11

Reishi Mushroom
(*Ganoderma lucidum*)

Lepakshi Md. Bhakshu,[1,*] *KV Ratnam,*[2] *Pulala Raghuveer Yadav,*[3]
C Meera Saheb,[1] *Anu Pandita*[4] *and Deepu Pandita*[5,*]

1. Introduction

The genus *Ganoderma* is represented by about 80 species of the Basidiomycotous (Ganodermataceae) macro-fungi and most of them are habituated in tropical parts of the globe (cosmopolitan distribution) and which were popularly used in the traditional system of medicine in Asia (Kirk et al., 2008). They are commonly known as shelf mushrooms of bracket fungi (Fig. 1). The important species of *Ganoderma* mentioned in *Pharmacopeia of Republic China* or *National Formulary*, are *G. applanatum*, *G. artum*, *G. formosanum*, *G. lucidum*, *G. sinensis*, *G. tsugae* (Gao et al., 2004a, b). *G. lucidum*, an ancient Chinese rejuvenating medicine, 'Fuzheng Guben', is referred as Lingzhi, marvellous herb preferably used in medicine in China, Japan, Korea and America (Yue et al., 2006; Meng et al., 2011) in addition to longevity (Noguchi et al., 2005, Sissi et al., 2011). It is considered as good in Channel tropism which boosts the health of essential organs (Yue et al., 2006).

It grows as a wood-decomposing fungus on trees with industrial application in bio-pulping process and bioremediation (Matos et al., 2007; FBRI, 2008; Cao et al., 2018). Its polysaccharide fraction has been found effective due to its anti-neurodegenerative, antioxidant and immunomodulation properties (Seweryn et al., 2021; Ji et al., 2007).

The polysaccharides, referred as Lz-8, are investigated widely and prove effective in immunomodulation and cancer treatment. Cao et al. (2018) demonstrated that the extract activates lymphocytes (T or B), NK cells, macrophages and various immune

[1] Department of Botany, PVKN. Government College (A), Chittoor-517 002, Andhra Pradesh, India.
[2] Department of Botany, Rayalaseema University, Kurnool-518 007, Andhra Pradesh, India.
[3] Department of Biotechnology, Indian Institute of Technology Hyderabad, Kandi-502 285, Telangana, India.
[4] Vatsalya Clinic, Krishna Nagar, New Delhi-110051, India.
[5] Government Department of School Education, Jammu, Jammu and Kashmir 180001, India.
* Corresponding authors: lmbakshu@gmail.com, deepupandita@gmail.com

Fig. 1. https://en.wikipedia.org/wiki/*Ganoderma_lucidum*.

cells for secretion of cytokines which are involved in control of tumour cells and have meagre toxicity on normal cells. It is also being practiced for domestic cultivation due to its medicinal properties and use as dietary supplements and as herbal tea which is effective in management of diabetes and hepato-protection (Sissi et al., 2011; Gao et al., 2004b; Hyun et al., 1990).

2. Cultivation

It often grows fully on sterilised wood or straw of paddy at 30–32°C optimum temperature in a duration of 25–30 days. For cultivation, it requires an optimum temperature of 30–32°C, good ventilation and 80–95% wetness. Fruit bodies are often harvested in two to three flushes once during the cultivation period of 120–150 days. The mushrooms have 25–30% biological potency. Due to its woody nature, the mushroom can be preserved in dried powdered form. Utmost care should be taken through the disposal of Reishi mushroom, because of its plant pathogenicity. The spent substrate is also burnt off to avoid its spread among the trees.

3. Phytochemicals of *Ganoderma*

G. lucidum is reported for its diverse secondary metabolites, such as triterpenoids, steroids, alkaloids, nucleotides, fatty acids and lactones besides polysaccharides (Sanodiya et al., 2009). These mushrooms consist of 90% of water and the remaining are macromolecules ranging between 10–40%, fats at 3–28% and others consisting of fibrous material, crude proteins and ash with vitamins, minerals, metallic components, inorganic substances, essential fatty acids, amino acids (Borchers et al., 1999; Mau et al., 2001) in addition to peptedo-glycans (Boh et al., 2007; Chang et al., 2007; Zhou et al., 2007) which are reported for varied beneficial biological activities.

3.1 Polysaccharides and Peptidoglycans

The high molecular-weight sugars and biologically active 'polyglycans' are reported from the mushrooms which show a wide range of physiochemical activities (Zhou et al., 2007). A plethora of polysaccharides were reported from all parts of *G. lucidum* in fermenters along with the amides (Ganoderan A-C). Its polysaccharides (GLPs) are found effective widely in anti-inflammatory, anti-mutagenic, antitumorogenic, hype-glycaemic, anti-ulcer along with the immunomodulatory effect (Miyazaki and Nishijima, 1981; Hikino et al., 1985; Tomoda et al., 1986; Bao et al., 2001; Wachtel-Galor et al., 2004a, b). The aldo-hexose is reported as a predominant component in the polysaccharides (Bao et al., 2001; Wang et al., 2002) besides the heteropolymers, such as mannose, galactose, saccharides, fructose with β and α-D (or L)-substitutions 1–4, 1–3, 1–6 conformational structures (Lee et al., 1999; Bao et al., 2002) which reported anti-tumerogenic effects (Bao et al., 2001; Zhang et al., 2001). The polysaccharide framework is structurally heavy and might be responsible for the hardness of *G. lucidum* (Upton, 2000). A broad array of sugar products from Lingzhi were marketed which were found beneficial in human illness, liver problems and cancers (Gao et al., 2005). The peptidoglycans have antiviral property (Li et al., 2005a), are immnomodulating (Wu and Wang, 2009) and have besides amides, a fucose-glycoprotein (Chien et al., 2004).

3.2 Triterpenes

In *G. lucidum*, the triterpenes chemical structure depend on the lanostane skeleton, especially lanosterol, which is synthesised through 'squalene cyclisation' (Haralampidis et al., 2002). The triterpenes obtained from *G. lucidum* are the Ganoderic acids A and B (Kubota et al., 1982) and several terpene-related compounds reported as new to science. The majority of biomolecules are ganoderic and lucidenic acids; however, various triterpenes products, like ganoderals, ganoderiols and ganodermic acids were reported in due course (Fig. 2) (Nishitoba et al., 1984; Sato et al., 1986; Budavari, 1989; Gonzalez et al., 1999; Ma et al., 2002; Akihisa et al., 2007; Zhou et al., 2007; Jiang et al., 2008; Chen et al., 2010). The purification methods for the extracts are standardised through traditional separation techniques as well as reverse-phase HPLC (Chen et al., 1999; 2008; Su et al., 2001). These classes of secondary metabolites offer a bitter taste and are found effective in biological properties. Some of the compounds were used as standard for the species marker compounds to confirm their identification (Chen et al., 1999; Liang et al., 2019; Su et al., 2001).

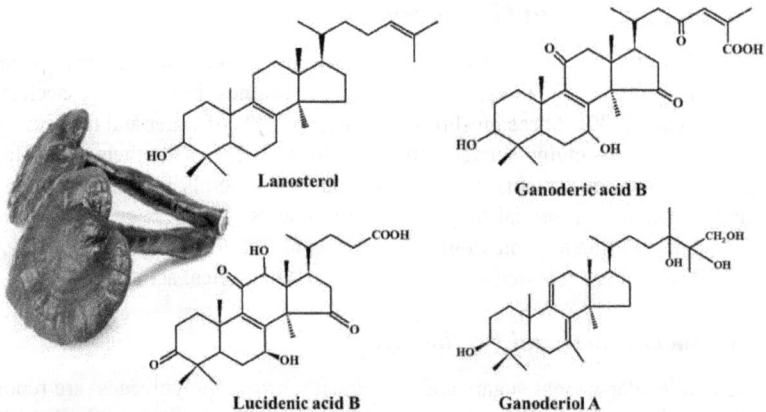

Fig. 2. Chemical structures of lanosterol and tree of many triterpenes isolated from *G. lucidum.*

3.3 Mineral Elements

Elemental analysis provided the information about the occurrence of magnesium, silica, sulphur, potassium, calcium and phosphorus to be the main mineral elements (Fig. 3). Iron, sodium, zinc, copper, manganese and strontium showed presence in

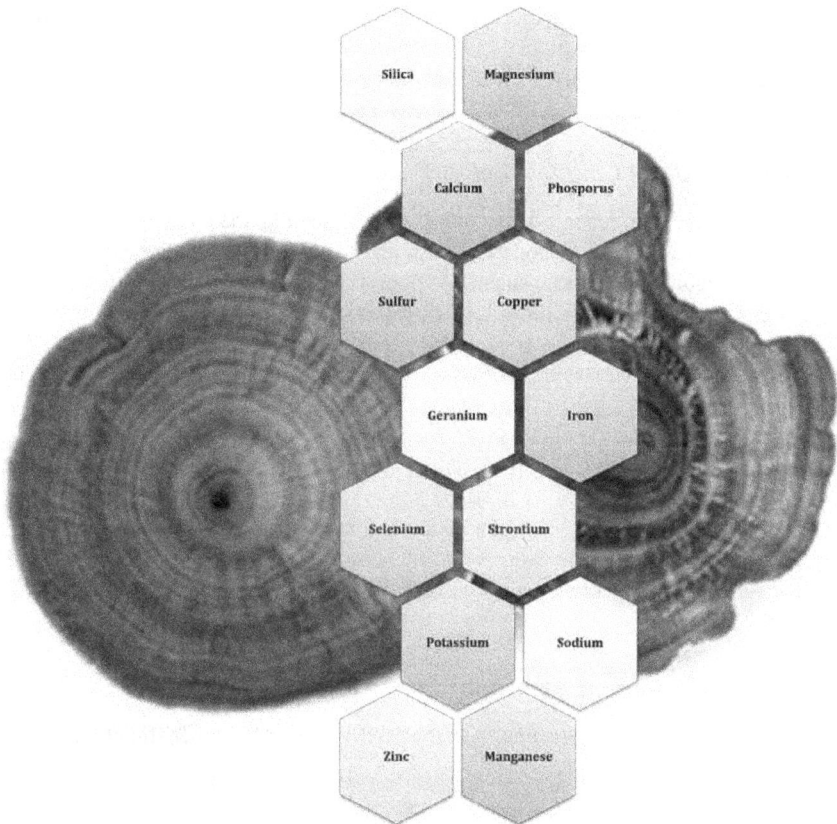

Fig. 3. Occurrence of mineral components in *G. lucidum*.

the fruit bodies of *G. lucidum* cultivated on the logs. Besides the heavy metals such as lead, cadmium and mercury were detected in lower amounts (Chen et al., 1998). Freeze-dried fruit bodies collected from the wild showed the mineral content (10.2%) with calcium, potassium and magnesium as the major mineral nutrients (Chiu et al., 2000). Instead, there are no reports on presence of cadmium or mercury though it possessed selenium up to 72 μg/g dry weight (Falandysz, 2008) and may bio-transform 20–30% into selenium-containing proteins (Du et al., 2008). Notably, a rare element with geranium (489 μg/g) has been reported from the fruiting body of wild origin (Chiu et al., 2000), while its presence appreciated in its potential as a medicine by being anti-tumour, anti-mutagenic, immune boosting (Mino et al., 1980; Kolesnikova et al., 1997). However, still certain proof is required for linking this component with the precise health edges related to this mushroom.

Besides the above components, the Lingzhi, possesses special types of metabolites, such as lectins and proteins from the dried fruiting bodies (Chang and Buswell, 1996; Mau et al., 2001). Bioactive proteins, including LZ-8, contribute to the medicinal effects (Van Der Hem et al., 1995)—a peptide fraction (GLP) demonstrated liver-protective effects (Sun et al., 2008) and a 15-kDa antifungal metabolite, ganodermin, obtained from the fruiting bodies (Wang and Ng, 2006). Certain carbohydrate-related metabolites, including lectins (114-KDa hexameric glycoprotein), reported the haemagglutinating property on human erythrocytes (Mau et al., 2001). Certain metabolites (sterol, nucleosides, nucleotides) and proteins were reported as specific inhibitory substances for α-glucosidase, SKG-3 from its fruiting structures (Wasser, 2005; Paterson, 2006; Kim and Nho, 2004; Fukuzawa et al., 2008).

3.4 Carbohydrate Composition

Diversified sugar components were characterised from all parts of Lingzhi, especially the polysaccharides (e.g., Ganoderans A, B and C or GL-PSs) and were reported to exert varied biological effects, such as hypoglycaemic, anti-inflammatory, antiulcer, immune-stimulating and antitumorigenic properties (Miyazaki and Nishijima, 1981; Hikino et al., 1985; Tomoda et al., 1986; Bao et al., 2001; Wachtel-Galor et al., 2004 a; b). Glucose has been reported as a major sugar component as indicated by structural analyses of GL-PSs (Bao et al., 2001; Wang et al., 2002). Besides, they are hetero-polymers and also possess mannose, xylose, galactose and fucose in numerous conformations, as well as 1–3, 1–4, and 1–6-linked β and α-D (or L)-substitutions (Lee et al., 1999; Bao et al., 2002) and attributed to its biological activities (Bao et al., 2001; Zhang et al., 2001). In addition, chitin is considered as the matrix for its hardness (Upton, 2000). The various formulations of *G. lucidum* were possessed with combinations of polysaccharides which may be used in chronic human ailments (Gao et al., 2005).

4. Pharmacological Applications

Various components of *G. lucidum* are reported for diversified beneficial biological effects (Fig. 4).

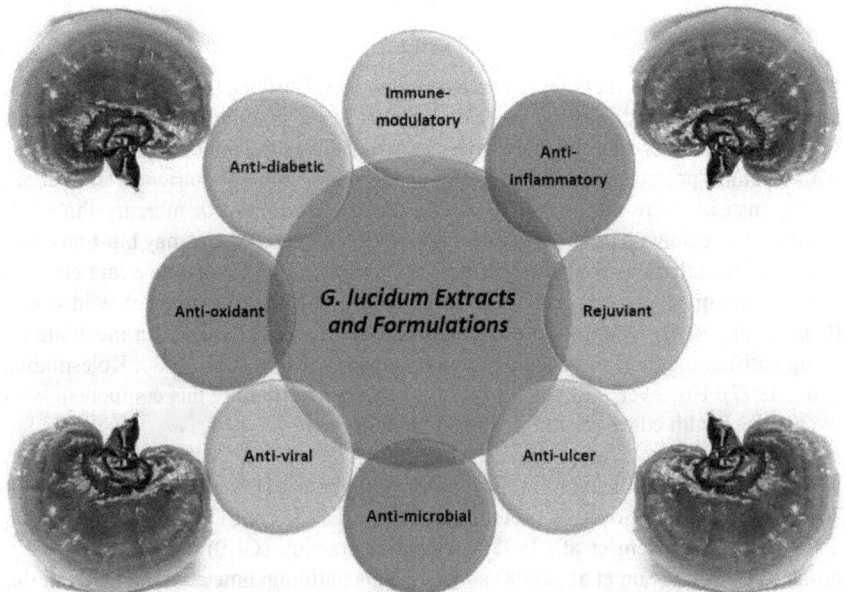

Fig. 4. Diversified biological effects of *G. lucidum*.

4.1 Antioxidant Activity

The extracts, polysaccharides and proteins of *G. lucidum* exhibited significant anti-oxidative properties which were thoroughly studied and reported. The antioxidant effect of its polysaccharide fraction bound to the protein (PBP) exhibited effective protection in the *in vivo* cancer-induced animal model (Ooi and Liu, 2000) besides the protective effect on RNA against fenton solution, UV irradiation and hydroxyl radical (Lee et al., 2001). The extract also reported the protective role against the oxide-induced injury to the DNA in the Raji cells (Shi et al., 2002). Hot aqueous extracts considerably provided protection to Raji cells from oxide (HO)-induced desoxyribonucleic acid injury (Shi et al., 2002).

Varied parts of *G. lucidum* and the extracts consisting triterpenoids and polysaccharides show antioxidant activity in *in vitro* systems (Mau et al., 2002; Shi et al., 2002; Gao et al., 2002a, b; Wachtel-Galor et al., 2005; Yuen and Gohel, 2005; 2008; Saltarelli et al., 2009; Wu and Wang, 2009). The extracts were found effective and showed antioxidant activity in human plasma during its consumption (Wachtel-Galor et al., 2004 a, b) and provided effective protection of DNA in human lymphocytes (Wachtel-Galor et al., 2005). The extracts oppositely affected the antioxidant status of HUC-PC cells (pre-malignant cells) in neoplastic diseases (Yuen and Gohel, 2005; 2008); in addition the extracts also showed bladder chemo-protection and injury of urinary organ in the cisplatin-induced animals and restored nephritic defence system (Sheena et al., 2003). Yue et al. (2006) showed that the triterpene fractions elicited protection in Hela cells against the doxorubicin-induced oxidative stress and DNA protection, apoptosis, as an effective scavenger of reactive oxygen species (ROS). Jia et al. (2009) reported stabilised antioxidants and lipid peroxidation in the diabetic animal system.

The ethyl alcohol extract of *G. lucidum* of 50 or 250 mg/kg (BW) was evaluated for its antioxidant effect on young and aged mice and determined the SOD, CAT activities and GSH, macromolecule peroxides (malondialdehydes) within the liver mitochondria in the tested mice groups apart from additional free radicals, like ABTS, DPPH, hydroxyl, FRAP, anion scavenging properties. The extract showed important useful effects within aged mice as compared to young mice selected as controls (Cherian et al., 2009). However, also found was a proportionate relation between the antioxidant effects, immunomodulatory and antineoplastic effects (Cherian et al., 2009). Inhibition effects of wild and domesticated *G. lucidum* samples were confirmed on different *in vitro* radicals, like DPPH, ABTS and chemical group radicals and ferric reducing antioxidants. The polyphenol contents were quantified to assess the effects. The methanol extracts showed effective scavenging property on the tested free radical species (Borchers et al., 2008; Mohsin et al., 2011; Rani et al., 2015).

4.2 Antiviral Potential

In view of the development of new antibiotics, the synthetic agents might possess the unsought severe effects than natural products from the medicinal plants with fungi being a good source for anti-infectious medicaments (Wasser and Weis, 1999; Zhong and Xiao, 2009) Kim et al. (2000) and *G. lucidum* is one among the natural antibiotic sources. Eo et al. (1999) and Oh et al. (2000) studied the effects of various fractions of *G. lucidum* on the virus strains, such as Herpes simplex viruses (HSV-1 & 2), New

Jersey strain of vesicular stomatitis virus (VSV) and reported significant inhibition in viral propagation in the *in vitro* studies. In addition the \tested samples exhibited non-toxic effects on the human cell lines. Further, the studies also demonstrated that PBP (protein bound polysaccharides) from *G. lucidum* exhibited synergetic effect when treated along with vidarabine and IFN-α during herpes treatment (Oh et al., 2000). The effects were further supported by Liu et al. (2004) and Li et al. (2005b) on herpes virus with protein-glucoside fraction. These studies confirmed that pre-treatment has been effective through interference in the events of viral replication (Li et al., 2005b). The triterpenes of *G. lucidum*, such as Ganoderiol F, *Ganoderma nontriol* demonstrated effective inhibition on anti-HIV replication (7.8 μg/ml) whereas Ganoderiol B, Ganoderic acid B, Ganoderic acid-α, B, C1 and H,3β-5α-dihydroxy-6β-methoxyergosta-7,22-diene were moderately significant in the inhibition of protease and reverse transcriptase of HIV (El-Mekkawy et al., 1998; Min et al., 1998).

The investigations of Li and Wang (2006) revealed that, ganoderic acid of *G. lucidum* effected significantly on the replication of hepatitis B virus (HBV) in HepG2215cells (HepG2-HBV-producing cell line) over eight days. The antigen production of HBV (HBsAg and HBeAg) was severally inhibited. Oral administration to elderly people affected by varicella zoster virus with the dried extract (36 and 72 g/day) of *G. lucidum* significantly reduced the post-herpetic pain and promoted healing of lesions; besides, the extracts did not show any toxic properties (Hijikata and Yamada, 1998). Similar was the case study affected with herpes labialis (a venereal disease) studied in human patients (Hijikata et al., 2007).

4.3 Antibacterial Role

The extract, when administered intra-peritonially (2 mg/mouse), enhanced the protection against the lethal *Escherichia coli* strain in ICR mice (Ohno et al., 1998). The chloroform extract found effective on the bacterial strains, such as *Bacillus subtilis, Staphylococcus aureus, Enterococcus faecalis* and *E. coli* as well as *P. aeruginosa* (Keypour et al., 2008) was also effective on the growth of *S. aureus* and *B. subtilis*. Besides, a study conducted by Yoon et al. (1994) found it effective on *E. coli, B. cereus, Micrococcus luteus, S. aureus, Proteus vulgaris* and typhoid bacillus, among the tested microorganisms.

4.4 Diabetes Mellitus

The intra-peritoneal application of Ganoderna (A and B) obtained from the Lingzhi fruiting bodies exhibited hypoglycemic effect in alloxan-affected diabetic mice and decrease in the glucose levels in blood plasma and the effect was persistent even after a day (Hikino et al., 1985). Tomodo (1986) and Hikino et al. (1989) reported that Ganoderan B has an hypoglycaemic phyto-component which has modulated carbohydrate metabolising enzymes and had the reversal effect in insulin levels in glucose-fed mice, revealing its antidiabetic potential; also Ganoderan C was proved as an hypoglycaemic agent.

Oral feeding of *G. lucidum* extract of 0.03 and 0.3 g/kg BW up to four weeks to the obese/diabetic (+db/+db) mice significantly lowered the aldohexose level (Seto et al., 2009) in addition to hypoglycaemic effect. The extracts wre found to reduce the content of hepatic PEPCK (phospho-enol pyruvic carboxy kinase) and proved its genes as a down-expression condition besides β-glucosidase inhibition (Kim et al., 1999). The polysaccharide-rich extracts, when treated to the streptozotocin-affected rats for thirty days, enhanced the glycemic profile, attenuated the aldose release in addition to normalidation of peroxidation, hydroperoxides and enzymatic or non-enzymatic antioxidant levels. It was noticed that the extracts were helpful in oxidative stress (Jia et al., 2009).

The studies on diabetic human patients were investigated for their glycemic profile after supplementing with Ganopoly, a polysaccharide-rich fraction, at 1800 mg thrice a day for 12 weeks. The result showed a significant betterment in diabetic conditions along with reduced levels of oxidation of haemoglobin and plasma aldohexose levels (Gao et al., 2004b). It was also recommended as an important nutrient for the diabetics under the purview of a physician since these aspects might need to be confirmed.

4.5 Gastrointestinal and Hepato-protective Effects

Hepato-protective efficiency of ether fraction (Ganodernic acid A) of *G. lucidum* was assessed against CCl_4-induced liver injury in animal models. Ganodernic acid A inhibits β-glucorinidase and this might be the way for hepato-protective activity. Liver markers, such as LDH, ALT and AST, were measured to find the liver injury (Kim et al., 1999; Lin et al., 1995). *G. lucidum* hot-water extract was administered to a mouse with free radicals induced by ethanol. The malondialdehyde level was high in kidney and liver homogenates of ethanol-administered group as compared to the treated group. The hot-water extract showed concentration-dependent antioxidant effect on kidney and liver lipid peroxidation and may be due to superoxide scavenging effect (Shieh et al., 2001).

In Chinese medicine, *G. lucidum* is widely used in treatment of chronic hepatopathy where liver injury was induced by d-galactosamine (d-GalN). The *G. lucidum* was given orally at three different concentrations, ranging between 60–180 mg/kg for a period of two weeks, to mice and the mice were exposed to d-galactosamine (750 mg/kg) which induced liver injury. The liver injury was observed with a significant increase in the AST, ALT concentrations in serum and the MDA levels in liver ($p < 0.01$) and by a prominent decrease in GSH and SOD levels in the liver. The pre-treatment with *G. lucidum* helped in the recovery of these increased markers to the normal level. The hepato-protective effect of *G. lucidum* was more effective at a concentration of 180 mg/kg (Shi et al., 2008). An aqueous extract of *G. formosanum*, *G. lucidum*, and *G. neo-japonicum* was administered in rats with liver toxicity induced by CCL_4. It resulted in recovery from liver toxicity (Lin et al., 1995). The benzo[α]pyrene (B[a]P) affects the hepatic enzymes. Animals given *G. lucidum* extract for 30 days before benzo[α]pyrene (B[α]P) showed inhibition of increase in SGOT, SGPT and ALP enzyme activities, boosted the reduced glutathione level and activities of other enzymes (Lakshmi et al., 2006). Ganoderic acid (8 μg/ml) from *G. lucidum* administered in HepG2215 cells over eight days inhibited the growth of hepatitis B virus (HBV) in the cells. Ganodermic acid (10 and 30 mg/kg/day) administration over seven days showed hepato-protective activity in the injured liver of moouse induced by CCL_4 and BCG plus lipopolysaccharide (Li and Wang, 2006).

In a study of liver damage induced by CCL_4, the medium of culture showed hepato-protective activity (Liu et al., 1998). The exudate of culture of *G. lucidum* was administered in animals. It showed hepato-protective activity as displayed by lowered AST and ALT levels in serum at 96 hours post injury. An exo-polymer (WK-003) produced from *G. lucidum* fermentation showed hepatoprotective activity. The exo-polymer consists of arabinose, rhamnose, xylose, mannose, galactose and glucose along with isoleucine, leucine, aspartic acid, glutamic acid, phenylalanine, valine, histidine, serine, glycine, arginine, alanine and tryptophan. The exo-polymer was administered for four consecutive days to rats. The glutamate pyruvic transaminase (GPT) activities in rats decreased from 871 to 263 (Song et al., 1998). *G. lucidum* has been investigated for polysaccharides with anti-fibrotic agents. The polysaccharides from *G. lucidum* decreased the collagen content in liver and enhanced the morphology while decreasing the total bilirubin, alkaline phosphatase (ALP), serum aspartate transaminase (AST) and alanine transaminase (ALT). The findings support the anti-fibrotic activity of polysaccharides in *G. lucidum* (Park et al., 1997).

The mice were treated with *G. lucidum* extract which decreased the hydroxyproline content in hepatic tissue and liver fibrosis in thio-acetamide-induced liver fibrosis in mice. After treatment, the extract significantly reduced the expression levels of mRNA in tissue inhibitor markers, like metalloproteinase 1, collagen (α1), metalloproteinase-13 and smooth muscle alpha-actin. Total collagenase activity decreased due to thioacetamide; the administration of *G. lucidum* extract reversed the total collagenase activity. Oral treatment of *G. lucidum* extract reversed the thio-acetamide-induced liver fibrosis (Wu et al., 2010). The stomach was treated with 100 μl (10 M) acetic acid for one minute and subsequently treated with polysaccharide of *G lucidum* in doses of 0.1, 0.5 or 1.0 g/kg intra-gastrically. The doses were given once daily for 14 consecutive days. The *G. lucidum* polysaccharide fraction was able to heal gastric ulcers in rats induced by acetic acid. Hence, polysaccharide of *G. lucidum* may be used as medicine to prevent or to treat ulcers (Gao et al., 2004c).

4.6 Anticancer Effects

The extracts or polysaccharides or proteins were reported to have anticancer properties (Chen et al., 2004; Kimura et al., 2002; Kim et al., 2007; Chung et al., 2001; Lin et al., 2003; Lu et al., 2001; 2003; 2004; Muller et al., 2006).

4.6.1 Myopathy

The products of *G. lucidum* were recommended for usage in myothaty after investigations and the Government of China approved polysaccharide of '*G. lucidum* Karst injection' for intramuscular application (Karst, 2011). The reports of qualified physicians of China suggested use of Lingzhi mushroom for prevention of cancers and for longevity (Zeng et al., 2018; Gao et al., 2009).

4.6.2 Prostate Cancer

Anticancer efficiency of *G. lucidum* was assessed in human breast and adenocarcinoma cell lines using *in vitro* studies. The study results revealed that the extracts significantly down-regulated the expression levels of nuclear kappa β and its related markers in both the tested cell lines depending on time and dose. It further arrested the cell cycle markers, such as cyclin B and Cdc2 proteins levels and enhanced p21 protein levels (Jiang et al., 2004a; Kuo et al., 2006). In another experiment, Jiang et al. (2004b) found that *G. lucidum* exhibited suppressive behaviour of MDA-MB-231 cells in breast cancer. The extract showed its anticancer activity by arresting cell cycle at G0/G1 phase and by inhibition of nuclear factor kappa β and related markers. Further Jiang et al. (2006) studied the anti-proliferative efficacy of *G. lucidum* against two cancer cell-lines models, i.e., MDA-MB-231 and MCF-7, revealing that *G. lucidum* suppressed the expression levels of oestrogen receptor in MCF-7 cells. However, no effect was observed on oestrogen response element in both the tested cell lines. It further significantly inhibited the THF-α-induced nuclear kappa-β mediated pro-inflammatory markers in both the cell lines. They concluded that *G. lucidum* can be used as an appropriate herb in chemotherapy of breast cancer (Hu et al., 2002).

A cancer study conducted by Sun et al. (2017) revealed that *G. lucidum* is associated closely with prostate cancer (PC) cells by enhancing the ROS production and causing aerobic stress. Wang et al. (2020) demonstrated that *G. lucidum* convinced cytotoxicity, accumulation of oxygen-generated free radicals and stimulation of apoptosis in PC-3 cells. In addition, it activated STAT-3, transcriptional factor which regularises apoptosis in cells and leads to hindering the prostate cancer. Further Wang et al. (2020) demonstrated the molecular mechanism in prostate cancer hampered through STAT-3 translocation by using the *G. lucidum* extracts. *G. lucidum* significantly attenuated NF-κβ and AP-1 pathway transcription factors, such as urokinase-type protease and its receptor proteins. Additionally, *G. lucidum* inhibited molecules involved in cell adhesion and cell migration in adenocarcinoma and breast cancer cell lines (Sliva, 2003).

4.6.3 Metastasis and Invasion

Weng and Yen (2010) summarised the reports on anticancer efficiency of *G. lucidum* and its constituents using *in vivo* and *in vitro* experimental studies. The reports revealed that *G. lucidum* significantly modulated the pro-inflammatory markers responsible for cancer invasion and metastasis (Weng and Yen, 2010). The polysaccharide and triterpene fractions of *G. lucidum* affected the growth and induced apoptosis in carcinoma cells of human A427 and A549. The molecular mechanism of action through gene-expression studies was thoroughly explained and the inhibition of cancer-related pathways responsible in partial or complete suppression of Mtorc1/2 path via activating AMPK and suppression of IGFR/PI3/Rheb among the tumour cells was helpful in understanding the role of GLPs (Sohretoglu et al., 2019).

Jin et al. (2016) investigated and surveyed the active constituents, such as polysaccharides and triterpenes isolated from *G. lucidum*, which showed beneficial effects on radiotherapy and chemotherapy in cancer treatment through several pathways. The extracts reported nausea and insomnia at marginal level apart from improvement in life (Jin et al., 2016). Liang et al. (2019) reviewed the pharmacological properties of two major groups of chemical constituents, i.e., polysaccharides and triterpenoids. Triterpenoids are pharmacologically effective in treating different diseases, such as fatigue syndrome, liver diseases and prostate cancer when combined with different medications. Clinical trials of seven genodendric acids (A, C2, D, F, DM, X and Y) are under progress. Anti-atherosclerotic activity of *G. lucidum* and its active ingredients, i.e., polysaccharides and triterpenoids were assessed using *in vivo* and *in vitro* models. Rabbits (*in vivo*) treated with *G. lucidum* and its active ingredients significantly reduced serum cholesterol and triglycerides levels to normal besides reducing plaque formation in arteries. In *in vitro* studies, *G. lucidum* strongly enhanced the antioxidant capacity by reducing the ROS and malondialdehyde in HUVECs activated with oxidised low-density lipoprotein. In phorbol myristate acetate and oxidised low-density lipoprotein challenged THP-1 cells, *G. lucidum* notably attenuated the higher levels of pro-inflammatory markers to normal levels in phorbol myristate acetate-challenged THP-1 cells. The extract significantly controlled the growth of foam cells in oxidised low-density lipoprotein challenged THP-1 cells (Li et al., 2021).

4.6.4 Ovarian Cancer

Hsiesh and Wu (2011) investigated antioxidant or detoxification and anticancer efficiency of *G. lucidum* in cancer (OVCAR-3) cells of ovaries, revealing that *G. lucidum* strongly suppressed the growth of cells and inhibited the regulation of cell cycle via suppression of cyclin D1. Further it enhanced the antioxidant capacity by increasing the SOD enzyme concentration in cells. Regarding detoxification, the extract significantly induced the enzymes involved in phase II-detoxification, such as glutathione S-transferase P1 (GSTP1) and NAD(P) H: quinone enzyme one (NQO1) through Nrf2 mediated signalling pathway Hsiesh and Wu (2011).

The hepato-protective mechanism of *G. lucidum* against formaldehyde-induced liver toxicity and inflammation was assessed in male Wistar rats at different concentrations. It indicated that rats treated with *G. lucidum* significantly altered the formaldehyde-induced body and organ reaction. Further, it prevented the inflammatory conditions by reducing the degree of hydrogen peroxide and MDA by enhancing the glutathione levels. Histopathology results revealed that *G. lucidum* restored the formaldehyde-induced microscopic anatomy changes (Oluwafemi Adetuyi et al., 2020).

4.6.5 Lung Cancer

Feng et al. (2013) assessed the repressive impact of triterpenes isolated from *G. lucidum* on cell growth and neoplasm development in A549 cells on Lewis mice. The chemical constituents strongly supressed growth of A549 cells (IC_{50} 24.63 µg/mL) and tumours in Lewis tumour-bearing mice (Feng et al., 2013). It enhanced the indices of immune organs, such as thymus and spleen. Reishi polysaccharide fraction of *G. lucidum* enriched with L-fucose notably inhibited the expansion of cancer cells and attenuated the tumour-associated inflammatory mediators, like MCP-1 in *in vitro* study (Gill and Kumar, 2017). The GLPs were investigated and found effective in lung cancer and found suppression of CD69, granzyme B and perforin in lymphocytes. The ganoderic acids and lanostene (triterpenoid) contributed to tolerance during lung cancer by inducing T-cell apoptosis and sequentially inducing suppression of CD8+T cell activation and improving Treg mediated disorders of immunity (Sun et al., 2014).

4.6.6 Liver Cancer

Antihepatoma studies, conducted on Hep 3B cells pre-incubated with 'lovastatin' of *G. lucidum*, confirmed liver protection in induced hepatome by carbon tetrachloride (CCl_4) in rats. The tested extracts were effective on hepatome without affecting normal cells and exhibited minimised LDH leakage, lipid peroxidation, improved internal antioxidants (Su et al., 2013). The liver cancer (human hepatocellular carcinoma cells)

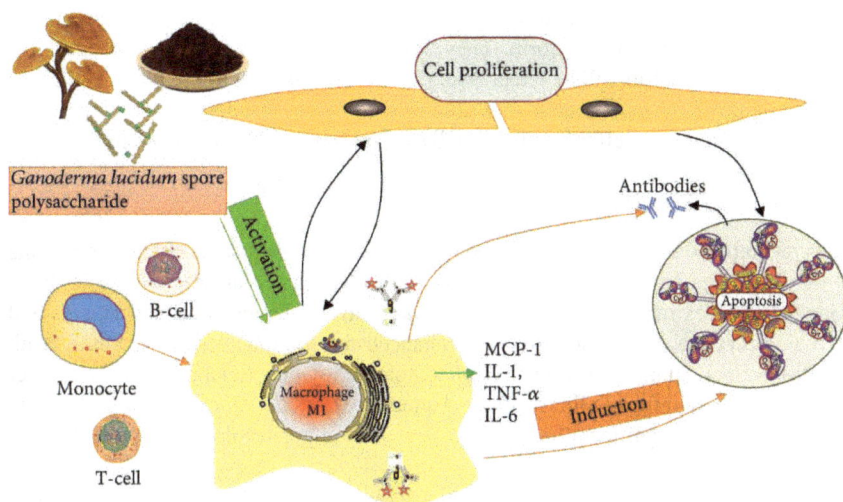

Fig. 5. GLSP enhances immunity and induces tumour cell apoptosis by activating macrophages (Song et al., 2021).

were significantly decreased by GLPS through decreased expression of miRNA in the liver carcinoma cells (Shen et al., 2014). In another study, the GLPS suppressed the hepatoma in mice and there was proliferation through multiplied IL-2 expression and secretion (Lin et al., 2003). GLSP (*Ganoderma lucidum* spore polysaccharide) enhances immunity and induces tumour cell apoptosis by activating macrophages and inhibiting the growth of hepatocellular carcinoma cells (HCC) by altering macrophage polarity and inducing apoptosis (Fig. 5) (Song et al., 2021).

4.6.7 Melanoma

The GLPS (polysaccharides of *G. lucidum*), when tested on skin cancer cells (B16F10), reported significantly repressed proliferation of lymphocyte and also the secretion of 'perforin and granzyme B' when activated with phytohemagglutinin and white corpuscle expansion within the mixed lymphocyte reaction (Sun et al., 2011a). The extracts of GLPS enhanced the power of B16F10 cells to switch on lymphocytes (Sun et al., 2011b). The GLPS increased the activity of molecules and costimulatory, histocompatibility and potency of immunity-mediated toxicity against B16F10 cells (Sun et al., 2012). Another study conducted by Sun et al. (2013) revealed that the GLPS showed the antagonistic effects of B16F10 cell-culture supernatants on lymphocytes and also reported Fas ligand and CD71 gene expression in lymphocytes. Barbieri et al., found that alcohol extract significantly repressed the discharge of IL-8, IL-6, MMP-2 and MMP-9 in carcinoma cells along with the pro-inflammatory conditions (Sohretoglu and Huang, 2018; Barbieri et al., 2017).

4.6.8 Leukaemia

Wang et al. (1997) discovered that lipopolysaccharides of *G. lucidum* play an indirect anti-tumour effect by enhancing the concentrations of IL-1 and IL-6 markers in *in vivo* models. Cao and Lin (2003) identified that lipopolysaccharides of *G. lucidum* strongly inhibited T-lymphocytes elicited by dendritic cells. Further, these polysaccharides promoted CIK cell growth and toxicity and enhanced the levels of TNF production, IL-2, supermolecule and perforin. They strongly decreased the levels of anti-CD3 and IL-2 (Zhu and Lin, 2006). Chan et al. (2008) reported that lipopolysaccharides of *G. lucidum* provoked monocytic leukemic cells to dendritic cells. Chang et al. (2014) reported the cytotoxic effects of these polysaccharides on natural killer cells.

4.7 Clinical Studies

Several authors reported anticancer potential of *G. lucidum* in different cancer patients through clinical studies. Gao et al. (2003 a, b) investigated anticancer efficiency of Ganopoly, an extract prepared from *G. lucidum* in 34 advanced cancer patients and revealed that the extract significantly enhanced the immune response. Shing et al. (2008) evaluated that *G. lucidum* exhibited noteworthy beneficial responses in cancer patients in six months of clinical study. Aqueous extract of *G. lucidum* fruit body and spores exhibited significant anticancer effect in five gynaecological cancer patients (Suprasert et al., 2014).

Different clinical experimental results showed the beneficial role of *G. lucidum* as an alternate adjuvant medical aid in patients with cancer with no obvious toxicity. Polysaccharide is the principle bioactive element within the soluble extracts of *G. lucidum*. Potent malignant neoplasm activity of polysaccharide was proved through anti-proliferative, immunomodulatory, pro-apoptotic and anti-angiogenic effects in *in vivo* and *in vitro* experimental models. Here, we tend to shortly summarise these malignant neoplasm effects of GLP and also the underlying mechanisms (Sohretoglu and Huang, 2018; Sohretoglu et al., 2019).

4.8 Toxicology

The pharmacology and medicine of *Ganoderma* are partly investigated in current studies. A sudden fatal disease was observed in a case who taken Lingzhii powder for a period of one to two months (Wanmuang et al., 2007). Additionally, non-Hodgkins malignant neoplastic disease was diagnosed and conferred with chronic watery looseness of the bowels while taking Lingzhi (Suprasert et al., 2014).

5. Conclusion

Species of Reishi mushrooms are historically used as a medicine in China and other geographic regions. Nutraceuticals prepared from *Ganoderma* are used to treat patients with vessel issues, leukaemia, leukopenia, hepatitis, asthma, nephritis, insomnia, gastritis, respiratory illness and for cholesterol-lowering. Trendy phytochemical analysis demonstrated that terpenes and polysaccharides are the chief active components. Recent medicine and clinical experiments recommend that *Ganoderma* may have a blood-thinner capacity and it exhibits anticancer/anti-tumour effects. It is effective against viruses causing hepatitis B and lowers blood sugar, anticancer, antimicrobial, immunomodulatory effect and in blood pressure. Therefore, the cultivation of Reishi mushroom was encouraging for the medicinal use during controlled conditions since in natural conditions it is going to have an effect on the tree vegetation.

References

Akihisa, T., Nakamura, Y., Tagata, M., Tokuda, H., Yasukawa, K., Uchiyama, E. Suzuki, T. and Kimura, Y. (2007). Anti-inflammatory and anti-tumour-promoting effects of triterpene acids and sterols from the fungus *Ganoderma lucidum*. *Chem. Biodivers.*, 4(2): 224–31.

Bao, X., Liu, C., Fang, J. and Li, X. (2001). Structural and immunological studies of a major polysaccharide from spores of *Ganoderma lucidum* (Fr.) Karst. *Carbohydr. Res.*, 332: 67–74.

Bao, X., Wang, X., Dong, Q., Fang, J. and Li, X. (2002). Structural features of immunologically active polysaccharides from *Ganoderma lucidum*. *Phytochemistry*, 59: 175–181.

Barbieri, A., Quagliariello, V., Del Vecchio, V., Falco, M., Luciano, A. and Amruthraj, N.J. et al. (2017). Anticancer and anti-inflammatory properties of *Ganoderma lucidum* extract effects on melanoma and triple-negative breast cancer treatment. *Nutrients*, 9: E210. Doi: 10.3390/nu9030210.

Boh, B., Berovic, M., Zhang, J. and Zhi-Bin, L. (2007). *Ganoderma lucidum* and its pharmaceutically active compounds. *Biotechnol. Annu. Rev.*, 13: 265–301.

Borchers, A.T., Krishnamurthy, A., Keen, C.L., Meyers, F.J. and Gershwin, M.E. (2008). The immunobiology of mushrooms. *Exp. Biol. Med.*, 233: 259–276.

Borchers, A.T., Stern, J.S. Hackman, R.M., Keen, C.L. and Gershwin, M.E. (1999). Mini review: Mushrooms, tumours and immunity. *Proc. Soc. Exp. Biol. Med.*, 221: 281–93.

Budavari, S. (1989). *The Merck Index*, 11. New Jersey: Merck & Co., p. 845.

Cao Yu, Xu Xiaowei, Liu Shujing, Huang, L. and Gu, J. (2018). *Ganoderma*: A cancer immunotherapy review. *Frontiers in Pharmacology*, 9(1217): 1–14.

Cao, L.Z. and Lin, Z.B. (2003). Regulatory effect of *Ganoderma lucidum* polysaccharides on cytotoxic T-lymphocytes induced by dendritic cells *in vitro*. *Acta Pharmacol. Sin.*, 24: 321–326.

Chan, W.K., Cheung, C.C., Law, H.K., Lau, Y.L. and Chan, G.C. (2008). *Ganoderma lucidum* polysaccharides can induce human monocytic leukemia cells into dendritic cells with immuno-stimulatory function. *Journal of Hematology and Oncology*, 1: 9. https://Doi.org/10.1186/1756-8722-1-9.

Chang, S.T. and Buswell, J.A. (1996). Mushroom nutraceuticals. *World J. Microbiol Biotechnol.*, 12: 473–6.

Chang, C.J., Chen, Y.Y.M., Lu, C.C. et al. (2014). *Ganoderma lucidum* stimulates NK cell cytotoxicity by inducing NKG2D/NCR activation and secretion of perforin and granulysin. *Innate. Immun.*, 20: 301–311. Doi: 10.1177/1753425913491789.

Chen, D.H., Wen-Yue Shiou, W.Y., Wang, K.C., Huang, S.Y., Shie, Y.T., Tsai, C.M. and Shie, J.F. (1999). Chemotaxonomy of triterpenoid pattern of HPLC of *Ganoderma lucidum* and *Ganoderma tsugae*. *J. Chin. Chem. Soc.*, 46: 47–51.

Chen, H.S., Tsai, Y.F., Lin, S., Lin, C.C., Khoo, K.H., Lin, C.H. and Wong, C.H. (2004). Studies on the immunomodulating and anti-tumour activities of *Ganoderma lucidum* (Reishi) polysaccharides. *Bioorg. Med. Chem.*, 12(21): 5595–601.

Chen, T.Q., Li, K.B., He, X.J., Zhu, P.G. and Xu, J. (1998). Micro-morphology, chemical components and identification of log-cultivated *Ganoderma lucidum* spore. *In*: Lu, M., Gao, K., Si, H.-F. and Chen, M.-J. (eds.). *Proc '98 Nanjing Intl Symp Science & Cultivation of Mushrooms*, 214, Nanjing, China, JSTC-ISMS.

Chen, Y., Bicker, W., Wu, J., Xie, M.Y. and Lindner, W. (2010). *Ganoderma* species discrimination by dual-mode chromatographic fingerprinting: A study on stationary phase effects in hydrophilic interaction chromatography and reduction of sample misclassification rate by additional use of reversed-phase chromatography. *J. Chromatogr.*, 1217(8): 1255–65.

Chen, Y., Zhu, S.B., Xie, M.Y., Nie, S.P., Liu, W., Li, C., Gong, X.F. and Wang, Y.X. (2008). Quality control and original discrimination of *Ganoderma lucidum* based on high-performance liquid chromatographic fingerprints and combined chemometrics methods. *Anal Chim. Acta*, 623(2): 146–56.

Cherian, E., Sudheesh, N.P., Janardhanan, K.K. and Patani, G. (2009). Free-radical scavenging and mitochondrial antioxidant activities of Reishi-*Ganoderma lucidum* (Curt: Fr) P. Karst and Arogyapacha-*Trichopus zeylanicus* Gaertn extracts. *J. Basic Clin. Physiol. Pharmacol.*, 20(4): 289–307.

Chien, C.M., Cheng, J.L., Chang, W.T., Tien, M.H., Tsao, C.M., Chang, Y.H., Chang, H.Y., Hsieh, J.F., Wong, C.H. and Chen, S.T. (2004). Polysaccharides of *Ganoderma lucidum* alter cell immunophenotypic expression and enhance CD56+ NK-cell cytotoxicity in cord blood. *Bioorg. Med. Chem.*, 12(21): 5603–9.

Chiu, S.W., Wang, Z.M., Leung, T.M. and Moore, D. (2000). Nutritional value of *Ganoderma* extract and assessment of its genotoxicity and antigenotoxicity using comet assays of mouse lymphocytes. *Food Chem. Toxicol.*, 38: 173–178.

Chung, W.T., Lee, S.H., Kim, J.D. et al. (2001). Effect of mycelial culture broth of *Ganoderma lucidum* on the growth characteristics of human cell lines. *J. Biosci. Bioeng.*, 92: 550–555.

Du, M., Wang, C., Hu, X.C. and Zhao, G. (2008). Positive effect of selenium on the immune regulation activity of Lingzhi or Reishi medicinal mushroom, *Ganoderma lucidum* (W. Curt.: Fr.) P. Karst (Aphyllophoromycetideae), proteins *in vitro. Int. J. Med. Mushrooms*, 10: 337–344.

El-Mekkawy, S., Meselhy, M.R., Nakamura, N. et al. (1998). Anti-HIV-1 and anti-HIV-1-protease substances from *Ganoderma lucidum. Phytochemistry*, 49(6): 1651–7.

Eo, S.K., Kim, Y.S., Lee, C.K. and Han, S.S. (1999). Antiviral activities of various water and methanol-soluble substances isolated from *Ganoderma lucidum. J. Ethnopharmacol.*, 68: 129–136. Doi: 10.1016/s0378-8741(99)00067-7.

Falandysz, J. (2008). Selenium in edible mushrooms. *J. Environ. Sci. Health C. Environ. Carcinog. Ecotoxicol. Rev.*, 26(3): 256–299.

FBRI. (2008). *New Enzymes for Biopulping*; archived from the original on 2009-01-04. Retrieved 2008-11-15.

Feng, L., Yuan, L., Du, M., Chen, Y., Zhang, M.H., Gu, J.F. et al. (2013). Anti-lung cancer activity through enhancement of immunomodulation and induction of cell apoptosis of total triterpenes extracted from *Ganoderma luncidum* (Leyss. ex Fr.) Karst. *Molecules*, 18: 9966–9981. Doi: 10.3390/molecules18089966.

Fukuzawa, M., Yamaguchi, R. Hide, I., Chen, Z., Hirai, Y., Sugimoto, A., Yasuhara, T. and Nakata, Y. (2008). Possible involvement of long chain fatty acids in the spores of *Ganoderma lucidum* (Reishi Houshi) to its anti-tumor activity. *Biol. Pharm. Bull.*, 31(10): 1933–7.

Gao, J.J., Min, B.S., Ahn, E.M., Nakamura, N., Lee, H.K. and Hattori, M. (2002a). New triterpene-aldehydes, lucialdehydes A-C, from *Ganoderma lucidum* and their cytotoxicity against murine and human tumour cells. *Chem. Pharm. Bull.*, 50: 837–840.

Gao, Y. and Zhou, S. (2009). Cancer prevention and treatment by *Ganoderma*, a mushroom with medicinal properties. *Food Rev. Int.*, 19: 275–325.

Gao, Y.H., Sai, X.H., Chen, G.L., Ye, J.X. and Zhou, S.F. (2003a). A randomized, placebo-controlled, multi-centre study of *Ganoderma lucidum* (W. Curt.: Fr.) Lloyd (Aphyllophoromycetideae) polysaccharides (Ganopoly) in patients with advanced lung cancer. *Int. J. Med. Mushrooms*, 5: 368–381.

Gao, Y.H., Zhou, S.F., Chen, G.L., Dai, X.H. and Ye, J.X.A. (2002b). Phase I/II study of a *Ganoderma lucidum* (Curr.: Fr.) P. Karst; Extract (Ganopoly) in patients with advanced cancer. *Int. J. Med. Mushrooms*, 4: 207–14.

Gao, Y.H., Zhou, S.F., Jiang, W.Q., Huang, M. and Sai, X.H. (2003b). Effects of Ganopoly (a *Ganoderma lucidum* polysaccharide extract) on immune functions in advanced-stage cancer patients. *Immunol. Invest.*, 32: 201–15.

Gao, Y., Chan, E. and Zhou, S. (2004a). Immunomodulating activities of *Ganoderma*, a mushroom with medicinal properties. *Food Rev. Int.*, 20:123–161. Doi: 10.1081/FRI-120037158.

Gao, Y., Gao, H., Chan, E., Tang, W. et al. (2005). Anti-tumour activity and underlying mechanisms of ganopoly, the refined polysaccharides extracted from *Ganoderma lucidum*, in mice. *Immunol. Invest.*, 34(2): 171–98.

Gao, Y., Lan, J., Dai, X., Ye, J. and Zhou, S. (2004b). A phase I/II study of Lingzhi mushroom *Ganoderma lucidum* (W. Curt.: Fr.) Lloyd (Aphyllophoromycetideae) extract in patients with type II diabetes mellitus. *Int. J. Med. Mushrooms*, 6: 33–40.

Gao, Y., Tang, W., Gao, H., Chan, E., Lan, J. and Zhou, S. (2004c). *Ganoderma lucidum* polysaccharide fractions accelerate healing of acetic acid-induced ulcers in rats. *J. Med. Food*, 7(4): 417–21.

Gill, B.S. and Kumar, S. (2017). *Ganoderma lucidum* targeting lung cancer signalling: A review. *Tumour Biology*. Doi:10.1177/1010428317707437.

Gonzalez, A.G., Leon, F., Rivera, A., Munoz, C.M. and Bermejo, J. (1999). Lanostanoid triterpenes from *Ganoderma lucidum. J. Nat. Prod.*, 62: 1700–1701.

Haralampidis, K., Trojanowska, M. and Osbourn, A.E. (2002). Biosynthesis of triterpenoid saponins in plants. *Adv. Biochem. Eng. Biotechnol.*, 75: 31–49.

Hijikata, Y. and Yamada, S. (1998). Effect of *Ganoderma lucidum* on postherpetic neuralgia. *Ame. J. Chin. Med.*, 26: 375–381.

Hijikata, Y., Yamada, S. and Yasuhara, A. (2007). Herbal mixtures containing the mushroom *Ganoderma lucidum* improve recovery time in patients with herpes genitalis and labialis. *J. Altern. Complement. Med.*, 13: 985–987.

Hikino, H., Ishiyama, M., Suzuki, Y. and Konno, C. (1989). Mechanisms of hypoglycemic activity of ganoderan B: A glycan of *Ganoderma lucidum* fruit body. *Planta Med.*, 55: 423–428.

Hikino, H., Konno, C., Mirin, Y. and Hayashi, T. (1985). Isolation and hypoglycemic activity of ganoderans A and B, glycans of *Ganoderma lucidum* fruit bodies. *Planta Med.*, 4: 339–340.

Hsieh, T.C. and Wu, J.M. (2011). Suppression of proliferation and oxidative stress by extracts of *Ganoderma lucidum* in the ovarian cancer cell line OVCAR-3. *Int. J. Mol. Med.*, 28(6): 1065–1069. Doi: 10.3892/ijmm.2011.788.

Hu, H., Ahn, N.S., Yang, X., Lee, Y.S. and Kang, K.S. (2002). *Ganoderma lucidum* extract induces cell cycle arrest and apoptosis in MCF-7 human breast cancer cell. *Int. J. Cancer*, 102: 250–3.

Hyun, J.W., Choi, E.C. and Kim, B.K. (1990). Studies on constituents of higher fungi of Korea (LXVII), antitumour components of the basidiocarp of *Ganoderma lucidum. Korean J. Mycol.*, 18: 58–69.

Ji, Z., Tang, Q., Zhang, J., Yang, Y., Jia, W. and Pan, Y. (2007). Immunomodulation of RAW264.7 macrophages by GLIS, a proteo-polysaccharide from *Ganoderma lucidum. J. Ethnopharmacol.*, 112: 445–50.

Jia, J., Zhang, X., Hu, Y.S. et al. (2009). Evaluation of *in vivo* antioxidant activities of *Ganoderma lucidum* poly-saccharides in STZ-diabetic rats. *Food Chem.*, 115: 32–36.

Jiang, J., Grieb, B., Thyagarajan, A. and Sliva, D. (2008). Ganoderic acids suppress growth and invasive behaviour of breast cancer cells by modulating AP-1 and NF-kappaB signalling. *Int. J. Mol. Med.*, 21: 577–584.

Jiang, J., Slivova, V. and Sliva, D. (2006). *Ganoderma lucidum* inhibits proliferation of human breast cancer cells by down-regulation of estrogen receptor and NF-kappaB signalling. *Int. J. Oncol.*, 29(3): 695–703.

Jiang, J., Slivova, V., Harvey, K., Valachovicova, T. and Sliva, D. (2004b). *Ganoderma lucidum* suppresses growth of breast cancer cells through the inhibition of Akt/NF-kappa-B signalling. *Nutr. Cancer*, 49(2): 209–216. Doi: 10.1207/s15327914nc4902_13.

Jiang, J., Slivova, V., Valachovicova, T., Harvey, K. and Sliva, D. (2004a). *Ganoderma lucidum* inhibits proliferation and induces apoptosis in human prostate cancer cells PC-3. *Int. J. Oncol.*, 24(5): 1093–9.

Jin, X., Ruiz Beguerie, J., Sze, D.M. and Chan, G.C. (2016). *Ganoderma lucidum* (Reishi mushroom) for cancer treatment. *Cochrane Database Syst. Rev.*, 4(4): CD007731.

Karst, P. (2011). Higher Basidiomycetes from central Himalayan hills of India. *Int. J. Med. Mushrooms*, 13(6): 535–44. Doi: 10.1615/intjmedmushr. v13.i6.50.

Keypour, S., Riahi, H., Moradali, M.F. and Rafati, H. (2008). Investigation of the antibacterial activity of a chloroform extract of Lingzhi or Reishi medicinal mushroom, *Ganoderma lucidum* (W.Curt.: Fr.) P. Karst. (Aphyllophoromycetideae). *Int. J. Med. Mushrooms*, 10(4): 345–349.

Kim, D.H., Shim, S.B., Kim, N.J. and Jang, I.S. (1999). β-Glucuronidase-inhibitory activity and hepatoprotective effect of *Ganoderma lucidum. Biol. Pharm. Bull.*, 22: 162–4.

Kim, K.C., Kim, J.S., Son, J.K. and Kim, I.G. (2007). Enhanced induction of mitochondrial damage and apoptosis in human leukemia HL-60 cells by the *Ganoderma lucidum* and *Duchesnea chrysantha* extracts. *Cancer Lett.*, 246: 210–17.

Kim, S.D. and Nho, H.J. (2004). Isolation and characterisation of alpha-glucosidase inhibitor from the fungus *Ganoderma lucidum. J. Microbiol.*, 42: 223–227.

Kim, Y.S., Eo, S.K., Oh, K.W., Lee, C.K. and Han, S.S. (2000). Antiherpetic activities of acidic protein bound polysaccharide isolated from *Ganoderma lucidum* alone and in combinations with interferons. *J. Ethnopharmacol.*, 72: 451–8.

Kimura, Y., Taniguchi, M. and Baba, K. (2002). Anti-tumour and antimetastatic effects on liver of triterpenoid fractions of *Ganoderma lucidum*: Mechanism of action and isolation of an active substance. *Anticancer Res.*, 22: 3309–18.

Kirk, P.M., Cannon, P.F., Minter, D.W. and Stalpers, J.A. (2008). J.A. *Dictionary of the Fungi* (10th ed.). Wallingford: CABI, pp. 272.

Kolesnikova, O.P., Tuzova, M.N. and Kozlov, V.A. (1997). Screening of immune-active properties of alkane carbonic acid derivatives and germanium-organic compounds *in vivo*. *Immunologiya.*, 10: 36–8.

Kubota, T., Asaka, Y., Miura, I. and Mori, H. (1982). Structures of ganoderic acids A and B, two new lanostane type bitter triterpenes from *Ganoderma lucidum* (Fr.) Karst. *Helv. Chim. Acta*, 65: 611–9.

Kuo, M.C., Weng, C.Y., Ha, C.L. and Wu, M.J. (2006). *Ganoderma lucidum* mycelia enhance innate immunity by activating NF-kappaB. *J. Ethnopharmacol.*, 103: 217–22.

Lakshmi, B., Ajith, T.A., Jose, N. and Janardhanan, K.K. (2006). Antimutagenic activity of methanolicextract of *Ganoderma lucidum* and its effect on hepatic damage caused by benzo[a] pyrene. *J. Ethnopharmacol.*, 107(2): 297–303.

Lee, J.M., Kwon, H., Jeong, H. et al. (2001). Inhibition of lipid peroxidation and oxidative DNA damage by *Ganoderma lucidum*. *Phytother. Res.*, 15(3): 245–9.

Lee, K.M., Lee, S.Y. and Lee, H.Y. (1999). Bistage control of pH for improving exopolysaccharide production from mycelia of *Ganoderma lucidum* in an air-lift fermenter. *J. Biosci. Bioeng.*, 88: 646–50.

Li, C.H., Chen, P.Y. et al. (2005a). Ganoderic acid X, a lanostanoid triterpene, inhibits topoisomerases and induces apoptosis of cancer cells. *Life Sci.*, 77(3): 252–65.

Li, Y.Q. and Wang, S.F. (2006). Anti-hepatitis B activities of ganoderic acid from *Ganoderma lucidum*. *Biotechnol Lett.*, 28(11): 837–41.

Li, Y., Tang, J., Gao, H., Xu, Y., Han, Y., Shang, H., Lu, Y. and Qin, C. (2021). *Ganoderma lucidum* triterpenoids and polysaccharides attenuate atherosclerotic plaque in high-fat diet rabbits. *Nutr. Metab. Cardiovasc. Dis.*, 31(6): 1929–1938.

Li, Z., Liu, J. and Zhao, Y. (2005b). Possible mechanism underlying the antiherpetic activity of a proteoglycan isolated from the mycelia of *Ganoderma lucidum in vitro*. *J. Biochem. Mol. Biol.*, 38(1): 34–40.

Liang, C., Tian, D., Liu, Y., Li, H., Zhu, J., Li, M., Xin, M. and Xia, J. (2019). Review of the molecular mechanisms of *Ganoderma lucidum* triterpenoids: Ganoderic acids A, C2, D, F, DM, X and Y. *Eur. J. Med. Chem.*, 174: 130–141.

Lin, J.M., Lin, C.C., Chen, M.F., Ujiie, T. and Takada, A. (1995). Radical scavenger and anti-hepatotoxic activity of *Ganoderma formosanum*, *Ganoderma lucidum* and *Ganoderma neo-japonicum*. *J. Ethnopharmacol.*, 47: 33–41.

Lin, S.B., Li, C.H., Lee, S.S. and Kan, L.S. (2003). Triterpene-enriched extracts from *Ganoderma lucidum* inhibit growth of hepatoma cells via suppressing protein kinase C, activating mitogen-activated protein kinases and G2-phase cell cycle arrest. *Life Sci.*, 72: 2381–90.

Liu, K.C., Phounsavan, S.F., Huang, R.L., Liao, C., Hsu, S.Y. and Wang, K.J. (1998). Pharmacological and liver functional studies on mycelium of *Ganoderma lucidum*. *Chin. Pharm. J.*, 40: 21–29.

Liu, J., Yang, F., Ye, L.B., Yang, X.J., Timani, K.A., Zheng, Y. and Wang, Y.H. (2004). Possible mode of action of antiherpetic activities of a proteoglycan isolated from the mycelia of *Ganoderma lucidum in vitro*. *J. Ethnopharmacol.*, 95: 265–272.

Lu, H., Kyo, E., Uesaka, T., Katoh, O. and Watanabe, H. (2003). A water-soluble extract from cultured medium of *Ganoderma lucidum* (Reishi) mycelia suppresses azoxymethane-induction of colon cancers in male F344 rats. *Oncol. Rep.*, 10: 375–9.

Lu, H., Uesaka, T., Katoh, O., Kyo, E. and Watanabe, H. (2001). Prevention of the development of preneoplastic lesions, aberrant crypt foci, by a water-soluble extract from cultured medium of *Ganoderma lucidum* (Rei-shi) mycelia in male F344 rats. *Oncol. Rep.*, 8: 1341–5.

Lu, Q.Y., Jin, Y.S., Zhang, Q., Zhang, Z., Heber, D., Go, V.L., Li, F.P. and Rao, J.Y. (2004). *Ganoderma lucidum* extracts inhibit growth and induce actin polymerization in bladder cancer cells *in vitro*. *Cancer Lett.*, 216(1): 9–20.

Ma, J., Ye, Q., Hua, Y., Zhang, D., Cooper, R., Chang, M.N., Chang, J.Y. and Sun, H.H. (2002). New lanostanoids from the mushroom *Ganoderma lucidum*. *J. Nat. Prod.*, 65(1): 72–75.

Matos, A.J., Bezerra, R.M. and Dias, A.A. (2007). Screening of fungal isolates and properties of *Ganoderma applanatum* intended for olive mill wastewater decolourisation and dephenolisation. *Lett. Appl. Microbiol.*, 45(3): 270–275.

Mau, J.L., Lin, H.C. and Chen, C.C. (2001). Non-volatile components of several medicinal mushrooms. *Food Res Int.*, 34: 521–6.

Mau, J.L., Lin, H.C. and Chen, C.C. (2002). Antioxidant properties of several medicinal mushrooms. *J. Agric. Food Chem.*, 50: 6072–7.

Meng, J., Hu, X., Shan, F., Hua, H., Lu, C., Wang, E. et al. (2011). Analysis of maturation of murine dendritic cells (DCs) induced by purified *Ganoderma lucidum* polysaccharides (GLPs). *Int. J. Biol. Macromol.*, 49: 693–699. Doi: 10.1016/j.ijbiomac.2011.06.029.

Min, B.S., Nakamura, N., Miyashiro, H., Bae, K.W. and Hattori, M. (1998). Triterpenes from the spores of *Ganoderma lucidum* and their inhibitory activity against HIV-1 protease. *Chem. Pharm. Bull.*, 46: 1607–12.

Mino, Y., Ota, N., Sakao, S. and Shi Momura, S. (1980). Determination of germanium in medicinal plants by atomic absorption spectrometry with electrothermal atomisation. *Chem. Pharm. Bull.*, 28: 2687–91.

Miyazaki, T. and Nishijima, M. (1981). Studies on fungal polysaccharides, XXVII. Structural examination of a water-soluble, antitumor polysaccharide of *Ganoderma lucidum*. *Chem. Pharm. Bull.*, 29: 3611–16.

Mohsin, M., Negi, P. and Ahmed, Z. (2011). Determination of the antioxidant activity and polyphenol contents of wild Lingzhi or Reishi medicinal mushroom, *Ganoderma lucidum* (W. Curt. Fr.) P. Karst, (higher Basidiomycetes) from central Himalayan hills of India. *Int. J. Med. Mushrooms*, 13(6): 535–44. Doi: 10.1615/intjmedmushr.v13.i6.50.

Muller, C.I., Kumagai, T., O'Kelly, J., Seeram, N.P., Heber, D. and Koeffler, H.P. (2006). *Ganoderma lucidum* causes apoptosis in leukemia, lymphoma and multiple myeloma cells. *Leuk. Res.*, 30: 841–8.

Nishitoba, T., Sato, H., Kasai, T., Kawagishi, H. and Sakamura, S. (1984). New bitter C27 and C30 terpenoids from fungus *Ganoderma lucidum* (reishi). *Agric. Biol. Chem.*, 48: 2905–7.

Noguchi, M., Kakuma, T., Tomiyasu, K., Konishi, F., Kumamoto, S., Kondo, R. and Matsuoka, K. (2005). Phase I study of a methanol extract of *Ganoderma lucidum* edible and medicine mushroom in men with mild symptoms of bladder outlet obstruction. *Urology*, 66: 21.

Oh, K.W., Lee, C.K., Kim, Y.S., Eo, S.K. and Han, S.S. (2000). Antiherpetic activities of acidic protein bound polysacchride isolated from *Ganoderma lucidum* alone and in combination with Acyclovir and Vidarabine. *J. Ethnopharmacol.*, 72: 221–227.

Ohno, N., Miura, N.N., Sugawara, N., Tokunaka, K., Kirigaya, N. and Yadomae, T. (1998). Immunomodulation by hot water and ethanol extracts of *Ganoderma lucidum*. *Pharm. Pharmacol. Lett.*, 4: 174–7.

Oluwafemi Adetuyi, B., Olamide Okeowo, T., Adefunke Adetuyi, O. et al. (2020). Biology (Basel). *Ganoderma lucidum* from red mushroom attenuates formaldehyde-induced liver damage in experimental male rat model. *Biology*, 9(10): 313. Doi: 10.3390/biology9100313.

Ooi, V.E. and Liu, F. (2000). Immunomodulation and anticancer activity of polysaccharide-protein complexes. *Curr. Med. Chem.*, 7: 715–29.

Park, E.J., Ko, G., Kim, J. and Dong, H.S. (1997). Antifibrotic effects of a polysaccharide extracted from *Ganoderma lucidum*, glycyrrhizin, and pentoxifylline in rats with cirrhosis induced by biliary obstruction. *Biol. Pharm. Bull.*, 20: 417–20.

Paterson, R.R. (2006). Ganoderma—A therapeutic fungal biofactory. *Phytochemistry*, 67(18): 1985–2001.

Rani, P., Lal, M.R., Maheshwari, U. and Krishnan, S. (2015). Antioxidant potential of Lingzhi or Eeishi medicinal Mushroom, *Ganoderma lucidum* (higher basidiomycetes) cultivated on *Artocarpus heterophyllus* sawdust substrate in India. *Int. J. Med. Mushrooms*, 17(12): 1171–7. Doi: 10.1615/intjmedmushrooms.v17.i12.70.

Saltarelli, R., Ceccaroli, P., Iotti, M., Zambonelli, A., Buffalini, M., Casadei, L., Vallorani, L. and Stocchi, V. (2009). Biochemical characterisation and antioxidant activity of mycelium of *Ganoderma lucidum* from Central Italy. *Food Chem.*, 116: 143–51.

Sanodiya, B.S., Thakur, G.S., Baghel, R.K., Prasad, G.B. and Bisen, P.S. (2009). *Ganoderma lucidum*: A potent pharmacological macrofungus. *Curr. Pharm. Biotechnol.*, 10(8): 717–42.

Sato, H., Nishitoba, T., Shirasu, S., Oda, K. and Sakamura, S. (1986). Ganoderiol A and B, new triterpenoids from the fungus *Ganoderma lucidum* (Reishi). *Agric. Biol. Chem.*, 50: 2887–2890.

Seto, S.W., Lam T.Y., Tam, H.L. et al. (2009). Novel hypoglycemic effects of *Ganoderma lucidum* water-extract in obese/diabetic (+db/+db) mice. *Phytomedicine*, 16(5): 426–36.

Seweryn, E., Ziała, A. and Gamian, A. (2021). Health-promoting of polysaccharides extracted from *Ganoderma lucidum*. *Nutrients*, 13(8): 2725. Doi:10.3390/nu13082725.

Sheena, M., Ajith, A. and Janardhanan, K. (2003). Prevention of nephrotoxicity induced by the anticancer drug Cisplatin, using *Ganoderma lucidum*, a medicinal mushroom occurring in South India. *Curr. Sci.*, 85: 478–482.

Shen, J., Park, H.S., Xia, Y.M., Kim, G.S. and Cui, S.W. (2014). The polysaccharides from fermented *Ganoderma lucidum* mycelia induced miRNAs regulation in suppressed HepG2 cells. *Carbohydr. Polym.*, 103: 319–324. Doi: 10.1016/j.carbpol.2013.12.044.

Shi, Y.L., James, A.E., Benzie, I.F. and Buswell, J.A. (2002). Mushroom-derived preparations in the prevention of H_2O_2-induced oxidative damage to cellular DNA. *Teratog. Carcinog. Mutagen*, 22: 103–11.

Shi, Y., Sun, J., He, H., Guo, H. and Zhang, S. (2008). Hepatoprotective effects of *Ganoderma lucidum* peptides against D-galactosamine-induced liver injury in mice. *J. Ethnopharmacol.*, 117: 415–19.

Shieh, Y.H., Liu, C.F., Huang, Y.K. et al. (2001). Evaluation of the hepatic and renal protective effects of *Ganoderma lucidum* in mice. *Ame. J. Chin. Med.*, 29: 501–507.

Shing, M.K., Leung, T.F., Chu, Y.L., Li, C.Y., Chik, K.W., Leung, P.C. et al. (2008). Randomized, double-blind and placebo-controlled study of the immunomodulatory effects of Lingzhi in children with cancers. *J. Clin. Oncol.*, 26(15 Suppl.): 14021–14021. Doi: 10.1200/jco.2008.26.15_suppl.14021.

Sissi, Wachtel-Galor, John, Y., John, A. and Buswell, I.F.F.B. (2011). *Ganoderma lucidum* (Lingzhi or Reishi) a medicinal Mushroom. *In*: Benzie, I.F.F. and Wachtel-Galor, S. (eds.). *Herbal Medicine: Biomolecular and Clinical Aspects* (2nd ed.). Boca Raton (FL): CRC Press/Taylor & Francis.

Sliva, D. (2003). *Ganoderma lucidum* (Reishi) in cancer treatment. *Integr. Cancer Ther.*, 2(4): 358–64. Doi: 10.1177/1534735403259066.

Sohretoglu, D. and Huang, S. (2018). *Ganoderma lucidum* polysaccharides as an anticancer agent. *Anticancer Agents Med Chem.*, 18(5): 667–674.

Sohretoglu, D., Zhang, C., Luo, J. and Huang, S. (2019). Reishi-Max inhibits mTORC1/2 by activating AMPK and inhibiting IGFR/PI3K/Rheb in tumour cells. *Signal. Transduct. Target Ther.*, 4: 21. Doi: 10.1038/s41392-019-0056-7.

Song, M., Li, Z.H., Gu, H.S., Tang, R.Y., Zhang, R., Zhu, Y.L., Liu, J.L., Zhang, J.J. and Wang, L.Y. (2021). Ganoderma lucidum Spore polysaccharide inhibits the growth of hepatocellular carcinoma cells by altering macrophage polarity and induction of apoptosis. *J. Immunology Res.*, vol., Article ID 6696606, p. 14. https://Doi.org/10.1155/2021/6696606.

Song, C.H., Yang, B.K., Ra, K.S., Shon, D.H., Park, E.J, Go, G.I. and Kim, Y.H. (1998). Hepatoprotective effect of extracellular polymer produced by submerged culture of *Ganoderma lucidum* WK-003. *J. Microbiol. Biotechnol.*, 8: 277–279.

Su, C.H., Yang, Y.Z., Ho, H., Hu, C.H. and Sheu, M.T. (2001). High-performance liquid chromatographic analysis for the characterisation of triterpenoids from *Ganoderma*. *J. Chromatogr. Sci.*, 39: 93–100.

Su, Z.Y., Sun Hwang, L., Chiang, B.H. and Sheen, L.Y. (2013). Antihepatoma and liver protective potentials of *Ganoderma lucidum* (ling zhi) fermented in a medium containing black soybean (hēi dòu) and astragalus membranaceus (shēng huáng qí). *J. Tradit. Complement. Med.*, 3(2): 110–8. Doi: 10.4103/2225-4110.110415.

Sun, L.X., Lin, Z.B., Duan, X.S., Lu, J., Ge, Z.H., Li, X.J. et al. (2011a). *Ganoderma lucidum* polysaccharides antagonize the suppression on lymphocytes induced by culture supernatants of

B16F10 melanoma cells. *J. Pharm. Pharmacol.*, 63: 725-735. Doi: 10.1111/j.2042-7158.2011. 01266.

Sun, L.X., Lin, Z.B., Li, X.J., Li, M., Lu, J., Duan, X.S. et al. (2011b). Promoting effects of *Ganoderma lucidum* polysaccharides on B16F10 cells to activate lymphocytes. *Basic Clin. Pharmacol. Toxicol.*, 108: 149–154. Doi: 10.1111/j.1742-7843.2010.00632.x.

Sun, L.X., Lin, Z.B., Duan, X.S., Lu, J., Ge, Z.H., Li, X.F. et al. (2012). Enhanced MHC class I and costimulatory molecules on B16F10 cells by *Ganoderma lucidum* polysaccharides. *J. Drug Target*, 20: 582–592. Doi: 10.3109/1061186X.2012.697167.

Sun, L.X., Li, W.D., Lin, Z.B., Duan, X.S., Li, X.F., Yang, N., Lan, T.F., Li, M., Sun, Y., Yu, M. and Lu, J. (2014). Protection against lung cancer patient plasma-induced lymphocyte suppression by *Ganoderma lucidum* polysaccharides. *Cell Physiol. Biochem.*, 33: 289–299. Doi: 10.1159/000356669.

Sun, L.X., Lin, Z.B., Duan, X.S., Lu, J., Ge, Z.H., Li, M., *et al.* (2013). *Ganoderma lucidum* polysaccharides counteract inhibition on CD71 and FasL expression by culture supernatant of B16F10 cells upon lymphocyte activation, *Exp. Ther. Med.*, 5: 1117-1122.

Sun, X.Z., Liao, Y., Li, W. and Guo, L.M. (2017). Neuroprotective effects of *Ganoderma lucidum* polysaccharides against oxidative stress-induced neuronal apoptosis. *Neural Regeneration Research*, 12(6): 953–958. https://Doi.org/10.4103/1673-5374.208590.

Suprasert, P., Apichartpiyakul, C., Sakonwasun, C., Nitisuwanraksa, P. and Phuackchantuck, R. (2014). Clinical characteristics of gynecologic cancer patients who respond to salvage treatment with Lingzhi. *Asian Pac. J. Cancer Prev.*, 15: 4193–4196. Doi:0.7314/APJCP.2014.15.10.4193.

Tomoda, M., Gonda, R., Kasahara, Y. and Hikino, H. (1986). Glycan structures of ganoderans B and C, hypoglycemic glycans of *Ganoderma lucidum* fruit bodies. *Phytochemistry*, 25: 2817–2820.

Upton, R. (2000). American Herbal Pharmacopeia and Therapeutic Compendium: Reishi Mushroom, *Ganoderma lucidum, Standards of Analysis, Quality Control, and Therapeutics*, U.S.A. Canada: Santa Cruz.

Van Der Hem, L., Van Der Vliet, A., Bocken, C.F.M., Kino, K., Hoitsma, A.J. and Tax, W.J.M. (1995). Lingzhi-8: Studies of a new immunomodulating agent. *Transplantation*, 60: 438–443.

Wachtel-Galor, S., Buswell, J.A., Tomlinson, B. and Benzie, I.F.F. (2004a). Lingzhi polyphorous fungus. *In: Herbal and Traditional Medicine: Molecular Aspects of Health*, New York: Marcel Dekker Inc., pp. 179–228.

Wachtel-Galor, S., Choi, S.W. and Benzie, I.F.F. (2005). Effect of *Ganoderma lucidum* on human DNA is dose dependent and mediated by hydrogen peroxide. *Redox Rep.*, 10(3): 145–149.

Wachtel-Galor, S., Szeto, Y.T., Tomlinson, B. and Benzie, F.I.F. (2004b). *Ganoderma lucidum* (Lingzhi): Acute and short-term biomarker response to supplementation. *Int. J. Food Sci. Nutr.*, 1: 75–83.

Wang, H. and Ng, T.B. (2006). Ganodermin, an antifungal protein from fruiting bodies of the medicinal mushroom *Ganoderma lucidum*. *Peptides*, 27: 27–30.

Wang, X., Wang, B., Zhou, L., Wang, X., Vishnu Priya, V., Krishna Mohan, S. and Feng, X. (2020). *Ganoderma lucidum* put forth antitumour activity against PC-3 prostate cancer cells via inhibition of Jak-1/STAT-3 activity. *Saudi Journal of Biological Sciences*, 27(10): 2632–2637. https://Doi.org/10.1016/j.sjbs.2020.05.044.

Wang, Y.Y., Khoo, K.H., Chen, S.T., Lin, C.C., Wong, C.H. and Lin, C.H. (2002). Studies on the immunomodulating and antitumor activities of *Ganoderma lucidum* (Reishi) polysaccharides: Functional and proteomic analyses of a fucose-containing glycoprotein fraction responsible for the activities. *Bioorg. Med. Chem.*, 10: 1057–1062.

Wang, S.Y., Hsu, M.L., Hsu, H.C. et al. (1997). The anti-tumor effect of *Ganoderma lucidum* is mediated by cytokines released from activated macrophages and T-lymphocytes. *Int. J. Cancer*, 70: 699–705.

Wanmuang, H., Leopairut, J., Kositchaiwat, C., Wananukul, W. and Bunyaratvej, S. (2007). Fatal fulminant hepatitis associated with *Ganoderma lucidum* (Lingzhi) mushroom powder. *J. Med. Assoc. Thai.*, 90(1): 179–181.

Wasser, S.P. and Weis, A.L. (1999). Medicinal properties of substances occurring in higher basidiomycetes mushrooms: Current perspectives. *Int. J. Med. Mushrooms*, 1: 31–62.

Wasser, S.P., Coates, P., Blackman, M., Cragg, G., Levine, M., Moss, J. and White, J. (2005). *Encyclopaedia of Dietary Supplements*, New York: Marcel Dekker, *Reishi or Lingzhi* (*Ganoderma lucidum*), pp. 680–90.

Weng, C.J. and Yen, G.C. (2010). The *in vitro* and *in vivo* experimental evidences disclose the chemo-preventive effects of *Ganoderma lucidum* on cancer invasion and metastasis. *Clin. Exp. Metastasis*, 27(5): 361–9. Doi: 10.1007/s10585-010-9334-z.

Wu, Y. and Wang, D. (2009). A new class of natural glycopeptides with sugar moiety-dependent antioxidant activities derived from *Ganoderma lucidum* fruiting bodies. *J. Proteome. Res.*, 8: 436–42.

Wu, Y.W., Fang, H.L. and Lin, W.C. (2010). Post-treatment of *Ganoderma lucidum* reduced liver fibrosis induced by thioacetamide in mice. *Phytother. Res.*, 24(4): 494–9.

Yoon, S.Y., Eo, S.K., Kim, Y.S., Lee, C.K. and Han, S.S. (1994). Antimicrobial activity of *Ganoderma lucidum* extract alone and in combination with some antibiotics. *Arch. Pharm. Res.*, 17: 438–442.

Yue, G.G., Fung, K.P., Tse, G.M., Leung, P.C. and Lau, C.B. (2006). Comparative studies of various *Ganoderma* species and their different parts with regard to their antitumour and immunomodulating activities *in vitro*. *J. Altern. Complement. Med.*, 12: 777–789. Doi: 10.1089/acm.2006.12.777.

Yuen, J.W. and Gohel, M.D. (2005). Anticancer effects of *Ganoderma lucidum*: A review of scientific evidence. *Nutr. Cancer*, 53: 11–17.

Yuen, J.W. and Gohel, M.D. (2008). The dual roles of *Ganoderma* antioxidants on urothelial cell DNA under carcinogenic attack. *J. Ethnopharmacol.*, 118: 324–30.

Yuen, J.W., Gohel, M.D. and Au, D.W. (2008). Telomerase-associated apoptotic events by mushroom *Ganoderma lucidum* on premalignant human urothelial cells. *Nutr. Cancer*, 60: 109–119.

Zeng, P., Guo, Z., Zeng, X., Hao, C., Zhang, Y., Zhang, M., Liu, Y., Li, H., Li, J. and Zhang, L. (2018). Chemical, biochemical, preclinical and clinical studies of *Ganoderma lucidum* polysaccharide as an approved drug for treating myopathy and other diseases in China. *J. Cell Mol Med.*, 22(7): 3278–3297. Doi: 10.1111/jcmm.13613.

Zhang, L., Zhang, M. and Chen, J. (2001). Solution properties of antitumor carboxy-methylated derivatives of a-(1→3)-D-Glucan from *Ganoderma lucidum*. *Chin. J. Polym. Sci.*, 19: 283–289.

Zhong, J.J. and Xiao, J.H. (2009). Secondary metabolites from higher fungi: Discovery, bioactivity and bio-production. *Adv. Biochem. Eng. Biotechnol.*, 113: 79–150.

Zhou, X., Lin, J., Yin, Y., Zhao, J., Sun, X. and Tang, K. (2007). Ganodermataceae: Natural products and their related pharmacological functions. *Ame. J. Chin. Med.*, 35: 559–74.

Zhu, X. and Lin, Z. (2006). Modulation of cytokines production, granzyme B and perforin in murine CIK cells by *Ganoderma lucidum* polysaccharides. *Carbohydr. Polym.*, 63: 188–97.

CHAPTER 12

Grifola frondosa
Nutraceutical and Medicinal Potential

Sudha Nandni,* Devanshu Dev and *Dayaram*

1. Introduction

Grifola frondosa comes under Grifolaceae family and Polyporales order edible Basidiomycetes fungus with a large pileus and overlapping caps. Maitake is the name used in Japan for the edible fruiting body. "Mai" denotes "dancing," while "takc" is used for "mushroom" in Japanese language. According to folklore, this mushroom was given the name "maitake" (dancing mushroom) because anyone who discovered it would dance with glee instead of telling his closest relatives where it was grown. Because of its appearance, *G. frondosa* is well-known in China by name "hui-shu-hua" (grey tree bloom). Sheep's head, king of mushrooms, hen-of-the-woods, and cloud

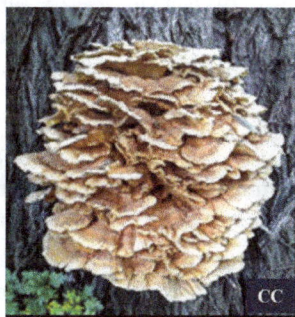
Fig. 1. Mature Maitake mushroom growing on a wooden log.

mushroom are some of the other names for it (Mayell, 2001). For optimal growth, it needs a cool temperature of 17–23°C. *G. frondosa* flourishes in the environment of Japan's north-eastern region. Temperate forests as are found in the eastern North America, Europe, and Asia are also conducive for the progression of this mushroom. Maitake is most usually seen on fallen or uprooted *Fagus crenata, Quercus serrata*, and *Quercus crispula* Fagaceae trees. Prior to the advent of artificially produced maitake onto the market, maitake was available at a reasonable price despite its rarity. The mushroom that has been artificially cultivated is currently accessible in large quantities in the market. Annual output has increased as a result of its outstanding flavour.

Dr. Rajendra Prasad Central Agricultural University, Pusa, Samastipur, 848125, Bihar.
Emails: dev9105@gmail.com; raudayaram@gmail.com
* Corresponding author: sudha.nandni@rpcau.ac.in

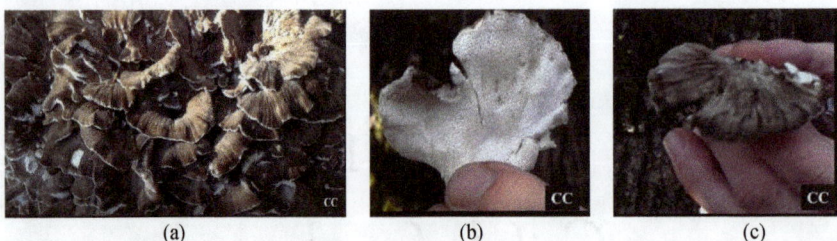

Fig. 2. (a) Showing mature Maitake mushroom growing on fallen tree (b) Showing polypores underside of the single Maitake mushroom (c) Showing upper side of the single Maitake mushroom in dark color.

2. Mycelium Growth and Artificial Cultivation

Mycelium and bioactive metabolites from *G. frondosa*, in addition to the fruiting body, are in great demand. There are two methods for cultivating mycelium:

a. Solid state fermentation (SSF) (Montoya et al., 2011) and

b. Submerged fermentation (Lee et al., 2003)

A typical substrate for SSF is sawdust enriched with rice or wheat bran (Takama et al., 1981). In most cases, submerged fermentation is more efficient, enabling further effective product quality control by producing more mycelial output in less time and utilising less space (Shih et al., 2008).

Japan was one among the first countries to develop an artificial culture of Maitake mushroom in the mid-1980s. The three main procedures for artificial production of the *G. frondosa* mushrooms are (1). Bottle culture (2). Bag culture and (3). Outdoor bed culture (Mayuzumi and Mizuno, 1997). Substrate preparation, substrate sterilisation, and bag culture are the three primary processes of bag cultivation.

a. Bottle culture: This approach is ideal for all-year cultivation and may be automated with little effort. The plucked mushroom is in smaller quantities due to the small bottles used (800–1000 ml).

b. Bag culture: This approach is most commonly employed in Japan. Sawdust is blended with rice bran and wheat bran. After adjusting the moisture of the mixture, it is placed in a polyethylene bag and moulded into a square-shaped culture bed.

c. Outdoor bed culture: Originally it was tried under natural environmental conditions in Japan. From inoculation to fruiting body production, this method takes around 6 months.

3. Bioactive Molecules

3.1 Polysaccharides

G. frondosa is an edible and considered a nutritious meal since it is high in protein, carbs, dietary fibre, vitamin D2 and minerals like - potassium phosphorus, sodium, calcium and magnesium and at the same time low in fat and calorie (Tabata et al., 2004). The high percentage of trehalose in addition to glutamic, and aspartic amino acid, as well as the 5′-nucleotide content, give *G. frondosa* a sweet and savory taste. (Ohno et al., 1984).

G. *frondosa* has roughly 83 to 96 per cent water content and 4 to 17 per cent dry matter in its newly grown fruiting body and fresh mycelium (Huang et al., 2011; Aljhouni et al., 1995), reflecting its soggy texture. The main components that contribute to the dry weight of Maitake mushroom are carbohydrates and protein, accounting for almost 70–80 per cent and 13–21 per cent of the fruiting body respectively.

Fruiting bodies, mycelia and fermentation broth may all be used to extract GFP (*G. frondosa* polysaccharides) using several extraction procedures (Takama et al., 1981). Diverse extraction procedures, like microwave extraction, as well as hot water extraction and chemical procedure like alkaline extraction and acid precipitation are examples of techniques that can be applied to acquire different water-miscible and water-immiscible components. A glycoprotein or a pure polysaccharide can be the active component (Montoya et al., 2011; Shih et al., 2008). A protein-bound polysaccharide or proteoglucan comprised of β-glucan is known as the D-fraction or GFP (either 1,6 linkage in glucose backbone with 1,3 linked glucose branches or 1,3 linked glucose backbone with 1,6 linked glucose branches) (Huang et al., 2011). Its glycans are commonly composed of d-glucose, d-xylose, as well as d-fucose, l-arabinose, d-mannose along with galactose and uronic acid (Tabata et al., 2004). The D-fraction, a β-glucan complex with approximately 30 per cent protein, is contained in this mushroom, which is increasingly being turned into commercially accessible supplemental medications and healthcare items (Hishida et al., 1988). A human experiment in its I/II phase done by the Memorial Sloan–Kettering, Cancer Centre in 2009 revealed that GFP might boost the immune systems in patients suffering from breast cancer (Deng et al., 2009).

Several different bioactive polysaccharide fractions including the MD-fraction, X-fraction, Grifolan, MZ-fraction and MT-α-glucan (Nanba and Kubo, 1998; Kubo et al., 1994; Adachi et al., 1994; Masuda et al., 2006; Lei et al., 2007) are produced from G. *frondosa* along with the D-fraction, which have various bioactive effects such as immune-modulation (Adachi et al., 1994), antitumor (Masuda et al., 2006) antivirus (Zhao et al., 2016), antidiabetic (Lei et al., 2007) and anti-inflammation (Su et al., 2020) respectively. Polysaccharides from G. *frondosa* have the ability to change gut microbiota, which are bacteria playing an essential role in the health and illness of humans. Especially, gut microbiota, have a function in the immunological homeostasis that may be related to polysaccharide anticancer effects (Liu et al., 2019). Polysaccharides found in G. *frondosa* regulate the makeup of the gut microbiota. G. *frondosa* polysaccharides are also found helpful in the treatment of metabolic diseases like NAFLD- Non-Alcoholic Fatty Liver Disease (Li et al., 2019) and diabetes (Chen et al., 2019), highlighting their impact for treating or avoiding hyperglycemia and hyperlipidemia.

3.2 Protein

Other molecular fractions having medical relevance, aside from polysaccharides, include protein components extracted from the fruiting bodies or mycelial biomass of G. *frondosa*. Glycoprotein of G. *frondosa* has anti-tumor (Cui et al., 2013) immune-enhancing (Tsao et al., 2013) antidiabetic, anti-hypertensive, anti-hyperlipidemic (Zhuang et al., 2007) and anti-viral effects (Gu et al., 2007). Being a medicinal mushroom, G. *frondosa* has many other medicinally important biomolecules other than polysaccharides and proteins.

3.3 Other Bioactive Molecules

G. frondosa contains bioactive small molecules in addition to some macro-molecular components like polysaccharides and proteins or peptides. Among the most important tiny molecules identified to have bioactivities are ergosterols, fatty acids, alkaloids, flavonoids, along with tocopherol and ascorbic acid. A furanone isolated from *G. frondosa*, Grifolaone A, displays particular anti-fungal action against *Pseudallescheria boydii*, a well-known opportunistic human pathogen, as well as several other plant infections (He et al., 2016).

4. Chemico-nutritional Constitution

4.1 Proximate Constitution

The fruiting body of *G. frondosa* has a crude fat content which is lesser than the crude fat (average) percentage of artificially grown mushrooms (4.3%), while the percentage of carbohydrates (17.2 per cent) and protein (70.3 per cent) is higher than other mushroom's average carbohydrates and protein to some extent, reflecting that *G. frondosa* has an excellent nutritional values.

4.2 Soluble Sugars

The total soluble sugar serving in Maitake mushroom fruiting bodies (90 to 190 mg/g) is greater than that found in the mycelium (70–90 mg/g). In a study, Sanmee et al., 2003 found that the total soluble sugar percentage of Maitake mushroom's fruiting body is more in comparison to that of several edible mushrooms like *Lactarius glaucescens* and *Craterellus odoratus*, suggesting an explanation for the mushroom's pleasant flavour. Growing period and cultivation environment have an impact on the total soluble sugar concentration and individual sugar content of different *G. frondosa* samples. The most prevalent sugar in *G. frondosa*'s fruiting body and mycelia is trehalose, a disaccharide composed of two molecules of glucose (Huang et al., 2011). As compared to the mycelium, the fruiting body contains about 40–60 mg/g trehalose, or between 50 and 160 mg/g dry weight. Glucose and mannitol are also present, but not at the same concentrations as in the mycelium. Arabitol and fructose are also present in the fruiting body.

4.3 Free Amino Acids

In the fruiting body of *G. frondosa*, there are approximately 15 to 60 milligrams of free amino acids per gram of dry weight, which is greater than a lot of other edible mushrooms, including *Tricholoma giganteum* and *Dictyophora indusiata* (Mau et al., 2001). *G. frondosa* holds a high level of free amino acids in its mycelium compared to its fruiting body. There are also many different amino acids in *G. frondosa*, including the fruiting body and mycelium.

5. Bioactivities and Pharmacological effects

Maitake is primarily consumed as a delicacy in Asia while its extracts are prescribed in the treatment of variety of ailments, including arthritis, hepatitis, and even the human immunodeficiency virus (HIV). Adaptogens are substances that conduct general activities in the body to boost immune system, develop the endurance, and reduce tiredness, according to some herbal specialists. Medicinal mushrooms are adaptogens. Maitake is occasionally prescribed to cancer patients because of its potential immunological and anticancer properties. Over the years, several fractions of the fungus have been separated in Japan with the goal of extracting β-glucans having significant therapeutic qualities.

Since maitake has long been thought of as a therapeutic fungus, therefore it has been the focus of considerable scientific investigation for over half a century. Researchers have discovered a variety of therapeutic characteristics that might bring significant health advantages as time advances and prepared the following timeline for the research advancement took place in GFP (Deng et al., 2009).

1955-First time *G. frondosa* was studied

1984-Anti tumour activity

1992-Anti-HIV effect

2002-Alcoholic oxidative injury and anti-radiation activity

2003-Anti-hyperlipidemia activity

2007-Anti-diabetic activity

2010-GFP launched as a SFDA drug named "Mai TeXiao" in China

G. frondosa is said to have vast range of pharmacological effects complementary to its high nutritional value which are discussed further in the chapter.

5.1 *Immunomodulatory Action*

The best well-known activity of *G. frondosa* components is immunomodulation, since the nutraceuticals have been demonstrated to have immunomodulatory characteristics. These immunomodulatory components boost the activities of macrophages and a variety of other cells related to immune system, including T-cells (cytotoxic) and NK (natural killer) cells (Kodama et al., 2004). Additionally, *G. frondosa* constituents may enhance the secretion of cytokines (signaling molecules), that includes interleukins (IL), interferons (IFN), tumour necrosis factors (TNF), and lymphokines with an anti-proliferative activity, triggering apoptosis and differentiation in tumour cells, and hence adding to the efficiency of cells related to immune system (Ishibashi et al., 2001).

G. frondosa polysaccharides have very earlier been identified as the primary components playing direct part in immunomodulation. Besides its direct antitumor activity, the D-fraction of *G. frondosa* is a major polysaccharide fraction with considerable immunomodulatory activity. By boosting the proliferation, differentiation, and activation of immune-competent cells, the D-fraction may be able to lower the effective dose of the chemo-therapeutic drug Mitomycin-C (MMC). It may also lessen MMC's immunosuppressive action. Aside from the D-fraction, Grifolan's (GRN) high molecular weight soluble forms have been identified. Both these components can trigger macrophages by activating cytokine release for creating TNF (Ishibashi et al., 2001; Wang et al., 2013).

5.2 Anti-Cancer Nature

Medicinal mushrooms are probably the most widely prescribed anticancer natural items, with results from controlled clinical trials indicating that they may be beneficial in cancer therapy. Among those utilized is the maitake mushroom (*Grifola frondosa*). Medicinal mushrooms include β-glucans, a kind of polysaccharide that promotes antitumor immunity via antibody–Fc interactions by activating complement receptors. Mouse models have indicated that β-glucans function synergistically with therapeutic antibodies such as trastuzumab or rituximab (Ajlouni et al., 1995). Human data are sparse and inadequate evidence is there to suggest for or against the oral use of Maitake mushroom for any reason at this time; nonetheless, they are regularly consumed as a meal and appear to be safe for general use. Maitake is a prominent cancer treatment. Although some modest studies have not identified any effects (Chen et al., 2019), In animal experiments, maitake extracts were shown to boost immune function and perhaps initiate host-mediated anticancer action (Adachi et al., 1994; Masuda et al., 2006; Su et al., 2020).

5.3 Antiviral and Antibacterial Properties

G. frondosa has several favourable benefits in the treatment of viral illnesses such as human immunodeficiency virus (HIV), hepatitis B virus (HBV), enterovirus 71 (EV71), and herpes simplex virus type 1 (HSV-1). *G. frondosa's* MD-fraction can combat HIV through a variety of mechanisms, including direct suppression of the virus, promotion of the natural defense system against the virus, and reduced sensitivity to opportunistic infections (Nanba et al., 2000). The GFP1 fraction is efficient in combating EV71, the pathogen responsible for foot-and-mouth disease. *G. frondosa* has the ability to inhibit viral replication of EV71 by reducing genomic RNA synthesis and protein production, making it a viable therapeutic molecule for EV71 therapy (Zhao et al., 2016). Besides polysaccharide components, the protein content extracted from *G. frondosa* shows anti-viral properties. It strongly suppresses HSV-1 multiplication in vitro and alleviates HSV-1-related symptoms including blepharitis (Gu et al., 2007). Because of the presence of D-fraction and its immune-stimulating activity, *G. frondosa* has antibacterial properties. When immunocompetent cells are stimulated, they encourage the production of cytokines, which leads to an increase in the activity of splenic T cells to destroy germs (Kodama et al., 2001).

5.4 Antidiabetic Effect

In numerous animal studies, the hypo-glycaemic properties of *G. frondosa* have been thoroughly researched and demonstrated. *G. frondosa* has an effect on fasting serum glucose (FSG). Polysaccharide fractions regulate insulin activity via the insulin signal pathway, where they promote glucose absorption, culminating the insulin receptor (IR) and insulin receptor substrate 1 (IRS-1) inactivation and ultimately, enhanced insulin secretion. In addition to increased insulin activity, *G. frondosa* produces hypoglycemic effects by inhibiting -glucosidase activity, since an anti—glucosidase effect limits starch breakdown into disaccharides and lowers blood glucose levels (Konno et al., 2013).

5.5 Lipid Metabolism Regulation

G. frondosa helps improving the metabolism of lipid and keeps in check the rise in liver lipid and serum lipid after the ingestion of high-fat feed. The fruiting body of *G. frondosa* suppressed cholesterol, triglyceride, and phospholipid levels in rats' blood by 30–80 per cent and enhanced the corresponding cholesterol excretion ratio in faeces by 1.8 times compared with those of the control group of animals (Nanba and Kubo, 1998).

5.6 Anti-Hypertension Effects

The antihypertensive effects of *G. frondosa*'s active components have mostly been investigated using systolic blood pressure (SBP) measurements in animal models. *G. frondosa* not only has preventive effect on hypertension, but also has treatment effect, i.e., it lowers the elevated blood pressure (Kabir and Kimura, 1989). There are commercially-available fractions of *G. frondosa* which when used as a baseline diet treated age-related hypertension partly via their effects on the renin-angiotensin system (Preuss et al., 2010).

5.7 Antioxidant Properties

Polysaccharides, proteins, fatty acids, and other constituents in *G. frondosa* have significant antioxidant properties. Reducing power and Fe2+ chelating activity as well as scavenging abilities of hydroxyl radicals, superoxide radicals, hydrogen peroxide are some of the most used antioxidant activity tests. Because of their radical scavenging action, antioxidant activity following UV irradiation, fibroblast proliferation, and collagen production, the polysaccharides from *G. frondosa* have the potential for cosmetic uses (Lee et al., 2003). The antioxidant activity of *G. frondosa* polysaccharides increases after the incorporation of zinc or selenium. The molecular weight, degree of branching, shape, as well as the synergistic action of polysaccharide, may influence antioxidant activity (Li et al., 2019). In addition to proteins, fatty acids and other compounds from *G. frondosa*, such as phenols and flavonoids have antioxidant properties as well.

5.8 Gut Micro-biota Regulating Properties

Mushrooms provide a variety of health advantages, owing to their polysaccharide content (Jayachandran et al., 2017). Bioactive polysaccharides are biological macromolecules of polysaccharides that cannot be absorbed directly by the body and are found in edible and medicinal mushrooms such as *G. frondosa* but are easily used by intestinal flora and control the gut microbiota (Cotillard et al., 2013). Because of their gut microbiota regulating mechanism, these polysaccharides have anti-diabetes, anti-cancer, anti-obesity, and antibacterial capabilities (Friedman, 2016). In particular, enhanced treatment of type 2 diabetic mellitus (T2DM) (Guo et al., 2020) and non-alcoholic fatty liver disease (NAFLD) is successful owing to gut microbiota homeostasis (Gangarapu et al., 2014). Treatment with *G. frondosa* polysaccharides (GFP-N) results in the relative abundance of some bacterial species such as *Bacteroides acidifaciens* (*B. acidifaciens*) and *Lactobacillus acidophilus* (*L. acidophilus*). Due to the gut microbiota regulation, immune system boosts which ultimately contributes to the anti-inflammatory and anti-tumor effect of the *G. frondosa* polysaccharides (GFP) (Zhang et al., 2018; Liu et al., 2019).

6. Conclusion

Maitake is being used from hundreds of years in traditional medicine, but there is no evidence of its usefulness in randomised, controlled studies. Preliminary evidence for cancer, diabetes, and immune-stimulation is now available, and it is based on animal and laboratory data as well as case studies. Maitake's appeal stems from its long history of usage in Chinese cultures for its putative immune-enhancing properties. There are inadequate evidences to recommend for or against the use of oral maitake for any indication due to a lack of human data. However, available animal and laboratory data, on the other hand, highlighted areas of ongoing study, as well as numerous speculative or historical uses for which there is insufficient proof. As a result, high-quality scientific information on maitake's safety and efficacy as an integrative treatment is required.

References

Adachi, Y. et al. (1994). Enhancement of cytokine production by macrophages stimulated with (1→3)-βD-glucan, grifolan (GRN), isolated from *Grifola frondosa*. *Biol. Pharm. Bull.*, 17: 1554–1560.

Aljhouni, S.O. et al. (1995). Changes in soluble sugars in various tissues of cultivated mushrooms, *Agaricus bisporus*, during postharvest storage. pp. 1865–1880. *In*: Charalambous, G. (ed.). *Developments in Food Science*, Elsevier, Amsterdam, The Netherlands, 37.

Chen, Y. et al. (2019). Hypoglycemic activity and gut microbiota regulation of a novel polysaccharide from *Grifola frondosa* in type 2 diabetic mice. *Food Chem. Toxicol.*, 126: 295–302.

Cotillard, A. et al. (2013). Dietary intervention impact on gut microbial gene richness. *Nature*, 500: 585–588.

Cui, F. et al. (2013). Purification and partial characterisation of a novel anti-tumour glycoprotein from cultured mycelia of *Grifola frondosa*. *Int. J. Biol. Macromol.*, 62: 684–690.

Deng, G. et al. (2009). A phase I/II trial of a polysaccharide extract from *Grifola frondosa* (maitake mushroom) in breast cancer patients: immunological effects. *J. Cancer Res. Clin. Oncol.*, 135(9): 1215–1221.

Friedman, M. (2016). Mushroom polysaccharides: Chemistry and anti-obesity, antidiabetes, anticancer, and antibiotic properties in cells, rodents, and humans. *Foods*, 5: 80.

Gangarapu, V. et al. (2014). Role of gut microbiota: Obesity and NAFLD. *Turk. J. Gastroenterol.*, 25: 133–140.

Gu, C.-Q. et al. (2007). Isolation, identification and function of a novel anti-HSV-1 protein from *Grifola frondosa*. *Antivir. Res.*, 75: 250–257.

Guo, W.-L. et al. (2020). Hypoglycemic and hypolipidemic activities of *Grifola frondosa* polysaccharides and their relationships with the modulation of intestinal microflora in diabetic mice induced by high-fat diet and streptozotocin. *Int. J. Biol. Macromol.*, 153: 1231–1240.

He, X. et al. (2016). Extraction, identification and antimicrobial activity of a new furanone, grifolaone A, from *Grifola frondosa*. *Nat. Prod. Res.*, 30: 941–947.

Hishida, I. (1988). Antitumour activity exhibited by orally administered extract from fruit body of *Grifola frondosa* (Maitake). *Chem. Pharm. Bull.*, 36: 1819–1827.

Huang, S.-J. et al. (2011). Nonvolatile taste components of culinary-medicinal maitake mushroom, *Grifola frondosa* (Dicks.:Fr.) S.F. Gray. *Int. J. Med. Mushrooms*, 13: 265–272.

Ishibashi, K.-I. et al. (2001). Relationship between solubility of Grifolan, a fungal 1,3-beta;- D-Glucan, and production of tumour necrosis factor by macrophages *in vitro. Biosci. Biotechnol. Biochem.*, 65: 1993–2000.

Jayachandran, M. et al. (2017). A critical review on health promoting benefits of edible mushrooms through gut microbiota. *Int. J. Mol. Sci.*, 18: 1934.

Kabir, Y. and Kimura, S. (1989). Dietary mushrooms reduce blood pressure in spontaneously hypertensive rats (SHR). *J. Nutr. Sci. Vitaminol.*, 35: 91–94.

Kodama, N. (2004). Administration of a polysaccharide from *Grifola frondosa* stimulates immune function of normal mice. *J. Med. Food*, 7: 141–145.

Kodama, N. et al. (2001). Addition of Maitake D-fraction reduces the effective dosage of Vancomycin for the treatment of Listeria-infected mice. *Jpn. J. Pharmacol.*, 87: 327–332.

Kodama, N. et al. (2005). Maitake D-Fraction enhances anti-tumour effects and reduces immunosuppression by mitomycin-C in tumour-bearing mice. *Nutrition*, 21: 624–629.

Konno, S. et al. (2013). Possible hypoglycemic action of SX-fraction targeting insulin signal transduction pathway. *Int. J. Gen. Med.*, 6: 181–187.

Kubo, K. et al. (1994). Anti-diabetic activity present in the fruit body of *Grifola frondosa* (Maitake), *I. Biol. Pharm. Bull.*, 17: 1106–1110.

Lee, B.C. et al. (2003). Biological activities of the polysaccharides produced from submerged culture of the edible Basidiomycete *Grifola frondosa. Enzym. Microb. Technol.*, 32: 574–581.

Lei, H. et al. (2007). Anti-diabetic effect of an α-glucan from fruit body of maitake (*Grifola frondosa*) on KK-Ay mice. *J. Pharm. Pharmacol.*, 59: 575–582.

Li, X. et al. (2019). The positive effects of *Grifola frondosa* heteropolysaccharide on NAFLD and regulation of the gut microbiota. *Int. J. Mol. Sci.*, 20: 5302.

Liu, L. et al. (2019). Natural polysaccharides exhibit anti-tumour activity by targeting gut microbiota. *Int. J. Biol. Macromol.*, 121: 743–751.

Masuda, Y. et al. (2006). Macrophage J774. 1 cell is activated by MZ-Fraction (Klasma-MZ) polysaccharide in *Grifola frondosa. Mycoscience*, 47: 360–366.

Mau, J.-L. et al. (2001). Non-volatile taste components of several speciality mushrooms. *Food Chem.*, 73: 461–466.

Mayell, M. (2001). Maitake extracts and their therapeutic potential—A review. *Altern. Med. Rev.*, 6: 48–60.

Mayuzumi, Y. and Mizuno, T. (1997). III. Cultivation methods of maitake (*Grifola frondosa*). *Food Rev. Int.*, 13: 357–364.

Montoya Barreto, S. et al. (2011). Modelling *Grifola frondosa* fungal growth during solid-state fermentation. *Eng. Life Sci.*, 11: 316–321.

Nanba, H. and Kubo, K. (1998). *Anti-tumour Substance Extracted from Grifola*. U.S. Patent 5,854,404.

Nanba, H. et al. (2000). Effects of maitake (*Grifola frondosa*) glucan in HIV-infected patients. *Mycoscience*, 41: 293–295.

Ohno, N. et al. (1984). Anti-tumour activity and structural characterisation of glucans extracted from cultured fruit bodies of *Grifola frondosa. Chem. Pharm. Bull.*, 32: 1142–1151.

Preuss, H.G. et al. (2010). Maitake mushroom extracts ameliorate progressive hypertension and other chronic metabolic perturbations in aging female rats. *Int. J. Med Sci.*, 7: 169.

Sanmee, R. et al. (2003). Nutritive value of popular wild edible mushrooms from northern Thailand. *Food Chem.*, 82: 527–532.

Shih, L. et al. (2008). Study of mycelial growth and bioactive polysaccharide production in batch and fed-batch culture of *Grifola frondosa. Bioresour. Technol.*, 99: 785–793.

Su, C.-H. et al. (2020). A (1→6)-Branched (1→4)-β-d-Glucan from *Grifola frondosa* inhibits lipopolysaccharide-induced cytokine production in RAW264. 7 macrophages by binding to TLR2 rather than Dectin-1 or CR3 Receptors. *J. Nat. Prod.*, 83: 231–242.

Tabata, T. et al. (2004). Comparison of chemical compositions of maitake (*Grifola frondosa* (Fr.) SF Gray) cultivated on logs and sawdust substrate. *Food Sci. Technol. Res.*, 10: 21–24.

Takama, F. et al. (1981). Parenchyma cells, chemical components of maitake mushroom (*Grifola frondosa* SF Gray) cultured artificially and their changes by storage and boiling. *In: Proceedings* of the Eleventh International Scientific Congress on the Cultivation of Edible Fungi, Sydney, Australia.

Tsao, Y.-W. et al. (2013). Characterisation of a novel maitake (*Grifola frondosa*) protein that activates natural killer and dendritic cells and enhances antitumor immunity in mice. *J. Agric. Food Chem.*, 61: 9828–9838.

Wang, Y. et al. (2013). Inducement of cytokine release by GFPBW2, a novel polysaccharide from fruit bodies of *Grifola frondosa*, through dectin-1 in macrophages. *J. Agric. Food Chem.*, 61: 11400–11409.

Zhang, B. et al. (2018). Anti-diabetic effect of baicalein is associated with the modulation of gut microbiota in streptozotocin and high-fat-diet induced diabetic rats. *J. Funct. Foods*, 46: 256–267.

Zhao, C. et al. (2016). Structural characterisation and antiviral activity of a novel heteropolysaccharide isolated from *Grifola frondosa* against enterovirus 71. *Carbohydr. Polym.*, 144: 382–389.

Zhuang, C. et al. (2007). *Glycoprotein with Antidiabetic, Antihypertensive, Antiobesity and Antihyperlipidemic Effects from Grifola frondosa, and a Method for Preparing Same.* U.S. Patent 7,214,778, 8 May.

CHAPTER 13

Lion's Mane (*Hericium erinaceus*)

Emeric Kochoni[1],* and Vincent Ezin[2]

1. Introduction

For about 200 years, scientific and technological progress has shaped our behaviour, especially our lifestyles (Grubler, 1990). The diseases which have been dreaded are being eradicated so that everywhere on Earth, life expectancy is greatly improving. Thanks to modern means of communication and transport, the Earth has become a global village, which has the effect of facilitating various exchanges (knowledge, goods, technology, etc.). However, these various developments and technological prowess are not without consequences and the global changes that are added to them pose new challenges to humanity.

For example, in the field of human health, one of the challenges is to deal with the many so-called metabolic symptoms or diseases which are consequences of modern lifestyle and aging, by preventing and curing them. This is what justifies in particular the enthusiasm of the industry in seeking alternative methods and/or new active principles or compounds that can be widely used in nutraceuticals and functional foods.

For this, the use of substances present in Nature (plants, algae, fungi) is an obvious choice. Medicinal mushrooms are therefore part of the natural products currently being researched to obtain promising bioactive and therapeutic compounds. Consumed for their organoleptic properties, mushrooms are also valued for their various nutritional and pharmacological properties (Khan et al., 2018; Liu et al., 2022; Reis et al., 2017). For this reason, they are increasingly interesting in maintaining the overall health and preventing disease and so are considered inherent nutraceuticals and functional foods (Reis et al., 2017; Yadav et al., 2020). They contain various compounds whose intrinsic bioactivities give them their positive effects or properties, like antitumour, antimicrobial and antifungal, anti-inflammatory, antiviral, anti-

[1] Institut national de recherche scientifique (INRS), Québec, Canada.
[2] Faculty of Agricultural Sciences, University of Aboemy-Calavi, Benin.
* Corresponding author: gbatchin.kochoni@gmail.com

diabetic and cardioprotective, hepatoprotective, nephroprotective and neuroprotective besides improving anxiety, cognitive function and depression (Khan et al., 2018; Reis et al., 2017; Wang et al., 2014).

The present work focuses on *Hericium erinaceus*, a culinary-medicinal mushroom used traditionally in Asian countries as medicine and as medicinal cuisine (Thongbai et al., 2015; Wang et al., 2014; Wang et al., 2015). *H. erinaceus* (also called Lion's Mane in the United States, Houtoujun in China and Yamabushitake in Japan) belongs to the phylum (division) Basidiomycota, subphylum Agaricomycotina, class Agaricomycetes, order Russulales, family Hericiaceae and genus *Hericium* (He et al., 2017; Thongbai et al., 2015). *H. erinaceus*, known to have various health benefits, is a luxury food in China and used as medicinal food to treat digestive system-associated disorders in China and Japan (Tiwari et al., 2022; Wang et al., 2019). Like many other mushrooms, *H. erinaceus* exhibits various pharmacological properties that justify its use as functional food and nutraceutical. It contains several bioactive compounds, like polysaccharides, proteins, terpenoids and hericenone, and many processed products and supplements are derived from them (Liu et al., 2016).

This study aims to review the state-of-the-art in the use of *H. erinaceus* and its extracts as nutraceuticals and functional foods with an overview of its main properties and bioactive compounds.

2. Bioactive Compounds of *H. erinaceus*

Lion's Mane mushroom contains an exceptionally high amount of structurally diverse and indubitably bioactive components, such as phenolic compounds, steroids, alkaloids, lactones, monounsaturated fatty acids, essential amino acids, polysaccharides and glycoproteins (Ma et al., 2021). These bioactive compounds, estimated to be more than 70, have also been classified into five organic compounds (erinacines, hericerins, steroids, alkaloids and lactones) extracted from both fruiting bodies (FB) and mycelia biomass (MB) (Friedman 2015; He et al., 2017). He et al., reported that hericerins (aromatic compounds) were only found in the FB of *H. erinaceus*, while erinacines were reported mainly in the mycelium and were found in trace amounts in fruiting body (He et al., 2017).

Among the bioactive compounds isolated from *H. erinaceus*, polysaccharides have been well studied and are well known as a major bioactive with high bioactivity (Khan et al., 2018; Wang et al., 2019). Polysaccharides are types of bioactive carbohydrates consisting of glucans comprising more than 10 monosaccharides linked by α (axial position of carbon) or β (equatorial position of carbon) glycosidic bonds involving carbons 1 and 3, 1 and 4 or 1 and 6 (Khan et al., 2018; Ma et al., 2021; Pandya et al., 2018). Some medicinal properties of *H. erinaceus* described below have been attributed to its polysaccharide fraction which mainly contains β-glucans (Khan et al., 2018; Khan et al., 2013). According to He et al. (2017) 100 g of *H. erinaceus* consist an average of 69.4 g dw [61.3–77.5] of total sugar, mainly polysaccharides (He et al., 2017). This type of polysaccharide or oligosaccharide has been shown to possess various therapeutic functions, such as antitumour, immunomodulation, anti-gastric ulcer, neuroprotection and neuroregeneration, anti-oxidation, hepatoprotection, anti-hyperlipidemia, anti-hyperglycemia, anti-fatigue and anti-aging (Gao et al., 2019; Khan et al., 2018; Khan et al., 2013; Liu et al., 2022).

Several articles have reviewed the recent advances in *H. erinaceus* research and study and have deciphered the possible health benefits of this mushroom and identified the different bioactive compounds responsible for these healing and curative properties (Friedman, 2015; He et al., 2017; Khan et al., 2018; Khan et al., 2013; Liu et al., 2022; Wang et al., 2014; Wang et al., 2019).

3. Health and Bioactive Properties

Lion's Mane mushroom has biologically active constituents with valuable effects on the brain, heart and intestines, among other things. Below are described some bioactive properties of Lion's Mane that give them it status of nutraceutical and functional food.

3.1 Nutritional and Digestive Functions

H. erinaceus is a well-known edible mushroom mainly used for its useful effects on health (He et al., 2017; Ma et al., 2021). However, Lion's Mane is also capable of nourishing the stomach. Its nutritional and digestive properties were studied and well recognised in the Asian culinary world. Lion's Mane mushroom is traditionally classified as a luxurious food in China (Liu et al., 2016). It is consumed raw, cooked, dried and steeped in tea. It has a stringy and meaty crab-like texture with the subtle 'seafood' flavour of lobster or crab that helps to create unique and delicious recipes.

Like many mushrooms, Lion's Mane contains important nutritional constituents, such as proteins, glucides and micronutrients. It is high in iron, vitamins, amino acids, carbohydrates, fatty acids and potassium. *H. erinaceus* contains a variety of bioactive compounds which are useful as a potential source of nutrients and energy (Friedman, 2015; Ma et al., 2021). According to Friedman, the nutritional quality of some mushrooms is quite similar to that of egg (Friedman, 2015).

Rodrigues et al., determined that *H. erinaceus* averages 21.4% protein, 58.8% sugar and 2.9% fat. It contains more monounsaturated and polyunsaturated fatty acids than saturated fatty acids. Lion's Mane's linoleic acid content makes up for over 30% of its fat content and it is a good source of nutrients (Rodrigues et al., 2015). Table 1 summarises the results of the composition of approximately 38 fruiting bodies and 19 mycelia since about two decades of research (Ulziijargal and Mau, 2011).

H. erinaceus also contains significant amounts of trace elements, like copper, iron and zinc, suggesting their micronutrient potential. More details on the composition of *H. erinaceus* are summarised in Table 2. Based on these findings, both FB and MB can serve as a resource for their use in functional foods and nutraceuticals.

Table 1. Fruiting bodies and mycelia composition in *H. erinaceus* (Ulziijargal and Mau, 2011).

	Ash (%)	Carbohydrate (%)					Fat (%)	Protein (%)	Energy[b] (kcal/100 g)
		RS[a]	SP[a]	Fiber	DF[a]	Total			
Fruiting Bodies	9.35	17.39	39.63	7.81	47.44	64.83	3.52	22.30	190.44
Mycelia	2.55	9.88	18.29	36.81	55.10	64.98	8.78	23.67	213.22

[a]RS: Reducing Sugar, SP: Soluble Polysaccharide, DF: Dietary Fiber = SP + Fiber. [b]kcal/100 g = RS × 4 + Fat × 9 + Protein × 4.

Table 2. Composition of nutrients and bioactive compounds in *H. erinaceus* (Friedman, 2015).

Compounds	Fruiting Body (FB)	Mycelia Biomass (MB)
Protein (%)	20.8	42.5
Total carbohydrate (%)	61.1	42.9
Fat (%)	5.1	6.3
Ash (%)	6.8	4.4
Water (%)	6.2	3.9
Energy (kcal/100 g)	374	398
Total free amino acid (mg/g dw)	14.3	30.6
The total fatty acid (#)[a]	11	7
γ-aminobutyric acid (GABA) (μg/g dw)	42.9	56.0
Ergothioneine (μg/g dw)	630.0	149.2
Lovostatin (μg/g dw)	14.4	NA

[a]individual unsaturated and saturated forms.

3.2 Antioxidant Properties

Antioxidant activity is assessed by different methods, such as free radical-scavenging activity assay, ferrous ion chelating assay, reducing power, etc. (Sánchez, 2017). Within the cell, reactive species have different causes and can target different macromolecules and cellular organelles. Cells are naturally equipped to prevent and relieve the effects of oxidative stress, but these antioxidant defences can also be acquired through diet (Kurutas, 2016; Reis et al., 2017). Natural substances are therefore obvious sources of antioxidant compounds and mushrooms are one of the most popular ones for having several antioxidants, such as phenolics, vitamins C and E and carotenoids (Reis et al., 2017).

According to numerous studies, the molecules of Lion's Mane have shown antioxidant capacities. *H. erinaceus* is composed of many elements, such as lectines, polysaccharides, proteins, hericenone, erinacol, erinacin and terpenoids, whose biological activities have been studied (Wang et al., 2014; Wang et al., 2019; Xu et al., 2010). Water-soluble oligosaccharides obtained from *H. erinaceus* FB revealed substantial antioxidant abilities, indicating that *H. erinaceus* oligosaccharide can serve as an effective health food and source of natural antioxidant substances (Gao et al., 2019; Hou et al., 2015).

The antioxidant expression of *H. erinaceus* has been proposed in the therapy of several ailments. For instance, studies have shown that Lion's Mane isolates can significantly accelerate wound healing (Abdulla et al., 2011), help prevent osteoporosis (Li et al., 2017), protect against alcohol-induced liver damage (Hao et al., 2015), slow skin aging (Xu et al., 2010) and relieve diabetes (Friedman, 2015). However, according to Friedman, the active ingredients responsible for the described effects remain unknown and that polysaccharide isolated from *H. erinaceus* (HEP) may help prevent many diseases (Friedman, 2015). Some studies have also associated the antioxidant activities of Lion's Mane extracts or compounds with many other effects, such as antihypolipidemic (Yang et al., 2003), prevention of oxidative stress-induced atherosclerosis pathogenesis (Rahman et al., 2014), antiplatelet therapy to prevent cardiovascular disease and stroke (Mori et al., 2010) and anti-hypertensivity (Abdullah et al., 2012). An interesting review by He et al. (2017) clearly establishes a link between the antioxidant activities of extracts (or compounds) of *H. erinaceus* and its hepatoprotective effects.

3.3 Immunomodulation Properties

Numerous studies reported that *H. erinaceus* has the ability to enhance the immune system function, thereby protecting the body against pathogen agents (Sheng et al., 2017). Immunity is, in fact, the ability of a living organism to recognise and destroy external harmful substances (Wu et al., 2018). Although these immune-boosting effects of Lion's Mane may be due to significant changes in the gut microbiota (Diling et al., 2017a; Diling et al., 2017b), research has primarily linked these effects to its bioactive compounds, particularly its polysaccharide content. According to research performed on mice, the immunomodulatory effects of the HEP could most likely be due to the effect of regulating the immune property of the intestinal mucosa by enhancing cell- and humoral-mediated immunity, macrophage phagocytosis and NK cell activity (Sheng et al., 2017). HEP has also been shown to increase SIgA production and trigger MAPK and AKT cellular communication pathways in the gut. A recent review by Liu et al. and Wang et al., summarised all the current advances on the immune-boosting activities of HEP, highlighting how HEP could be an immunoregulatory agent in nutraceuticals and functional foods (Liu et al., 2022; Wang et al., 2019).

3.4 Antitumour Properties

A number of works suggest that *H. erinaceus* has beneficial anticancer effects (Lee et al., 2015; Li et al., 2014a; Li et al., 2010). These anticancer effects are linked to its bioactive compounds as well as its antioxidant and immunoregulatory properties (Assemie and Abaya, 2022; He et al., 2017). According to Khan et al., immune regulation is the major mechanism that governs anticancer activities (Khan et al., 2013). Extracts of Lion's Mane would have a therapeutic potential against leukemia (Kim et al., 2011) and gastrointestinal cancers (Li et al., 2014a). *H. erinaceus* has been widely studied for polysaccharides with anticancer properties (Li et al., 2014a; Mizuno et al., 1995). By reviewing the available literature, besides directly inducing tumour cell apoptosis and cell cycle arrest, HEP can also act on several cell receptors and produce effective antitumour effects (He et al., 2017). Numerous reports indicate that HEP also has shown anticancer activities against gastrointestinal cancers (Liu et al., 2022). However, Khan et al., reported that polysaccharides promote the increase in T cells and macrophages, thus suggesting the existence of a link between the immunoregulatory properties of HEP and its antitumour activities (Khan et al., 2013). The resistances of HEP on various types of cancers have been summarised recently by Wang et al., while in their review, Pandya et al., gave a brief state-of-the-art on bioactive polysaccharides as being potentially antitumour (Pandya et al., 2018; Wang et al., 2019). Besides HEP, a series of Lion's Mane metabolomes have been reported for their anticancer activities (Cui et al., 2014; Lee et al., 2015). All of these findings clearly show the potential of *H. erinaceus* bioactive compounds as nutraceuticals and functional foods in the fight against cancer.

3.5 Cardiovascular Disorders Risk Improvement

It is well recognised that the main danger to heart disease is associated with obesity, high triglyceride levels, high content of oxidised cholesterol and a significant propensity to generate blood clots. It was reported that Lion's Mane mushroom may also help to prevent heart disease by influencing some of these factors. Like other medicinal mushrooms, *H. erinaceus* has a long tradition of use as a nutraceutical in the dysregulation of lipid metabolism associated with cardiovascular diseases (Khan et al., 2013). It has been found that *H. erinaceus* extracts can prevent increases in LDL cholesterol (known as 'bad' cholesterol), proliferate HDL (or 'good' cholesterol) (Yang et al., 2003) and reduce blood triglycerides, an early indicator of heart disease (Choi et al., 2013; Rahman et al., 2014). These extracts may also prevent blood clots from forming and help reduce the threat of stroke (Mori et al., 2010). This is because the walls of the arteries are prone to attach the oxidised cholesterol molecules, leading to the threat of stroke and heart attack as a result of their hardening and increase. Therefore, lowering oxidation should benefit heart health (Mori et al., 2010; Rahman et al., 2014). And once again, polysaccharides have been widely reported to be one of the bioactive compounds extracted from *H. erinaceus* that exerts these lipid-lowering effects (Friedman, 2015; Wang et al., 2019).

3.6 Sugar Homeostasis Improvement (Anti-diabetic Properties)

Numerous reports show that *H. erinaceus* can also help manage the symptoms of diabetes when the body can no longer maintain blood sugar homeostasis (He et al., 2017; Khan et al., 2013; Khursheed et al., 2020). Results obtained from rat studies suggest that *H. erinaceus* extracts possess hypoglycemic properties (Liang et al., 2013). According to Khan et al., the concentrated methanol substance from *H. erinaceus* had significantly lower rates of blood sugar elevation (Khan et al., 2013). A mechanism that governs Lion's Mane's ability to improve blood sugar homeostasis has been described to inhibit the activity of alpha-glucosidase, which catalyses the breakdown of carbohydrates in the small intestine. Then, the body becomes unable to digest and absorb carbohydrates as efficiently, leading to a drop in blood sugar levels (Khursheed et al., 2020; Wu and Xu, 2015). Furthermore, *H. erinaceus* has been shown to relieve neuropathic pain in humans and is thought to be related to its antioxidant properties (Friedman, 2015; Yi et al., 2015). *H. erinaceus* has real potential as a therapeutic supplement for diabetes and supports its use as nutraceutical and functional food.

3.7 Neurogenerative Potential and Neurodegenerative Diseases Improvement

The neuroprotective potential of *H. erinaceus* has been demonstrated through many researches and several results highlight its promising role in the treatment of degenerative brain and other neural system diseases (Wang et al., 2019; Wong et al., 2015). The anti-neuroinflammatory expressions of the bioactive components of Lion's Mane mushrooms have been reviewed by Kushairi et al., with particular emphasis on their pathological effects on degenerative diseases, like Parkinson's and Alzheimer's disease, brain ischemia, epilepsy and depression. The beneficial properties of mushrooms on the nervous system are mainly attributed to bioactive compounds in the mycelium and other organs (Kushairi et al., 2020). Researchers have shown that extracts from Lion's Mane mushroom can help speed brain or spinal cord recovery by triggering the growth and repair of nerve cells (Wang et al., 2014). According to other research, *H. erinaceus* may affect the brain function by improving neurite outgrowth in the brain and in related organs (Samberkar et al., 2015). Wong et al., found that daily oral administration of fresh *H. erinaceus* FB extract could promote regeneration of injured rat peroneal nerves during early recovery (Wong et al., 2011). *H. erinaceus* extract may also help reduce the severity of brain damage after stroke (Lee et al., 2014). The effects of *H. erinaceus* supplementation on patients with mild cognitive impairment have been studied, showing a significant increase in scores on cognitive function. These results suggest the efficacy of *H. erinaceus* in the treatment of mild cognitive impairment (Mori et al., 2009). A review by Wang et al., summarised the available information on the neuroprotective effects of HEP and highlighted its potential role in the treatment or prevention of Alzheimer's disease (Wang et al., 2019). The effectiveness of HEP has also been reported in restoring sensory dysfunction after nerve injury (Wong et al., 2015). Moreover, another review specified that only *erinacine A* would have pharmacological action on the central nervous system in rats (Li et al., 2018). Bailly and Gao recently reviewed the molecular diversity of *erinacine A* and the mechanisms underlying their neuroprotective ability (Bailly and Gao, 2020). Research remains very active to study the full therapeutic potential of *H. erinaceus* in the treatment of neurodegenerative diseases, reflecting the hope of its use as a nutraceutical and functional food.

3.8 Other Properties and Side Effects of H. erinaceus

In addition to their abilities described above, extracts and bioactive compounds of *H. erinaceus* are also used to treat several other diseases, based largely on their antioxidant properties. For example, *H. erinaceus* is established for its potent anti-inflammatory properties which play a key role in reducing fatty tissue inflammation, considered to be a determining factor in the onset of metabolic syndrome (Li et al., 2014a; Mori et al., 2015). These anti-inflammatory properties also allow *H. erinaceus* to improve the functioning of the stomach and the digestive system (Liu et al., 2022; Thongbai et al., 2015; Wong et al., 2013).

Lion's Mane was also used for its hepatoprotective effects (Liu et al., 2022). Studies demonstrated that HEP has the potential to prevent and mitigate liver damage (Liu et al., 2022; Wang et al., 2019).

Lion's Mane may also exhibit antibacterial effects (Liu et al., 2016; Shang et al., 2013; Zhu et al., 2014). It was reported that *H. erinaceus* exhibits antimicrobial activity by inhibiting the growth of several bacteria, such as *Bacillus cereus, B. subtilis, Salmonella* sp., *Enterococcus faecalis, Shigella* sp., and *Plesiomonas shigelloides* (Khan et al., 2013; Thongbai et al., 2015). These effects depend on the type of *H. erinaceus* isolate studied (Friedman, 2015).

It has also been reported to improve mental health and general well-being. Overall, consumption of Lion's Man can help improve mood and sleep (Okamura et al., 2015; Vigna et al., 2019), fight fatigue (He et al., 2017; Liu et al., 2015), reduce appetite (Khursheed et al., 2020), return circadian rhythms to normal (Furuta et al., 2016) and treat depression, anxiety and menopause (Chong et al., 2019; Nagano et al., 2010; Yao et al., 2015).

H. erinaceus has been shown to have the potential to inhibit the loss of trace elements in rat skulls (He et al., 2017). It has also been found to improve plant growth (Thongbai et al., 2015). Most of these bioactive properties reported for *H. erinaceus* are summarised in Table 3.

Table 3. Wide-range of bioactive properties in *H. erinaceus* (Reis et al., 2017).

Sample/Extract/Compound	Bioactivity
Lectin	Antiviral
Aqueous extract	Anti-inflammatory
	Hypoglycemic/Anti-diabetic Hypolipidemic
Methanol extract	Hypoglycemic/Anti-diabetic Hypocholesterolemic
Chloroform extract	Anti-inflammatory
Chloroform fraction from ethanolic extract	Anti-inflammatory
Exo-polymers	Hypolipidemic (including hypocholesterolemic)
Hericirine (ergosterol conjunction-type alkaloid)	Anti-inflammatory

H. erinaceus bioactive metabolomes are believed to be safe for use as nutraceuticals and functional foods (Pandya et al., 2018). Animal studies suggest that this fungus and its isolates are relatively nontoxic, even in high doses (Lakshmanan et al., 2016; Li et al., 2014b; Pandya et al., 2018). However, allergic reactions (difficulty in breathing, skin rashes) have been reported in people with a known allergy to mushrooms (Maes et al., 1999; Nakatsugawa et al., 2003).

4. Conclusion and Perspective

H. erinaceus, a well-known medicinal mushroom, has been used for centuries in Asia as a beneficial food and in traditional medicine. It contains bioactive substances that have a variety of health benefits and is widely used in nutraceuticals and functional foods. Research has shown that Lion's Mane mushroom and its extracts can prevent and protect against several symptoms or diseases. It is a very promising source of active ingredients in the modern nutritional industry. This review presents the state of scientific knowledge on the current uses of *H. erinaceus* according to their real or supposed potential properties. Since most of this knowledge has emanated from animal research, future prospects should aim to understand their harmlessness and improve their bioavailability, particularly in humans.

References

Abdulla, M.A., Fard, A.A., Sabaratnam, V., Wong, K.H., Kuppusamy, U.R., Abdullah, N. and Ismail, S. (2011). Potential activity of aqueous extract of culinary-medicinal Lion's Mane mushroom, *Hericium erinaceus* (Bull.: Fr.) Pers. (Aphyllophoromycetideae) in accelerating wound healing in rats. *Int. J. Med. Mushrooms*, 13(1): 33–39. 10.1615/intjmedmushr.v13.i1.50.

Abdullah, N., Ismail, S.M., Aminudin, N., Shuib, A.S. and Lau, B.F. (2012). Evaluation of selected culinary-medicinal mushrooms for antioxidant and ACE inhibitory activities. *Evid. Based Complement. Alternat. Med.*, 2012: 464238. 10.1155/2012/464238.

Assemie, A. and Abaya, G. (2022). The effect of edible mushroom on health and their biochemistry. *Int. J. Microbiol.*, 2022: 8744788–8744788. 10.1155/2022/8744788.

Bailly, C. and Gao, J.-M. (2020). Erinacine A and related cyathane diterpenoids: Molecular diversity and mechanisms underlying their neuroprotection and anticancer activities. *Pharmacol. Res.*, 159: 104953. 10.1016/j.phrs.2020.104953.

Choi, W.-S., Kim, Y.-S., Park, B.-S., Kim, J.-E. and Lee, S.-E. (2013). Hypolipidaemic effect of *Hericium erinaceum* grown in *Artemisia capillaris* on obese rats. *Mycobiology*, 41(2): 94–99. 10.5941/MYCO.2013.41.2.94.

Chong, P.S., Fung, M.-L., Wong, K.H. and Lim, L.W. (2019). Therapeutic potential of *Hericium erinaceus* for depressive disorder. *Int. J. Mol. Sci.*, 21(1): 163. 10.3390/ijms21010163.

Cui, F.J., Li, Y.-H., Zan, X.-Y., Yang, Y., Sun, W.-J., Qian, J.-Y., Zhou, Q. and Yu, S.-L. (2014). Purification and partial characterisation of a novel hemagglutinating glycoprotein from the cultured mycelia of *Hericium erinaceus*. *Process Biochem.*, 49(8): 1362–1369. 10.1016/j.procbio.2014.04.008.

Diling, C., Chaoqun, Z., Jian, Y., Jian, L., Jiyan, S., Yizhen, X. and Guoxiao, L. (2017a). Immunomodulatory activities of a fungal protein extracted from *Hericium erinaceus* through regulating the gut microbiota. *Front. Immunol.*, 8: 666–666. 10.3389/fimmu.2017.00666.

Diling, C., Xin, Y., Chaoqun, Z., Jian, Y., Xiaocui, T., Jun, C., Ou, S. and Yizhen, X. (2017b). Extracts from *Hericium erinaceus* relieve inflammatory bowel disease by regulating immunity and gut microbiota. *Oncotarget*, 8(49): 85838–85857. 10.18632/oncotarget.20689.

Friedman, M. (2015). Chemistry, nutrition, and health-promoting properties of *Hericium erinaceus* (Lion's Mane) mushroom fruiting bodies and mycelia and their bioactive compounds. *J. Agric. Food Chem.*, 63(32): 7108–7123. 10.1021/acs.jafc.5b02914.

Furuta, S., Kuwahara, R., Hiraki, E., Ohnuki, K., Yasuo, S. and Shimizu, K. (2016). *Hericium erinaceus* extracts alter behavioural rhythm in mice. *Biomed. Res.*, 37(4): 227–232. 10.2220/biomedres.37.227.

Gao, Y., Zheng, W., Wang, M., Xiao, X., Gao, M., Gao, Q. and Xu, D. (2019). Molecular properties, structure, and antioxidant activities of the oligosaccharide Hep-2 isolated from cultured mycelium of *Hericium erinaceus*. *J. Food Biochem.*, 43(9): e12985. 10.1111/jfbc.12985.

Grubler, A. (1990). *The Rise and Fall of Infrastructures: Dynamics of Evolution and Technological Change in Transport*. Physica-Verlag, Heidelberg.

Hao, L., Xie, Y., Wu, G., Cheng, A., Liu, X., Zheng, R., Huo, H. and Zhang, J. (2015). Protective effect of *Hericium erinaceus* on alcohol-induced hepatotoxicity in mice. *Evid. Based Complement. Alternat. Med.*, 2015: 418023–418023. 10.1155/2015/418023.

He, X., Wang, X., Fang, J., Chang, Y., Ning, N., Guo, H., Huang, L., Huang, X. and Zhao, Z. (2017). Structures, biological activities and industrial applications of the polysaccharides from *Hericium erinaceus* (Lion's Mane) mushroom: A review. *Int. J. Biol. Macromol.*, 97: 228–237. 10.1016/j.ijbiomac.2017.01.040.

Hou, Y., Ding, X. and Hou, W. (2015). Composition and antioxidant activity of water-soluble oligosaccharides from *Hericium erinaceus*. *Mol. Med. Rep.*, 11(5): 3794–3799. 10.3892/mmr.2014.3121.

Khan, A.A., Gani, A., Khanday, F.A. and Masoodi, F.A. (2018). Biological and pharmaceutical activities of mushroom β-glucan discussed as a potential functional food ingredient. *Bioact. Carbohydr. Diet. Fibre*, 16: 1–13. 10.1016/j.bcdf.2017.12.002.

Khan, M.A., Tania, M., Liu, R. and Rahman, M.M. (2013). *Hericium erinaceus*: An edible mushroom with medicinal values. *J. Complement Integr. Med.*, 10(1): 253–258. 10.1515/jcim-2013-0001.

Khursheed, R., Singh, S.K., Wadhwa, S., Gulati, M. and Awasthi, A. (2020). Therapeutic potential of mushrooms in diabetes mellitus: Role of polysaccharides. *Int. J. Biol. Macromol.*, 164: 1194–1205. 10.1016/j.ijbiomac.2020.07.145.

Kim, S.P., Kang, M.Y., Choi, Y.H., Kim, J.H., Nam, S.H. and Friedman, M. (2011). Mechanism of *Hericium erinaceus* (Yamabushitake) mushroom-induced apoptosis of U937 human monocytic leukemia cells. *Food Funct.*, 2(6): 348–356. 10.1039/C1FO10030K.

Kurutas, E.B. (2016). The importance of antioxidants which play the role in cellular response against oxidative/nitrosative stress: Current state. *Nutr. J.*, 15(1): 71–71. 10.1186/s12937-016-0186-5.

Kushairi, N., Tarmizi, N.A.K.A., Phan, C.W., Macreadie, I., Sabaratnam, V., Naidu, M. and David, P. (2020). Modulation of neuroinflammatory pathways by medicinal mushrooms, with particular relevance to Alzheimer's disease. *Trends Food Sci. Technol.*, 104: 153–162. 10.1016/j. tifs.2020.07.029.

Lakshmanan, H., Raman, J., David, P., Wong, K.-H., Naidu, M. and Sabaratnam, V. (2016). Haematological, biochemical and histopathological aspects of *Hericium erinaceus* ingestion in a rodent model: A sub-chronic toxicological assessment. *J. Ethnopharmacol.*, 194: 1051–1059. 10.1016/j.jep.2016.10.084.

Lee, K.-F., Chen, J.-H., Teng, C.-C., Shen, C.-H., Hsieh, M.-C., Lu, C.-C., Lee, K.-C., Lee, L.-Y., Chen, W.-P., Chen, C.-C., Huang, W.-S. and Kuo, H.-C. (2014). Protective effects of *Hericium erinaceus* mycelium and its isolated erinacine A against ischemia-injury-induced neuronal cell death via the inhibition of iNOS/p38 MAPK and nitrotyrosine. *Int. J. Mol. Sci.*, 15(9): 15073–15089. 10.3390/ijms150915073.

Lee, S.R., Jung, K., Noh, H.J., Park, Y.J., Lee, H.L., Lee, K.R., Kang, K.S. and Kim, K.H. (2015). A new cerebroside from the fruiting bodies of *Hericium erinaceus* and its applicability to cancer treatment. *Bioorg. Med. Chem. Lett.*, 25(24): 5712–5715. 10.1016/j.bmcl.2015.10.092.

Li, G., Yu, K., Li, F., Xu, K., Li, J., He, S., Cao, S. and Tan, G. (2014a). Anticancer potential of *Hericium erinaceus* extracts against human gastrointestinal cancers. *J. Ethnopharmacol.*, 153(2): 521–530. 10.1016/j.jep.2014.03.003.

Li, I.C., Chen, Y.-L., Lee, L.-Y., Chen, W.-P., Tsai, Y.-T., Chen, C.-C. and Chen, C.-S. (2014b). Evaluation of the toxicological safety of erinacine A-enriched *Hericium erinaceus* in a 28-day oral feeding study in Sprague–Dawley rats. *Food Chem. Toxicol.*, 70: 61–67. 10.1016/j. fct.2014.04.040.

Li, I.C., Lee, L.-Y., Tzeng, T.-T., Chen, W.-P., Chen, Y.-P., Shiao, Y.-J. and Chen, C.-C. (2018). Neurohealth properties of *Hericium erinaceus* mycelia enriched with Erinacines. *Behav. Neurol.*, 2018: 5802634–5802634. 10.1155/2018/5802634.

Li, W., Lee, S.H., Jang, H.D., Ma, J.Y. and Kim, Y.H. (2017). Antioxidant and anti-osteoporotic activities of aromatic compounds and sterols from *Hericium erinaceum*. *Molecules*, 22(1): 108. 10.3390/molecules22010108.

Li, Y., Zhang, G., Ng, T.B. and Wang, H. (2010). A novel lectin with antiproliferative and HIV-1 reverse transcriptase inhibitory activities from dried fruiting bodies of the monkey head mushroom *Hericium erinaceum*. *J. Biomed. Biotechnol.*, 2010: 716515–716515. 10.1155/2010/716515.

Liang, B., Guo, Z., Xie, F. and Zhao, A. (2013). Antihyperglycemic and antihyperlipidemic activities of aqueous extract of *Hericium erinaceus* in experimental diabetic rats. *BMC Complement. Altern. Med.*, 13: 253–253. 10.1186/1472-6882-13-253.

Liu, J., Du, C., Wang, Y. and Yu, Z. (2015). Anti-fatigue activities of polysaccharides extracted from *Hericium erinaceus*. *Exp. Ther. Med.*, 9(2): 483–487. 10.3892/etm.2014.2139.

Liu, J., Wang, W., Hu, Q., Wu, X., Xu, H., Su, A., Xie, M. and Yang, W. (2022). Bioactivities and molecular mechanisms of polysaccharides from *Hericium erinaceus*. *J. Fut. Foods*, 2(2): 103–111. 10.1016/j.jfutfo.2022.03.007.

Liu, J.-H., Li, L., Shang, X.-D., Zhang, J.-L. and Tan, Q. (2016). Anti-*Helicobacter pylori* activity of bioactive components isolated from *Hericium erinaceus*. *J. Ethnopharmacol.*, 183: 54–58. 10.1016/j.jep.2015.09.004.

Ma, B., Feng, T., Zhang, S., Zhuang, H., Chen, D., Yao, L. and Zhang, J. (2021). The inhibitory effects of *Hericium erinaceus* β-glucan on *in vitro* starch digestion. *Front. Nutr.*, 7: 621131. 10.3389/fnut.2020.621131.

Maes, M.F.J., van Baar, H.M.J. and van Ginkel, C.I.W. (1999). Occupational allergic contact dermatitis from the mushroom White Pom Pom® (*Hericium erinaceum*). *Contact Dermatitis*, 40(5): 289–290. 10.1111/j.1600-0536.1999.tb06073.x.

Mizuno, T., Saito, H., Nishitoba, T. and KaWagishi, H. (1995). Antitumour-active substances from mushrooms. *Food Rev. Int.*, 11(1): 23–61. 10.1080/87559129509541018.

Mori, K., Inatomi, S., Ouchi, K., Azumi, Y. and Tuchida, T. (2009). Improving effects of the mushroom Yamabushitake (*Hericium erinaceus*) on mild cognitive impairment: A double-blind placebo-controlled clinical trial. *Phytother. Res.*, 23(3): 367–372. 10.1002/ptr.2634.

Mori, K., Kikuchi, H., Obara, Y., Iwashita, M., Azumi, Y., Kinugasa, S., Inatomi, S., Oshima, Y. and Nakahata, N. (2010). Inhibitory effect of hericenone B from *Hericium erinaceus* on collagen-induced platelet aggregation. *Phytomedicine*, 17(14): 1082–1085. 10.1016/j.phymed.2010.05.004.

Mori, K., Ouchi, K. and Hirasawa, N. (2015). The anti-inflammatory effects of Lion's Mane culinary-medicinal mushroom, *Hericium erinaceus* (higher basidiomycetes) in a coculture system of 3T3-L1 adipocytes and RAW264 macrophages. *Int. J. Med. Mushrooms*, 17(7): 609–618. 10.1615/IntJMedMushrooms.v17.i7.10.

Nagano, M., Shimizu, K., Kondo, R., Hayashi, C., Sato, D., Kitagawa, K. and Ohnuki, K. (2010). Reduction of depression and anxiety by 4 weeks *Hericium erinaceus* intake. *Biomed. Res.*, 31(4): 231–237. 10.2220/biomedres.31.231.

Nakatsugawa, M., Takahashi, H., Takezawa, C., Nakajima, K., Harada, K., Sugawara, Y., Kobayashi, S., Kondo, T. and Abe, S. (2003). *Hericium Erinaceum* (Yamabushitake) extract-induced acute respiratory distress syndrome monitored by serum surfactant proteins. *Intern. Med.*, 42(12): 1219–1222. 10.2169/internalmedicine.42.1219.

Okamura, H., Anno, N., Tsuda, A., Inokuchi, T., Uchimura, N. and Inanaga, K. (2015). The effects of *Hericium erinaceus* (Amyloban® 3399) on sleep quality and subjective well-being among female undergraduate students: A pilot study. *Pers. Med. Universe*, 4: 76–78. 10.1016/j.pmu.2015.03.006.

Pandya, U., Dhuldhaj, U. and Sahay, N.S. (2018). Bioactive mushroom polysaccharides as antitumour: An overview. *Nat. Prod. Res.*, 33(18): 2668–2680. 10.1080/14786419.2018.1466129.

Rahman, M.A., Abdullah, N. and Aminudin, N. (2014). Inhibitory effect on *in vitro* LDL oxidation and HMG Co-A reductase activity of the liquid-liquid partitioned fractions of *Hericium erinaceus* (Bull.) Persoon (Lion's Mane Mushroom). *Biomed. Res. Int.*, 2014: 828149. 10.1155/2014/828149.

Reis, F.S., Martins, A., Vasconcelos, M.H., Morales, P. and Ferreira, I.C.F.R. (2017). Functional foods based on extracts or compounds derived from mushrooms. *Trends Food Sci. Technol.*, 66: 48–62. 10.1016/j.tifs.2017.05.010.

Rodrigues, D.M., Freitas, A.C., Rocha-Santos, T.A., Vasconcelos, M.W., Roriz, M., Rodríguez-Alcalá, L.M., Gomes, A.M. and Duarte, A.C. (2015). Chemical composition and nutritive value of *Pleurotus citrinopileatus var cornucopiae*, *P. eryngii*, *P. salmoneo stramineus*, *Pholiota nameko* and *Hericium erinaceus*. *J. Food Sci. Technol.*, 52(11): 6927–6939. 10.1007/s13197-015-1826-z.

Samberkar, S., Gandhi, S., Naidu, M., Wong, K.-H., Raman, J. and Sabaratnam, V. (2015). Lion's Mane, *Hericium erinaceus* and Tiger Milk, *Lignosus rhinocerotis* (Higher Basidiomycetes) medicinal mushrooms stimulate neurite outgrowth in dissociated cells of brain, spinal cord, and retina: An *in vitro* study. *Int. J. Med. Mushrooms*, 17(11): 1047–1054. 10.1615/IntJMedMushrooms.v17.i11.40.

Sánchez, C. (2017). Reactive oxygen species and antioxidant properties from mushrooms. *Synth. Syst. Biotechnol.*, 2(1): 13–22. 10.1016/j.synbio.2016.12.001.

Shang, X., Tan, Q., Liu, R., Yu, K., Li, P. and Zhao, G.-P. (2013). *In vitro* anti-helicobacter pylori effects of medicinal mushroom extracts, with special emphasis on the Lion's Mane mushroom, *Hericium erinaceus* (Higher Basidiomycetes). *Int. J. Med. Mushrooms*, 15(2): 165–174. 10.1615/IntJMedMushr.v15.i2.50.

Sheng, X., Yan, J., Meng, Y., Kang, Y., Han, Z., Tai, G., Zhou, Y. and Cheng, H. (2017). Immunomodulatory effects of *Hericium erinaceus* derived polysaccharides are mediated by intestinal immunology. *Food Funct.*, 8(3): 1020–1027. 10.1039/C7FO00071E.

Thongbai, B., Rapior, S., Hyde, K.D., Wittstein, K. and Stadler, M. (2015). *Hericium erinaceus*, an amazing medicinal mushroom. *Mycol. Progress*, 14(10): 91. 10.1007/s11557-015-1105-4.

Tiwari, H., Deshmukh, H., Wagh, N.S. and Lakkakula, J. (2022). Biological macromolecules as anticancer agents. pp. 243–272. *In*: Nayak, A.K., Dhara, A.K. and Pal, D. (eds.). *Biological Macromolecules*, Academic Press. 10.1016/B978-0-323-85759-8.00011-7.

Ulziijargal, E. and Mau, J.-L. (2011). Nutrient compositions of culinary-medicinal mushroom fruiting bodies and mycelia. *Int. J. Med. Mushrooms*, 13(4): 343–349. 10.1615/IntJMedMushr.v13.i4.40.

Vigna, L., Morelli, F., Agnelli, G.M., Napolitano, F., Ratto, D., Occhinegro, A., Di Iorio, C., Savino, E., Girometta, C., Brandalise, F. and Rossi P. (2019). *Hericium erinaceus* improves mood and sleep disorders in patients affected by overweight or obesity: Could circulating Pro-BDNF and BDNF be potential biomarkers? *Evid.-based Complement. Alternat. Med.*, 2019: 7861297. 10.1155/2019/7861297.

Wang, M., Gao, Y., Xu, D., Konishi, T. and Gao, Q. (2014). *Hericium erinaceus* (Yamabushitake): A unique resource for developing functional foods and medicines. *Food Funct.*, 5(12): 3055–3064. 10.1039/C4FO00511B.

Wang, M., Konishi, T., Gao, Y., Xu, D. and Gao, Q. (2015). Anti-gastric ulcer activity of polysaccharide fraction isolated from mycelium culture of Lion's Mane medicinal mushroom, *Hericium erinaceus* (Higher Basidiomycetes). *Int. J. Med. Mushrooms*, 17(11): 1055–1060. 10.1615/IntJMedMushrooms.v17.i11.50.

Wang, X.-Y., Zhang, D.-d., Yin, J.-Y., Nie, S.-P. and Xie, M.-Y. (2019). Recent developments in *Hericium erinaceus* polysaccharides: Extraction, purification, structural characteristics and biological activities, *Crit. Rev. Food Sci. Nutr.*, 59(sup1): S96–S115. 10.1080/10408398.2018.1521370.

Wong, J.-Y., Abdulla, M.A., Raman, J., Phan, C.-W., Kuppusamy, U.R., Golbabapour, S. and Sabaratnam, V. (2013). Gastroprotective effects of Lion's Mane mushroom *Hericium erinaceus* (Bull.:Fr.) Pers. (Aphyllophoromycetideae) extract against ethanol-induced ulcer in rats. *Evid.-based Complement. Alternat. Med.*, 2013: 492976. 10.1155/2013/492976.

Wong, K.-H., Kanagasabapathy, G., Bakar, R., Phan, C.-W. and Sabaratnam, V. (2015). Restoration of sensory dysfunction following peripheral nerve injury by the polysaccharide from culinary and medicinal mushroom, *Hericium erinaceus* (Bull.: Fr.) Pers. through its neuroregenerative action. *Food Sci. Technol.*, 35(4): 712–721. 10.1590/1678-457x.6838.

Wong, K.-H., Naidu, M., David, P., Abdulla, M.A., Abdullah, N., Kuppusamy, U.R. and Sabaratnam, V. (2011). Peripheral nerve regeneration following crush injury to rat peroneal nerve by aqueous extract of medicinal mushroom *Hericium erinaceus* (Bull.: Fr) Pers. (Aphyllophoromycetideae). *Evid.-based Complement. Alternat. Med.*, 2011: 580752–580752. 10.1093/ecam/neq062.

Wu, F., Zhou, C., Zhou, D., Ou, S., Zhang, X. and Huang, H. (2018). Structure characterisation of a novel polysaccharide from *Hericium erinaceus* fruiting bodies and its immunomodulatory activities. *Food Funct.*, 9(1): 294–306. 10.1039/C7FO01389B.

Wu, T. and Xu, B. (2015). Antidiabetic and antioxidant activities of eight medicinal mushroom species from China. *Int. J. Med. Mushrooms*, 17(2): 129–140. 10.1615/IntJMedMushrooms.v17.i2.40.

Xu, H., Wu, P.-R., Shen, Z.-Y. and Chen, X.-D. (2010). Chemical analysis of *Hericium erinaceum* polysaccharides and effect of the polysaccharides on derma antioxidant enzymes, MMP-1 and TIMP-1 activities. *Int. J. Biol. Macromol.*, 47(1): 33–36. 10.1016/j.ijbiomac.2010.03.024.

Yadav, S.K., Ir, R., Jeewon, R., Doble, M., Hyde, K.D., Kaliappan, I., Jeyaraman, R., Reddi, R.N., Krishnan, J., Li, M. and Durairajan, S.S.K. (2020). A mechanistic review on medicinal mushrooms-derived bioactive compounds: Potential mycotherapy candidates for alleviating neurological disorders. *Planta Med.*, 86(16): 1161–1175. 10.1055/a-1177-4834.

Yang, B.-K., Park, J.-B. and Song, C.-H. (2003). Hypolipidemic effect of an exo-biopolymer produced from a submerged mycelial culture of *Hericium erinaceus*. *Biosci. Biotechnol. Biochem.*, 67(6): 1292–1298. 10.1271/bbb.67.1292.

Yao, W., Zhang, J.-C., Dong, C., Zhuang, C., Hirota, S., Inanaga, K. and Hashimoto, K. (2015). Effects of amycenone on serum levels of tumor necrosis factor-α, interleukin-10, and depression-like behavior in mice after lipopolysaccharide administration. *Pharmacol. Biochem. Behav.*, 136: 7–12. 10.1016/j.pbb.2015.06.012.

Yi, Z., Shao-Long, Y., Ai-Hong, W., Zhi-Chun, S., Ya-Fen, Z., Ye-Ting, X. and Yu-Ling, H. (2015). Protective effect of ethanol extracts of *Hericium erinaceus* on alloxan-induced diabetic neuropathic pain in rats. *Evid.-based Complement. Alternat. Med.*, 2015: 595480–595480. 10.1155/2015/595480.

Zhu, Y., Chen, Y., Li, Q., Zhao, T., Zhang, M., Feng, W., Takase, M., Wu, X., Zhou, Z., Yang, L. and Wu, X. (2014). Preparation, characterisation, and anti-*Helicobacter pylori* activity of Bi^{3+}-*Hericium erinaceus* polysaccharide complex. *Carbohydr. Polym.*, 110: 231–237. 10.1016/j.carbpol.2014.03.081.

CHAPTER 14

Shiitake (*Lentinula edodes*)

Ariadna Thalía Bernal-Mercado,[1]
Francisco Rodríguez-Félix,[1] Carlos Gregorio Barreras-Urbina,[1,2]
Tomás Jesús Madera-Santana,[2] Karla Hazel Ozuna-Valencia,[1]
Maria Jesús Moreno-Vásquez,[3] Lorena Armenta-Villegas,[3]
Carmen Lizette Del-Toro-Sánchez,[1] Miguel Ángel Urías-Torres[1]
and *José Agustín Tapia-Hernández[1,*]*

1. Introduction

Lentinula edodes, commonly known as shiitake, is the most widely consumed edible fungus in the world and is prized for its distinctive flavour in both eastern and, more recently, western cuisine (Morales et al., 2020). Also, it is recognised for its sensory, nutritional and functional attributes and is the second most cultivated edible fungus on the planet, preceded by the white mushroom (*Agaricus bisporus*) (Rivera et al., 2017). Among the main countries where this mushroom can be found in the wild is China and Japan. In China alone, in 2003, production amounted to 2 million tons, accounting for two-thirds of global manufacture (Xiang et al., 2016). Figure 1 illustrates the physical appearance of two varieties of *L. edodes*—Donko and Koshin.

L. *edode* cultivation has economic advantages in its production as it is used as agricultural residue and in forestry, as well as its price in the global market. In countries, such as Mexico, techniques have been adapted and modified traditional methods are used to reduce the crop cycle and lower the production costs when substrate materials previously not considered, such as shavings and sawdust of different trees, such as oak (*Quercus* sp.), pine (*Pinus* sp.), among others are used

[1] Departamento de Investigación y Posgrado en Alimentos (DIPA). Blvd. Luis Encinas y Rosales S/N, Col. Centro, Hermosillo, Sonora, México. C.P. 83000.
[2] Centro de Investigación en Alimentación y Desarrollo, A.C., Coordinación de Tecnología de Alimentos de Origen Vegetal, Carretera Gustavo Enrique Astiazarán Rosas Núm. 46. La Victoria, C.P. 83304, Hermosillo, Sonora, Mexico.
[3] Departamento de Ciencias Químico Biológicas, Universidad de Sonora, Blvd. Luis Encinas y Rosales, S/N, Colonia Centro, 83000 Hermosillo, Sonora, Mexico.
* Corresponding author: Joseagustin.tapia@unison.mx

A   **B**

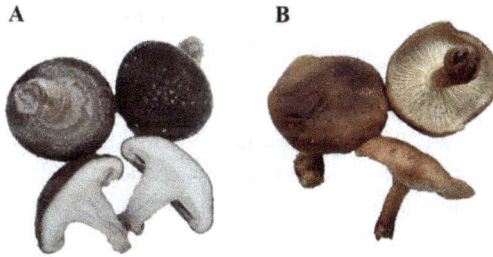

Fig. 1. Physical appearance of two varieties of *L. edodes* (A) Donko and (B) Koshin (García et al., 2021).

(Romero-Arenas et al., 2015). Also, different substrates for mushroom culture have been investigated, including wheat straw and some by-products from the processing of sunflower, flax, maize, cotton, coffee, grapes and other plants. Philippoussis et al. (2007) formulated a substrate for mushroom with agricultural by-products (wheat, oak-wood sawdust, corn-cobs) and added these with 25% of supplements, including wheat bran (10%), millet (10%), soybean flour (4%) and $CaCO_3$ (1%). The results showed higher performance of the fungus with the formulations containing wheat straw and corncobs, than the formulation with oak-wood sawdust, which, in terms of sporophores' size and quantity, proved least effective.

On the other hand, this mushroom's structure contains components including phenolic compounds, lenthionine, glucans, ergothioneine, ergosterol, eritadenine, and other substances. Also, it has important nutritional and medicinal properties for humans, as well as culinary and industrial applications (Rincão et al., 2012). Most mushrooms, sucb as *Lentinula edode*, comprise vitamins, mainly riboflavin, niacin, biotin thiamine, and vitamin C. The biological activity of this mushroom is antioxidant, hypocholesterolemic, antithrombotic and antihypertensive (Morales et al., 2010). These activities are mainly determined by the following compounds: nucleotides, terpenoids, phenols, steroids, the glycoprotein derivatives and polysaccharides and promote health benefits (Sasidharan et al., 2010).

Edible mushrooms take precedence in our daily food regime and are appropriate for people of all ages and specific groups, such as diabetics and children in comparison to other foods that have been proved to manage intestinal microorganisms, such as fruits, tea and cereals (Xue et al., 2020). Therefore, the aim of this chapter is to demonstrate how *Lentinula edodes* may be used as a functional food due to its composition and its nutraceutical qualities.

2. Structure and Taxonomy of *L. edodes*

The creation of dikaryotic mycelia by matin compatible monokaryotic strains and, later, the formation of fruiting bodies on the dikaryon are both dependent on the mating-type genes of mushroom-forming fungus from the Agaricomycetes (Au et al., 2014). *Lentinula* mushrooms correspond to a class of higher fungi that break down wood. The genus is well-known to produce shiitake mushrooms. The Shiitake species' delineation is a contentious topic, according to Kobayashi et al. (2020). *L. edodes* from continental and northeast Asia, *L. lateritia* from tropical Asia and Australia, and *L. novaezelandieae* from New Zealand, were the first three species that shiitake were divided into. However, it has been shown that these species are interfertile, which is why they are currently considered a single species. The taxonomy of *L. edodes* it is presented in Table 1.

L. edodes has a typical sexual life cycle with meiotic basidiospores evolving into monokaryotic mycelium. By fusing two monokaryons with various mating alleles at the A and B mating-type loci, *L. edodes*, a tetrapolar heterothallic basidiomycete, creates a dikaryon that may mature into a fruiting body. The structure of *L. edodes* is shown in Fig. 2.

Table 1. Taxonomic tree of *L. edodes*.

Division	Basidiomycota
Class	Basidiomycetes
Order	Agaricales
Family	Agaricaceae
Genus	Lentinula
Species	Edodes

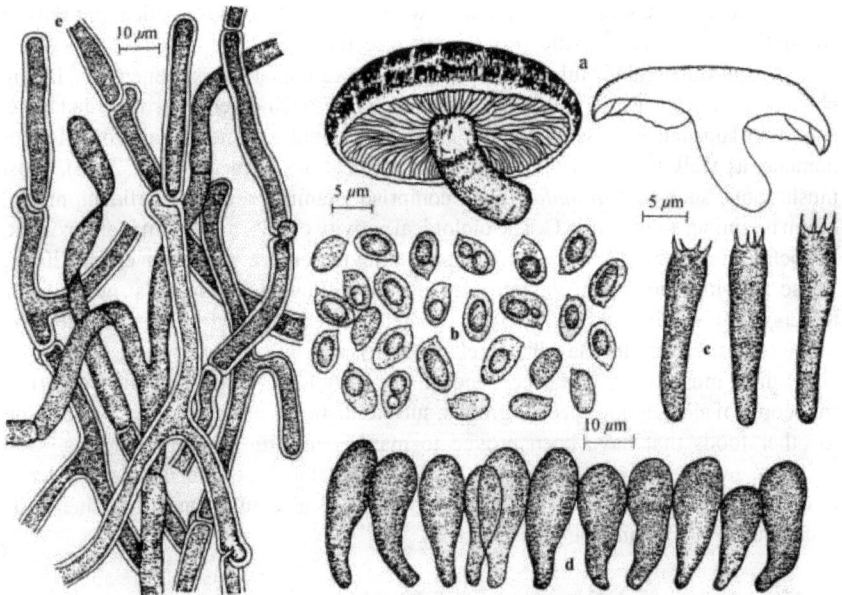

Fig. 2. *L. edodes* (a) fruit body, (b) spores, (c) basidia, (d) cheilocystidia, (e) elements of pileal cuticle (Wasser, 2005).

3. Chemical Composition of *L. edodes*

The nutritional profile and bioactive components of *L. edodes* have recently increased their appeal. In traditional cuisine of Asia and mainly China, it is used as a culinary ingredient for its characteristic flavour and smell. Protein and carbs are abundant in dried shiitake mushrooms. They are mostly composed of 58–60% carbs, 20–23% protein (digestibility 80–87%), 9–10% fibre, 3–4% fats and 4–5% ash. The chemical

Table 2. Chemical composition of *L. edodes*.

Group	Compounds
Fatty acid	
	Linoleic
	Palmitic
	Tetradecenoic
	Oleic
	Stearic
	Myristic
Free sugars	
	Trehalose
	Glycerol
	Mannitol
	Arabinol
	Mannose
	Arabinose
Soluble polysaccharides	
	Heteroglycans
	Heterogalactans
	Heteromannans
	Xyloglucans
Insoluble polysaccharides	
	Heteroglycan
	Polyuronide
	β-glucan
	Chitin

composition is determined by macromolecules, such as fatty acids, sugars and soluble and insoluble polysaccharides, as well as trace elements, such as vitamins and volatile compounds of different organic nature (Finimundy et al., 2014). Table 2 shows the chemical composition of macromolecules present in *L. edodes*.

Shiitake mushroom is a rich resource of minerals and vitamins, such as vitamin B_1, B_2, B_{12}, provitamin D_2, and niacin (Fukushima-Sakuno, 2020; Finimundy et al., 2014). Table 3 shows the trace elements present in *L. edodes*. Concentrations of B complex vitamins in *L. edodes* (mg or µg/100 g) are: B_1 0.05–0.07 (mg/100 g); B_2 0.15–0.2 (mg/100 g); folates 21.51 (µg/100 g); B_{12} 0.07–0.05 (µg/100 g), also, vitamin C 2.1–1.6 (mg/100 g) (Çağlarırmak, 2007). All these types of vitamins have health benefits as nutraceutical compounds.

Additionally, one of the key factors in classifying all mushrooms is their flavour, which is a combination of taste and aroma. Shiitake mushrooms are widely liked by customers worldwide and account for a significant portion of the world's mushroom consumption, mostly due to their flavour and nutritional value (Hou et al., 2021). Lu

Table 3. Trace elements present in *L. edodes*.

Trace Element
Vitamins
Pro-vitamin D_2 (calciferol)
Pantothenic acid
B_1 (thiamine)
B_2 (rivoflavin)
B_6 (pyridoxine)
B_{12} (niacin)
Minerals
Fe
Mn
Ca
K
Zn
Cd

et al. (2021) performed a study to determine the volatile compounds of this mushroom, using modified headspace solid-phase microextraction (HS-SPME) coupled with gas chromatography-mass spectrometry (GC-MS) and electronic nose (Enose) during vacuum freeze drying. Some identified compounds had the aroma of different foods – dimethyl disulfide (onion), benzaldehyde (almond), 1-Octen-3-one (mushroom), 3-octanone (herb), (E)-2-octenal (green and fat), octanoic acid, ethyl ester (fruit and fat) and 2,4,5-tetrathiane (mushroom). Some volatile compounds derived from alcohols present in *L. edodes* are shown below (Table 4).

4. Bioactive Properties of *L. edodes* for use as Nutraceutical

L. edodes is one of the most farmed mushrooms globally and is appreciated for its medicinal and edible properties. It has also gained interest due to its economic importance and organoleptic merits (Sharif et al., 2018). Shiitake mushroom has several therapeutic applications in Eastern countries as it can cure tumours, flu, heart disease, obesity, high blood pressure, diabetes, immune system, cancer and microbial infection (Bisen et al., 2010). There are even shiitake food supplements in the market and are used for different purposes. *L. edodes* has favourable qualities for human health and are mostly linked to their composition. This mushroom includes polyphenols with antioxidant and antimicrobial properties, ergosterol, β-glucans, eritadenine associated with hypocholesterolemic properties, lentinan related to antiviral, immunomodulation and anticancer activities among others (Table 5) (Morales et al., 2020). As a result, shiitake mushrooms are a desirable source for the recovery of nutraceuticals that pharmaceutical industries can valorise. A nutraceutical is a food or food component that offers health or medical advantages, such as illness prevention and treatment. This chapter aims to review the existing literature concerning the nutraceutical properties of *L. edodes*.

Table 4. Alcohol volatile compounds present in *L. edodes* mushrooms (Hu et al., 2021).

Volatile Compounds	(µg/kg db) Fresh
1-Pentanol dimer	1.2567
1-Propanol	1.2467
1-Octen-3-ol	1.1601
1-Pentanol	1.5228
2-Methyl-1-butanol	1.2346
3-Methyl-3-butenol	1.248
Ethanol	1.0443
1-Butanol	1.1775
Isobutanol	1.1705
3-Methyl-1-butanol	1.4908
1-Octen-3-ol dimer	1.5985
2-Methyl-1-pentanol	1.2888
Isobutanol dimer	
1-Propanol	1.3696
3-Methyl-3-butenol dimer	1.1127
1-Hexanol	1.5116
Trans-3-hexenol	1.3295
1-Octanol	1.2612
2-Hexanol	1.4738
	1.2805

Table 5. Main *L. edodes* compounds and their nutraceutical effect.

Activity	Compounds	References
Antioxidant	Phenolic compounds Polysaccharides	(Chikari et al., 2020) (Song et al., 2020) (Zhang et al., 2018)
Antimicrobial	Lentinan (β-D-glucans) Phenolic compounds Lenthionine Oxalic acid	(Santoyo et al., 2012) (Kwak et al., 2016) (Hatvani, 2001)
Antitumor	Mannogalactoglucan Lentinan (β-D-glucans) Lentin	(Jeff et al., 2013)´ (Li et al., 2018a) (Zhang et al., 2015)
Immunomodulation	Mannogalactoglucan Lentinan (β-D-glucans)	(Jeff et al., 2013) (Chen et al., 2020)
Cardiovascular protect	Eritadenine	(Yang et al., 2013) (Asada et al., 2019)
Antidiabetic	Lentinan	(Zhang et al., 2016)
Anti-inflammatory	Polysaccharides	(Song et al., 2020)

4.1 Antioxidant Activity

Free radicals are produced naturally during normal cellular function as part of the metabolism. Free radical overproduction can harm cell components such as of lipids membranes, proteins and nucleic acids. The body contains a complete antioxidant protection system to counteract their harmful action. However, oxidative stress arises when an imbalance exists between an excessive free radical amount and the antioxidant defences (Lobo et al., 2010). Chronic oxidative stress has been linked to many diseases, including chronic inflammation, neurodegenerative illness (Alzheimer's and Parkinson's diseases), cardiovascular disorders, cancer, diabetes, pulmonary disease and premature aging (Reuter et al., 2010; Maritim et al., 2003). Several studies have revealed that the consumption of antioxidant molecules in dietary intake could help to reduce oxidative damage and prevent and treat such diseases. Antioxidants are stable compounds that may donate an electron to free radicals and neutralise it, limiting or delaying its ability to cause harm (Lobo et al., 2010).

Several edible mushroom species, including shiitake mushrooms (*L. edodes*), present bioactive compounds with antioxidant activity (Yin et al., 2018). For example, Carneiro et al. (2013) demonstrated the antioxidant potential of dried powder of *L. edodes* using the FRAP (ferric reducing antioxidant power), ABTS (2,2'-azino-bis(3-ethylbenzothiazoline-6-sulfonic acid), and DPPH (2,2-diphenyl-1-picryl-hydrazyl-hydrate) methods (E_{50} = 64.79, 24.84, 26.32 mg/ml, respectively). The authors suggested its use as an antioxidant source to prevent diseases related to oxidative stress due to the phenolic content. In another study, the antioxidant activity of an aqueous extract of dried shiitake was EC_{50} = 0.191 mg/mL for the ABTS assay, 0.400 mg/mL for the DPPH and 1.488 mg/mL for the ferrous ion chelating activity (de Melo Macoris et al., 2017). Moreover, the aqueous extract of *L. edodes* showed potential antioxidant activities with an IC_{50} of 48.30 µg/mL for DPPH, 15.92 µg/mL for ABTS and 30.10 µg/mL for ORAC assays; however, its activity was lower than for the *Pleurotus ostreatus* (oyster mushroom) (Elhusseiny et al., 2021).

Extraction of shiitake mushrooms has been done using a variety of procedures. Xiaokang et al. (2020) found that the extraction process (hot-water extraction, microwave-assisted extraction, or organic-solvent extraction) had a significant effect on the total phenolic compound and antioxidant capacity of shiitake mushrooms. The results showed that hot-water extraction possessed 23 and 26% more antioxidant activity than microwave-assisted and organic solvent extraction. Similarly, different extraction methods to obtain shiitake mushroom extracts were assessed for antioxidant activity by DPPH (Kitzberger et al., 2007). The procedures adopted were high-pressure techniques, utilising pure CO_2 and CO_2 with co-solvent in pressures up to 30 MPa and low-pressure methods, using organic solvents (dichloromethane, n-hexane and ethyl acetate). The results demonstrated that the highest antioxidant activity (AA) was achieved with dichloromethane (64.83% AA) and ethyl acetate (92.93% AA) solvents with medium polarity in a traditional organic solvent extraction. The antioxidant activity of supercritical extracts with pure CO_2 between 30–50°C and 15–30 MPa was limited (11% AA) (Kitzberger et al., 2007). According to the authors, these results are likely related to polar chemicals with higher antioxidant activity extracted using polar solvents, whereas non-polar solvent properties resulted in extracting non-polar constituents with minimal antioxidant activity. Because of its polar character, ethanol is a potential co-solvent for supercritical fluid extraction to get antioxidant compounds. In this study, the antioxidant activity increased with greater ethanol concentration in the supercritical fluid extraction at 40°C and 20 MPa, up to 72.97% AA, for 15% ethanol and 63.96% AA with 10% ethanol (Kitzberger et al., 2007).

Some of the positive effects of mushrooms on human health are related to the presence of β-glucans. Morales et al. (2020) reported the antioxidant capacity of D-glucans with β-(1→6) and β-(1→3), (1→6) linkages fractions isolated from *L. edodes* extracts (IC_{50} of 183.8 µg/mL). In diverse *in vitro* and *in vivo* situations, polysaccharides from *L. edodes* are demonstrated to influence antioxidant enzymes, such as superoxide dismutase, peroxidase, catalase and glutathione peroxidase (Tang et al., 2020). The interaction of hydrogen with radicals is involved in hydroxyl radical scavenging by polysaccharides, although the detailed mechanism needs to be clarified (Shi et al., 2013). The antioxidant polysaccharide capacity of spent *L. edodes* substrate (SLPS), a residue generated in mushroom industries that pollutes the environment, was investigated. The freeze-dried SLSP had the best polysaccharide yield (13.00%) and antioxidant activity with EC_{50} values of 0.051 mg/mL, 0.379 mg/mL, and 0.719 mg/mL on ABTS, DPPH and superoxide anion radicals, respectively and these were much lower than freeze-dried *L. edodes* polysaccharide (Zhu et al., 2018).

Some studies have explored the antioxidant potential in animal models. Nisar et al. (2017) investigated the hepatoprotective, immunomodulatory and antioxidant properties of dried powder of *L. edodes* fruiting bodies in hypercholesterolemic rats. Three groups of albino rats were created: control, hypercholesteremic and the shitake group. The group without treatment was fed with only a chow maintenance diet. Cholic acid and cholesterol-based feeding in a chow maintenance diet for 24 days caused hypercholesterolemia in rats. For 42 days, one group of rats was supplemented with *L. edodes*, whereas the hypercholesterolemia group continued with the high-cholesterol diet. Following the inclusion of *L. edodes* in the diet, the total oxidant status decreased significantly, the cell-mediated immune reaction significantly improved and the liver enzymes reduced substantially.

Similarly, the influence of *L. edodes* polysaccharides on oral ulceration was investigated by Yu et al. (2009). Three experimental groups (each with 10 rats) were allocated at random to the rats: the control group received no treatment for oral ulcers, the low-dose polysaccharides group received treatment for oral ulcers and the polysaccharides group received treatment for oral ulcers with a high-dose. An amount of 0.3 ml of the polysaccharide's solution or physiological saline solution was administered to treatment groups and control group, respectively. A total of 18 days was spent administering the therapy. The results showed that *L. edodes* polysaccharides supplementation significantly increased serum antioxidant enzyme activity and lowered serum mucosal interleukin-2 levels.

4.2 Antimicrobial Effect

Microbial infections have dramatically grown worldwide due to an alarming increase in multidrug resistance being one of the leading global health concerns. Antibiotic resistance has been increased by the overuse of antibiotics in both people and animals (WHO, 2021). Currently, this problem is one of the biggest dangers to development, food security and global health. Infections are becoming more challenging to treat and antibiotics are becoming less effective, prolonging hospital stay, increasing costs and mortality rates. Therefore, there is an urgency in searching for novel and effective antimicrobial agents. Among the most widely cultivated mushrooms, *L. edodes* has a promising antibacterial, antifungal and antiviral activity.

Scientific literature has reported the efficacy of *L. edodes* to inhibit the growth of Gram-positive and Gram-negative bacteria. For example, the antibacterial activity of ethanolic *L. edodes* extract was reported against *Staphylococcus aureus*, *Acinetobacter baumannii*, *Klebsiella pneumoniae*, and *Enterococcus faecalis* with minimum inhibitory concentrations (MIC) in a range of 5.1–6.01 mg/ml. The antibacterial activity of this extract was attributed to the main compounds – isosorbide/dianhydromannitol, chlorogenic acid, rutin, syringic acid, ferulic acid, p-coumaric acid, 2-hydroxy cinnamic acid, abscisic acid, protocatechuic acid and trans-cinnamic acid. The antibacterial mechanisms proposed for this mushroom were DNA and protein leakage and destruction of bacterial cell membrane permeability (Erdoğan Eliuz, 2021). The greatest concentration of total phenolic compounds in *L. edodes* var. Koshin aqueous extracts resulted in the highest ABTS scavenging ability, which in turn produced the best antibacterial effectiveness in preventing the development of methicillin-resistant *Staphylococcus aureus*. Additionally, the combination of commercial antibiotics and mushroom extracts had positive synergistic effects on the tested microorganisms (Garcia et al., 2021).

Moreover, an aqueous extract of dried shiitake mushrooms obtained from local trade was evaluated against bacteria and yeast pathogens. *Candida albicans*, *K. pneumoniae* and *Saccharomyces cerevisiae* were the most sensitive microorganisms, within 2–3 mg/mL MIC range. This extract showed a MIC result of 12.5 mg/mL against *Pseudomonas aeruginosa*, *Escherichia coli*, *Salmonella enterica*, and *S. aureus* (de Melo Macoris et al., 2017). Similarly, Parola et al. (2017) reported the antibacterial capacity of crude water extract of shiitake mushroom against *P. aeruginosa* and *S. aureus*. Another study described the effectiveness of a protein extracted from the *L. edodes* fruiting body against *E. coli* and *S. aureus* with an inhibition area of 2.48 cm and 2.68 cm, respectively (Sánchez-Minutti et al., 2016). Lenthionine, a cyclic organosulphur molecule that contributes to the shiitake's flavour, is another antibacterial agent detected in the *L. edodes* mycelium cultured in submerged fluid culture. This chemical inhibited the growth of *S. aureus*, *Bacillus subtilis* and *E. coli* (Hatvani, 2001).

The activity against microorganisms of *L. edodes* was exploited for the control of bacterial spots caused by *Xanthomonas campestris* pv. *vesicatoria* in tomato cultivar, Agriset 761. The foliar spray of mycelia culture filtrate greatly decreased the occurrence of bacterial spots in tomato leaves. The treatment application did not affect tomato germination, height and flowering. However, phytotoxicity signs were detected under natural circumstances attributed to oxalic acid in the filtrate. After removing oxalic acid from the mycelia culture filtrate, the authors suggested that the product could be used as a biopesticide (Kaur et al., 2016). In a similar approach, 15 commercially accessible *L. edodes* strains were evaluated for their effect against *X. campestris* pv. *versicatoria* and *Erwinia amylovora*. The majority of the *L. edodes* strains suppressed the growth of both bacteria with a similar effect to 100 mg/ml of streptomycin sulphate (Kaur et al., 2019). A similar study carried out by Kwak et al. (2016) indicated that the culture filtrate of *L. edodes* is a promising ecological control strategy for plant diseases. The authors demonstrated that organics acids, mainly oxalic acid, exhibited antibacterial activity against *Ralstonia solanacearum*, *Pectobacterium carotovorum* subsp. *Carotovorum*, *X. campestris* pv. *Campestris*, *Pseudomonas tolaasii* KACC10365, *E. coli*, *Agrobacterium tumefaciens*, *Xanthomonas axonopodis* and *B. subtilis*.

Polysaccharides from mushrooms have demonstrated antibacterial action; however, the action mode is still unclear. There might be multiple possible polysaccharide targets against bacteria, including cell wall, cytoplasmic membrane and DNA, resulting in bacteria being unable to develop resistance (He et al., 2010). These polysaccharides possess a β-(1→3)-linked backbone with single β-(1→6)-linked glucose side chains, when completely oxidised, produce a significant number of aldehyde groups with antibacterial characteristics (Zi et al., 2018). In addition, it has been reported that the lentinan (β-glucan) antibacterial mechanism can be related to the non-specific modulation of the immune system.

Scientific evidence has been obtained to support the antiviral properties of *L. edodes*. For example, an aqueous extract of *L. edodes* showed promising inhibitory activity against *Herpes simplex virus* (IC_{50} = 27.32 µg/mL) and *Adenovirus* (IC_{50} = 34.83 µg/mL) (Elhusseiny et al., 2021). Aqueous and ethanol extracts, as well as polysaccharides from the fruiting body of *L. edodes*, were tested for their ability to inhibit the reproduction of poliovirus type 1 (PV-1) and bovine herpes virus type 1 (BoHV-1) (Rincão et al., 2012). The aqueous and polysaccharide extracts were more effective when administered at 0 hours of infection, but the ethanol extract was more beneficial after one and two hours of infection. The three extracts had minimal virucidal action and there was no significant suppression of viral adsorption. The extracts, according to the authors, block the early replication phase of both viruses; nevertheless, more research is needed to clarify the inhibitory mechanisms.

Ren et al. (2018a) evaluated the compound lentinan from *L. edodes* mycelia for antiviral action against infectious hematopoietic necrosis virus, explaining the role of this polysaccharide in the formation of a stable virion-polysaccharide complex (Nguyen et al., 2012). Furthermore, due to the affinity and competition between lentinan and herpes virus for cell receptors, lentinan could selectively bind with cell surface receptors, further inhibiting viral penetration in cells. In addition, after post-cell infection, lentinan had direct virucidal and therapeutic actions, preventing viral replication. Furthermore, polysaccharides could represent the immunomodulating activity by stimulating a wide range of immune responses (Galindo-Villegas and Hosokawa, 2004). Polysaccharide-induced immune improvement may restrict virus proliferation by creating an antiviral microenvironment (Lee et al., 2014). TNF-α, which is produced by various cell types, may play an essential role in the organism's immune response, which is linked to the activation of growth and transcription factors. The active contributors in the cytokine network and the human immunological response to infection are IL-2 and IL-11 (Zelová and Hošek, 2013; Sogo et al., 2009). Lentinan was found to have immune-stimulatory properties, including stimulating interferon production *in vitro*, reducing the cytotoxic result of human immunodeficiency virus and suppressing herpes simplex virus type 1 release from cells (Ren et al., 2018a; Shalhoub et al., 2015).

4.3 Anti-tumour and Anticancer Properties

Cancer is among the most important reasons of death globally, with 19.3 million new cases and 9.9 million deceased in 2020 (GLOBOCAN, 2020). Cancer is a multifactorial disease characterised by acquired genetic abnormalities by the exposition to external and internal factors (90–95%) besides inherited genetic aberrations (5–10%) (Maru et al., 2016). Conventional therapy to treat cancer is not wholly effective, has severe side effects due to non-selectivity action on cells (cancerous and non-cancerous), has a greater rate of multidrug resistance and is highly costly (Fekrazad et al., 2017). Therefore, developing and searching for novel agents with minimal side effects and high anticancer effect is urgently needed. In the last decades, researchers have been interested in the pharmacological effects of natural products, especially anticancer agents. Several mushroom species have anti-proliferative, anti-tumour and anticancer activity attributed to their main compounds, including polysaccharides, proteins, polysaccharide-protein complexes, dietary fibre, terpenoids, steroids and phenols (Joseph et al., 2021). In this approach, water-soluble polysaccharides with anti-tumour drugs seem appealing for further use in clinical oncology (Vetvicka et al., 2021).

Polysaccharides are the most active mushroom compounds with anticancer, antiviral, anti-inflammatory, immunomodulatory, anti-carcinogenic, antioxidant and neuroprotective properties (Singh et al., 2021; Ho et al., 2020). Their chemical structures mainly determine polysaccharide anti-tumour action. Glycans with anti-tumour action range from homopolymers to complex heteropolymers, isolated and identified from fruiting bodies, mycelia and culture media. Specifically, *L. edodes* has a potent anticancer activity mainly due to the presence of polysaccharides, such as β-glucans. The glucan lentinan is made up of a core chain of (13)-linked -D-glucopyranose units that are replaced at O-6 by -D-glucopyranose at an occurrence of two branches for each five core chain units (Fig. 3) (Morales et al., 2020). This polysaccharide is widely interesting due to its *in vitro* and *in vivo* antitumor activity (Zhang et al., 2011). Other α-D-glucans have also been found in *L. edodes* extracts; however, their bioactivity is not thoroughly studied (Gil-Ramírez et a

Morales et al. (2020) reported that *L. edodes* polysaccharide-rich extract possesses a wide range of bioactive glucans, which were effectively extracted by utilising a

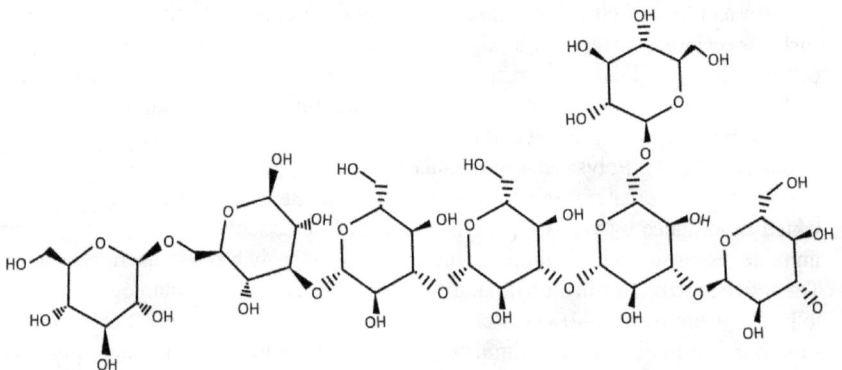

Fig. 3. Lentinan is a β-1,3 beta-glucan with β-1,6 branching polysaccharides found in the fruiting body of shiitake mushroom and has immunomodulating and anti-tumour properties.

simple and effective technique. The glucans found in this extract consisted of a linear (1→6)-β-D-glucan with little quantities of (1→3)-β-D-glucan (G-1), a branched (1→3), (1→6)-β-D-glucan (G-2) and a linear (1→3)-α-D-glucan (G-3). The glucans showed cytotoxic properties on tumoural breast cells but had no effect on normal breast cell growth. When measured by MTT, the G-3 fraction had the maximum cytotoxic action, eliminating 73% viability afterward 48 hours of incubation. Also, the G-3 showed a cytotoxic effect when analysed by Live/Dead® Viability. The authors suggested that distinct structures, like branching grade and α/β-configuration, can significantly impact their conformation, solubility and cell contact, resulting in varied biological results.

The polysaccharides extract from *L. edodes* obtained under high-pressure cooking showed high inhibited HepG2 (hepatocellular carcinoma) and HeLa (cervical carcinoma) cells. This extract inhibited tumour development and increased organ index levels in H22 tumour-bearing mice *in vivo*. In addition, the extract increased the actions of antioxidant enzymes (glutathione peroxidase, superoxide dismutase and catalase) and demonstrated immunomodulatory effects by increasing IL-2 and TNF-α values (Li et al., 2018a). A homogenous polysaccharide obtained from the fruiting bodies of *L. edodes* showed a promising *in vitro* anticancer efficacy on HeLa cells. The polysaccharides greatly reduced the proliferation of HeLa cells in a concentration-dependent way by activating apoptosis-inducing caspase-3 and 9 after the mitochondria were disrupted and cytochrome C was released (Ya, 2017). Similarly, lentinan, a water-extracted polysaccharide from *L. edodes*, effectively reduced HT-29 colon cancer cell proliferation and lowered tumour development in nude mice by inducing cell death via ROS-mediated intrinsic and TNF-mediated extrinsic mechanisms. The authors proposed that polysaccharides from *L. edodes* arc a promising option for colon cancer prevention and therapy (Wang et al., 2017).

Jeff et al. (2013) suggested that mannogalactoglucan of *L. edodes* can be investigated as possible anticancer drug and used as tumour cell growth inhibitor in food and medicinal sector. The preliminary results showed that (1 → 6)-β-d-glucan (WPLE-N-1) and two mannogalactoglucans (WPLE-N-2 and WPLE-N-3) obtained from the *L. edodes* basidiocarps by hot water extraction, ethanol precipitation, anion exchange chromatography and further purified by gel-permeation chromatography, exhibited *in vitro* anti-tumour capacity against Sarcoma S-180, Carcinoma HCT-116 and HT-29. Similarly, the anti-tumour activity of a branched β-(1, 3)-glucan obtained from *L. edodes* was investigated by Xu et al. (2016). The glucan suppressed 75% of S-180 tumour development *in vivo*, which was more than the positive control of cytoxan (54%). The anti-tumour mechanism proposed is that shiitake glucan stimulates immunological responses to cause cell apoptosis via a caspase 3-dependent signalling route and suppresses cell growth via a p53-dependent signalling route. It also slows tumour development by inhibiting angiogenesis by reducing VEGF expression.

The exact mechanism of the anti-tumour effect of glucan mushrooms is still unclear. As was previously described, glucans can inhibit the proliferation of some cancer cells (liver, breast, colon) and can directly attack cancer cells. Furthermore, through attaching to lymphocyte surfaces or serum-specific proteins, these polysaccharides activated the natural killer cells (NK), T cells, macrophages and other effector cells (Li et al., 2018b; Bisen et al., 2010). The anticancer mechanism of action

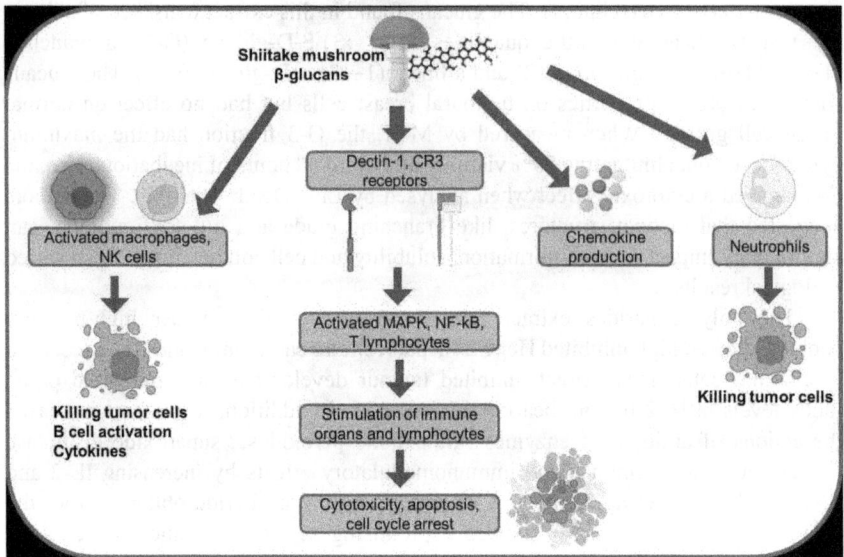

Fig. 4. Immunomodulatory mode of action proposed of β-glucans of *L. edodes* mushroom (adapted from Vetvicka et al., 2021).

of β-glucan is based on its binding to dectin-1, which stimulates the activation of many components during immunomodulation. Following binding, β-glucan triggers spleen-associated tyrosine kinase, which in turn activates the nuclear factor kappa-beta (NF-kB) transcription factor through the family proteins of caspase recruitment domain (CARD9) or mitogen-activated protein kinases (NIK), resulting in the production of interleukins (10, 23, 2, 6) and TNF. When these cytokines are triggered, T and B cells, as well as DCs, multiply. When declin-1 and MyD88 are engaged, several additional signalling pathways are also activated. As a result, numerous cytotoxic substances are produced by T, B and DC cells, which recruit tumour-specific immune replies in the host (Vetvicka et al., 2021; Jin et al., 2018). Figure 4 summarises the immunomodulatory mode of action proposed by β-glucans mushroom for the anti-tumour activity.

4.4 *Immunomodulating Activity*

The ability of edible mushrooms to modulate the immune system has sparked considerable attention. In terms of anticancer qualities, polysaccharides have an important role in immunomodulatory activity in mushrooms owing to their influence on dendritic cells, macrophages, lymphocytes and natural killer cells, which are engaged in both innate and adaptive immunity (Meng et al., 2016; Wasser, 2002). For example, two mannogalactoglucan-type polysaccharides (WPLE-N-2 and WPLE-A0.5-2) isolated from the fruiting bodies of *L. edodes* were evaluated for their impact on the cellular immunological reply of mice with Sarcoma 180. Mice were given 100 mg/kg body weight polysaccharides for 10 days. The mice in the polysaccharide group had significantly less tumours than the group without treatment. These polysaccharides have been shown to increase nitrite oxide formation in peritoneal macrophages, improve macrophage phagocytosis in animals with tumour and enhance splenocyte proliferation caused by concanavalin and lipopolysaccharide (Jeff et al., 2016).

A study conducted with the same approach reported that three polysaccharide parts (F1, F2, F3) obtained from *L. edodes* aqueous extracts protected the host from immunosuppression in mice (Chen et al., 2020). Only cellular immunity was increased by F1, while cellular, humoural and innate immunities were all improved by F2 and F3. The molecular weight had a substantial effect on their bioactivity mechanism. The conclusions of the study achieved by Murphy et al. (2020) demonstrated a significant physicochemical difference between their novel in-house (IH-Lentinan) and a commercial (Carbosynth-Lentinan) lentinan extract, which exhibited various immunomodulatory and pulmonary cytoprotective actions *in vitro*. In a lung epithelial model, both lentinan products decreased inflammation, but IHL had this effect in lesser dosages. The composition of the lentinan extracts differed significantly according to physicochemical analysis, with CL comprising more alpha-glucans and less beta-glucans. Both lentinan extracts inhibited cytokine-induced NF-B start in human alveolar epithelial A549 cells, observing that the IHL extract produced a higher effect at lower concentrations. In activated THP-1-derived macrophages, the CL extract suppressed pro-inflammatory cytokine generation (TNF-, IL-6, IL-8, IL-22 and IL-2) more efficiently. The CL extract decreased oxidative stress-induced early apoptosis, but the IHL extract reduced late apoptosis. All of these may have interesting implications for COVID-19 therapy targeting cytokine storms (Murphy et al., 2020).

4.5 Anti-inflammatory Effects

The inflammatory process is an essential part of the immune response because it serves as a defence mechanism against pathogenic microorganisms or chemical agents harmful to the body (Rock and Kono, 2008). When this response is exaggerated and unregulated, it can result in tissue injury, physiological decompensation, organ dysfunction and death (Sherwood and Toliver-Kinsky, 2004). Several diseases are caused by a chronic inflammation process, such as rheumatoid arthritis, Crohn's disease, tuberculosis, chronic lung disease, Alzheimer's and Parkinson's disease (Rea et al., 2018; Gupta et al., 2018). Alternative therapies based on natural compounds are a promising treatment for many diseases' complications caused by the inflammatory process. Polysaccharides from *L. edodes* mushrooms have caught research attention owing to their anti-inflammatory properties (Du et al., 2018).

The enzymatic residue polysaccharide of *L. edodes* mushroom was proposed as a functional food and potential candidate for treating lipopolysaccharide-induced sepsis in mice due to its anti-inflammatory effect (Ren et al., 2018b). Sepsis is a syndrome characterised by systemic inflammation and the overproduction of pro-inflammatory cytokines, resulting in diverse organic dysfunctions, and even is the principal reason of decease in the intensive care unit. Ren et al. (2018b) demonstrated that the enzymatic residue polysaccharide of *L. edodes* showed a lower effect at the levels of TNF-α, IL-6 and IL-1β on treated mice (600 mg/kg bw) than that in the non-treated induced mice. On the other hand, shiitake mushroom was used to formulate a mouthwash to evaluate its antibacterial and anti-inflammatory efficacy in 45 participants with moderate to severe gingivitis (Pillai et al., 2021). When compared to the reference point, the group considered for the treatment mouthwash demonstrated an important reduction in gingival and plaque index at 15 days and one-month later. It also showed a significant reduction in gingival inflammation. However, the anti-inflammatory mechanisms were not studied or proposed.

4.6 Antidiabetic Activity

Nowadays, diabetes is found among the most frequent and severe chronic illnesses affecting human health. Diabetes occurs when cells and tissues are unable to use blood glucose because of the abnormal metabolism of insulin, resulting in hyperglycemia and various metabolic and functional consequences (Lo and Wasser, 2011). The prevalence of diabetes has increased significantly over the last decades, rendering it a significant public health challenge (Cheng et al., 2021). Hence, the prevention and treatment of diabetes is a necessary concern to be resolved. As stated before, natural products have attracted interest in preventing or relieving some diseases. Bioactive compounds of *L. edodes* have demonstrated a crucial antidiabetic activity due to their effect on anti-glycation capacity, α-glucosidase reduction and antioxidant activity, which can alleviate diabetes and its difficulties (Cao et al., 2020).

The polysaccharide of *L. edodes* mycelia inhibit α-glucosidase action reversibly in a mixed-type way (Cao et al., 2020). This enzyme is involved in carbohydrate metabolism and its inhibition can slow down the release and absorption of glucose, with the consequent reduction in postprandial and fasting blood glucose values. The polysaccharides inhibit the development of advanced glycation end products and high glucose-induced stimulation of MAPK signalling routes. The author suggested that *L. edodes* polysaccharides can be a promising candidate to control diabetes and its application in the functional food industry (Cao et al., 2020). In a similar approach, Cao et al. (2019) investigated the protective effect of *L. edodes* mycelia polysaccharide on islet β cells harmed by glucose. This study demonstrated that the mushrooms' polysaccharides reduced glucose-induced oxidative stress and prevented apoptosis in islet β cells. The inhibitory mechanism to reduce glucose toxicity involved cell signalling routes – p38 MAPK, JNK, NF-κB, and Nrf2.

Hyperglycemia is a symptom of diabetes and inhibitors of glucose transporters in the intestine and proximal tubule cells are one type of treatment. In this regard, at 0.1 mg/ml, the *L. edodes* ethanol extract significantly reduced glucose uptake by the monolayer of differentiated Caco-2 cells; nevertheless, at the same dosage, neither water nor the 50% (v/v) ethanol extract had an impact on glucose transport through these cells. In this study, the *L. edodes* extract reduced plasma glucose levels by 64.0% in mice whose hyperglycemia was brought on by eating a high-fat diet (Hata, 2021). Another study examined the impact of *L. edodes* on a streptozotocin-induced diabetes mellitus in rats in pre-clinical testing under two gestational situations (before and after fetal implantation). The results demonstrated that the shiitake mushroom induced improved maternal glucose tolerance and an increase in insulin levels but did not reduce mother-fetal hyperglycemia (Laurino et al., 2019). Additionally, it repaired liver damage, reduced lipid and cholesterol levels and shielded the mice against post-implantation losses. All these effects can be associated with the antioxidant activity of β-glucans and phenolic compounds in *L. edodes*.

4.7 Effect on Cardiovascular Diseases

Cardiovascular diseases are an important reason for death worldwide and have been recognised as an essential contributor to the cost of medical care (Amini et al., 2021). These diseases are related to heart failure, hypertension, platelet aggregation, cardiac problems and high low-density cholesterol and triglycerides. Hence, various scientific investigations have sought new therapeutic agents to prevent and treat cardiovascular diseases. β-glucans from edible mushrooms have been marketed as functional foods to reduce hypercholesterolemia because these chemicals inhibit the absorption of exogenous cholesterol (Morales et al., 2018).

Specifically, eritadenine (Fig. 5) obtained from the *L. edodes* fruiting bodies presented hypocholesterolemic influence associated with methionine metabolism (Asada et al., 2019). *L. edodes* may function in lipid metabolism regulation, as evidenced by reduced hypercholesterolemia in mice through regulating cholesterol 7αhydroxylase 1 (CYP7A1) expression in the liver (Yang et al., 2013). This research investigated how *L. edodes*-affected rats had hypercholesterolemia caused by eating a rich-fat diet. Rats were administered *L. edodes* doses of 5, 10 and 20% over a period of four weeks. Shiitake lowered total blood cholesterol, low-density lipoprotein and triglyceride in a dose-dependent way in mice treated at all concentrations. Additionally, supplementing with eritadenine and *L. edodes* increased the expression of CYP7A1 mRNA in hypercholesterolemic animals, decreased lipid accumulation in liver tissues and inhibited the formation of atherosclerotic plaques caused by a diet rich in fats (Yang et al., 2013).

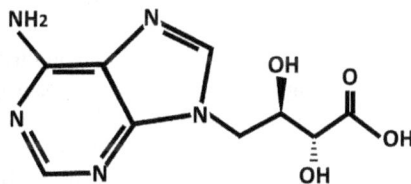

Fig. 5. Eritadenine, is a compound found in shiitake mushrooms responsible for its hypocholesterolemic activity.

Asada et al. (2019) reported that shiitake and medium-chain-triglyceride-rich plant oil acted simultaneously to decrease lipid levels and lecithin-cholesterol acyltransferase activity in blood serum. Similarly, shiitake mushroom bars were evaluated on oxidative stress and cholesterolemia values in participants with borderline high cholesterol, low-density lipoprotein or triglycerides. A total of 68 people was randomly assigned to the placebo or intervention group. Following 66 days of eating the mushroom bars, the shiitake group had a 10% reduction in triglycerides. In addition, this group had higher endogenous antioxidants, lower glutathione levels and lower lipid peroxidation. According to the researchers, *L. edodes* could be a valuable functional food that improves redox status and prevents dyslipidemia (Spim et al., 2021b).

5. Use of *L. edodes* for Functional Foods Formulation

In recent decades, consumers have demanded food products with greater health benefits. Marketers have ridden the tide of rising health awareness and realised the significant potential of adding functional ingredients to everyday items (Aguiar et al., 2019; Aggett et al., 1999). Functional food is defined as any natural food or a food to which a constituent has been added or eliminated by a technological procedure that presents a good impact on an individual's health besides its nutritious benefits (Granato et al., 2020). Food must be taken as part of a regular diet to qualify as a functional food; it cannot be a capsule, pill, or powder made from natural components.

Mushrooms are used as a bioactive component in functional foods and it has attracted the scientific community's interest due to several reports on their health advantages. Humans have consumed mushrooms, including *L. edodes*, for centuries for their nutritional benefits and organoleptic values and currently, people are eating mushrooms as functional foods because of their medicinal properties (Ho et al., 2020). Mushrooms possess high nutritional value; they are a great source of protein with all essential amino acids, even higher than that of most vegetables. Mushrooms are high in dietary fibre, so their calorific value is low; they also contain vitamins and a broad variety of bioactive characteristics (Feeney et al., 2014; Chang, 2006). Mushroom-derived bioactive compounds have immunomodulating, antioxidant, ant-itumour and antibacterial activities that can be employed as functional food additives for enhancing nutritional benefits. Specifically, *L. edodes* contains many medicinal substances, such as polysaccharides, terpenoids, sterols, polyphenols and lipids that are beneficial in treating various diseases.

Nowadays, many mushroom products and compounds are available on the global pharmaceutical and food markets, referred to as nutraceuticals or functional food additives (Shamtsyan, 2016). The extracts can be consumed as a beverage, freeze-dried, or turned into powder before being added to food products, like bakery products, pasta, patties and snacks, making them easier to handle, package, transport, consume, etc. Also, dried powdered mushroom extracts can be put into capsules and used as dietary supplements or mushroom nutraceuticals with health advantages (Morris et al., 2017). Many of these bioactive compounds can also be employed to enhance the functional characteristics of typical foods.

Mushrooms can be employed in the functional food formulations due to their nutritional diversity and positive effects. Functional bars, for example, are a simple product to market because of their stability, ease of transportation, low production cost and high acceptance, as well as the flexibility of different functional ingredients that may be added to the bar. In recent decades, the shiitake mushroom has become more widely used in biotechnology and in the prevention and treatment of chronic diseases, particularly hypercholesterolemia and diabetes. Spim et al. (2021b) have reported that shiitake bars can improve redox status and prevent dyslipidemia. In this context, dried shiitake was incorporated into cereal bars to prepare two sweet and two salty bars and assessed for their sensory acceptability and maintenance of their antioxidants and nutritional values. Sweet bar 1 (oats, brown and white sugar, dry plum, nuts, flaxseed, glucose, lecithin, quinoa, peanut, chia, coconut oil, glycerine) performed better in terms of sensory analysis, purchasing intention and nutritional qualities. These bars maintained the shiitake concentrations of proteins, fibres, glucans and phenolic compounds. Sweet bars are a simple marketing option because of their stability, low production costs, high acceptance and the opportunity to include other functional components that are excellent for human health, such as shiitake mushrooms (Spim et al., 2021a).

Furthermore, shiitake has been used to partially substitute wheat flour in creating biscuits deficient in various components, such as minerals, vitamins, proteins and dietary fibre. Essential amino acids, such as lysine and tryptophan, are lacking in wheat flour (Prodhan et al., 2015). To generate functional bread, partial substitutions or wheat flour supplementation with other nutritional elements are strongly suggested. Because the addition of shiitake mushroom powder to other culinary items imparts a distinct and pleasant flavour and aftertaste to meals, many countries use mushrooms in various goods, including soups, sauces and pickles. With this in mind, Van Toan and Anh (2018) substituted shiitake mushroom powder for 5–10% of the wheat flour in biscuits, resulting in a considerable boost in fibre and protein content compared to wheat flour alone.

On the other hand, considering the global demand for healthier products, low-meat sausages tend to use mainly meat replacements in the product. Commonly they have used rice bran, tomato pomace, non-meat proteins and hydrocolloids; however, the addition of mushrooms is still being explored (Savadkoohi et al., 2014). Liyan Wang et al. (2019) employed *L. edodes* as a 25%, 50%, 75% and 100% pork lean, mean substitute in sausages. *L. edodes* improved the sausage's moisture content, overall dietary fibre content, methionine, glutamic acid, cysteine and total phenolic content, cooking loss and antioxidant properties; but also, *L. edodes* lowered the sausage's protein, ash, pH, energy content and texture. The findings showed that *L. edodes* might be utilised as a meal to decrease the amount of lean pork meat used in sausages.

Other studies on functional fungal properties for pharmacological treatment reported that adding shiitake powder (cap, stem, whole) to noodles improved the protein content, depressed the glycaemic reply and augmented the antioxidant activity (Liwen Wang et al., 2020). Shiitake noodles also showed antioxidant qualities at the cellular level based on the IEC-6 intestinal epithelial cell line, suggesting that the insoluble conjugated phenolic acids were released after a simulated digestion process. This is the base for using shiitake as a functional food to prevent chronic diseases.

Recent findings have demonstrated the use of shiitake components not as ingredients added to food, but also as part of a food design in the preparation of functional coatings. For example, soluble dietary fibre from *L. edodes* was used as a coating material to improve the viability and release efficiency of the probiotic *Lactobacillus plantarum* LP90 (Xue et al., 2021). The results showed efficient microencapsulation and uniform distribution of *L. plantarum*. *L. edodes* fibre from high and low molecular weight showed greater stability under temperature, humidity, storage and gastrointestinal digestion. This microencapsulation can be applied in the package and as a delivery system of probiotics in functional foods. Table 6 shows more studies using shiitake as an active ingredient in the formulation of functional foods.

Table 6. Studies reporting the use of *L. edodes* as functional foods.

Food System	*L. edodes* Incorporation	Results	References
Pickled product	Sun-dried shiitake	• Shower organoleptic acceptation • Increased protein content • Increased polyphenol content • Showed good nutritional profile	(Singh et al., 2016)
Muffin	Shiitake powder (5, 10, 15%)	• Increased polyphenol content • Increased antioxidant potential • Showed better consumer acceptability than the control rice muffin • Improve properties of rice muffin, making it more suitable for producing quality gluten-free muffin	(Olawuyi and Lee, 2019)
Fermented sausages	Shiitake stipes extracted by ethanol/water at 50/50 ratio (0.6%)	• Increased antioxidant activities • Increased inhibitory capacity against lipid oxidation • Decreased the rate of total aerobic count • Showed strong antimicrobial activity against *S. aureus, L. monocytogenes, Salmonella typhimurium,* and *E. coli.* • Improved the shelf-life stability	(Van Ba et al., 2017)
Non-sucrose gluten-free cookie	Shiitake powder (5, 10, 15, 20, 25%)	• The quality features varied significantly to different degrees • Consumer's acceptance decreased with the concentration increase	(Feng et al., 2018)
Biscuit	Shiitake powder (5–15%)	• Increased contents of fibre and protein	(Van Toan and Thu, 2018)
Pasta	Shiitake powder	• Protein and soluble and insoluble dietary fibre-enriched • Decreased the starch digestibility of pasta • Enhanced the antioxidant activity • Increased total phenolic content • Proposed for lowering the potential glycaemic response	(Lu et al., 2018)
Semolina-extruded snack products	Shiitake powder (5, 10, 15%)	• Reduced product expansion, decreased water absorption index, increased water solubility index and altered microstructure characteristics • Reduced the degree of starch gelatinisation and digestibility • Increased levels of phenolic compounds and antioxidants	(Lu et al., 2020)

6. Conclusion

In several Asian countries, shiitake has economic potential both in their own countries and export to Western countries. Bioactive compounds of *L. edodes* have been drawing interest to formulate dietary supplements and medicines with pharmacological impact. The incorporation of this fungus in different foods provides technological and bioactive properties. More studies of the compounds present in *L. edodes* must be carried out, both in extract and isolated and above all, in different chronic-degenerative diseases, in addition, to seeing the technological potential of its incorporation in different foods, monitoring flavour, colour, texture and digestibility of the other components of the food.

References

Aggett, P., Ashwell, M., Bornet, F., Diplock, A., Fern, E. and Roberfroid, M. (1999). Scientific concepts of functional foods in Europe consensus document. *Br. J. Nutr.*, 81: 1–27.

Aguiar, L.M., Geraldi, M.V., Betim Cazarin, C.B. and Maróstica Junior, M.R. (2019). Functional food consumption and its physiological effects. pp. 205–225. *In*: Segura Campos, M.R. (ed.). *Bioactive Compounds*, Elsevier, New York, USA.

Amini, M., Zayeri, F. and Salehi, M. (2021). Trend analysis of cardiovascular disease mortality, incidence and mortality-to-incidence ratio: Results from global burden of disease study 2017. *BMC Public Health*, 21: 1–12.

Asada, N., Kairiku, R., Tobo, M. and Ono, A. (2019). Effects of shiitake intake on serum lipids in rats fed different high-oil or high-fat diets. *J. Diet. Suppl.*, 16: 345–356.

Bisen, P., Baghel, R.K., Sanodiya, B.S., Thakur, G.S. and Prasad, G. (2010). *Lentinus edodes*: A macrofungus with pharmacological activities. *Curr. Med. Chem.*, 17: 2419–2430.

Çağlarırmak, N. (2007). The nutrients of exotic mushrooms (*Lentinula edodes* and *Pleurotus* species) and an estimated approach to the volatile compounds. *Food Chem.*, 105: 1188–1194.

Cao, X.-Y., Liu, D., Bi, R.-C., He, Y.-l., He, Y. and Liu, J.-l. (2019). The protective effects of a novel polysaccharide from *Lentinus edodes* mycelia on islet β (INS-1) cells damaged by glucose and its transportation mechanism with human serum albumin. *Int. J. Biol. Macromol.*, 134: 344–353.

Cao, X., Xia, Y., Liu, D., He, Y., Mu, T., Huo, Y. and Liu, J. (2020). Inhibitory effects of *Lentinus edodes* mycelia polysaccharide on α-glucosidase, glycation activity and high glucose-induced cell damage. *Carbohydr. Polym.*, 246: 116659.

Carneiro, A.A., Ferreira, I.C., Dueñas, M., Barros, L., Da Silva, R., Gomes, E. and Santos-Buelga, C. (2013). Chemical composition and antioxidant activity of dried powder formulations of *Agaricus blazei* and *Lentinus edodes*. *Food Chem.*, 138: 2168–2173.

Chang, S.-T. (2006). The world mushroom industry: Trends and technological development. *Int. J. Med. Mushrooms*, 8.

Chen, S., Liu, C., Huang, X., Hu, L., Huang, Y., Chen, H. and Tang, W. (2020). Comparison of immunomodulatory effects of three polysaccharide fractions from *Lentinula edodes* water extracts. *J. Funct. Foods*, 66: 103791.

Cheng, H.-T., Xu, X., Lim, P.S. and Hung, K.-Y. (2021). Worldwide epidemiology of diabetes-related end-stage renal disease, 2000–2015. *Diabetes Care*, 44: 89–97.

Chikari, F., Han, J., Wang, Y. and Ao, W. (2020). Synergised subcritical-ultrasound-assisted aqueous two-phase extraction, purification, and characterization of *Lentinus edodes* polysaccharides. *Process Biochem.*, 95: 297–306.

de Melo Macoris, J.D., Brugnari, T., Boer, C.G., Contato, A.G., Peralta, R.M. and de Souza, C.G.M. (2017). Antioxidant properties and antimicrobial potential of aqueous extract of basidioma from *Lentinus edodes* (Berk.) Sing.(Shiitake). *Int. J. Curr. Microbiol. Appl. Sci.*, 6: 3757–3767.

Du, B., Zhu, F. and Xu, B. (2018). An insight into the anti-inflammatory properties of edible and medicinal mushrooms. *J. Funct. Foods*, 47: 334–342.

Elhusseiny, S.M., El-Mahdy, T.S., Awad, M.F., Elleboudy, N.S., Farag, M., Yassein, M.A. and Aboshanab, K.M. (2021). Proteome analysis and *in vitro* antiviral, anticancer and antioxidant capacities of the aqueous extracts of *Lentinula edodes* and *Pleurotus ostreatus* edible mushrooms. *Molecules*, 26: 4623.

Erdoğan Eliuz, E.A. (2021). Antibacterial activity and antibacterial mechanism of ethanol extracts of *Lentinula edodes* (Shiitake) and *Agaricus bisporus* (button mushroom). *Int. J. Environ. Health Res.*, 1–14.

Feeney, M.J., Miller, A.M. and Roupas, P. (2014). Mushrooms—biologically distinct and nutritionally unique: Exploring a 'third food kingdom'. *Nutrition Today*, 49: 301.

Fekrazad, R., Afzali, M., Pasban-Aliabadi, H., Esmaeili-Mahani, S., Aminizadeh, M. and Mostafavi, A. (2017). Cytotoxic effect of *Thymus caramanicus* Jalas on human oral epidermoid carcinoma KB cells. *Braz. Dent. J.*, 28: 72–77.

Feng, T., Wang, W., Zhuang, H., Song, S., Yao, L., Sun, M. and Xu, Z. (2018). *In vitro* digestible properties and quality characterisation of nonsucrose gluten-free *Lentinus edodes* cookies. *J. Food Process. Preserv.*, 42: e13454.

Finimundy, T.C., Dillon, A.J.P., Henriques, J.A.P. and Ely, M.R. (2014). A review on general nutritional compounds and pharmacological properties of the *Lentinula edodes* mushroom. *Food Nutr. Sci.*, 5: 1–11.

Fukushima-Sakuno, E. (2020). Bioactive small secondary metabolites from the mushrooms *Lentinula edodes* and *Flammulina velutipes*. *J. Antibiot.*, 73: 687–696.

Galindo-Villegas, J. and Hosokawa, H. (2004). Immunostimulants: Towards temporary prevention of diseases in marine fish. *Advances en Nutrición Acuicola*. Recovered from: https://nutricionacuicola.uanl.mx/index.php/acu/article/view/202.

Garcia, J., Afonso, A., Fernandes, C., Nunes, F.M., Marques, G. and Saavedra, M.J. (2021). Comparative antioxidant and antimicrobial properties of *Lentinula edodes* Donko and Koshin varieties against priority multidrug-resistant pathogens. *S. Afr. J. Chem. Eng.*, 35: 98–106.

Gil-Ramírez, A., Smiderle, F.R., Morales, D., Iacomini, M. and Soler-Rivas, C. (2019). Strengths and weaknesses of the aniline-blue method used to test mushroom $(1\rightarrow3)$-β-d-glucans obtained by microwave-assisted extractions. *Carbohydr. Polym.*, 217: 135–143.

GLOBOCAN. (2020). *Global Cancer Observatory: Cancer Today*. Retrieved from https://www.uicc.org/news/globocan-2020-new-global-cancer-data.

Granato, D., Barba, F.J., Bursać Kovačević, D., Lorenzo, J.M., Cruz, A.G. and Putnik, P. (2020). Functional foods: Product development, technological trends, efficacy testing, and safety. *Annu. Rev. Food Sci. Technol.*, 11: 93–118.

Gupta, S.C., Kunnumakkara, A.B., Aggarwal, S. and Aggarwal, B.B. (2018). Inflammation, a double-edge sword for cancer and other age-related diseases. *Front. Immunol.*, 9: 2160.

Hata, K. (2021). *In vitro* and *in vivo* antidiabetic effects of the ethanol extract from *Lentinula edodes* (Shiitake). *International Journal on Nutraceuticals, Functional Foods and Novel Foods*, 15: 279–284.

Hatvani, N. (2001). Antibacterial effect of the culture fluid of *Lentinus edodes* mycelium grown in submerged liquid culture. *Int. J. Antimicrob. Agents*, 17: 71–74.

He, F., Yang, Y., Yang, G. and Yu, L. (2010). Studies on antibacterial activity and antibacterial mechanism of a novel polysaccharide from *Streptomyces virginia* H03. *Food Control*, 21: 1257–1262.

Ho, L.-H., Zulkifli, N.A. and Tan, T.-C.J.A.I.T.M. (2020). *Edible Mushroom: Nutritional Properties, Potential Nutraceutical Values and its Utilisation in Food Product Development*, 19–38.

Hou, H., Liu, C., Lu, X., Fang, D., Hu, Q., Zhang, Y. and Zhao, L. (2021). Characterization of flavor frame in shiitake mushrooms (*Lentinula edodes*) detected by HS-GC-IMS coupled with electronic tongue and sensory analysis: Influence of drying techniques. *LWT*, 146: 111402.

Jeff, I.B., Fan, E., Tian, M., Song, C., Yan, J. and Zhou, Y. (2016). *In vivo* anticancer and immunomodulating activities of mannogalactoglucan-type polysaccharides from *Lentinus edodes* (Berkeley) Singer. *Cent. Eur. J. Immunol.*, 41: 47.

Jeff, I.B., Yuan, X., Sun, L., Kassim, R.M., Foday, A.D. and Zhou, Y. (2013). Purification and in vitro anti-proliferative effect of novel neutral polysaccharides from *Lentinus edodes*. *Int. J. Biol. Macromol.*, 52: 99–106.

Jin, Y., Li, P. and Wang, F. (2018). β-glucans as potential immunoadjuvants: A review on the adjuvanticity, structure-activity relationship and receptor recognition properties. *Vaccine*, 36: 5235–5244.

Joseph, T.P., Zhao, Q., Chanda, W., Kanwal, S., Fang, Y., Zhong, M. and Huang, M. (2021). Expression and *in vitro* anticancer activity of Lp16-PSP, a member of the YjgF/YER057c/UK114 protein family from the mushroom *Lentinula edodes* C 91-3. *Arch. Microbiol.*, 203: 1047–1060.

Kaur, H., Nyochembeng, L.M., Banerjee, P., Cebert, E. and Mentreddy, S. (2019). Optimisation of fermentation conditions of *Lentinula edodes* (berk.) pegler (shiitake mushroom) mycelia as a potential biopesticide. *J. Agric. Sci.*, 11.

Kaur, H., Nyochembeng, L.M., Mentreddy, S.R., Banerjee, P. and Cebert, E. (2016). Assessment of the antimicrobial activity of *Lentinula edodes* against *Xanthomonas campestris* pv. vesicatoria. *Crop. Prot.*, 89: 284–288.

Kitzberger, C.S.G., Smânia Jr, A., Pedrosa, R.C. and Ferreira, S.R.S. (2007). Antioxidant and antimicrobial activities of shiitake (*Lentinula edodes*) extracts obtained by organic solvents and supercritical fluids. *J. Food Eng.*, 80: 631–638.

Kobayashi, T., Oguro, M., Akiba, M., Taki, H., Kitajima, H. and Ishihara, H. (2020). Mushroom yield of cultivated shiitake (*Lentinula edodes*) and fungal communities in logs. *J. For. Res.*, 25: 269–275.

Kwak, A.-M., Lee, I.-K., Lee, S.-Y., Yun, B.-S. and Kang, H.-W. (2016). Oxalic acid from *Lentinula edodes* culture filtrate: Antimicrobial activity on phytopathogenic bacteria and qualitative and quantitative analyses. *Mycrobiol.*, 44: 338–342.

Laurino, L.F., Viroel, F.J., Caetano, E., Spim, S., Pickler, T.B., Rosa-Castro, R.M. and Grotto, D. (2019). *Lentinus edodes* exposure before and after fetus implantation: Materno-fetal development in rats with gestational diabetes mellitus. *Nutrients*, 11: 2720.

Lee, M.H., Seo, D.J., Kang, J.-H., Oh, S.H. and Choi, C. (2014). Expression of antiviral cytokines in Crandell-Reese feline kidney cells pretreated with Korean red ginseng extract or ginsenosides. *Food Chem. Toxicol.*, 70: 19–25.

Li, W., Wang, J., Chen, Z., Gao, X., Chen, Y., Xue, Z. and Chen, H. (2018a). Physicochemical properties of polysaccharides from Lentinus edodes under high pressure cooking treatment and its enhanced anticancer effects. *Int. J. Biol. Macromol.*, 115: 994–1001.

Li, W., Wang, J., Hu, H., Li, Q., Liu, Y. and Wang, K. (2018b). Functional polysaccharide Lentinan suppresses human breast cancer growth via inducing autophagy and caspase-7-mediated apoptosis. *J. Funct. Foods*, 45: 75–85.

Lo, H.-C. and Wasser, S.P. (2011). Medicinal mushrooms for glycemic control in diabetes mellitus: History, current status, future perspectives, and unsolved problems. *Int. J. Med. Mushrooms*, 13.

Lobo, V., Patil, A., Phatak, A. and Chandra, N. (2010). Free radicals, antioxidants and functional foods: Impact on human health. *Pharmacogn. Rev.*, 4: 118.

Lu, X., Hou, H., Fang, D., Hu, Q., Chen, J. and Zhao, L. (2022). Identification and characterization of volatile compounds in *Lentinula edodes* during vacuum freeze-drying. *J. Food Biochem.*, 46: e13814.

Lu, X., Brennan, M.A., Narciso, J., Guan, W., Zhang, J., Yuan, L. and Brennan, C.S. (2020). Correlations between the phenolic and fibre composition of mushrooms and the glycaemic and textural characteristics of mushroom enriched extruded products. *LWT*, 118: 108730.

Lu, X., Brennan, M.A., Serventi, L., Liu, J., Guan, W. and Brennan, C.S. (2018). Addition of mushroom powder to pasta enhances the antioxidant content and modulates the predictive glycaemic response of pasta. *Food Chem.*, 264: 199–209.

Maritim, A., Sanders, A. and Watkins Iii, J. (2003). Diabetes, oxidative stress and antioxidants: A review. *J. Biochem. Mol. Toxicol.*, 17: 24–38.

Maru, G.B., Hudlikar, R.R., Kumar, G., Gandhi, K. and Mahimkar, M.B. (2016). Understanding the molecular mechanisms of cancer prevention by dietary phytochemicals: From experimental models to clinical trials. *World J. Biol. Chem.*, 7: 88–99. Doi:10.4331/wjbc.v7.i1.88.

Meng, X., Liang, H. and Luo, L. (2016). Anti-tumour polysaccharides from mushrooms: A review on the structural characteristics, antitumor mechanisms and immunomodulating activities. *Carbohydr. Res.*, 424: 30–41.

Morales, D., Piris, A.J., Ruiz-Rodriguez, A., Prodanov, M. and Soler-Rivas, C. (2018). Extraction of bioactive compounds against cardiovascular diseases from *Lentinula edodes* using a sequential extraction method. *Biotechnol. Prog.*, 34: 746–755.

Morales, D., Rutckeviski, R., Villalva, M., Abreu, H., Soler-Rivas, C., Santoyo, S. and Smiderle, F.R. (2020). Isolation and comparison of α-and β-D-glucans from shiitake mushrooms (*Lentinula edodes*) with different biological activities. *Carbohydr. Polym.*, 229: 115521.

Morris, H.J., Llauradó, G., Beltrán, Y., Lebeque, Y., Bermúdez, R.C., García, N. and Moukha, S. (2017). The use of mushrooms in the development of functional foods, drugs, and nutraceuticals. *Wild Plants, Mushrooms Nuts: Functional Food Properties and Applications*, 5: 123–159.

Murphy, E.J., Masterson, C., Rezoagli, E., O'Toole, D., Major, I., Stack, G.D. and Rowan, N.J. (2020). β-Glucan extracts from the same edible shiitake mushroom *Lentinus edodes* produce differential *in-vitro* immunomodulatory and pulmonary cytoprotective effects—Implications for coronavirus disease (COVID-19) immunotherapies. *Sci. Total Environ.*, 732: 139330.

Nguyen, T.L., Chen, J., Hu, Y., Wang, D., Fan, Y., Wang, J. and Chen, X. (2012). *In vitro* antiviral activity of sulfated *Auricularia auricula* polysaccharides. *Carbohydr. Polym.*, 90: 1254–1258.

Nisar, J., Mustafa, I., Anwar, H., Sohail, M.U., Hussain, G., Ullah, M.I. and Basit, A. (2017). Shiitake culinary-medicinal mushroom, *Lentinus edodes* (*Agaricomycetes*): A species with antioxidant, immunomodulatory, and hepatoprotective activities in hypercholesterolemic rats. *Int. J. Med. Mushrooms*, 19.

Olawuyi, I.F. and Lee, W.Y. (2019). Quality and antioxidant properties of functional rice muffins enriched with shiitake mushroom and carrot pomace. *Int. J. Food Sci. Technol.*, 54: 2321–2328.

Parola, S., Chiodaroli, L., Orlandi, V., Vannin, C. and Panno, L. (2017). *Lentinula edodes* and *Pleurotus ostreatus*: Functional food with antioxidant-antimicrobial activity and an important source of Vitamin D and medicinal compounds. *Funct. Foods Health Dis.*, 7: 773–794.

Philippoussis, A., Diamantopoulou, P. and Israilides, C. (2007). Productivity of agricultural residues used for the cultivation of the medicinal fungus *Lentinula edodes*. *International Biodeterioration and Biodegradation*, 59(3): 216–219.

Pillai, S., Shetty, N., Mathur, A., Bali, A. and Pal, P. (2021). Evaluation of *Lentinula edodes* (Shiitake Mushroom) mouthwash in patients with gingivitis. *Journal of Oral and Hygiene Health*, 9: 2.

Prodhan, U.K., Linkon, K.M.M.R., Al-Amin, M.F. and Alam, M.J. (2015). Development and quality evaluation of mushroom (*Pleurotus sajor-caju*) enriched biscuits. *Emir. J. Food Agric.*, 27: 542–547. Doi:10.9755/ejfa.2015.04.082.

Raudaskoski, M. and Kothe, E. (2010). Basidiomycete mating type genes and pheromone signaling. *Eukaryot. Cell*, 9(6): 847–59. Doi: 10.1128/EC.00319-09.

Rea, I.M., Gibson, D.S., McGilligan, V., McNerlan, S.E., Alexander, H.D. and Ross, O.A. (2018). Age and age-related diseases: Role of inflammation triggers and cytokines. *Front. Immunol.*, 9: 586.

Ren, G., Xu, L., Lu, T. and Yin, J. (2018a). Structural characterisation and antiviral activity of lentinan from *Lentinus edodes* mycelia against infectious hematopoietic necrosis virus. *Int. J. Biol. Macromol.*, 115: 1202–1210.

Ren, Z., Liu, W., Song, X., Qi, Y., Zhang, C., Gao, Z. and Jia, L. (2018b). Antioxidant and anti-inflammation of enzymatic-hydrolysis residue polysaccharides by *Lentinula edodes*. *Int. J. Biol. Macromol.*, 120: 811–822.

Reuter, S., Gupta, S.C., Chaturvedi, M.M. and Aggarwal, B.B. (2010). Oxidative stress, inflammation, and cancer: How are they linked? *Free Radic. Biol. Med.*, 49: 1603–1616.

Rincão, V.P., Yamamoto, K.A., Ricardo, N.M.P.S., Soares, S.A., Meirelles, L.D.P., Nozawa, C. and Linhares, R.E.C. (2012). Polysaccharide and extracts from *Lentinula edodes*: Structural features and antiviral activity. *Virol. J.*, 9: 1–6.

Rivera, O.A., Albarracín, W. and Lares, M. (2017). *Componentes Bioactivos del Shiitake (Lentinula edodes Berk. Pegler) y su impacto en la salud*, *Archivos Venezolanos de Farmacología y Terapéutica*, 36(3): 67–71.

Rock, K.L. and Kono, H. (2008). The inflammatory response to cell death. *Annu. Rev. Pathol.*, 3: 99–126.

Romero-Arenas, O., Martínez Guerrero, M.A., Damián Huato, M.A., Ramírez Valverde, B. and López-Olguín, J. (2015). Producción del hongo Shiitake (Lentinula edodes Pegler) en bloques sintéticos utilizando residuos agroforestales. *Revista Mexicana de Ciencias Agrícolas*, 6: 1229–1238.

Sánchez-Minutti, L., Téllez-Téllez, M., Tlecuitl-Beristain, S., Santos-López, G., Díaz, R., Gupta, V. K. and Díaz-Godínez, G. (2016). Antimicrobial activity of a protein obtained from fruiting body of *Lentinula edodes* against *Escherichia coli* and *Staphylococcus aureus*. *J. Environ. Biol.*, 37: 619.

Santoyo, S., Ramírez-Anguiano, A.C., Aldars-García, L., Reglero, G. and Soler-Rivas, C. (2012). Antiviral activities of *Boletus edulis*, *Pleurotus ostreatus* and *Lentinus edodes* extracts and polysaccharide fractions against Herpes simplex virus type 1. *J. Food Nutr. Res.*, 51: 225–235.

Sasidharan, S., Aravindran, S., Latha, L.Y., Vijenthi, R., Saravanan, D. and Amutha, S. (2010). *In vitro* antioxidant activity and hepatoprotective effects of *Lentinula edodes* against paracetamol-induced hepatotoxicity. *Molecules*, 15: 4478–4489.

Savadkoohi, S., Hoogenkamp, H., Shamsi, K. and Farahnaky, A. (2014). Colour, sensory and textural attributes of beef frankfurter, beef ham and meat-free sausage containing tomato pomace. *Meat Sci.*, 97: 410–418.

Shalhoub, S., Farahat, F., Al-Jiffri, A., Simhairi, R., Shamma, O., Siddiqi, N. and Mushtaq, A. (2015). IFN-α2a or IFN-β1a in combination with ribavirin to treat Middle East respiratory syndrome coronavirus pneumonia: A retrospective study. *J. Antimicrob. Chemother.*, 70: 2129–2132.

Shamtsyan, M. (2016). Potential to develop functional food products from mushroom bioactive compounds. *Journal of Hygienic Engineering and Design*, 15: 51–59.

Sharif, S., Atta, A., Huma, T., Shah, A.A., Afzal, G., Rashid, S. and Mustafa, G. (2018). Anticancer, antithrombotic, antityrosinase, and anti-α-glucosidase activities of selected wild and commercial mushrooms from Pakistan. *Food Sci. Nutr.*, 6: 2170–2176.

Sherwood, E.R. and Toliver-Kinsky, T. (2004). Mechanisms of the inflammatory response. *Best Pract. Res. Clin. Anaesthesiol.*, 18: 385–405.

Shi, M., Zhang, Z. and Yang, Y. (2013). Antioxidant and immunoregulatory activity of *Ganoderma lucidum* polysaccharide (GLP). *Carbohydr. Polym.*, 95: 200–206.

Singh, J., Sindhu, S.C. and Sindhu, A. (2016). Development and evaluation of value added pickle from dehydrated shiitake (*Lentinus edodes*) Mushroom. *Development* 1.

Singh, M.P., Rai, S.N., Dubey, S.K., Pandey, A.T., Tabassum, N., Chaturvedi, V.K. and Singh, N.B. J.C.R.I.B. (2021). Biomolecules of mushroom: A recipe of human wellness, 1–18.

Sogo, T., Kawahara, M., Ueda, H., Otsu, M., Onodera, M., Nakauchi, H. and Nagamune, T. (2009). T cell growth control using hapten-specific antibody/interleukin-2 receptor chimera. *Cytokine*, 46: 127–136.

Song, X., Ren, Z., Wang, X., Jia, L. and Zhang, C. (2020). Antioxidant, anti-inflammatory and renoprotective effects of acidic-hydrolytic polysaccharides by spent mushroom compost (*Lentinula edodes*) on LPS-induced kidney injury. *Int. J. Biol. Macromol.*, 151: 1267–1276.

Spim, S.R.V., Castanho, N.R.C.M., Pistila, A.M.H., Jozala, A.F., Júnior, J.M.O. and Grotto, D. (2021a). *Lentinula edodes* mushroom as an ingredient to enhance the nutritional and functional properties of cereal bars. *J. Food Sci. Technol.*, 58: 1349–1357.

Spim, S.R.V., Pistila, A.M.H., Pickler, T.B., Silva, M.T. and Grotto, D. (2021b). Effects of shiitake culinary-medicinal mushroom, *Lentinus edodes* (*Agaricomycetes*), bars on lipid and antioxidant profiles in individuals with borderline high cholesterol: A double-blind randomized clinical trial. *Int. J. Med. Mushrooms*, 23.

Tang, W., Liu, C., Liu, J., Hu, L., Huang, Y., Yuan, L. and Bian, S. (2020). Purification of polysaccharide from *Lentinus edodes* water extract by membrane separation and its chemical composition and structure characterisation. *Food Hydrocoll.*, 105: 105851.

Van Ba, H., Seo, H.-W., Cho, S.-H., Kim, Y.-S., Kim, J.-H., Ham, J.-S. and Pil-Nam, S. (2017). Effects of extraction methods of shiitake by-products on their antioxidant and antimicrobial activities in fermented sausages during storage. *Food Control*, 79: 109–118.

Van Toan, N. and Anh, V.Q. (2018). Preparation and improved quality production of flour and the made biscuits from purple sweet potato. *J. Food Nutr. Res.*, 4: 1–14.

Van Toan, N. and Thu, L.N.M. (2018). Preparation and improved quality production of flour and the made biscuits from shitake mushroom (*Lentinus edodes*). *Clinical Journal of Nutrition and Dietetics*, 1: 1–9.

Vetvicka, V., Teplyakova, T.V., Shintyapina, A.B. and Korolenko, T.A.J.J.O.F. (2021). Effects of medicinal fungi-derived β-glucan on tumor progression. *J. Fungi*, 7: 250.

Wang, J., Li, W., Huang, X., Liu, Y., Li, Q., Zheng, Z. and Wang, K. (2017). A polysaccharide from *Lentinus edodes* inhibits human colon cancer cell proliferation and suppresses tumour growth in athymic nude mice. *Oncotarget*, 8: 610.

Wang, L., Guo, H., Liu, X., Jiang, G., Li, C., Li, X. and Li, Y. (2019). Roles of *Lentinula edodes* as the pork lean meat replacer in production of the sausage. *Meat Sci.*, 156: 44–51.

Wang, L., Zhao, H., Brennan, M., Guan, W., Liu, J., Wang, M. and Brennan, C. (2020). *In vitro* gastric digestion antioxidant and cellular radical scavenging activities of wheat-shiitake noodles. *Food Chem.*, 330: 127214.

Wasser, S. (2002). Medicinal mushrooms as a source of antitumor and immunomodulating polysaccharides. *Applied Microbiology and Biotechnology Reports*, 60: 258–274.

Wasser, S.P. (2005). Shiitake (*Lentinus edodes*). *Encyclopedia of Dietary Supplements*, 653–664.

WHO. (2021). *Antimicrobial Resistance*. Retrieved from https://www.who.int/news-room/fact-sheets/detail/antimicrobial-resistance.

Xiang, X., Li, C., Li, L., Bian, Y., Kwan, H.S., Nong, W., Cheung, M.K. and Xiao, Y. (2016). Genetic diversity and population structure of Chinese *Lentinula edodes* revealed by InDel and SSR markers. *Mycological Progress*, 15(4): 1–13.

Xiaokang, W., Lyng, J.G., Brunton, N.P., Cody, L., Jacquier, J.-C., Harrison, S.M. and Papoutsis, K. (2020). Monitoring the effect of different microwave extraction parameters on the recovery of polyphenols from shiitake mushrooms: Comparison with hot-water and organic-solvent extractions. *Biotechnol. Rep.*, 27: e00504.

Xu, H., Zou, S., Xu, X. and Zhang, L. (2016). Anti-tumour effect of β-glucan from *Lentinus edodes* and the underlying mechanism. *Sci. Rep.*, 6: 1–13.

Xue, Z., Chen, Z., Gao, X., Zhang, M., Panichayupakaranant, P. and Chen, H. (2021). Functional protection of different structure soluble dietary fibres from *Lentinus edodes* as effective delivery substrate for *Lactobacillus plantarum* LP90. *LWT*, 136: 110339.

Xue, Z., Ma, Q., Chen, Y., Lu, Y., Wang, Y., Jia, Y., and Chen, H. (2020). Structure characterization of soluble dietary fiber fractions from mushroom *Lentinula edodes* (Berk.) Pegler and the effects on fermentation and human gut microbiota *in vitro*. *Food Res. Int.*, 129: 108870.

Ya, G. (2017). A *Lentinus edodes* polysaccharide induces mitochondrial-mediated apoptosis in human cervical carcinoma HeLa cells. *Int. J. Biol. Macromol.*, 103: 676–682.

Yang, H., Hwang, I., Kim, S., Hong, E.J. and Jeung, E.B. (2013). *Lentinus edodes* promotes fat removal in hypercholesterolemic mice. *Exp. Ther. Med.*, 6: 1409–1413.

Yin, C., Fan, X., Fan, Z., Shi, D. and Gao, H. (2018). Optimisation of enzymes-microwave-ultrasound assisted extraction of Lentinus edodes polysaccharides and determination of its antioxidant activity. *Int. J. Biol. Macromol.*, 111: 446–454.

Yu, Z., LiHua, Y., Qian, Y. and Yan, L. (2009). Effect of *Lentinus edodes* polysaccharide on oxidative stress, immunity activity and oral ulceration of rats stimulated by phenol. *Carbohydr. Polym.*, 75: 115–118.

Zelová, H. and Hošek, J. (2013). TNF-α signalling and inflammation: Interactions between old acquaintances. *Inflamm. Res.*, 62: 641–651.

Zhang, J., Wen, C., Qin, W., Qin, P., Zhang, H. and Duan, Y. (2018). Ultrasonic-enhanced subcritical water extraction of polysaccharides by two steps and its characterisation from *Lentinus edodes*. *International, Journal of Biological Macromolecules*, 118: 2269–2277.

Zhang, Y., Li, Q., Shu, Y., Wang, H., Zheng, Z., Wang, J. and Wang, K. (2015). Induction of apoptosis in S180 tumour bearing mice by polysaccharide from *Lentinus edodes* via mitochondria apoptotic pathway. *J. Funct. Foods*, 15: 151–159.

Zhang, Y., Li, S., Wang, X., Zhang, L. and Cheung, P.C. (2011). Advances in lentinan: Isolation, structure, chain conformation and bioactivities. *Food Hydrocoll.*, 25: 196–206.

Zhang, Y., Mei, H., Shan, W., Shi, L., Chang, X., Zhu, Y. and Han, X. (2016). Lentinan protects pancreatic β cells from STZ-induced damage. *J. Cell. Mol. Med.*, 20: 1803–1812.

Zhu, H., Tian, L., Zhang, L., Bi, J., Song, Q., Yang, H. and Qiao, J. (2018). Preparation, characterization and antioxidant activity of polysaccharide from spent *Lentinus edodes* substrate. *Int. J. Biol. Macromol.*, 112: 976–984.

Zi, Y., Zhu, M., Li, X., Xu, Y., Wei, H., Li, D. and Mu, C. (2018). Effects of carboxyl and aldehyde groups on the antibacterial activity of oxidised amylose. *Carbohydr. Polym.*, 192: 118–125.

CHAPTER 15

Tiger's Milk Mushroom (*Lignosus rhinocerus*)

Phaniendra Alugoju[1,2] and *Tewin Tencomnao*[1,2,*]

1. Introduction

Historically mushrooms have been widely used in culinary as an appetiser and nutritious food, as well as for medicinal purposes to treat various human ailments in different parts of the world. A mushroom is a macrofungus with distinct fruiting bodies that can be hypogeous or epigeous, large enough to be seen with the naked eye and harvested with hands (Chang and Miles, 1992). *Lignosus rhinocerus* (Cooke) Ryvarden (Tiger's Milk mushroom) (synonym: *Polyporus rhinocerus, Lignosus rhinocerotis*) belongs to order Polyporales, family Polyporaceae in the phylum Basidiomycota and has been considered as one of the most potent mushrooms. *L. rhinocerus*, hereafter referred as TMM (Tiger's Milk Mushroom) is a white-rot fungus and it bears a central stipitate pilei originating from the underground tuber-like sclerotium (Fig. 1). TMM is used as a folk-medicine by indigenous people in Southeast Asian countries, such as Thailand, Philippines, Malaysia, Indonesia and also found in South China, Papua New Guinea, New Zealand and Australia (Ryvarden, 2001). In recent years, TMM has become increasingly popular due to its wide variety of ethno-botanical applications. The name Tiger's Milk Mushroom comes from the fact that this mushroom emerged at the spot where tigress' milk accidentally dribbled during lactation (Nallathamby et al., 2018). TMM is also known as *Cendawan susu rimau* in Malay language and is also refereed as Malaysia's national treasure mushroom. In Malaysia, people use this mushroom as a health tonic, antipruritic, antipyretic and additionally to treat breast cancer, food poisoning, cough, asthma and wounds (Tan et al., 2010; Lai et al., 2011). In general, sporophore (fruiting body), the above-ground part of the mushrooms is the part with

[1] Natural Products for Neuroprotection and Anti-Ageing Research Unit, Chulalongkorn University, Bangkok 10330, Thailand.
[2] Department of Clinical Chemistry, Faculty of Allied Health Sciences, Chulalongkorn University, Bangkok 10330, Thailand.
* Corresponding author: tewin.t@chula.ac.th

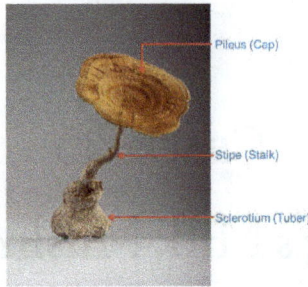

Fig. 1. The morphology of Tiger's Milk Mushroom (*Lignosus rhinocerus*).

medicinal value. In contrast, sclerotium (a compact mass of hardened mycelia) of TMM is the underground part which is the edible part with medicinal importance, and it is 1–5 cm in diameter, having a thickness of 1–3 cm. The outer greyish brown layer of the sclerotium is rough, whereas the inner part has ivory granular texture with a slight milky odour (Yang and Fang, 2008).

2. Nutritional Composition of TMM

Nutrient composition analysis revealed that sclerotium of wild TMM is abundant in carbohydrate and dietary fibre content but low-fat content, moderate protein content and all essential amino acids, except tryptophan (Lau et al., 2013), whereas higher amounts of protein and water-soluble substances were reported in the TMM cultivar TM02 (Yap et al., 2013). It was also reported that, not only the sclerotium, but other parts, including fruiting body and mycelium, are also composed of carbohydrates, fibres and essential fatty acids (Lau et al., 2013). Both wild type and cultivated strains exhibited significant superoxide anion radical scavenging activity, indicating the strong antioxidant activity of TMM; however, cultivated strain (TMM cultivar TM02) showed stronger antioxidant capacity and antiproliferative activity than the wild type (Yap et al., 2013). Aqueous methanol extract of the mycelium TMM was also reported to exhibit higher antioxidant activities compared to that of sclerotium (Lau et al., 2014).

Yap et al. (2014) sequenced the genome of TMM to unravel the molecular and genetic basis of its therapeutic properties and found that its genome encodes several enzymes involved in primary metabolism (e.g., carbohydrate and glycoconjugate metabolism) and secondary metabolism (triterpenoid and sesquiterpenoid metabolic pathways), along with cytochrome P450s, putative bioactive proteins (lectins and fungal immunomodulatory proteins), polyketide, nonribosomal peptide and laccases (Yap et al., 2014). Mushrooms are rich sources of an array of pharmacologically active proteins and peptides, such as antimicrobial proteins, ribosome inactivating proteins (RIP), fungal immunomodulatory proteins (FIP), lectins, ribonucleases, and laccases (Xu et al., 2011). Yap et al., have performed proteomic analysis of TMM sclerotium using two-dimensional electrophoresis coupled with mass spectrometry and revealed that proteome of sclerotium is rich in some of the differentially expressed proteins (e.g., lectins, immunomodulatory proteins and aegerolysin) with pharmaceutical value and some with antioxidant potential (e.g., superoxide dismutase, catalase and glutathione transferase) (Yap et al., 2015). A hyperbranched β-d-glucan was also isolated from the TMM sclerotia (Hu et al., 2017b).

Wild sclerotium of TMM is very difficult obtain to obtain as it grows underground. In a study, researchers have grown sclerotium of TMM through a cultivation technique and analysed its nutritional contents as well as antioxidant and cytotoxic capacity. Research results revealed that cultivated sclerotium contains higher contents of carbohydrate and protein and low-fat content when compared to wild type sclerotium. Total flavonoids and total β-D-glucan levels were also found to be higher in cultivated sclerotium as compared to wild type. This points to the fact that cultivated tuber exhibits higher antioxidant activity than the wild type (Jamil et al., 2017).

3. Pharmacological Activities

Owing to its wide and traditional use in folklore medicine for the treatment of various ailments, the pharmacological activities of TMM including anti-inflammatory, neuroprotective, anticancer, immunomodulatory, anti-asthmatic, anti-ulcer, antidiabetic, antiviral and anti-aging activities are discussed below (Fig. 2). Lists of studies on the protective effects of TMM in different *in vitro* and *in vivo* are summarised in Table 1.

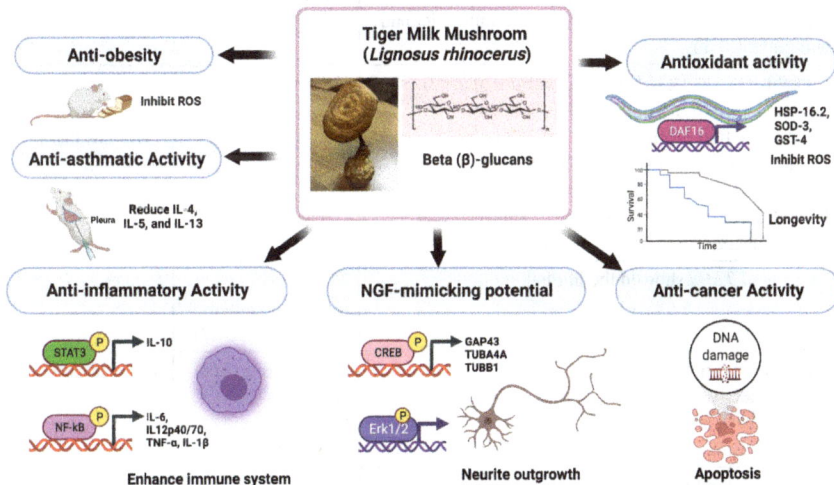

Fig. 2. Schematic representation of the pharmacological activities of TMM.

4. Anti-inflammatory Activity

Inflammation is the protective response of the body against microbiologic agents, physical trauma, harmful chemicals and is characterised by redness, swelling, pain and heat. Lee et al., have conducted experiments to demonstrate the anti-inflammatory effects of TMM cultivar TM02 sclerotium against carrageenan-induced inflammation in rat models. Authors found that three different extracts including cold water, hot water and methanol extracts possess anti-inflammatory activity, particularly protein component of the heavy molecular weight fraction (HMWF) of cold-water extract showed potent anti-inflammatory action by inhibiting TNF-alpha production. This indicates the possible anti-inflammatory activity of TMM sclerotial powder (Lee et al., 2014). *In vitro* studies also reported that the ethyl acetate fraction and hydroethanolic extract of selerotia powder of TMM can diminish inflammatory burden via down

Table 1. List of studies on the protective effects of TMM in different *in vitro* and *in vivo* models.

Protective Effects of TMM	*In Vitro* or *In Vivo* Model Used	References
Cold water extract and its fraction could prevent cancer via TNF signalling pathway leading to apoptosis and cell cycle arrest.	ORL-204 oral cancer cell lines	Yap et al., 2022
Cold water extract of TMM inhibited dengue virus replication and infection.	A monkey kidney epithelial cell line (Vero cells)	Khazali et al., 2021
Extracts of *TMM* showed functional antioxidant and anti-aging properties via the DAF-16/FOXO signalling pathway.	*Caenorhabditis elegans*	Kittimongkolsuk et al., 2021
Ethanol extract showed neuroprotection through the alleviation of glutamate-induced ROS in HT22 cells and the prevention of neurotoxicity in *C. elegans* in an Alzheimer and a Huntington model.	HT-22 mouse hippocampal cells and *C. elegans*	Kittimongkolsuk et al., 2021
Oral supplementation of *L. rhinocerus* powder effectively improved respiratory health, immunity and antioxidant status.	Healthy volunteers	Tan et al., 2021
Ethanol and water extracts of sclerotium of *TMM* inhibited the activity of both HIV-1 protease (PR) and reverse transcriptase (RT) enzymes, while hexane extract suppressed only the HIV-1 PR activity. *In silico* study revealed that phytochemicals, heliantriol F and 6 alpha-fluoroprogesterone, displayed great binding energies with HIV-1 PR and HIV-1 RT.	Human T lymphoblast cell line (MOLT-4 cells) *In silico* study	Sillapachaiyaporn and Chuchawankul, 2020
Extracts of *TMM* showed the immunomodulating properties.	Monocyte/ macrophage-like cells (RAW 264.7)	Sum et al., 2020
Aqueous and methanol extracts exhibited the neuroprotective property by promoting neurite outgrowth. In addition, methanol extract attenuated dexamethasone-induced neurotoxicity and apoptosis.	Human embryonic stem cells (hESCs) SH-SY5Y human neuroblastoma cells	Yeo et al., 2019
Fractionated extract of sclerotial powder showed apoptosis-inducing activities against breast cancer cells.	MCF-7 human breast adenocarcinoma cells	Yap et al., 2018b
Cold water extract exerted the prevention of AGE-mediated diabetic complications through glycation inhibition.	Cell free *in vitro* system	Yap et al., 2018a
Hot-water extract exerted the anti-asthmatic activity on ovalbumin-induced airway inflammation.	Sprague Dawley rats	Johnathan et al., 2016
Polysaccharide from hot-water extract exhibited anticancer activity by inhibition of growth of human lung carcinoma.	A549 human lung cancer cell line	Lai et al., 2014
Extracts of sclerotial powder exhibited antioxidant and anti-proliferative activity against human breast carcinoma cells	Human breast adenocarcinoma cells (MCF-7)	Yap et al., 2013

Table 1 contd. ...

...Table 1 contd.

Protective Effects of TMM	*In Vitro* or *In Vivo* Model Used	References
Oral feeding of sclerotial powder (at 100 mg/kg) could not induce chronic toxicity, anti-fertility and teratogenic effects as well as genotoxicity.	Sprague Dawley rats	Lee et al., 2013
Cold-water extract of sclerotia exhibited anti-proliferative activity against human breast and lung carcinoma cells.	Human breast adenocarcinoma cells (MCF-7), human lung carcinoma (A549), human breast cell (184B5) and human lung cell (NL 20).	Lee et al., 2012
Ethanol and water extracts possessed significant antimicrobial activity.	Cell free *in vitro* system	Mohanarji et al., 2012
Aqueous extract exerted the neuroprotection by stimulating neurite outgrowth.	PC-12 cells (derived from a pheochromocytoma of rat's adrenal medulla)	Eik et al., 2012

regulation of neuroinflammatory iNOS and COX2 genes and subsequent nitric oxide (NO) production in brain microglial (BV2) cells (Nallathamby et al., 2016). In addition, hot aqueous extract and its n-butanol fraction as well as ethyl acetate fraction were shown to significantly inhibit NO production in lipopolysaccharide-stimulated BV2 microglia cells (Seow et al., 2017). Taken together, all these studies indicate the potent anti-inflammatory potential of TMM.

5. Neuroprotective Activity

A neurite (or neuronal process) represents the external projections, such as axons and dendrites of the neuronal cell body. Neurite outgrowth is a fundamental process in the differentiation of the neurons. Moreover, neurite outgrowth is a critical step in the formation of synaptic connections, thus essential for the development of central nervous system network as well as for the regeneration following any disease or trauma (Miller and Suter, 2018; Meldolesi, 2011). Therefore, therapeutic strategies are essential for inducing neurite development and conserving the neurite architecture and synaptic connections. Eik et al., have shown neurite-stimulating activities of TMM. The aqueous sclerotium TMM extract was demonstrated to induce neurite outgrowths. Interestingly, a combination of aqueous extract and NGF has been reported to have additive effects, thereby significantly enhancing neurite outgrowth compared to aqueous extract alone (Eik et al., 2012). Phan et al., have also reported that aqueous extracts prepared from both sclerotium and mycelium can promote neurite outgrowth in mouse neuroblastoma (Neuro-2a) cells, indicating that TMM could be developed as a potent dietary food additive for improving brain health (Phan et al., 2013). Neuritin is also a neurotrophin involved in dendritic outgrowth, maturation and axonal regeneration and it can also inhibit neuronal and behavioural defects (Son et al., 2012; Yao et al., 2018). Tan et al.'s experimental results revealed that co-treatment of aqueous extract of TMM with NGF significantly enhanced neuritin levels in PC12 cells and neuritin abundance is positively proportional to the average neurite length (Tan et al., 2021).

Indigenous people prepare a decoction of sclerotium of TMM and consume it to increase their alertness during hunting. It was also documented that consumption of TMM can improve stamina and alertness in healthy people (Tan et al., 2012). Both neuritogenic activity and neuronal network are essential for improving mental alertness. Nerve growth factor (NGF), a neurotrophin, is essential for maintenance of neuron functions by promoting neuronal survival, proliferation, development and neuritogenesis. NGF was believed to stimulate neuritogenesis in PC12 cell through the activation of mitogen-activated protein kinase kinase/extracellular signal-regulated kinase (MEK/ERK) pathway (Kao et al., 2001; Vaudry et al., 2002). In view of this, Seow et al. (2015) have investigated neuritogenic activity of TMM cultivar TM02 sclerotium in rat pheochromocytoma (PC-12) cells and they found that hot aqueous and ethanol extracts as well as crude polysaccharides exhibited significant neuritogenic activity. However, except hot aqueous extract, ethanol extract and crude polysaccharides did not induce NGF production in PC-12 cells. This study also reported that similar to NGF, only the hot aqueous extract of TMM sclerotium induced neuritogenesis in PC-12 cells via activation of TrkA-MEK1/2-ERK1/2 signaling pathway (Seow et al., 2015). This study indicates that hot aqueous extract might mimic the NGF in stimulating the neuritogenesis. Samberkar et al., conducted *in vitro* experiments to investigate neuritogenic potential of TMM sclerotium in dissociated cells of brain, spinal cord and retina from chick embryo. Treatment with TMM sclerotium extract (50 µg/mL) triggered maximum neurite outgrowth of 20.77% and 24.73% in brain and spinal cord, respectively. On the other hand, 20.77% of neurite outgrowth was observed in retinal cells at 25 µg/mL. These results point to the fact that TMM possess anti-neurogenerative properties against different neurogenerative diseases, including Alzheimer's disease, Parkinson's disease, etc.; however future clinical studies are still needed to further warrant the neuroprotective effects of TMM (Samberkar et al., 2015). It was also shown that methanolic extract of TMM attenuated the dexamethasone-induced apoptosis and reduction in phospho-Akt level in hESC-derived neural stem cells. Thus, it can be suggested that the neuroprotective effects of TMM are partly linked through the activation of Akt signalling (Yeo et al., 2019). *In vitro* experimental findings from our research lab reported that only ethanolic extract of TMM significantly reduced glutamate-induced burden of ROS and apoptosis in HT22 cells. Furthermore, treatment with ethanolic extract improved Chemotaxis Index (CI) and decreased PolyQ40 aggregation in *C. elegans*. From these *in vitro* and *in vivo* studies, it can be suggested that ethanolic extract of TMM sclerotium might also offer neuroprotection (Kittimongkolsuk et al., 2021a).

6. Anticancer Activities

A hot water-soluble polysaccharide, isolated from the sclerotium of TMM, exhibited antiproliferative effects by significantly inhibiting the growth of different leukemic cell lines, such as HL-60 human acute promyelocytic leukemia cell line, K562 chronic myelogenous leukemia cell line and THP-1 human acute monocytic leukemia cell line (Lai et al., 2008). The cytotoxic activity of this polysaccharide extract was attributed to its ability to induce apoptosis in cancer cell lines. Also, the cold-water extract of sclerotia showed antiproliferative activity against MCF-7 human breast carcinoma and A549 human lung carcinoma but did not show cytotoxicity against 184B5 human breast normal cell line as well as NL 20 human lung normal cell line (Lee et al., 2012; Lau et al., 2013; Yap et al., 2013). Yap et al., isolated a partially purified carcinogenic protein portion (termed F5) from the cold-water extract of TMM TM02 sclerotia and evaluated its cytotoxic activity in breast cancer MCF7 cell line. In this *in vitro* study, it was shown that cytotoxic protein fraction (termed F5) can induce apoptosis in MCF7 cells via upregulation of Bax, BID, cleaved BID and caspase-8 and -9 activities and downregulation of Bcl-2. Through LC-MS/MS analysis, the cytotoxic substance in the fraction F5 was identified to be a serine protease (Yap et al., 2018b). The high molecular weight fraction (HMW) of cold-water extract of sclerotial powder of TMM cultivar TM02 significantly induced apoptosis in oral cancer ORL-204 cells by arresting cells in G_0/G_1 phase and increasing caspase-3/7 cleavage. Also, HMW fraction inhibited the activities of MIP2 and COX-2 in ORL-204 cells. Altogether, these studies indicate that sclerotial powder can be developed as a natural anticancer dietary therapy (Yap et al., 2022).

6.1 Immunomodulatory Properties

Several studies have investigated the immunomodulatory activities of TMM (Sum et al., 2020; Wong et al., 2009; Hu et al., 2017a). *In vitro* studies demonstrated that polysaccharides of sclerotium can stimulate human immune cells, such as CD56+ natural killer (NK) cells and human normal spleen monocytes/macrophages (Wong et al., 2009). Spleen and thymus are the chief immune organs involved in humoral and cellular immune response. Hu et al., isolated four polysaccharide fractions from aqueous extract of TMM sclerotia and evaluated their immunomodulatory effects against cyclophosphamide (Cy)-induced immunosuppression in mice models. Treatment with cyclophosphamide leads to a decline in spleen and thymus indices (which are calculated, based on the weight of each organ compared to the total body weight). However, intraperitoneal administration of four polysaccharide fractions for eight days significantly increased the Cy-induced decline in spleen and thymus indices. Polysaccharide fractions also increased the number of splenocytes, lymphocytes and megakaryocytes as well as enhanced the serum levels of cytokines TNF-α and INF-γ in Cy-treated mice group. Researchers have found that two of the polysaccharide fractions (LRP-1 and LRP-2) were polysaccharide-protein complexes (46–68% β-d-glucan and 27–48% protein), while the other two fractions (LRP-3 and LRP-4) were composed of only β-d-glucose. Despite their composition, all fractions significantly reversed the Cy-induced immunosuppression (Hu et al., 2017a). As immunomodulation is a key factor in the enhancement of host's own defence mechanism and to recognide and selectively eliminate cancerous cells, it can be suggested that both the polysaccharides and polysaccharide-protein complexes of TMM sclerotia can be developed as a potential food- or drug-based anticancer therapies.

7. Anti-asthmatic Activity

TMM has been traditionally used to treat asthma, which is a lung disorder characterised by chronic inflammation of lower inflammatory tract. Chronic air way inflammation in asthma leads to elevated T helper cell type-2 (Th2) immune response (i.e., elevation of Th2 cells) and subsequent release of specific cytokines, such as interleukin (IL-)-4, IL-5, IL-9 and IL-13 which in turn stimulate eosinophil infiltration into bronchoalveolar lavage fluid and immunoglobulin E (IgE) production (Song et al., 2019; Quirt et al., 2018). IgE causes the production of other pro-inflammatory cytokines, such as histamine and cysteinyl leukotrienes, which cause bronchospasm (smooth muscle contraction in the bronchioles), edema and increased mucus secretion, leading to asthmatic symptoms (Quirt et al., 2018). A study was conducted to investigate anti-asthmatic effects of sclerotia of TMM cultivar TM02 using ovalbumin-induced allergic asthmatic rat model and the study reported that hot water extract of sclerotia significantly inhibited elevation in serum total IgE levels as well as IL-4, IL-5 and IL-13 levels in BALF. Additionally, the hot-water extract of sclerotia markedly suppressed eosinophil count in BALF as well as attenuated eosinophil infiltration into the lungs. Thus, these results provide scientific evidence for the traditional asthma treating potential of TMM (Johnathan et al., 2016). Similarly, yet interestingly, Muhamad et al.'s research demonstrated that intranasal administration of hot-water extract of sclerotia of TMM cultivar TM02 could significantly attenuate airway inflammation in ovalbumin-induced allergic asthmatic rat model (Muhamad et al., 2019).

Airways are adversely affected and bronchial constriction is usually seen in many lung diseases, including asthma, COPD and cystic fibrosis. Research evidence indicates that sclerotial of TMM might exert antagonistic effects on the bronchial constriction. It was demonstrated that the high-molecular-weight fraction (HMW; composed of 71% carbohydrates and 4% proteins) and medium-molecular-weight fraction (MMW; composed of 35% carbohydrates and 1% proteins) of sclerotial of TMM have substantially relaxed pre-contracted airways compared to the cold-water sclerotial extract and low-molecular-weight fraction (LMW). Moreover, both the HMW fraction and cold-water extract significantly attenuated carbachol- and calcium-induced airway contractions and the bronchorelaxation efficacy of reported fractions and cold-water extract was attributed to their ability to inhibit extracellular Ca^{2+} influx. These results suggest that the polysaccharide-protein complex or proteins present in the HMW and MMW fractions are expected to cause bronchorelaxation of cold-water extract of TMM sclertoia. Moreover, the bronchorelaxation effect was reported to be both concentration- and incubation time-dependent (Lee et al., 2018a, b, c). Asthma is characterised by airway hyperresponsiveness (AHR), which is a common symptom. AHR is described as an increase in the airways' sensitivity and response to a variety of stimuli (Ritz, 2012). Treatment with different doses of hot-water extract of TMM (125, 250 and 500 mg/kg) significantly repressed airway hyperresponsiveness (AHR) in house dust mite (HDM)-induced asthma in mice. These studies conclude that TMM not only attenuates airway inflammation, but also alleviates AHR in asthmatics (Johnathan et al., 2021).

Recently, an open-labelled prospective study was conducted to investigate the effects of TMM on respiratory health in humans. In this study, 50 healthy participants, aged 30–50, were supplemented with TMM at a dose of 300 mg two time daily for three months. Results revealed that supplementation of TMM significantly attenuated the level of IL-1β, IL-8, malondialdehyde (MDA), as well as respiratory symptoms. Additionally, a significant increase in the level of IgA, total antioxidant capacity and pulmonary function were observed. Taken together, it can be suggested that TMM supplementation efficiently enhances respiratory health via ameliorating the immunity and antioxidant status (Tan et al., 2021).

8. Anti-ulcer Activities

The anti-ulcer activities of TMM were also evaluated in different models of experimentally-induced gastric ulcer models. All ulcer groups showed significant damage and haemorrhage of gastric mucous membrane. In contrast, oral administration of TMM powder (250 and 500 mg/kg) offered significant protection against gastric ulceration in all ulcer-induced models and the protection was dose dependent (Nyam et al., 2016).

9. Antidiabetic Activity

Diabetes mellitus, or diabetes, is a metabolic disease that is the leading risk factor for mortality in the world. Nyam et al., reported that TMM also possess antidiabetic effects. They reported that treatment with freeze dried powder of TMM for 60 days significantly reduced streptozotocin (STZ)-induced increase in blood glucose levels to a normal range (3.0–7.0 mmol/L) in rats and also reduced the body weight of rats. In addition, treatment with freeze-dried powder enhanced the endogenous antioxidant enzyme activities (superoxide dismutase and catalase) and reduced glutathione levels, whereas decreased lipid peroxidation in the liver of STZ induced-diabetic rat model. This suggests the possible use of TMM as a food supplement to reduce the burden of diabetes-associated hyperglycemia and oxidative stress (Nyam et al., 2017). Advanced glycation end products (AGEs), such as carboxymethyl-hydroxylysine, Nε-(carboxymethyl)lysine, pentosidine, and pyrraline have been considered as the pathological hallmarks of diabetic-associated complications. These AGEs can increase inflammation and oxidative stress in aging as well as in some age-associated metabolic diseases, including diabetes (Chaudhuri et al., 2018). Yap et al.'s studies indicate that TMM sclerotia can exhibit anti-glycation activity. A medium-molecular-weight fraction isolated from the cold-water extract of sclerotial powder of TMM TM02 cultivar inhibited the production of AGEs, including Nε-(carboxymethyl)lysine and pentosidine in a human serum albumin-glucose system (Yap et al., 2018a). Therefore, TMM sclerotia powder might be also used for ameliorating diabetes complications as well as for reducing the burden of AGEs-mediated oxidative stress in aging.

10. Antiviral Effects

Antiviral activities of TMM have also been reported against human immunodeficiency virus type-1 and dengue virus serotype 2 (DENV2) (Ellan et al., 2019; Sillapachaiyaporn and Chuchawankul, 2020). Ethanolic and water extracts of TMM significantly inhibited the activity of two important enzymes, HIV-1 protease and reverse transcriptase that are essential for the replication of HIV (Sillapachaiyaporn and Chuchawankul, 2020). In another study, hot aqueous extract and aqueous soluble extracts prepared from TMM showed anti-DENV2 activity by significantly decreasing the expression of dengue virus envelope (ENV) and non-structural protein 5 (NS5) genes *in vitro* (Ellan et al., 2019).

11. Anti-aging Activity

Laboratory experiments conducted by our research group suggest the anti-aging potential of TMM in *Caenorhabditis elegans*. Different solvent extracts (ethanol, cold and hot water) of powder of sclerotium of TMM TM02 (a cultivated strain) decreased the ROS levels, lipofuscin, and subsequent oxidative stress in worms. All the extracts extended the lifespan of worms. However, further studies are needed to establish the potential anti-aging role of TMM (Kittimongkolsuk et al., 2021b).

12. TMM as Probable Antihypertensive Drug Source

Hypertension is a major risk factor in a variety of cardiovascular disorders, including coronary disease, heart failure and stroke. Angiotensin-converting enzyme (ACE) inhibitors (ACEIs) are the most widely used for treatment of hypertension and associated cardiovascular and renal diseases (Herman et al., 2022). Multiple clinical trials have suggested that ACE inhibitors can be used as antihypertensive drugs (Messerli et al., 2018). Synthetic ACE inhibitors have been effectively employed clinically to decrease the mortality rate in patients with high blood pressure and cardiac failure. As these chemical ACE inhibitors are associated with side effects, there has been increasing attention on the discovery of natural ACE inhibitors. Genomic data mining studies revealed that cystathionine beta-synthase (CBS) protein of *L. rhinocerus* (LrCBS) shares a significant structural similarity (> 80%) with the *Ganodema lucidum* (GlCBS), which can function as an inhibitor of Angiotensin-Converting Enzyme (ACE). *In silico* protein-docking study discovered that both GlCBS and LrCBS might control the C-terminal domain of ACE (C-ACE) activity and both GlCBS and LrCBS interact with ACE at the same region that leads to the inhibition of ACE function (Goh et al., 2022). However, more research is needed to prove TMM's antihypertensive effects.

13. Sub-acute and Acute Toxicity of Sclerotium of TMM

In vivo sub-acute toxicity studies revealed that daily oral administration of sclerotial powder of TMM at a dose of up to 1,000 mg/kg b.w (for 28 days) had no detrimental impact on growth rate, hematological and biochemical parameters in rats. Moreover, treatment with sclerotial powder did not cause any histopathological changes in many organs, including the kidneys, liver, lungs, heart and spleen. The NOAEL dosage is set at 1,000 mg/kg b.w. since the maximal test dose (1,000 mg/kg) was not linked with any toxicity (Lee et al., 2011). Acute toxicity studies on rats showed that daily oral treatment with sclerotial powder at levels up to 1000 mg/kg for 180 days had no negative effects. Furthermore, sclerotial powder had no influence on the fertility of rats or caused teratogenic effects in their progeny (Lee et al., 2013). From the Ames test and *in vitro* chromosome aberration test, it was shown that TMM mycelium (up to a dose of 100 mg/mL) does not induce mutagenicity and genotoxicity (Lee et al., 2013; Chen et al., 2013). Studies also demonstrate that TMM mycelium had no effect on erythropoiesis in mice (Chen et al., 2013). Furthermore, it was shown that oral treatment of TMM mycelium (at dosages of 850, 1700 and 3400 mg/kg/day) had no deleterious effects on pregnant rats, showing that the NOAEL dose (3400 mg/kg/day) had no developmental toxicity in rats (Jhou et al., 2017).

Li et al. (2021) also conducted experiments to assess the safety of oral toxicity of TMM mycelium powder in male and female Sprague-Dawley rats. Rats were orally given different doses of TMM mycelium powder (850, 1700 and 3400 mg/kg b.w) for 13 weeks. This study reported that there were no significant changes in clinical symptoms, body weight, ophthalmological tests, hematological, or serum biochemistry parameters, and no animals died following 13 weeks of treatment. Furthermore, necropsy and histopathological examination showed that treatment with TMM mycelium powder did not cause adverse effects. Based on these results, the NOAEL dose of TMM mycelium was identified to be greater than 3400 mg/kg in rats (Li et al., 2021).

14. Conclusion

Medicinal mushrooms have gained significant interest in recent years as a source of therapeutically active substances. Nowadays there have been an increasing number of both *in vitro* and *in vivo* experimental studies to unravel the composition and therapeutic potential of sclerotium as well as mycelium of Tiger's Milk Mushroom (TMM). The polysaccharide β-glucans or polysaccharide–protein complexes content of Tiger's Milk Mushroom exerted a plethora of pharmacological properties, such as anticancer, antidiabetic, neuroprotective, anti-asthmatic, antiviral, anti-ulcer and anti-aging activities. Acute and sub-acute toxicity studies also indicated that sclerotium and mycelium of TMM do not exhibit any toxic effects up to a dose of 3400 mg/kg body weight in rats. Future clinical trials are required to further evaluate the therapeutic potential of TMM as a dietary supplement and/or its chemical constituents as drugs for the treatment of various diseases.

Abbreviations

L. rhinocerus-Lignosus rhinocerus, TMM-Tiger's Milk Mushroom, FIP-fungal immunomodulatory protein, RIP-ribosome inactivating proteins, TNF-alpha-Tumour necrosis factor alpha, iNOS-inducible nitric oxide synthase, COX2-cycloxygenase-2, PC12-phaeocytochroma cells, NGF-Nerve growth factor, p-Akt-phospho-Akt, CI-index-Chemotaxis Index (CI), HMW-High molecular weight fraction, NK cells- natural killer cells, Cy-cyclophosphamide, IFN- γ-interferon gamma, LRP-1/2/3/4-polysaccharide fractions of *L. rhinocerus*, Th2-T helper cell type-2, IL-4/5/9/136-Interleukin-4/5/9/13, Ig E/A- Immunoglobulin E/A, BALF-bronchoalveolar lavage fluid, COPD-chronic obstructive pulmonary disease, Ca^{+2}-calcium ion, MMW fraction-medium-molecular-weight fraction, AHR-Airway hyperresponsiveness, MDA-malondialdehyde, STZ-streptozotocin, SOD-superoxide dismutase, CAT-catalase, AGEs-advanced glycation end products, HIV-1-Human immunodeficiency virus type-1, DENV2-dengue virus serotype 2 (DENV2), ENV gene-envelope gene, NS-5 gene-non-structural protein 5 gene, ROS-reactive oxygen species, ACE-Angiotensin-converting enzyme, CBS-cystathionine beta-synthase (CBS), NOAEL-no-observed-adverse-effect level.

References

Chang, S.T. and Miles, P.G. (1992). Mushroom biology—A new discipline. *Mycologist*, 6(2): 64–65.

Chaudhuri, Jyotiska, Yasmin Bains, Sanjib Guha, Arnold Kahn, David Hall, Neelanjan Bose, Alejandro Gugliucci and Pankaj Kapahi (2018). The role of advanced glycation end products in aging and metabolic diseases: bridging association and causality. *Cell Metabolism*, 28(3): 337–352.

Chen, T.I., Zhuang, H.W., Chiao, Y.C. and Chen, C.C. (2013). Mutagenicity and genotoxicity effects of *Lignosus rhinocerotis* mushroom mycelium. *J. Ethnopharmacol.*, 149(1): 70–74.

Eik, L.F., Naidu, M., David, P., Wong, K.H., Tan, Y.S. and Sabaratnam, V. (2012). *Lignosus rhinocerus* (Cooke) Ryvarden: A medicinal Mushroom that stimulates neurite outgrowth in PC-12 cells. *Evid-based Complement. Alternat. Med.*, 2012: 320308.

Ellan, K., Thayan, R., Raman, J., Kipj Hidari, Ismail, N. and Sabaratnam, V. (2019). Antiviral activity of culinary and medicinal mushroom extracts against dengue virus serotype 2: An *in-vitro* study. *BMC Complement, Altern. Med.*, 19(1): 260.

Goh, N.Y., Mohamad Razif, M.F., Yap, Y.H., Ng, C.L. and Fung, S.Y. (2022). *In silico* analysis and characterisation of medicinal mushroom cystathionine beta-synthase as an angiotensin converting enzyme (ACE) inhibitory protein. *Comput. Biol. Chem.*, 96: 107620.

Herman, L.L., Padala, S.A., Ahmed, I. and Bashir, K. (2022). Angiotensin Converting Enzyme Inhibitors (ACEI). *In: StatPearls*, Treasure Island (FL): StatPearls Publishing, StatPearls Publishing LLC.

Hu, T., Huang, Q., Wong, K. and Yang, H. (2017a). Structure, molecular conformation, and immunomodulatory activity of four polysaccharide fractions from *Lignosus rhinocerotis sclerotia*. *Int. J. Biol. Macromol.*, 94(Pt A): 423–430.

Hu, T., Huang, Q., Wong, K., Yang, H., Gan, J. and Li, Y. (2017b). A hyperbranched β-d-glucan with compact coil conformation from *Lignosus rhinocerotis sclerotia*. *Food Chem.*, 225: 267–275.

Jamil, N.A.M., Rashid, N.M.N., Hamid, M.H.A., Rahmad, N. and Al-Obaidi, J.R. (2017). Comparative nutritional and mycochemical contents, biological activities and LC/MS screening of tuber from new recipe cultivation technique with wild type tuber of Tiger's Milk mushroom of species *Lignosus rhinocerus*. *World J. Microbiol. Biotechnol.*, 34(1): 1.

Jhou, B.Y., Liu, H.H., Yeh, S.H. and Chen, C.C. (2017). Oral reproductive and developmental toxicity of *Lignosus rhinocerotis* mycelium in rat. *J. Ethnopharmacol.*, 208: 66–71.

Johnathan, M., Gan, S.H., Wan Ezumi, M.F., Faezahtul, A.H. and Nurul, A.A. (2016). Phytochemical profiles and inhibitory effects of Tiger Milk mushroom (*Lignosus rhinocerus*) extract on ovalbumin-induced airway inflammation in a rodent model of asthma. *BMC Complementary and Alternative Medicine*, 16: 167–167.

Johnathan, M., Muhamad, S.A., Gan, S.H., Stanslas, J., Mohd Fuad, W.E., Hussain, F.A., Wan Ahmad, W.A.N. and Nurul, A.A. (2021). *Lignosus rhinocerotis* Cooke Ryvarden ameliorates airway inflammation, mucus hypersecretion and airway hyperresponsiveness in a murine model of asthma. *PLoS ONE*, 16(3): e0249091.

Kao, S., Jaiswal, R.K., Kolch, W. and Landreth, G.E. (2001). Identification of the mechanisms regulating the differential activation of the mapk cascade by epidermal growth factor and nerve growth factor in PC12 cells. *J. Biol. Chem.*, 276(21): 18169–18177.

Khazali, Ahmad Suhail, Nurshamimi Nor Rashid, Shin-Yee Fung and Rohana Yusof. (2021). *Lignosus rhinocerus* TM02® sclerotia extract inhibits dengue virus replication and Infection. *Journal of Herbal Medicine*, 30: 100505.

Kittimongkolsuk, P., Pattarachotanant, N., Chuchawankul, S., Wink, M. and Tencomnao, T. (2021a). Neuroprotective effects of extracts from tiger milk mushroom *Lignosus rhinocerus* against glutamate-induced toxicity in HT22 hippocampal neuronal cells and neurodegenerative diseases in Caenorhabditis elegans. *Biology (Basel)*, 10(1).

Kittimongkolsuk, P., Roxo, M., Li, H., Chuchawankul, S., Wink, M. and Tencomnao, T. (2021b). Extracts of the Tiger's Milk Mushroom (*Lignosus rhinocerus*) enhance stress resistance and extend lifespan in Caenorhabditis elegans via the DAF-16/FoxO signalling pathway. *Pharmaceuticals (Basel)*, 14(2).

Lai, Connie, Ka Hing Wong and Peter Cheung. (2008). Antiproliferative effects of sclerotial polysaccharides from polyporus rhinocerus Cooke (Aphyllophoromycetideae) on different kinds of leukemic cells. *International Journal of Medicinal Mushrooms – Int. Med. Mushrooms,* 10: 255–264.

Lai, W.H., Siti Murni, M.J., Fauzi, D., Abas Mazni, O. and Saleh, N.M. (2011). Optimal culture conditions for mycelial growth of *Lignosus rhinocerus. Mycobiology,* 39(2): 92–95.

Lai, Wei, Zamri Zainal and Fauzi Daud. (2014). *Preliminary Study on the Potential of Polysaccharide from Indigenous Tiger's Milk Mushroom (Lignosus rhinocerus) as Anti-lung Cancer Agent,* vol. 1614.

Lau, B.F., Abdullah, N. and Aminudin, N. (2013). Chemical composition of the Tiger's Milk Mushroom, *Lignosus rhinocerotis* (Cooke) Ryvarden, from different developmental stages. *J. Agric. Food Chem.,* 61(20): 4890–4897.

Lau, B.F., Abdullah, N., Aminudin, N. and Lee, H.B. (2013). Chemical composition and cellular toxicity of ethnobotanical-based hot and cold aqueous preparations of the Tiger's Milk Mushroom (*Lignosus rhinocerotis*). *J. Ethnopharmacol.,* 150(1): 252–262.

Lau, B.F., Abdullah, N., Aminudin, N., Lee, H.B., Yap, K.C. and Sabaratnam, V. (2014). The potential of mycelium and culture broth of *Lignosus rhinocerotis* as substitutes for the naturally occurring sclerotium with regard to antioxidant capacity, cytotoxic effect and low-molecular-weight chemical constituents. *PLoS ONE,* 9(7): e102509.

Lee, M.K., Li, X., Yap, A.C.S., Cheung, P.C.K., Tan, C.S., Ng, S.T., Roberts, R., Ting, K.N. and Fung, S.Y. (2018a). Airway relaxation effects of water-soluble sclerotial extract from *Lignosus rhinocerotis. Front. Pharmacol.,* 9: 461.

Lee, M.K., Lim, K.H., Millns, P., Mohankumar, S.K., Ng, S.T., Tan, C.S., Then, S.M., Mbaki, Y. and Ting, K.N. (2018b). Bronchodilator effects of *Lignosus rhinocerotis* extract on rat isolated airways is linked to the blockage of calcium entry. *Phytomedicine,* 42: 172–179.

Lee, M.K., Millns, P., Mbaki, Y., Ng, S.T., Tan, C.S., Lim, K.H., Then, S.M., Mohankumar, S.K. and Ting, K.N. (2018). Data on the *Lignosus rhinocerotis* water-soluble sclerotial extract affecting intracellular calcium level in rat dorsal root ganglion cells. *Data Brief,* 18: 1322–1326.

Lee, M.L., Tan, N.H., Fung, S.Y., Tan, C.S. and Ng, S.T. (2012). The antiproliferative activity of sclerotia of *Lignosus rhinocerus* (Tiger's Milk mushroom). *Evidence-based Complementary and Alternative Medicine,* 2012: 697603.

Lee, S.S., Enchang, F.K., Tan, N.H., Fung, S.Y. and Pailoor, J. (2013). Preclinical toxicological evaluations of the sclerotium of *Lignosus rhinocerus* (Cooke), the Tiger Milk mushroom. *J. Ethnopharmacol.,* 147(1): 157–163.

Lee, S.S., Tan, N.H., Fung, S.Y., Sim, S.M., Tan, C.S. and Ng, S.T. (2014). Anti-inflammatory effect of the sclerotium of *Lignosus rhinocerotis* (Cooke) Ryvarden, the Tiger's Milk mushroom. *BMC Complement. Altern. Med.,* 14: 359.

Lee, Sookshien, Nget Tan, Shin Fung, Jayalakshmi Pailoor and Si Mui Sim. (2011). Evaluation of the sub-acute toxicity of the sclerotium of *Lignosus rhinocerus* (Cooke), the Tiger's Milk mushroom. *Journal of Ethnopharmacology,* 138: 192–200.

Li, I.C., Yang, B.H., Lin, J.Y., Lin, S. and Chen, C.C. (2021). Nutritional and 13-week subchronic toxicological evaluation of *Lignosus rhinocerotis* mycelium in sprague-dawley rats. *Int. J. Environ. Res. Public Health,* 18(3).

Meldolesi, J. (2011). Neurite outgrowth: This process, first discovered by Santiago Ramon y Cajal, is sustained by the exocytosis of two distinct types of vesicles. *Brain Res. Rev.,* 66(1-2): 246–255.

Messerli, F.H., Bangalore, S., Bavishi, C. and Rimoldi, S.F. (2018). Angiotensin-converting enzyme inhibitors in hypertension: To use or not to use? *J. Am. Coll. Cardiol.,* 71(13): 1474–1482.

Miller, Kyle E. and Daniel M. Suter. (2018). An integrated cytoskeletal model of neurite outgrowth. *Frontiers in Cellular Neuroscience,* 12.

Mohanarji, S., Dharmalingam, S. and Kalusalingam, A. (2012). Screening of *Lignosus rhinocerus* extracts as antimicrobial agents against selected human pathogens. *Journal of Pharmaceutical and Biomedical Sciences,* 18(18): 1–4.

Muhamad, S.A., Muhammad, N.S., Ismail, N.D.A., Mohamud, R., Safuan, S. and Nurul, A.A. (2019). Intranasal administration of *Lignosus rhinocerotis* (Cooke) Ryvarden (Tiger's Milk mushroom) extract attenuates airway inflammation in murine model of allergic asthma. *Exp. Ther. Med.,* 17(5): 3867–3876.

Nallathamby, N., Serm, L.G., Raman, J., Malek, S.N.A., Vidyadaran, S., Naidu, M., Kuppusamy, U.R. and Sabaratnama, V. (2016). Identification and *in vitro* evaluation of lipids from sclerotia of *Lignosus rhinocerotis* for antioxidant and anti-neuroinflammatory activities. *Nat. Prod. Commun.*, 11(10): 1485–1490.

Nallathamby, Neeranjini, Chia-Wei Phan, Syntyche Ling-Sing Seow, Asweni Baskaran, Hariprasath Lakshmanan, Sri N. Abd Malek and Vikineswary Sabaratnam. (2018). A status review of the bioactive activities of Tiger Milk Mushroom *Lignosus rhinocerotis* (Cooke) Ryvarden. *Frontiers in Pharmacology*, 8: 998–998.

Nyam, K.L., Chang, C.Y., Tan, C.S. and Ng, S.T. (2016). Investigation of the Tiger's Milk medicinal mushroom, *Lignosus rhinocerotis* (Agaricomycetes), as an antiulcer agent. *Int. J. Med. Mushrooms*, 18(12): 1093–1104.

Nyam, K.L., Chow, C.F., Tan, C.S. and Ng, S.T. (2017). Antidiabetic properties of the Tiger's Milk medicinal mushroom, *Lignosus rhinocerotis* (Agaricomycetes), in streptozotocin-induced diabetic rats. *Int. J. Med. Mushrooms*, 19(7): 607–617.

Phan, C.W., David, P., Naidu, M., Wong, K.H. and Sabaratnam, V. (2013). Neurite outgrowth stimulatory effects of culinary-medicinal mushrooms and their toxicity assessment using differentiating Neuro-2a and embryonic fibroblast BALB/3T3. *BMC Complement. Altern. Med.*, 13: 261.

Quirt, Jaclyn, Kyla J. Hildebrand, Jorge Mazza, Francisco Noya and Harold Kim. (2018). Asthma. *Allergy, Asthma, and Clinical Immunology: Official Journal of the Canadian Society of Allergy and Clinical Immunology*, 14(Suppl 2): 50–50.

Ritz, Thomas. (2012). Airway responsiveness to psychological processes in asthma and health. *Frontiers in Physiology*, 3.

Ryvarden, Maria Núñez and Leif. (2001). *East Asian Polypores 2, Polyporaceae s. lato.*

Samberkar, S., Gandhi, S., Naidu, M., Wong, K.H., Raman, J. and Sabaratnam, V. (2015). Lion's mane, *Hericium erinaceus* and Tiger's Milk, *Lignosus rhinocerotis* (Higher Basidiomycetes) medicinal mushrooms stimulate neurite outgrowth in dissociated cells of brain, spinal cord and retina: An *in vitro* study. *Int. J. Med. Mushrooms*, 17(11): 1047–1054.

Seow, S.L., Eik, L.F., Naidu, M., David, P., Wong, K.H. and Sabaratnam, V. (2015). *Lignosus rhinocerotis* (Cooke) Ryvarden mimics the neuritogenic activity of nerve growth factor via MEK/ERK1/2 signaling pathway in PC-12 cells. *Sci. Rep.*, 5: 16349.

Seow, S.L., Naidu, M., Sabaratnam, V., Vidyadaran, S. and Wong, K.H. (2017). Tiger's Milk medicinal mushroom, *Lignosus rhinocerotis* (Agaricomycetes) sclerotium inhibits nitric oxide production in LPS-stimulated BV2 microglia. *Int. J. Med. Mushrooms*, 19(5): 405–418.

Sillapachaiyaporn, C. and Chuchawankul, S. (2020). HIV-1 protease and reverse transcriptase inhibition by tiger milk mushroom (*Lignosus rhinocerus*) sclerotium extracts: *In vitro* and *in silico* studies. *J. Tradit. Complement. Med.*, 10(4): 396–404.

Son, H., Banasr, M., Choi, M., Chae, S.Y., Licznerski, P., Lee, B., Voleti, B. et al. (2012). Neuritin produces antidepressant actions and blocks the neuronal and behavioural deficits caused by chronic stress. *Proc. Natl. Acad. Sci. USA*, 109(28): 11378–11383.

Song, Min, Shan Cai, Hong Luo, Yi Jiang, Min Yang, Yan Zhang, Hong Peng and Ping Chen. (2019). Short-term pulmonary infiltrate with eosinophilia caused by asthma: A phenotype of severe, eosinophilic asthma? Five cases and a review of the literature. *Allergy, Asthma and Clinical Immunology: Official Journal of the Canadian Society of Allergy and Clinical Immunology*, 15: 48–48.

Sum, A.Y.C., Li, X., Yeng, Y.Y.H., Razif, M.F.M., Jamil, A.H.A., Ting, N.S., Seng, T.C., Cheung, P.C.K. and Fung, S.Y. (2020). The immunomodulating properties of Tiger's Milk medicinal mushroom, *Lignosus rhinocerus* TM02® cultivar (Agaricomycetes) and its associated carbohydrate composition. *Int. J. Med. Mushrooms*, 22(8): 803–814.

Tan, C.S., Ng, S.T., Vikineswary, S., Lo, F.P. and Tee, C.S. (2010). Genetic markers for identification of a malaysian medicinal mushroom, *Lignosus rhinocerus* (cendawan susu rimau).

Tan, C.S., Ng, S.T., Yap, Y.H.Y., Lee, S.S. and Lee, M.L. (2012). Breathing new life to a Malaysia lost national treasure—The Tiger's Milk mushroom (*Lignosus rhinocerotis*) in Mushroom science. Paper presented at the XVIII: *Proceedings of the 18th Congress of the International Society for Mushroom Science.*

Tan, E.S.S., Leo, T.K. and Tan, C.K. (2021). Effect of tiger milk mushroom (*Lignosus rhinocerus*) supplementation on respiratory health, immunity and antioxidant status: an open-label prospective study. *Sci Rep.*, 11(1): 11781.

Tan, Y.H., Lim, C.S., Wong, K.H. and Sabaratnam, V. (2021). Neuritin protein expression is positively correlated with neurite outgrowth induced by the Tiger's Milk Mushroom, *Lignosus rhinocerotis* (Agaricomycetes), in PC12 cells., *Int. J. Med. Mushrooms*, 23(6): 1–11.

Vaudry, D., Stork, P.J., Lazarovici, P. and Eiden, L.E. (2002). Signalling pathways for PC12 cell differentiation: Making the right connections. *Science*, 296(5573): 1648–1649.

Wong, Ka Hing, Connie Lai and Peter Cheung. (2009). Stimulation of human innate immune cells by medicinal mushroom sclerotial polysaccharides. *International Journal of Medicinal Mushrooms*, 11: 215–223.

Xu, X., Yan, H., Chen, J. and Zhang, X. (2011). Bioactive proteins from mushrooms. *Biotechnol. Adv.*, 29(6): 667–674.

Yang, L. and Fang, Z.J. (2008). Identification of the sclerotium of *Pleurotus tuber-regium* (Fr.) Sing and *Lignosus rhinocerus* (Cooke) Ryv. *Lishizhen Med. Mater. Med. Res.*, 19: 178–179.

Yao, Jin-jing, Qian-ru Zhao, Jun-mei Lu and Yan-ai Mei. (2018). Functions and the related signaling pathways of the neurotrophic factor neuritin. *Acta Pharmacologica Sinica*, 39(9): 1414–1420.

Yap, H.Y., Chooi, Y.H., Firdaus-Raih, M., Fung, S.Y., Ng, S.T., Tan, C.S. and Tan, N.H. (2014). The genome of the Tiger's Milk mushroom, *Lignosus rhinocerotis*, provides insights into the genetic basis of its medicinal properties. *BMC Genomics*, 15(1): 635.

Yap, H.Y., Fung, S.Y., Ng, S.T., Tan, C.S. and Tan, N.H. (2015). Genome-based proteomic analysis of *Lignosus rhinocerotis* (Cooke) Ryvarden sclerotium. *Int. J. Med. Sci.*, 12(1): 23–31.

Yap, H.Y., Tan, N.H., Ng, S.T., Tan, C.S. and Fung, S.Y. (2018a). Inhibition of protein glycation by Tiger's Milk Mushroom [*Lignosus rhinocerus* (Cooke) Ryvarden] and search for potential antidiabetic activity-related metabolic pathways by genomic and transcriptomic data mining. *Front Pharmacol.*, 9: 103.

Yap, H.Y.Y., Kong, B.H., Yap, C.S.A., Ong, K.C., Zain, R.B., Tan, S.H., Zaini, Z.M., Ng, S.T., Tan, C.S. and Fung, S.Y. (2022). Immunomodulatory effect and an intervention of TNF signalling leading to apoptotic and cell cycle arrest on ORL-204 oral cancer cells by Tiger's Milk Mushroom, *Lignosus rhinocerus*. *Food Technol. Biotechnol.*, 60(1): 80–88.

Yap, H.Y.Y., Tan, N.H., Ng, S.T., Tan, C.S. and Fung, S.Y. (2018b). Molecular attributes and apoptosis-inducing activities of a putative serine protease isolated from Tiger's Milk mushroom (*Lignosus rhinocerus*) sclerotium against breast cancer cells *in vitro*. *Peer J.*, 6: e4940.

Yap, Y.H., Tan, N., Fung, S., Aziz, A.A., Tan, C. and Ng, S. (2013). Nutrient composition, antioxidant properties and anti-proliferative activity of *Lignosus rhinocerus* Cooke sclerotium. *J. Sci. Food Agric.*, 93(12): 2945–2952.

Yeo, Y., Tan, J.B.L., Lim, L.W., Tan, K.O., Heng, B.C. and Lim, W.L. (2019). Human embryonic stem cell-derived neural lineages as *in vitro* models for screening the neuroprotective properties of *Lignosus rhinocerus* (Cooke) Ryvarden. *Biomed. Res. Int.*, 2019: 3126376.

CHAPTER 16

Gucchi (*Morchella esculenta*)

Anu Pandita,[1] Deepu Pandita,[2,] Jyothi Chaitanya Pagadala,[3]*
Sugumari Vallinayagam,[4] Bhoomika Inamdar,[5] Manohar MV,[5]
Amogha G Paladhi,[6] Siji Jacob[6] and Devananda Devegowda[7]

1. Introduction

Mushrooms are spore-bearing fleshy fruiting bodies of fungus often present above the ground. Greeks and Romans included mushrooms in their diet. Romans considered mushrooms as the food of supernatural beings, despite the Chinese contemplating them as the elixir of the human being (Bashir et al., 2014). Mushrooms have attracted people due to their versatile activities, viz., functional food and resource for drug development and as nutraceuticals with medicinal properties (O'Donnell et al., 2011; Mohmand et al., 2011). In addition, due to their pharmacological qualities, mushrooms are becoming more important in daily diet for their nutritional value, which is connected with protein and low fat/energy contents (Khatun et al., 2012; Surcek, 1988). Many mushroom species exist in Nature, but still, merely some are used as food. Nowadays, people are deeply interested in food bio-active compounds that are beneficial and intended for health promotions and disease-threat reduction. Collecting comprehensive information concerned with food bio-actives is essential to attain suitable functional food products (Manzi et al., 2001). *Morchella esculenta* is commonly known as 'morel mushroom' or sponge mushroom or Gucchi in Indian locality of Jammu & Kashmir and Himachal Pradesh. *M.*

[1] Vatsalya Clinic, Krishna Nagar, New Delhi, 110051, India.
[2] Government Department of School Education, Jammu, Jammu and Kashmir 180001, India.
[3] Department of Plant Sciences, University of Hyderabad Gachibowli, Hyderabad, Telangana, India-500046.
[4] Department of Biotechnology, Vel Tech Rangarajan Dr. Sagunthala R&D Institute of Science and Technology, Chennai, India.
[5] JSS Medical College (Deemed to be University), Mysuru, Karnataka, India.
[6] CHIRST (Deemed to be University), Bengaluru, Karnataka, India.
[7] Centre of Excellence in Molecular Biology & Regenerative Medicine, Department of Biochemistry, JSS Medical College, JSS Academy of Higher Education & Research, Mysuru, 15, India.
* Corresponding author: deepupandita@gmail.com

esculenta is among the economically valuable wild morel mushroom species and cannot be cultivated commercially. It is appreciated for its spongy, honeycomb texture and unique flavour. *M. esculenta* is well known for its distinctive flavour, texture, aroma and taste and belongs to order Pezizales and family Morchellaceae. *M. esculenta* is one kind of highly-priced forest mushroom (over Rs 20,000/kg) (Thakur et al., 2021; Paliwal, 2013; Prasad et al., 2002). Due to its elevated price, it plays a significant role in the country's economy (Vivek et al., 2018).

Functional foods that are prepared from morel mushrooms are of high medicinal properties (Prasad et al., 2002). Prasad et al. (2002) and Wasser and Weis (1999) described utilisation of *M. esculenta* as preventive medicine for various medicinal purposes by traditional hill region societies, and in China and Japan, as folk-medicine since ancient times (Jong and Birmingham, 1992). *M. esculenta* is one of the best members of the genus which is edible, tasty and possesses excellent medicinal and functional properties in India (Duncan et al., 2002). Litchfield et al. (1963) identified *M. esculenta* comprises numerous bioactive substances with polysaccharides, trace elements, proteins, vitamins, and dietary fibres. Bioactive compounds provide high health benefits and longevity of life by protecting against regenerative diseases due their highly significant and antioxidant property (Alves et al., 2012; Thakur et al., 2021). Pharmacological properties are imparted due to the vast range of bioactive components present in them (Manikandan et al., 2011; Meng et al., 2010a, b). *M. esculenta* is popularly known for its nutraceutical and medicinal/health benefits, due to the bioactive compounds present in it which include proteins, vitamins, polysaccharides, micronutrients, macronutrients, dietary fibres, etc. (Litchfteld et al., 1963). According to studies, compounds, that were isolated, exhibited anticancer, antiviral, antibacterial, antiparasitic, hepatoprotective, cardiovascular, antioxidant, antimicrobial, antitumour, anti-inflammatory and antidiabetic activities (Nitha et al., 2010; Meng et al., 2010a, b). Due to its various active constituents, it also acts as an immune stimulant. As it is proven to be a rich source of B complex vitamins and proteins, studies have been conducted using its fruiting bodies for the treatment of illness, cold, headache and diseases like hepatitis B. It is also used for the treatment of various health conditions, like fatigue, blood cholesterol, insomnia, blood pressure issues and also in maintaining blood-glucose levels (Mohmand and Goldfarb, 2011). Elmastas et al. (2006) recognised an antioxidant activity exhibited by the fruiting body of *M. esculenta*. Besides, Mau et al. (2004) predicted antioxidant activities by the mycelia of Gucchi that contained beta-carotene and linoleic acid. Derivatives from *M. esculenta* ingested to prevent diseases caused by oxidative stress and consequences were caused due to the same (Alves et al., 2012). Various number of synthetic antioxidants have been showcased to cause damage to the health of an individual. Thus, extracts or products of *M. esculenta* are of better beneficial and nutraceutical value to help in the prevention of diseases and has cognitive effect on maintenance of health (Yadav et al., 1998; Kamath and Rajini, 2007; Nitha et al., 2010). *M. esculenta* is identified as an antitumour and anti-inflammatory agent (Nitha et al., 2013; Nitha et al., 2007), which is recognised as possessing polysaccharides (Yang et al., 2014). Li et al. (2013) noted that a few polysaccharides extracted from *M. esculenta* reasonably showed antitumour capacity. Heleno et al. (2013) described that some mushroom extracts demonstrate antibacterial activity adversely on several pathogens, viz., *Staphylococcus aureus*, *Salmonella typhimurium*, *E. coli*, *Listeria monocytogenes*, and *Enterobacter cloacae*. It could be used as a body tonic, laxative, purgative, or emollient for stomach problems, wound healing and general weakness (Shameem et al., 2017). The proteins present in it consist of nine vital amino acids that humans require. Besides their excellent protein component, *M. esculenta* is an excellent resource of nutrients similar to phosphorus and iron. *M. esculenta* has important bioactive secondary metabolites, viz., alkaloids, glycoproteins, amino acids, sterols, polyphenols, lactones, terpenoids, chelating agents, sesquiterpenes, nucleotide analogs, metals, vitamins, polysaccharides, along with β-glucans (Taskin, 2013; Mattila et al., 2001). Erjavec et al. (2012) reported the presence of new proteins in *M. esculenta* with biological activities for use in various biotechnological approaches and future drug improvement. It also includes hydrophobins, protease inhibitors, proteins that inactivate ribosomes and lignocellulose-degrading enzyme proteases (Erjavec et al., 2012). In addition, it may be used as a purgative, laxative, body tonic, emollient for stomach troubles, wound healing agent and to reduce general weakness. It can be poisonous and may produce many adverse reactions if not appropriately used or if eaten raw (Vivek et al., 2018). This chapter mainly focuses on *Morchella esculenta* as a nutraceutical and functional food, its habit, habitat, general characteristics, availability, biologically active compounds present and pharmacological and medicinal value.

2. Classification, Habit, Habitat and General Characteristics of *Morchella esculenta*

Kingdom: Fungi
Division: Ascomyscota
Subdivision: Pezizomycotina
Class: Pezizomycetes
Order: Pezizales
Family: Morchellaceae
Genus: *Morchella*
Species: *esculenta*

Morchella esculenta consists of a cylindrical structure (Prasad et al., 2002). The pileus, the upper part which retains the total plant weight of 70–80%, is about 3–9 cm long, 2–5 cm broad, has honeycombed heads that are spongy and round or asymmetrical pits/irregular pits present on it (Ajmal et al., 2015). It is pale or black, brown and yellow in colour. The edible fruiting bodies of *M. esculenta* bloom in the months from March to June (Raman, 2018). The lower part is composed of the stalk or stipe which constitutes about 20–30% of the total weight. It has size of about 1–4 cm, width of 0.5–3 cm and a hollow body structure (Raman et al., 2018). Hamayun et al. (2003) predicted the general appearance of gucchi as being white to pale grey in colour and turning greyish-brown at ripeness. The stripes are slightly distended at the base, narrowing towards the upper part and support it. When fresh, their sizes range from 2–25 cm and the length decreases up to 0.1–10 cm on drying (Negi, 2006; Raman, 2018).

The word *esculenta* is derived from Latin language and means 'edible'. *M. esculenta* is known by different names in different localities as indigenous people of particular places possess different ethnological aspects, especially the people of

Kullu, Mandi, Kinnaur, Shimla and Chambal districts call it *gucchi, chaeu, dhunghloo, bhuntu, jamchu, chuahar khukh, chunchroo, jangmuts* (Semwal, 2014). *M. esculenta* are edible mushrooms famous as true morel mushroom, gucchi, morel, sponge morel, common morel, yellow morel, etc., and are distinctive, significant and efficiently advantageous species of wild mushroom (Dorfelt, 2013). Besides, the country of origin for *M. esculenta* is Himachal Pradesh, Kullu district, part of the western Himalayas. In India, *M. esculenta* is naturally found in high-altitude hills with cold habitats and thick coniferous forests across temperate zones of Jammu & Kashmir, Uttaranchal and Himachal Pradesh with humus and nutrient-rich soil. These grow best during the rains. They consist of a sponge cap with a rigid conical body. Many have found these morels in Himalayan ranges and it is known that the climate of Mt. Abu in Rajasthan is also compatible for its growth. They are also found under pear trees growing naturally. Therefore, cool and high-altitude places or hilly land environments with greater slopes than 33% during the months of June and July are compatible for their growth (Manikandan et al., 2011; Thakur et al., 2021. They are habitually mature on logs of decayed wood and leaves, stumps and humus-rich loamy soil (Wagay and Vyas, 2011). When they are small with low height, they are commonly black (Nautiyal et al., 2001). According to Ali et al. (2011) and Hamayun et al. (2006), the habitat of *M. esculenta* maintains a mycorrhizal or saprobic association with hardwood of coniferous forests at a height of about 2500–3500 m above sea level (Tahoun and Ali, 1986; Ajmal et al., 2015). They are very found to possess a high market due to which they are subsequently known as the 'growing gold of mountains' (Tahoun and Ali, 1986; Thakur et al., 2021).

3. Collection and Production of *Morchella esculenta*

The production of *M. esculenta* worldwide is 1.5 million tonnes of fresh weight and 150 tonnes of dry weight. India and Pakistan are the major morel-producing countries and each country has about 50 tonnes of dry morels (FAO, 2002). The collection of *M. esculenta* is challenging and it also requires concentration and zeal. It is effortless to collect during spring and early summer, when it is collected after the ascocarp reaches 6.5–8 cm in height and 4.4–7.5 cm in diameter (Hamayun et al., 2006). The processing processes, like drying, storage and marketing are the major challenges in the commercialisation of *M. esculenta.* Its high moisture content is the reason for its reduced life. Therefore, it should be appropriately dried and stored and preserved in a closed chamber. The most acceptable way of storage for morel is to keep them to dry with a bit of appropriate ventilation. It is mainly traded internationally to countries like France, Germany, Switzerland, Austria, Belgium and the Middle East (Hamayun et al., 2006).

4. Bioactive Compounds/Constituents of *Morchella esculenta*

Chemical composition of *M. esculenta* is reported to be of high nutritional value due to primary and secondary metabolites present in it. Carbohydrates are the most abundant macronutrients and proteins are secondary to it. They contain very low or negligible amounts of fat. Mannitol and trehalose were found in these mushrooms and the presence of fructose was also detected. They contain high levels of total sugars and polyol is the most abundant among them (Kalač, 2009), and trehalose is found to be the storage disaccharide (Koide et al., 2000). About 24 types of fatty acids were identified and quantified by assessing the samples of these mushrooms. *Morchella esculenta* is not only a resource of nutrients but also has been described as a therapeutic food, helpful in hindering diseases, viz., hypercholesterolemia, hypertension, diabetes and cancer (Fig. 1). Biologically active compounds of morels strengthen the immune system and shield against pre-radicals, like carcinogens (Ramesh and Pattar, 2010). Various previous studies have shown the active components present in gucchi are therapeutic agents against cancer, they are immunomodulating and known to cure chronic bronchitis (Seth et al., 2014). Several studies concluded that the functional characteristics of mushrooms are mainly their dietary fibre, predominantly chitin and beta-glucans. This shows the properties like anti-tumoor, antiviral, antithrombotic, lowered blood sugar levels and immunomodulating (Perera and Li, 2011). *M. esculenta* contains outstanding components with outstanding properties to prevent or treat diverse diseases. Because of their low-fat percentage, *M. esculenta* can be used in low-calorie diets. They possess all the essential amino acids necessary for humans. Organic acids are also known to be found in these mushrooms. Experiments were performed and researchers succeeded in quantifying oxalic and fumaric acids and they were assessed for their toxic effects (Nagarajkumar et al., 2005), in which fumaric acid possessed biologically positive effects, such as neuroprotective, chemo-preventive and anti-inflammatory and anti-microbial activity (Baati et al., 2011). It is found that phenolic compounds are present in a wide range and possess physiological properties, anti-atherogenic, anti-thrombotic, cardio-protective, anti-inflammatory, anti-allergic, anti-microbial and vasodialatory effects which are related to antioxidant activity of

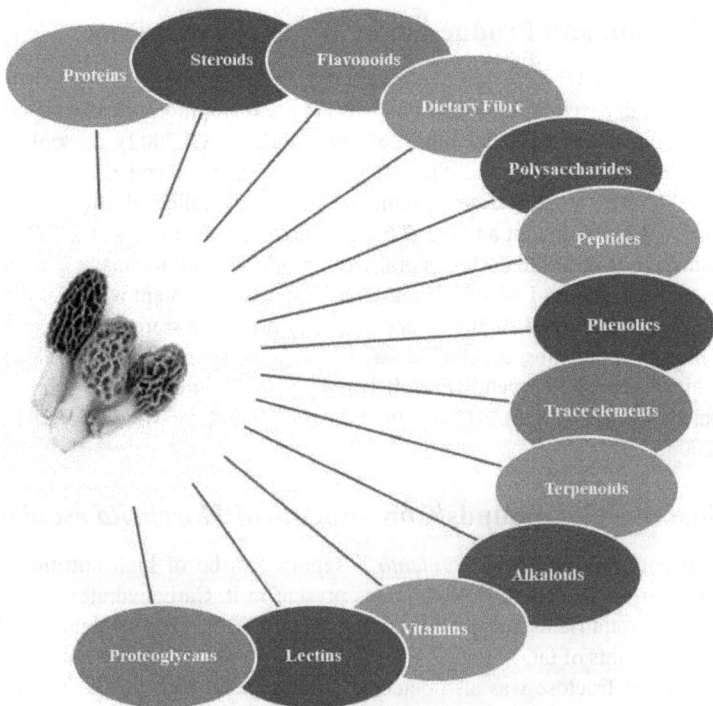

Fig. 1. Bioactive compounds of *M. esculenta*.

M. esculenta (Ferreira et al., 2009). Furthermore, they are a comparatively high-quality source of nutrients similar to phosphorus and iron and also include vitamins, thiamine, riboflavin, ascorbic acid, ergosterol and niacin. It has also been reported that *M. esculenta* is a therapeutic food, helpful in preventing hypertension, diabetes, hypercholesterolemia and cancer. Certain efficient distinctiveness is mainly due to dietary fibres, bioactive components, antioxidants, lectins and antimicrobial agents in *M. esculenta* and also have polysaccharides whilch are immunomodulating and are used as health-promoting food supplements (nutraceuticals). The activity method of a range of secondary metabolites isolated from the remedial and wild edible mushroom is yet to be discovered.

The fruiting body of *M. esculenta* is edible and can be used as flavouring agent in various food items (Prasad et al., 2002). The fruiting body of *M. esculenta* contains a wide range of active constituents, including carotenoids, tocopherols, phenolic compounds and organic acids (Fig. 2). Badshah et al. (2021) reported the presence of carotenoids resembling β-carotene and lycopene. Tocopherols contain δ-to copherol, α-tocopherol and γ-tocopherol. Phenolic compounds are protocatechuic acid, P-coumaric acid, and p-hydroxybenzoic acid and contain organic acids like citric acid, quinic acid, fumaric acid, oxalic acid, and malic acid (Heleno et al., 2013). Badshah et al. (2021) revealed the presence of several vital nutrients in morchella, viz., proteins, polyunsaturated fatty acids and carbohydrates. Also noted was the presence of phenolic compounds and isolated and purified polysaccharides from *M. esculenta* mushroom, which were also tested for specific bioactivities. Correspondingly, Lee et al. (2018) proved the presence of some sterols and fatty acids with their antitumour activity from the methanolic extracts of fruiting bodies of *M. esculenta*. Yang et al. (2015) purified a polysaccharide from *M. esculenta* with 43.6 kDa containing monomer units of glucose, mannose, galactose and arabinose. Likewise, Huo et al. (2020) found that a polysaccharide with an 81 kDa from *M. esculenta* also proved that its anticancer activity is dosage and time-dependent. Hu et al. (2013) confirmed the presence of polysaccharides isolated from the mycelium of *M. esculenta* and established its anti-proliferative activity against hepatoma cell lines. Similarly, Cai et al. (2018a) isolated a heteropolysaccharide from *M. esculenta* and recognised its anti-melanogenesis action; also these heteropolysaccharides can be used in skin cancer treatment. Polysaccharides, predominantly β-glucans, builds up the immune system. β-glucans are found to be more in mushrooms, chiefly in *M. esculenta*. Hence, *M. esculenta* is a precious dietary resource and a promising curative source to treat

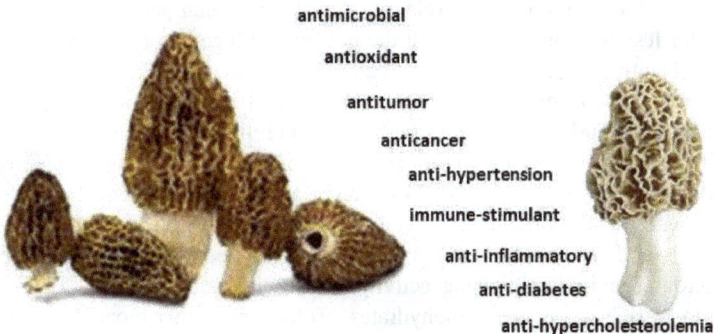

antimicrobial

antioxidant

antitumor

anticancer

anti-hypertension

immune-stimulant

anti-inflammatory

anti-diabetes

anti-hypercholesterolemia

Fig. 2. Pharmacological properties of *M. esculenta*.

widespread disorders, for instance, inflammation, cancer and diabetes (Guillamon et al., 2010).

Concerning their high nutritional and therapeutic potential, *M. esculenta* come across in different applications, specifically as functional foods or as a resource of nutraceuticals for maintaining and promoting well-being and standard of life.

4.1 Nutritional Significance

M. esculenta fruiting body is fit for human consumption. It is composed of extremely nourishing, appetising and healthy constituents (Elmastas et al., 2006). It abounds in protein, vitamins, carbohydrates and predominantly vitamin B besides containing traces of vitamin A, D and C. It contains minerals, including calcium, copper, sodium, zinc, manganese, iron, magnesium, phosphorus, selenium and potassium and has less fat and less calories (Negi, 2006; Mattila et al., 2001). Cheung (2010) noted 38% carbohydrates, 32.7% protein, 17.6% fibre, 2.0% fat and 9.7% ash in *M. esculenta*. Wahid et al. (1988) reported the presence of 195 mg/g iron, 62.6 mg/g copper, 98.9 mg/g zinc, 54.7 mg/g manganese, 23.5 mg/g potassium, 3.49 mg/g phosphorus, 0.85 mg/g calcium, 1.82 mg/g magnesium and 0.18 mg/g sodium (Table 1). Previous reports elucidate the presence of various aromatic compounds, for instance, aldehydes, phenol, acids, alcohol, esters, ketones and terpene. The major aromatic compounds in *M. esculenta* reported by Genccelep et al. (2009) are phenol, which includes 15.55% of ester, 50.88% alcohol and 11.37% of carbamic acid. Proteins acquired from the mycelia of *M. esculenta* are equivalent to vegetative protein and are a good source of supply as a protein supplement (Taskin, 2013). *M. esculenta* can be more easily digested than other vegetables due to its high protein content and also as it is rich in vitamins, B-complex and minerals. It has been revealed that *M. esculenta* helps treat diseases like colds, stomachache, headaches and hepatitis B. It can diminish fatigue, sleeping troubles and also reduce blood cholesterol levels. It is an excellent alternative source for anemia. It also aids in regulating the blood sugar level (Janardhanan et al., 1970; Ying et al., 1987; Nitha and Meera, 2006; Wahid et al., 1988; Kumar et al., 2000, Collins, 1999; Mattila et al., 2001; Duncan et al., 2002; Genccelep et al., 2009; Mahmood et al., 2011, Sher et al., 2011; Sharma et al., 2016; Halder and Sharma, 2017; Rai et al., 2017; Sharma et al., 2017; Mohmand et al., 2011; Sharma and Arora, 2017). *M. esculenta* is a delicious and edible mushroom (Sharma et al., 2016). Jander-Shagug and Masaphy (2010) reported its toxicity when raw or freshly collected and eaten in big amounts. Heleno et al. (2013) elucidated that the implicit neurotoxin is volatile. For intoxication, the morels may be fit for human consumption when it is cooked for less time and removes all the poison and then can be consumed in huge amounts (Nautiyal et al., 2001; Fayaz et al., 2012).

Nutritional value of *M. esculenta* and its fruiting bodies were assessed against the specimens from Turkey (Elmastas et al., 2006) and that of Spain (Jové et al., 2017), to learn the comparative anti-microbial activity of mycelia specimens obtained from the regions of Turkey and Spain (Kalyoncu et al., 2010). Samples from Serbia and Portugal were first reported to contain a wide range of phenolic compounds, organic compounds and fatty acids. Thus the bioactive compounds are proven to possess anti-microbial and demyelinising activity while wild samples of *M. esculenta* are known to be rich sources of carbohydrates, proteins and other bioactive compounds

Table 1. Nutritional value of raw *M. esculenta* (morel mushroom) (*Source*: Collected from USDA nutritional and nutrient database).

Nutrient (For Serving Size of 1 Cup, 66 grams)	Amount in Grams	% of Daily Values
Water	59.14	N/D
Energy	85 kJ	N/D
Protein	2.06	4.12
Total fat (lipid)	0.38	1.09
Ash	1.04	N/D
Carbohydrate	3.37	2.59
Dietary fibre	1.8	4.74
Sugar	0.4	N/D
Sucrose	0	N/D
Glucose (dextrose)	0.4	N/D
Minerals	**Quantity in mg/μg**	**% of Daily Values**
Calcium	28	2.80
Iron	8.04	100.50
Magnesium	13	3.10
Phosphorus	128	18.29
Potassium	271	5.77
Sodium	14	0.93
Zinc	1.34	12.18
Copper	0.412	45.78
Manganese	0.387	16.83
Selenium	1.5 μg	2.73
Vitamins	**Amount mg/μg**	**% of Daily Values**
Vitamin B_1	0.046 mg	3.83
Vitamin B_2	0.135 mg	10.38
Vitamin B_3	1.486 mg	9.29
Vitamin B_5	0.29 mg	5.80
Vitamin B_6	0.09 mg	6.92
Vitamin B_9	6 μg	1.50
Folic acid	0 μg	N/D
Folate, food	6 μg	N/D
Folate, DEF	6 μg	N/D
Vitamin D ($D_2 + D_3$)	3.4 μg	22.67
Vitamin D_2	3.4 μg	N/D
Vitamin D_3	0 μg	N/D
Vitamin D	136 IU	N/D

Table 1 contd. ...

...Table 1 contd.

Fatty Acids, Total Saturated	0.043 g	N/D
Capric acid 10:00 (Decanoic acid)	0.001 g	N/D
Myristic acid 14:00 (Tetradecanoic acid)	0.001 g	N/D
Palmitic acid 16:00 (Hexadecanoic acid)	0.034 g	N/D
Stearic acid 18:00 (Octadecanoic acid)	0.006 g	N/D
Lignoceric acid (tetracosanoic acid) 24:00	0.001 g	N/D
Fatty Acids, Total Monounsaturated	0.034 g	N/D
Palmitoleic acid 16:1 (hexadecenoic acid)	0.001 g	N/D
16:1 c	0.001 g	N/D
16:1 t	0 g	N/D
Oleic acid 18:1 (Octadecenoic acid)	0.015 g	N/D
18:1 c	0.015 g	N/D
18:1 t	0 g	N/D
Nervonic acid (Tetracosenoic acid) 24:1 c	0.002 g	N/D
Fatty Acids, Total Polyunsaturated	0.286 g	N/D
Linoleic acid 18:2 (Octadecadienoic acid)	0.142 g	N/D
Alpha-linolenic acid 18:3 n-3 c,c,c (Octadecatrienoic acid)	0.142 g	N/D
Eicosadienoic acid 20:2 n-6 c,c	0.001 g	N/D
Campesterol	2 mg	N/D

The aforementioned percent of daily values (% DVs) are established bases on 2,000 calorie diet consumption. Depending on a person's daily calorie requirements, daily values (DVs) might be varied. The U.S. Department of Agriculture suggests cited values and www.healthbenefitstimes. com also recommends. Also made the calculations based on average age of 19–50 years and a body weight minimum of 194 lbs.

including tocopherols and organic acids. Polyunsaturated fatty acids are present in high proportion than that of mono and unsaturated fatty acids (Heleno et al., 2013).

4.2 Pharmacological Properties of M. esculenta

The pharmacological properties of Morchella species show its use in Chinese traditional medicine since 2,000 years and in Malaysia and Japan to cure several diseases (Hobbs, 1995). They possess a wide diversity of bio-compounds that play bioactive and nutritional important roles in preventive ailments in the traditional hill communities (Wasser and Weis, 1999; Prasad et al., 2002; Ferreira et al., 2009; Ferreira et al., 2010; Alves et al., 2012). *M. esculenta* is known for its pharmaceutical and nutritional values due to the antitumour, anticancerous, antioxidant, antimicrobial and anti-inflammatory activities and is also used as a flavouring agent (Mau et al., 2004; Nitha and Janardhanan, 2008; Nitha et al., 2010; Alves et al., 2012; Nitha et al., 2013).

In studies reported by Kalyoncu et al. (2010) and Alves et al. (2012), mycelia of *M. esculenta* has antimicrobial activities. Preceding reports appraise that chloroform, methanol and ethanol extracts hold antibacterial properties of *M. esculenta* (Badshah et al., 2012). *M. esculenta* exhibits antibacterial activity against *Staphylococcus aureus, Salmonella typhimurium, Enterobacter cloacae, Escherichia coli*, and *Listeria monocytogenes* (Heleno et al., 2013). Biological and anti-inflammatory activities of *M. esculenta* are reported against *E. coli, Bacillus mesenteric* and *Bacillus subtilis* (Sher et al., 2011). Inflammation occurs due to a range of reasons, including insect bites, toxic drugs, or numerous chronic diseases (Collins, 1999). Different compounds present in *M. esculenta* show intense anti-inflammatory activity against insect bites on entire body. Methanolic extracts of *M. esculenta* exhibit an anti-inflammatory response and reduce pain (Kumar et al., 2000; Nitha et al., 2006).

Pharmacological and nutritional activities of *M. esculenta* are reported due to the composition of the biological compounds that are present (Mau et al., 2004, Tsai et al., 2006). Studies reveal that the percentage composition of compounds in *M. esculenta* is as: 38% of carbohydrates, 32.7% of proteins, 17.6% fibre, 2% of fat and lastly 9.7% of ash. Polysaccharides are the major components in the extracts of *M. esculenta*. Many polysaccharide and polysaccharide-protein complexes were extracted and isolated from these mushrooms for use as therapeutic agents which are macromolecules that are efficient and potential immunomodulators with a wide range of clinical applications (Tzianabos, 2000). The compound isolated from the fruiting body of *M. esculenta* exhibits the platelet aggregation inhibitor that has been patented (Christine et al., 2002). The polysaccharide galactomannan, isolated from the fruiting bodies of *M. esculenta*, has been described to display immunostimulatory activity (Duncan et al., 2002). *M. esculenta* polysaccharide, named galactomannan, is one of the compounds of high economic value (Duncan et al., 2002) and increases immune response of the system to various diseases and also has excellent immunomodulatory mechanism (Duncan et al., 2002; Cui et al., 2011). Polysaccharides prove effects of antimicrobial activity, anti-actinomycete properties and antibacterial activity (Heleno et al., 2013). Heleno et al. (2013) found the presence of antioxidant properties and antibacterial activities in *M. esculenta* methanolic extracts against some bacteria (Figs. 1 and 2). Heteropolysaccharide, isolated from the extract of protein bodies of *M. esculenta*, had shown antimelanogenesis or demelanising property (Cai et al., 2018a). A skin-lightening cosmetic extracted from cultured *M. esculenta* contains melanin formation inhibitor (Sharma et al., 2016; Halder et al., 2017; Halder and Sharma, 2017; Rai et al., 2017; Sharma et al., 2017; Sharma et al., 2017). These properties were investigated *in vitro* and *in vivo* which exhibited low productions of melanin without any toxicity to the cell (Cai et al., 2018b). By the virtue of this property, polysaccharides obtained from *M. esculenta* can be utilised as therapeutic agents in skin cancer treatments. It is universally used to treat indigestion, excessive phlegm and cardiac diseases (Ying et al., 1987). Powder extracts of *M. esculenta* can be used as an antiseptic to treat wounds and heal stomach aches and as a laxative and an emollient (Mahmood et al., 2011).

These also have antitumour activity. The ethanol extract of *M. esculenta* mycelium exhibits noteworthy antitumour activity. It has both curative and preventive properties against both superficial and solid tumours in a dose-dependent manner, has been reported by Nitha et al. (2013). *M. esculenta* mycelia are composed of compounds that might have to modulate tumorigenesis at diverse stages or act at the exact location. *M. esculenta* also contains important nutrients, like phenolic compounds, polyunsaturated fatty acids and other carbohydrate extracts which were fed to mice and acted as useful source of microbiota as well as for obtaining short chain fatty acids in the body of mice (Tietel and Masaphy, 2018). The monosaccharide units obtained from the purified form of polysaccharide of molecular weight 43.6 Kda from *M. esculenta* is composed of mannose, glucose, galactose, arabinose as monomeric subunits (Yang et al., 2015). *M. esculenta* also has a polysaccharide of weight 81 Kda and is a potential inhibitor of growth and prognosis of human colorectal cancer HT29 which is dose-dependent and time-dependent in nature (Liu et al., 2016). Various fatty acids and sterol compounds obtained from methanolic extract from the fruiting bodies of *M. esculenta* showed anticancer and antitumour activity against human lung cancer cells IC_{50} in the range of 157–278 µM (Lee et al., 2018). One among the polysaccharides obtained from mycelium of *M. esculenta* exhibited anti-proliferative activity against liver cancer cell lines (Hu et al., 2013). Tables 2 and 3 summarizes the medicinal properties of *M. esculenta.*

4.3 Effects of M. esculenta *Polysaccharides on Mice in* In Vivo *Experiments*

M. esculenta polysaccharides (MEP), being nutraceutical components in extracts of *M. esculenta*, were analysed for their immunomodulatory effects on vital organs of thymus and spleen in mice in *in vivo* conditions (Kyo et al., 2001). The analysis of immunity indexes of these organs revealed that the immunosuppressive agents induce variations in the weight of spleen and thymus, causing decrease in immune activity. The effects of polysaccharides are positive towards immune expression. MEP compounds showed a significant increase in relative weight of thymus and spleen when compared with the control group. The experimental analysis in mice showed slight increase in immunity of mice concluding that MEP in *M. esculenta* can rejuvenate MRP under *in vivo* conditions. The results are due to the inductive activity of MEP present in *M. esculenta* to stimulate the proliferation of T-lymphocytes and also purified lymphocytes, but, no significant increase in B-lymphocytes was observed. A significant increase in proliferating activity of MEP at 10–30 mg/kg doses on T-lymphocytes and purified lymphocytes was reported, but the activity decreased in high doses of 60 mg/kg (Cui et al., 2011). Simultaneously, MEP also influences the activity of lymphocytes under *in vitro* conditions at high concentrations. So the concentration of 30 mg/kg MEP showed maximum proliferation which is demonstrated by the dose-dependent studies of polysaccharides on mice models (Cui et al., 2011).

4.4 Effects of MEP on Macrophage Nitric Oxide

Nitric oxide (NO) is a signalling molecule which is involved in enhancement of phagocytic activity of macrophages. The MEP from the extracts of morels impact NO, which alters the immunity, regulates apoptosis and anti-proliferative activity against tumour cells. NO as a gaseous molecule is synthesised from amino acid L-arginine in the presence of nitric acid synthase enzyme and is known as signalling molecule because of the key role to alarm the body of infections (Huang et al., 2012). Experiments have proven that *M. esculenta* (MEP) showed a significant stimulation of macrophages to produce NO, thus proving it as a principle effector molecule which is produced by macrophages by virtue of cytotoxic activity (Badshah et al., 2021).

Table 2. Phytoconstituents of *M. esculenta*, activity and medicinal properties.

S. No.	Phytoconstituents	Activity	Medicinal Properties	References
1.	Polysaccharides	Hyperglycemic activity.	Promote serum insulin secretion (INS) in diabetic rats.	Liu et al., 2018
2.	Phenolics	Antioxidant.	Delay oxidative damage in the human body.	Thakur et al., 2014
3.	Glycosylated Morchella protein hydrolysate (MPH)	Antioxidant.	Break the free radical chain, chelating metal ions, providing hydrogen and electrons, and scavenging active oxygen.	Zhang et al., 2018a
4.	Ethanolic extract of mycelium	Anti-inflammatory.	Inflammation-associated diseases like arthritis.	Nitha et al., 2007
5.	Ethanolic extract of mycelium	Antitumour.	Preventive properties against as cites and solid tumour.	Nitha et al., 2007
6.	Ethanolic extract of mycelium	Antioxidant.	Inhibit free radical formation and can also scavenge them.	Nitha et al., 2010
7.	Glycated *M. esculenta* protein isolate (MPI–xylose conjugates)	Antioxidant.	Chain-breaking, electron and hydrogen atom donation, metal-chelating, and oxygen-scavenging mechanisms.	Zhang et al., 2018b
8.	Aqueous-ethyl alcohol extract	Radioprotective effect.	Protect mitochondria from oxidative stress and DNA from deleterious effects induced by radiation.	Nitha et al., 2020
9.	*Morchella esculenta* polysaccharide (MEP) extracted by pulsed electric field (xylose, glucose, mannose, rhamnose and galactose)	Anti-proliferating.	Inhibit the proliferation and growth of human colon cancer HT-29 cells.	Liu et al., 2016
10.	Exopolysaccharides (EPS)	Antioxidant.	Scavenging of superoxide radicals.	Meng et al., 2010a,b
11.	Heteropolysaccharide from the fruiting bodies of *M. esculenta*	Antioxidant.	Protect zebrafish embryos against AAPH-induced oxidative damage, reduction of ROS production; NO production and cell death.	Cai et al., 2018a

Table 2 contd. ...

...Table 2 contd.

S No.	Phytoconstituents	Activity	Medicinal Properties	References
12.	Arabitol, glucose, mannitol and trehalose, monosodium glutamate-like components, umami	Non-volatile taste components.	Sweet and bitter components and intense umami taste.	Tsai et al., 2006
13.	Polysaccharopeptides and polysaccharides	Antioxidant, acetylcholinesterase (AChE) and butyryl cholinesterase (BChE) inhibition activities.	Can be used for the treatment of Alzheimer's and Parkinson's diseases.	Badshah et al., 2021
14.	Methanolic extract of *M. esculenta*	Anti-microbial property.	Inhibited *L. monocytogenes, S. typhimurium, S. aureus, E. cloacae* and *E. coli.*	Heleno et al., 2013
15.	Methanolic extract of *M. esculenta*	Demelanising properties.	Demelanising activity against micromycetes (*Aspergillus fumigatus, A. flavus, Penicillium funiculosum* and *P. ochrochloron*).	Heleno et al., 2013
16.	Polysaccharide MEP-II (from the fermentation broth of *Morchella esculenta*)	Antitumour agent.	Apoptosis of HepG2 cells through ROS generation.	Hu et al., 2013

Table 3. Health-stimulating effects of *Morchella esculenta* (Tietel and Masaphy, 2018).

M. esculenta Extract	Activity	References
Methylene chloride	Anti-inflammatory activity.	Kim et al., 2011
Ethanolic and methanolic	Immunostimulatory.	Duncan et al., 2002
Methanol	Antimutagenic, antimitotic and radical scavenging activity.	Stojkovic et al., 2013; Heleno et al., 2013; Jander-Shagug and Masaphy, 2010
Ethanol	Antiscavenging activity, antitumour and anti-inflammatory.	Elmastas et al., 2006; Gursoy et al., 2009; Nitha et al., 2007; Nitha et al., 2010; Nitha et al., 2013
Water	Scavenging effects.	Ramirez-Anguiano et al., 2007
Aqueous ethanolic	Nephroprotective activity.	Nitha and Janardhanan, 2008

Dose-dependent response of MEP was observed at a dose of 10 µg/ml. NO synthesis by macrophages increased with an increase in the concentration of MEP. In comparison to the increased activity, the production of NO was very low in non-activated macrophage. When compared with other immunomodulatory polysaccharides, MEP increases the production of NO because of the variations in the structure or molecular weight. These polysaccharides are carbohydrates composed of monomeric subunits called monosaccharides which are responsible for biological activity due to their chemical nature. The MEP is proven to have anticancerous activity due to its branching at $\beta(1 \rightarrow 3)$ position and also has a main chain of glucan and an additional $\beta(1 \rightarrow 6)$ branch points (Cui et al., 2011; Yang et al., 2015). β glycosidic linkage plays a very prominent role in antitumour and is also known to show immunomodulatory activities. By C13-NMR, the polysaccharides of *M. esculenta* are analysed and it is known that MEP are rich in mannose, glucose and galactose and also contain very little quantities of rhamnose, arabinose and xylose that are arranged at 1-3, 1-4 β linkages (Li et al., 2017). The proteinized MEP from the extracts of *M. esculenta* possess a low modulatory potential to block AChE when compared to the standard galanthamine, thus inferring that *M. esculenta* polysaccharides (MEP) show better therapeutic potential and also with the conjugation of other drugs show excellent control in Alzheimer's diseases which is also a neurological or cerebral syndrome (Yang et al., 2015; Liu et al., 2016).

5. Conclusion

M. esculanta is known to be a non-timber source of phytochemicals. Since ages it is used as food amongst the local population of India and other countries of the world. Morels are globally highly valued for their outstanding organoleptic characteristics with tremendous nutritional significance in preventing or treating different diseases. Furthermore, morels are considered as functional food. The bioactive components of morel's have nutritional, nutraceutical and medicinal and pharmacological importance. Morels have been scientifically proved to be hyperglycemic, antioxidant, anti-inflammatory, immunostimulatory, anti-proliferating, anti-microbial, demelanising, antitumour and radioprotective. They possess immunomodulatory property as experimented by the analysis using lymphoid organs, like spleen and thymus. For their elevated nutritional and therapeutic prospects, *M. esculenta* can find diverse applications as functional foods or as a resource of nutraceuticals meant for preservation and endorsement of well-being and quality of human life.

References

Ajmal, M., Akram, A., Ara, A., Akhund, S. and Nayyar, B.G. (2015). *Morchella esculenta*: An edible and health beneficial mushroom, Pak. *J. Food Sci.*, 25(2): 71–78.

Ali, H., Sannai, J., Sher, H. and Rashid. A. (2011). Ethnobotanical profile of some plant resources in Malam Jabba Valley of Swat, Pakistan. *J. Med. Plants Res.*, 5(17): 4171–4180.

Alves, M., Ferreira, I.F.R., Dias, J., Teixeira, V., Martins, A. and Pintado, M. (2012). A review on antimicrobial activity of mushroom (basidiomycetes) extracts and isolated compounds. *Planta Med.*, 78(16): 1707–1718. https://Doi.org/10.1055/s-0032-1315370.

Baati, T., Horcajada, P., Gref, R., Couvreur, P. and Serre, C. (2011). Quantification of fumaric acid in liver, spleen, and urine by high-performance liquid chromatography coupled to photodiode-array detection. *J. Pharm. Biomed. Anal.*, 56(4): 758–762. https://Doi.org/10.1016/j.jpba.2011.07.011.

Badshah, S.L., Riaz, A., Muhammad, A., Tel Çayan, G., Çayan, F., Emin Duru, M., Ahmad, N., Emwas, A.-H. and Jaremko, M. (2021). Isolation, characterisation, and medicinal potential of polysaccharides of *Morchella esculenta*. *Molecules*, 26(5): 1–12. https://Doi.org/10.3390/molecules26051459.

Bashir, A., Vaida, N. and Dar, M.A. (2014). Medicinal Importance of mushrooms—A review. *International Journal of Advanced Research*, 2(12): 1–4.

Cai, Z.N., Li, W., Mehmood, S., Pan, W.J., Wu, Q.X., Chen, Y. and Lu, Y.M. (2018a). Effect of polysaccharide FMP-1 from *Morchella esculenta* on melanogenesis in B16F10 cells and zebrafish. *Food Funct.*, 9(9): 5007–5015. https://Doi.org/10.1039/c8fo01267a.

Cai, Z.N., Li, W., Mehmood, S., Pan, W.J., Wang, Y., Meng, F.J., Wang, X.F., Lu, Y.M. and Chen, Y. (2018b). Structural characterisation, *in vitro* and *in vivo* antioxidant activities of a heteropolysaccharide from the fruiting bodies of *Morchella esculenta*. *Carbohydr. Polym.*, 195: 29–38. https://Doi.org/10.1016/j.carbpol.2018.04.069.

Cheung, P. (2010). The nutritional and health benefits of mushrooms. *Nutrition Bulletin*, 35: 292–299.

Christine, J.G.D., Pugh, N., Pasco, D.S. and Ross, S.A. (2002). Isolation of a galactomannan that enhances macrophage activation from the edible fungus *Morchella esculenta*. *J. Agric. Food Chem.*, 50(20): 5683–5685.

Collins, T. (1999). Acute and chronic inflammation. pp. 50–51. *In*: Cotran, R.S., Kumar, V. and Collins, T. (eds.). *Textbook of Robbins Pathologic Basis of Diseases*. W.B. Sounders Company, Philadelphia, 6th ed.

Cui, H.L., Chen, Y., Wang, S.S., Kai, G.Q. and Fang, Y.M. (2011). Isolation, partial characterisation and immunomodulatory activities of polysaccharide from *Morchella esculenta*. *J. Sci. Food Agric.*, 91(12): 2180–2185. https://Doi.org/10.1002/jsfa.4436.

Demoulin, V. and Merriott, J.V.R. (1981). Key to the gasteromycetes of great Britain. *Bull. Brit. Mycol. Soc.*, 15(1): 37–43.

Dorfelt, H. (2013). Morchellaceae. *In*: Hanelt P. Mansfeld (ed.). *Encyclopedia of Agricultural and Horticultural Crops*: (*Except Ornamentals*).

Duncan, C., Pugh, J., Pasco, G., David, N., Ross, S. and Samir, A. (2002). Isolation of a galactomannan that enhances macrophage activation from the edible fungus *Morchella esculenta*. *J. Agric. Food Chem.*, 50(20): 5683–5685. https://Doi.org/10.1021/jf020267c.

Elmastas, M., Turkekul, I., Ozturk, L., Glucin, I., Isildak, O. and Aboul-Enein, H.Y. (2006). Antioxidant activity of two wild edible mushrooms (*Morchella vulgaris* and *Morchella esculenta*) from north Turkey. *Comb. Chem. High Throughput Screen*, 9(6): 443–448. https://Doi.org/10.2174/138620706777698544.

Erjavec, J., Kos, J., Ravnikar, M., Dreo, T. and Sabotic, J. (2012). Proteins of higher fungi—From forest to application. *Trends Biotechnol.*, 30: 259–273.

FAO. (2002). Non-wood forest products from temperate broad-leaved trees, by W M Ciesla, FAO NWFP Series No. 15, Rome, pp. 125.

Fayaz, A.L., Lone, S., Aziz, M.A. and Malla, F.A. (2012). Ethnobotanical studies in the tribal areas of district Kupwara, Kashmir, India. *Int. J. Pharma Bio. Sci.*, 3(4): 399–411.

Ferreira, I., Barros, L. and Abreu, R. (2009). Antioxidants in wild mushrooms. *Curr. Med. Chem.*, 16(12): 1543–1560. https://doi.org/10.2174/092986709787909587.

Ferreira, I.C.F.R., Vaz, J.A., Vasconcelos, M.H. and Martins, A. (2010). Compounds from wild mushrooms with antitumour potential. *Anticancer Agents Med. Chem.*, 10(5): 424–436.

Genccelep, H., Uzun, Y., Tunccturk, Y. and Demirel, K. (2009). Determination of mineral contents of wild-grown edible mushrooms. *Food Chem.*, 113(4): 1033–1036.

Guillamon, E., Lafuente, A.G., Lozano, M., Arrigo, M.D., Rostagno, M.A., Villares, A. and Martinez, J.A. (2010). Edible mushrooms: Role in the prevention of cardiovascular diseases. *Fitoterapia*, 81: 715–723.

Gursoy, N., Sarikurkcu, C., Cengiz, M. and Solak, M.H. (2009). Antioxidant activities, metal contents, total phenolics and flavonoids of seven *Morchella* species. *Food and Chemical Toxicology*, 47: 2381–2388.

Hamayun, M., Khan, M.A. and Begum, S. (2003). Marketing of medicinal plants of Utror-Gabral Valleys, Swat, Pakistan. *J. Ethnobot. Leaflets*, SIUC, USA.

Hamayun, M., Khan, S.A., Ahmad, H., Shin, D.H. and Lee, I.J. (2006). Morel collection and marketing: A case study from the Hindu-Kush mountain region of Swat, Pakistan. *Lyonia J. Ecol. Appl.*, 11(2): 7–13.

Heleno, S.A., Stojković, D., Barros, L., Glamočlija, J., Soković, M., Martins, A., Queiroz, M.J.R.P. and Ferreira, I.C.F.R. (2013). A comparative study of chemical composition, antioxidant and antimicrobial properties of *Morchella esculenta* (L.) Pers. from Portugal and Serbia. *Food Res. Int.*, 51(1): 236–243. https://Doi.org/10.1016/j.foodres.2012.12.020.

Hobbs, C. (1995). *Medicinal Mushrooms.* Bot Press, CA, USA.

Hu, M., Chen, Y., Wang, C., Cui, H., Duan, P., Zhai, T., Yang, Y. and Li, S. (2013). Induction of apoptosis in HepG2 cells by polysaccharide MEP-II from the fermentation broth of *Morchella esculenta. Biotechnol. Lett.*, 35(1): 1–10. https://Doi.org/10.1007/s10529-012-0917-4.

Hu, M., Chen, Y., Wang, C., Cui, H., Duan, P., Zhai, T., Yang, Y. and Li, S. (2013). Induction of apoptosis in HepG2 cells by polysaccharide MEP-II from the fermentation broth of *Morchella esculenta. Biotechnol. Lett.*, 35(1): 1–10. https://doi.org/10.1007/s10529-012-0917-4.

Huang, M., Zhang, S., Zhang, M., Ou, S. and Pan, Z. (2012). Effects of polysaccharides from Morchella conica on nitric oxide production in lipopolysaccharide-treated macrophages. *Appl. Microbiol. Biotechnol.*, 94(3): 763–771. https://Doi.org/10.1007/s00253-011-3711-7.

Huo, W., Qi, P., Cui, L., Zhang, L., Dai, L., Liu, Y., Hu, S., Feng, Z., Qiao, T. and Li, J. (2020). Polysaccharide from wild morels alters the spatial structure of gut microbiota and the production of short-chain fatty acids in mice. *Biosci. Microbiota Food Health*, 39: 219–226.

Janardhanan, K.K., Kaul, T.N. and Husain, A. (1970). Use of vegetable waste for the production of fungal protein from *Morchella* species. *J. Food Sci. Technol.*, 7: 197–199.

Jander-Shagug, G. and Masaphy, S. (2010). Free radical scavenging activity of culinary/medicinal morel Mushrooms, *Morchella Dill.* ex Pers. (Ascomycetes): Relation to colour and phenol contents. *International Journal of Medicinal Mushrooms*, 12: 299–307.

Jong, S.C. and Birmingham, J.M. (1992). Medicinal benefits of the Mushroom ganoderma. *Adv. Appl. Microbiol.*, 37(C): 101–134. https://Doi.org/10.1016/S0065-2164(08)70253-3.

Jové, M., Collado, R., Quiles, J.L., Ramírez-Tortosa, M.C., Sol, J., Ruiz-Sanjuan, M., Fernandez, M., Cabrera, C., de la, T., Ramírez-Tortosa, C., Granados-Principal, S. et al. (2017). A plasma metabolomic signature discloses human breast cancer. *Oncotarget.*, 8(12): 19522–19533. https://Doi.org/10.18632/oncotarget.14521.

Kalač, P. (2009). Chemical composition and nutritional value of European species of wild growing mushrooms: A review. *Food Chem.*, 113(1): 9–16. https://Doi.org/10.1016/j.foodchem.2008.07.077.

Kalyoncu, F., Oskay, M., Saglam, H., Erdogan, T.F. and Tamer, A.U. (2010). Antimicrobial and antioxidant activities of mycelia of 10 wild mushroom species. *J. Med. Food.*, 13(2): 415–419. https://Doi.org/10.1089/jmf.2009.0090.

Kamath, V. and Rajini, P.S. (2007). Altered glucose homeostasis and oxidative impairment in pancreas of rats subjected to dimethoate intoxication. *Toxicology*, 231(2-3): 137–146. https://Doi.org/10.1016/j.tox.2006.11.072.

Khatun, S., Islam, A., Cakilcioglu, U. and Chatterjee, N.C. (2012). Research on Mushroom as a potential source of nutraceuticals: A review on Indian perspective. *American Journal of Experimental Agriculture*, 2(1): 47–73.

Kim, J.A., Lau, E., Tay, D. and De Blanco, E.J.C. (2011). Antioxidant and NF-kappa B inhibitory constituents isolated from *Morchella esculenta. Natural Product Research*, 25: 1412–1417.

Koide, R.T., Shumway, D.L. and Stevens, C.M. (2000). Soluble carbohydrates of red pine (*Pinus resinosa*) mycorrhizas and mycorrhizal fungi. *Mycol. Res.*, 104(7): 834–840. https://Doi.org/10.1017/S0953756299002166.

Kumar, S., Zeireis, K., Wiegrebe, W. and Mullar, K. (2000). Medicinal plants from Nepal: Evaluation as inhibitors of leukotriene biosynthesis. *J. Ethnopharmacol.*, 70(3): 191–195.

Kyo, E., Uda, N., Kasuga, S. and Itakura, Y. (2001). Recent advances on the nutritional effects associated with the use of garlic as a supplement immunomodulatory effects of aged garlic. *Extract*, 1: 1075–1079.

Lee, S.R., Roh, H.S., Lee, S., Park, H.B., Jang, T.S., Ko, Y.J., Baek, K.H. and Kim, K.H. (2018). Bioactivity-guided isolation and chemical characterisation of antiproliferative constituents from morel mushroom (*Morchella esculenta*) in human lung adenocarcinoma cells. *J. Funct. Foods*, 40: 249–260. https://Doi.org/10.1016/j.jff.2017.11.012.

Li, S., Gao, A., Dong, S., Chen, Y., Sun, S., Lei, Z. and Zhang, Z. (2017). Purification, antitumor and immunomodulatory activity of polysaccharides from soybean residue fermented with *Morchella esculenta*. *Int. J. Biol. Macromol.* [Internet], 96: 26–34. https://Doi.org/10.1016/j.ijbiomac.2016.12.007.

Li, S.H., Sang, Y.X., Zhu, D., Yang, Y.N., Lei, Z.F. and Zhang, Z.Y. (2013). Optimisation of fermentation conditions for crude polysaccharides by *Morchella esculenta* using soybean curd residue. *Ind. Crop Prod.*, 50: 666–672.

Litchfield, J.H., Vely, V.G. and Overbeck, C.R. (1963). Nutrient content of morel mushroom mycelium: Amino acid composition of the protein. *J. Food Sci.*, 28(6): 741–743. https://Doi.org/10.1111/j.1365-2621.1963.tb01682.x.

Liu, C., Sun, Y., Cui, W. and Xu, N. (2018). Effects of *Morchella esculenta* acidic polysaccharide on nerve growth factor of diabetes mellitus rats. *NeuroQuantology*, 16(6): 816–821. https://Doi.org/10.14704/nq.2018.16.6.1609.

Liu, C., Sun, Y., Mao, Q., Guo, X., Li, P., Liu, Y. and Xu, N. (2016). Characteristics and antitumor activity of *Morchella esculenta* polysaccharide extracted by pulsed electric field. *Int. J. Mol. Sci.*, 17(6). https://Doi.org/10.3390/ijms17060986.

Mahmood, A., Malik, R.N., Shinwari, Z.K. and Mahmood, A. (2011). Ethnobotanical survey of plants from Neelum, Azad Jammu & Kashmir, Pakistan. *Pak. J. Bot.*, 43: 105–110.

Manikandan, K., Sharma, V.P., Kumar, S., Kamal, S. and Shirur, M. (2011). Edaphic conditions of natural sites of *Morchella* and *Phellorinia*. *Mushroom Res.*, 20(2): 117–120.

Manzi, P., Aguzzi, A. and Pizzoferrato, L. (2001). Nutritional value of mushrooms widely consumed in Italy. *Food Chemistry*, 73: 321–325.

Mattila, P., Konko, K., Eurola, M., Pihlawa, J.M., Astola, J., VahteristoLietaniemi, V., Kumpulainen, J., Valtonen, M. and Piironen, V. (2001). Contents of vitamins, mineral elements, and some phenolic compounds in cultivated mushrooms. *J. Agric. Food Chem.*, 49: 2343–2348.

Mau, J.L., Chang, C.N., Huang, S.J. and Chen, C.C. (2004). Antioxidant properties of methanolic extracts from *Grifola frondosa*, *Morchella esculenta* and *Termitomyces albuminosus* mycelia. *Food Chemistry*, 87(1): 111–118. https://Doi.org/10.1016/j.foodchem.2003.10.026.

Meng, F., Liu, X., Jia, L., Song, Z., Deng, P. and Fan, K. (2010a). Optimisation for the production of exopolysaccharides from *Morchella esculenta* SO-02 in submerged culture and its antioxidant activities *in vitro*. *Carbohydr. Polym.*, 79(3): 700–704. https://Doi.org/10.1016/j.carbpol.2009.09.032.

Meng, F., Zhou, B., Lin, R., Jia, L., Liu, X., Deng, P., Fan, k., Wang, G., Wang, L. and Zhang, J. (2010b). Extraction optimisation and *in vivo* antioxidant activities of exopolysaccharide by *Morchella esculenta* SO-01. *Bioresour. Technol.*, 101(12): 4564–4569.

Mohmand, H. and Goldfarb, S. (2011). Renal dysfunction associated with intra-abdominal hypertension and the abdominal compartment syndrome. *J. Am. Soc. Nephrol.*, 22(4): 615–621. https://Doi.org/10.1681/ASN.2010121222.

Mohmand, A.Q.K., Kousar, M.W., Humaira Zafar, Bukhari, K.T. and Khan, M.Z. (2011). Medical importance of fungi with special emphasis on mushrooms. *Isra Medical Journal*, 3(1): 31–37.

Nagarajkumar, M., Jayaraj, J., Muthukrishnan, S., Bhaskaran, R. and Velazhahan, R. (2005). Detoxification of oxalic acid by *Pseudomonas fluorescens* strain PfMDU2: Implications for the biological control of rice sheath blight caused by *Rhizoctonia solani*. *Microbiol. Res.*, 160(3): 291–298. https://Doi.org/10.1016/j.micres.2005.02.002.

Nautiyal, S., Maikhuri, R.K., Rao, K.S. and Saxena, K.G. (2001). Medicinal plant resources in Nanda Devi biosphere reserve in the central himalayas. *J. Herb. Spices Med. Plants*, 8(4): 47–64.

Negi, C.S. (2006). Morels (*Morchella* spp.) in Kumaon Himalaya. *Nat. Prod. Rad.*, 5(4): 306–310.

Nitha, B. and Janardhanan, K.K. (2008). Aqueous-ethanolic extract of morel mushroom mycelium *Morchella esculenta*, protects cisplatin and gentamicin induced nephrotoxicity in mice. *Food Chem. Toxicol.*, 46(9): 3193–3199. https://Doi.org/10.1016/j.fct.2008.07.007.

Nitha, B., De, S., Adhikari, S.K., Devasagayam, T.P.A. and Janardhanan, K.K. (2010). Evaluation of free radical scavenging activity of morel mushroom, *Morchella esculenta* mycelia: A potential source of therapeutically useful antioxidants. *Pharm. Biol.*, 48(4): 453–460. https://Doi.org/10.3109/13880200903170789.

Nitha, B., De, S., Adhikari, S., Devasagayam, T. and Janardhanan, K. (2010). Evaluation of free radical scavenging activity of morel mushroom, *Morchella esculenta* mycelia: A potential source of therapeutically useful antioxidants. *Pharmaceutical Biology*, 48: 453–460.

Nitha, B., De, S., Devasagayam, T.P.A. and Janardhanan, K.K. (2020). Edible mushroom *Morchella esculenta* (L.) Pers. mycelium protects DNA and mitochondria from radiation induced damages. 58(December): 842–847.

Nitha, B., Fijesh, P.V. and Janardhanan, K.K. (2013). Hepatoprotective activity of cultured mycelium of morel mushroom, *Morchella esculenta*. *Exp. Toxicol. Pathol.* [Internet], 65(1-2): 105–112. https://Doi.org/10.1016/j.etp.2011.06.007.

Nitha, B., Fijesh, P. and Janardhanan, K.K. (2013). Hepatoprotective activity of cultured mycelium of Morel mushroom, *Morchella esculenta*. *Exp. Toxicol. Pathol.*, 65(1-2): 105–112.

Nitha, B., Meera, C.R. and Janardhanan, K.K. (2006). *Anti-inflammatory and Antitumour Activities of Cultured Mycelium of Morel Mushroom, Morchella esculenta.* Department of Microbiology, Amala, Can. Res. Centre, Thrissur, 680555.

Nitha, B., Meera, C.R. and Janardhanan, K.K. (2007). Anti-inflammatory and antitumour activities of cultured mycelium of morel mushroom, *Morchella esculenta*. *Curr. Sci.*, 92: 235–239.

O'Donnell, K., Rooney, A.P., Mills, G.L., Kuo, M., Weber, N.S. and Rehner, S.A. (2011). Phylogeny and historical biogeography of true morels (Morchella) reveals an early Cretaceous origin and high continental endemism and provincialism in the Holarctic. *Fungal Genet. Biol.*, 48: 252–265. 10.1016/j.fgb.2010.09.006.

Paliwal, A., Bohra, A., Pillai, U. and Purohit, D.K. (2013). First report of *Morchella*—An edible morel from mount Abu, Rajasthan. *Middle East Journal of Scientific of Research*, 18(3): 327–329.

Perera, P.K. and Li, Y. (2011). Mushrooms as a functional food mediator in preventing and ameliorating diabetes. *Functional Foods in Health and Disease*, 4: 161–171.

Prasad, P., Chauhan, K., Kandari, L.S., Maikhuri, R.K., Purohit, A., Bhatt, R.P. and Rao, K.S. (2002). *Morchella esculenta* (Gucchi): Need for scientific intervention for its cultivation in Central Himalayas. *Curr Sci.*, 82(9): 1098–1100.

Rai, A., Sharma, A. and Parashar, B. (2017). Cannabis Sativa: Boon or Curse. *World Journal of Pharmacy and Pharmaceutical Sciences*, 6(10): 332–338.

Raman, V.K., Sharma, A. and Parashar, B. (2018). Probiotic properties of lactic acid bacteria isolated from animal sources. *International Journal of Development Research*, 8(03): 19660–19665.

Ramesh, C. and Pattar, M.G. (2010). Antimicrobial properties, antioxidant activity and bioactive compounds from six wild edible mushrooms of Western Ghats of Karnataka, India. *Pharmacognosy Res.*, 2(2): 107–112. https://Doi.org/10.4103/0974-8490.62953.

Ramirez-Anguiano, A.C., Santoyo, S., Reglero, G. and Soler-Rivas, C. (2007). Radical scavenging activities, endogenous oxidative enzymes, and total phenols in edible mushrooms are commonly consumed in Europe. *Journal of the Science of Food and Agriculture*, 87: 2272–2278.

Semwal, K. (2014). Edible mushrooms of the Northwestern Himalaya, India: A study of indigenous knowledge, distribution and diversity. *Mycosphere*, 5(3): 440–461. https://Doi.org/10.5943/mycosphere/5/3/7.

Seth, R., Zafar Haider, S. and Mohan, M. (2014). Pharmacology, phytochemistry and traditional uses of *Cordyceps sinensis* (Berk.) Sacc: A recent update for future prospects. *Indian J. Tradit. Knowl.*, 3(3): 551–556.

Shameem, N., Kamili, A.N., Ahmad, M., Masoodi, F.A. and Parray, J.A. (2017). Antimicrobial activity of crude fractions and morel compounds from wild edible mushrooms of north-western Himalaya. *International Journal of Microbial Pathogenesis*, 105: 356–360.

Sharma, A. and Parashar, B. (2017). *Trillium govanianum*: A boon to medicinal World. *Der. Pharma. Chemica*, 9(14): 14–30.

Sharma, A., Sharma, S., Chandel, S., Vatsa, E. and Parashar, B. (2016). A review on *Morchella esculanta*: Therapeutically potent plant. *World Journal of Pharmacy and Pharmaceutical Sciences*, 5(9): 685–699.

Sharma, A., Sharma, S., Naresh, R. and Parashar, B. (2017). *Mesua ferrae* Linn: A review of the Indian medical herb. *Sys. Rev. Pharm.*, 8(1): 19–23.

Sher, H., Elyemeni, M., Hussain, K. and Sher, H. (2011). Ethnobotanical and economic observations of some plant resources from the northern parts of Pakistan. *Ethnobot. Res. Appl.*, 9: 27–41.

Stojkovic, D.S., Davidovic, S., Zivkovic, J., Glamoclija, J., Ciric, A., Stevanovic, M., Ferreira, I. and Sokovic, M. (2013). Comparative evaluation of anti-mutagenic and antimitotic effects of *Morchella esculenta* extracts and protocatechuic acid. *Frontiers in Life Science*, 7: 218–223.

Surcek, M. (1988). *The Illustrated Book of Mushrooms and Fungi*. Octopus Book, London, 311 pp.

Tahoun, M.K. and Ali, H.A. (1986). Specificity and glyceride synthesis by mycelial lipases of Rhizopus delemar. *Enzyme Microb. Technol.*, 8(7): 429–432. https://Doi.org/10.1016/0141-0229(86)90152-3.

Taskin, H. (2013). Detection of volatile aroma compounds of *Morchella* by Headspace Gas Chromatography Mass Spectrometry (HS-GC/MS). *Not. Bot. Horti. Agrobo.*, 41(1): 122–125.

Thakur, M., Paul, A. and Chawla, S. (2014). Qualitative phytochemical screening, total phenolic content and antioxidant activity in methanolic extracts of Myristica fragrans Houtt. (Mace). *Food Sci Res J.*, 5(2): 135–138. https://Doi.org/10.15740/has/fsrj/5.2/135-138.

Thakur, M., Sharma, I. and Tripathi, A. (2021). Ethnomedicinal aspects of morels with special reference to *Morchella esculenta* (Gucchi) in Himachal Pradesh (India): A review. *Current Research in Environmental & Applied Mycology* (*Journal of Fungal Biology*), 11(1): 284–293. https://Doi.org/10.5943/cream/11/1/21.

Tietel, Z. and Masaphy, S. (2018). True morels (*Morchella*)—Nutritional and phytochemical composition, health benefits and flavour: A review. *Crit. Rev. Food Sci. Nutr.*, 58(11): 1888–1901. https://Doi.org/10.1080/10408398.2017.1285269.

Tsai, S.Y., Weng, C.C., Huang, S.J., Chen, C.C. and Mau, J.L. (2006). Non-volatile taste components of *Grifola frondosa, Morchella esculenta* and *Termitomyces albuminosus* mycelia. *LWT – Food Sci. Technol.*, 39(10): 1066–1071. https://Doi.org/10.1016/j.lwt.2005.07.017.

Tzianabos, A.O. (2000). Polysaccharide immunomodulators as therapeutic agents: Structural aspects and biologic function. *Clin. Microbiol. Rev.*, 13(4): 523–533. https://Doi.org/10.1128/CMR.13.4.523-533.2000.

Vivek, K.R., Manish, S., Amit, S. and Bharat, P. (2018). *Morchella esculenta*: A herbal boon to pharmacology. *International Journal of Development Research*, 8(3): 19660–19665.

Wagay, J.A. and Vyas, D. (2011). Phenolic quantification and antioxidant activity of *Morchella esculenta. Int. J. Pharma Bio. Sci.*, 2(1): 188–197.

Wahid, M., Sattar, A. and Khan, S. (1988). Composition of wild and cultivated mushrooms of Pakistan. *Mushroom, J. Trop.*, 8(2): 47–51.

Wasser, S.P. and Weis, A. (1999). Medicinal properties of substances occurring in higher Basidiomycetes mushrooms: Current perspectives (review). *Int. J. Med., Mushrooms*, 1: 31–62.

Yadav, R.P., Saxena, R.K., Gupta, R. and Davidson, S. (1998). Lipase production by Aspergillus and Penicillium species. *Folia Microbiol.* (Praha), 43(4): 373–378. https://Doi.org/10.1007/BF02818576.

Yang, H., Yin, T. and Zhang, S. (2014). Isolation, purification and characterisation of polysaccharides from wide *Morchella esculenta* (L.) Pers. *Int. J. Food Prop.*, 18: 1385–1390.

Yang, H., Yin, T. and Zhang, S. (2015). Isolation, purification and characterisation of polysaccharides from wide *Morchella esculenta* (L.) Pers. *Int. J. Food Prop.*, 18(7): 1385–1390. https://Doi.org/10.1080/10942912.2014.915849.

Ying, J., Mao, X., Ma, Q., Zong, Y. and Wen, H. (1987). *Icons of Medicinal Fungi from China.* Xu, Y. (Trans.)., Science Press: Beijing, 38–45.

Zhang, Q., Wu, C., Fan, G., Li, T. and Sun, Y. (2018a). Improvement of antioxidant activity of *Morchella esculenta* protein hydrolysate by optimized glycosylation reaction. *CYTA – J. Food* [Internet], 16(1): 238–246. https://Doi.org/10.1080/19476337.2017.1389989.

Zhang, Q., Wu, C., Fan, G., Li, T. and Wen, X. (2018b). Characteristics and enhanced antioxidant activity of glycated *Morchella esculenta* protein isolate. *Food Sci Technol.*, 38(1): 126–133. https://Doi.org/10.1590/1678-457x.01917.

CHAPTER 17

Cordyceps
(*Ophiocordyceps sinensis*)

Parthasarathy Seethapathy,[1,]* *Anu Pandita*,[2] *Deepu Pandita*,[3,]*
Subbiah Sankaralingam,[4] *Harinathan Balasundaram*[5]
and *Kousalya Loganathan*[6]

1. Introduction

Fungi are one of the most prevalent types of creatures (Araújo and Hughes, 2016). The fungal kingdom constitutes a six-phylum within the domain of eukaryotes (Lücking et al., 2021). However, it is hypothesised and partially supported that there are between one and four million species of fungi worldwide. Fungi inhabit a variety of ecological niches and are a source of enormous ecological diversity. Moreover, fungi serve numerous environmental tasks, similar to macrofungi, such as mushrooms. These fungi decompose dead organic debris and develop mutually advantageous partnerships, but they can also be pathogenic to plants and animals. Rotting dead matter mushrooms preserve soil nutrients by recycling, but mutualistic or pathogenic relationships may prevent specific organisms from dominating ecosystems to an excessive degree. Numerous cultures around the world have long acknowledged mushrooms as medicinal and nutritive food since 18,700 years ago (Spain), 6,000 years ago (China) and 4,600 years ago (Egypt). Mushroom use has been documented (Li et al., 2021a). Researchers do not even know how ancient humans discovered safe and edible mushroom species. Nevertheless, it was believed that it did, as with natural plants and other life forms gathered or bought for sustenance. Then,

[1] Department of Plant Pathology, Amrita School of Agricultural Sciences, Amrita Vishwa Vidyapeetham, Coimbatore, 642109, Tamil Nadu, India.
[2] Vatsalya Clinic, Krishna Nagar, New Delhi, 110051, India.
[3] Government Department of School Education, Jammu, Jammu and Kashmir 180001, India.
[4] PG and Research Department of Botany, Saraswathi Narayanan College, Madurai - 625022, India.
[5] PG and Research Department of Microbiology, Virudhunagar Hindu Nadars' Senthikumara Nadar College, Virudhunagar - 626001, India.
[6] Department of Botany, Nirmala College for Women, Coimbatore, 641018, Tamil Nadu, India.
* Corresponding authors: spsarathyagri@gmail.com, deepupandita@gmail.com

flavour, appearance and the absence of anaphylaxis indicate which mushrooms are safe to consume (Niazi and Ghafoor, 2021). According to a comprehensive literature assessment and reputable sources providing metabarcoding data, around 1,600,000 of the 6.2 million fungal species classified as Macromycetes produce sporocarps. Approximately 3,283 mushroom species have been classified as edible, representing approximately 20% of all mushroom species recorded in extant data. About 700 are considered functional medicinal mushrooms, with pharmacological qualities of 2,000 known, safe species and 10 of them have reached the level of commercial production many countries (Baldrian et al., 2021).

The mushroom industry encompasses edible, medicinal, and wild varieties. In the past 60 years, the global production of edible mushrooms has multiplied by 30. According to the FAO's Statistical Database, China produces the most mushrooms, followed by the United States and the Netherlands. It is anticipated that mushroom cultivation will continue its upward trend. In the past four decades, the field of medicinal mushrooms has achieved significant progress. It has been decided that a new area of science that is successful has emerged (Niazi and Ghafoor, 2021). Many mushrooms create secondary metabolites with antifungal, antibacterial and antiviral effects and nutritional and culinary characteristics during their life cycle. In Asia, its use for health promotion and disease treatments has a long history, whereas it is relatively recent in the rest of the world. Frequently, insects and macrofungi form mutualistic, parasitic and commensal partnerships (Araújo and Hughes, 2016). Despite their significance, these interactions are among the least-studied areas of fungal biodiversity. *Ophiocordyceps* sp., is the most beneficial entomopathogenic fungus due to its therapeutic capabilities and the price of its fruiting bodies. Recent technical advancements have considerably contributed to profiling the varying clinical importance of this mushroom and other closely related mushrooms with comparable therapeutic potential.

Ophiocordyceps sinensis, an Ascomycete caterpillar-parasitizing fungus, is one of several medicinal mushrooms that have been used in China, India, Tibet, Nepal and Bhutan for over two millennia to prolong vigour and vitality (Jiraungkoorskul and Jiraungkoorskul, 2016). It possesses three essential characteristics for disease management and cure in the present conventional medical system. More than 3,600 articles, monographs and online databases about *O. sinensis*, usually called DongChongXiaCao (DCXC) in Chinese, were examined in a study to find its cytotoxic properties. It has no cytotoxicity or adverse reactions in mammals; secondly, it is not limited to a single organ; and thirdly, its use has enhanced benefits in regulating organ functions (Smiderle et al., 2014). It is the most frequently grown medicinal mushroom globally, particularly in China, Europe, Ghana, India, Japan, Kenya, Korea, South East Asia, Taiwan, Tanzania and the United States. It is revered as the most potent natural treatment in Asia and many African countries due to its distinctive pharmacological properties and perceived absence of adverse effects. *O. sinensis* is a precious and exotic medicinal mushroom with a rich heritage in China. The traditional Chinese and Tibetan medical systems recognise *O. sinensis* as a potent restorative and aphrodisiac. In addition, *O. sinensis* has been associated with diarrhoea, headaches, colds, rheumatism, asthma, allergic rhinitis, irregular menstrual cycles, immunomodulating effects, cholesterol reduction, and enhancement of stamina and libido. Indeed, *O. sinensis* and its compounds have demonstrated positive effects on cardiovascular, obesity, type 2 diabetes, tiredness, neurological, renal, gastrointestinal, hepatic, respiratory, and immunological disorders as well as erectile problems (Xu et al., 2016; Belwal et al., 2018; Ashraf et al., 2020). This mushroom is potent in therapeutic activities including adaptogenic, aphrodisiac, antiaging, anticancer, antidiabetic, antifungal, anti-inflammatory, antimicrobial, antinociceptive, anti-proliferative, antiosteoporotic, antioxidant, antitumour agent, antiviral, hepatoprotective, hypoglycemic, hypolipidemic, nootropic and immunomodulatory effects. It improves physical performance, reminiscence and learning capabilities (Lo et al., 2013; Liu et al., 2015; Das et al., 2021). This report will analyse the commercial aspects of its nutritional and healing properties.

2. History and Significance

Caterpillar fungus (*O. sinensis*), usually referred to as the Chinese caterpillar mushroom or Yartsa gunbu, DongChongXiaCao, bu, or kira ghass, is a parasitic mushroom indigenous to the Sino-Tibetan Plateau and the Indo-Nepalese Himalayas. *Ophiocordyceps* has more arthropods and other fungal parasites than any other kingdom of fungi, with over 500 species. Numerous *Ophiocordyceps* species are insect pest diseases and are exciting options for biological management. These pathogenic fungi create several physiologically active chemicals that contribute to their pathogenicity. This fungus can only infect a specific insect species or a small group of insects closely related to one another. With the medical characteristics of this mushroom in mind, numerous groups have undertaken research on its various features. To date, physiological investigations, an enzyme profile and an antioxidant assay have been conducted, and attempts are being made to investigate biomass and fruiting body-development potential. In addition, recent research has shown that fungus can be used to cure a vast array of diseases (Zhou et al., 2014; Ashraf et al., 2020; Rathi et al., 2021). It has been used as a medicinal mushroom in China since the Qing dynasty (Bhetwal et al., 2021). It has been recognised as an oral medicine in China and Tibet regions since 1964. In recent years, the price of *O. sinensis* natural goods has skyrocketed and is now offered at or above the price of gold. This fungus, sometimes called 'Himalayan gold', grows in large patches in several high-altitude areas of the Nepalese Himalayas. Every year, many *Ophiocordyceps* are taken from these areas and illegally exported. However, this mushroom biodiversity is threatened due to its stringent host range for lepidopteran caterpillars, limited global dispersal and the decades-long overharvesting performed by people. It is listed in the *Red Book* as a threatened species due to its exceptional medicinal properties and higher economic returns soon after harvesting, which have led to its massive overexploitation (Smith-Hall and Bennike, 2022).

3. Taxonomy

Kingdom: Fungi
Phylum: Ascomycota
Class: Sordariomycetes
Order: Hypocreales
Family: Ophiocordycipitaceae

According to MycoBank (https://www.mycobank.org; May 12, 2022), *Ophiocordyceps* is a vast and diversified genus within the Ascomycetous family Ophiocordycipitaceae with 336 identified species. They are parasitic fungi, with the majority being endoparasitoids of insects and other arthropods, and the remainder parasitic on other fungi. The phylogenetic analysis of Cordyceps revealed that genetically it was grouped into four different genera: *Cordyceps*, *Elaphocordyceps*, *Metacordyceps* and *Ophiocordyceps*. It has been demonstrated that *O. sinesis* belongs to the related-group, depending on the proposed concept of *Ophiocordyceps* Petch; at present, the preferred name is *Ophiocordyceps siensis* (Berk.) G.H. Sung, J.M. Sung, Hywel Jones and Spatafora [Obligate synonym: *Cordyceps sinensis* (Berk.) Sacc. or Basionym: *Sphaeria sinensis* Berk.] (Sung et al., 2007). The term *Cordyceps sinensis* is derived from Latin words. A *cord* refers to 'club', *ceps* to 'head', and *sinensis* to 'manufactured in China'. The origin of the Chinese term, Dong Chong Xia Cao (DCXC), is the growth process of *O. sinensis*, which decodes into 'winter worm, summer plant, or summer grass, winter insect'. The fungus infects Hepialidae worms in springtime and midsummer, parasitising and transforming them into 'stiff caterpillars' in midwinter, hence the name 'Dong Chong', which refers to the 'winter worm' (Fig. 1). The stroma germinates in the following summer months and then sprouts over the head capsule of Xia Cao larvae (Belwal et al., 2018). In recent decades, the controlled cultivation of the fungal species derived from *Ophiocordyceps* through the fermentation process has become the subject of intensive research. China has found more than 30 different kinds of fungi and 13 of them are linked to the asexual forms of *Ophiocordyceps*. Such varieties have been recorded as *Cephdosporium sinensis*, *Chrysosporium sinensis*, *Hirsutella hepiali*, *H. sinensis*, *Mortierella hepiali*, *Paecilomyces hepiali*, *P. sinensis*, *Scytalidium hepiali*, *Stachybotrys elegans*, and *Tolypocladium sinensis* were discovered to be non-anamorphic strains among them (Zhou et al., 2014; Rathi et al., 2021). A disagreement has arisen over the uniformity of classification of anamorphs in the *Ophiocordyceps* due to the plethora of strains and diverse nomenclature. Currently, there is evidence suggesting that *Hirsutella sinensis* is an asexual life form of the *Ophiocordyceps* (Cannon et al., 2009; Jiraungkoorskul and Jiraungkoorskul, 2016). Nevertheless, the link between the anamorph and the

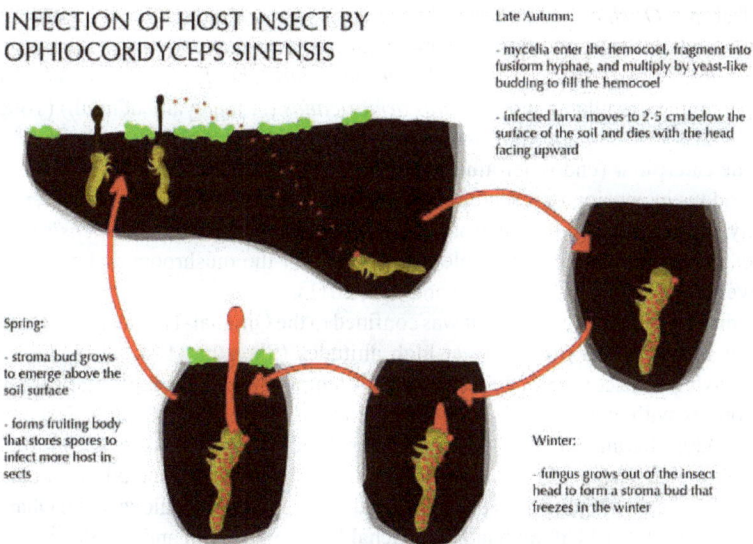

INFECTION OF HOST INSECT BY OPHIOCORDYCEPS SINENSIS

Late Autumn:

- mycelia enter the hemocoel, fragment into fusiform hyphae, and multiply by yeast-like budding to fill the hemocoel

- infected larva moves to 2-5 cm below the surface of the soil and dies with the head facing upward

Spring:

- stroma bud grows to emerge above the soil surface

- forms fruiting body that stores spores to infect more host insects

Winter:

- fungus grows out of the insect head to form a stroma bud that freezes in the winter

Fig. 1. https://microbewiki.kenyon.edu/index.php/Ophiocordyceps_sinensis.

Ophiocordyceps requires further investigation. According to previous accounts, over 1200 entomopathogenic fungus species are known, with *Ophiocordyceps* being a larger genus with roughly 500 species. Several species of *Ophiocordyceps* and related genera of *Cordyceps*, including *Cordyceps liangshanesis*, *Cordyceps militaris*, *Cordyceps ophioglossoides*, *Ophiocordyceps cicadicola*, *Ophiocordyceps sinensis* and *Ophiocordyceps sobolifera*, have been cultivated for their therapeutic characteristics (Tuli et al., 2017; Deshmukh et al., 2020).

4. Diversity

O. sinensis is an entomopathogenic fungus whose life cycle is multifaceted. The majority of these fungi grow from mummified larvae covered in mycelia with a prominent primordium fruit body or stroma with a length of > 1 cm that erupt ascus and ascospores from the corpse of their arthropod host (Fig. 1). Generally, the stromata generate a stipe that assists in elevating the fruitful or spore-producing part above the insect host, which is typically buried in the soil (Fig. 1). Hypocreales are typically found in regions with a hot and humid climate, except for *O. sinensis*, which is discovered at a high altitude of 3,000–5,000 meters MSL, making it extremely expensive on the international market of USD 60,000–75,000 per kg at earlier periods (Zhou et al., 2014; Qin et al., 2018) and national and continental market with a price range of USD 12,000–40,000 per kg (Negi et al., 2020). Lepidopteran larvae are the favoured hosts for *Ophiocordyceps* infection, but Coleoptera, Orthoptera, Hemiptera and Hymenoptera insects are also susceptible to infection (Kumar et al., 2016; Araújo and Hughes, 2016; Wang and Yao, 2011). The most common species infect immature Coleoptera (*Ophiocordyceps stylophora*), Diptera (*Ophiocordyceps dipterigena*), Hemiptera (*Ophiocordyceps nutans*), Hymenoptera (*Ophiocordyceps sphecocephala*) and Lepidoptera (*Cordyceps militaris*) stages (www.cordyceps.us). Another related species, *C. militaris* (L.) Fr. is a predominant endoparasitoid on pupae and frequently on larvae and is extensively dispersed throughout America, Asia, and Europe. *O. sinensis* is also multifaceted parasitism consisting of a anamorphic fungus and underground-dwelling moth larvae (Lepidoptera), particularly species of *Thitarodes xiaojinensis*, *Thitarodes armoricanus*, and *Thitarodes jianchuanensis* (ghost moth caterpillars) and *Hepialus armoricanus* (Himalayan bat moth) (Tong and Guo, 2022). It is composed of two different parts – the fungal fruiting body (stroma) and the caterpillar (endosclerotium) (Fig. 1). In China, during the 1980s, there was a tremendous expansion of research on the richness and classification of the Hepialidae family. Of the 91 listed insects from 13 different genera linked to *O. sinensis* as a host insect, 57 are known to be possible host species for the mushroom and can be found all over the Tibetan Plateau (Wang and Yao, 2011).

Initial reports indicated that it was confined to the Qinghai-Tibetan Plateau's Golog Tibetan Autonomous Prefecture at high altitudes (Shrestha et al., 2010). Recently, *O. sinensis* was discovered in India in the chilly temperatures of the high highlands, along the borders with Nepal and China, at elevations of between 3,300 and 4,600 m (Negi et al., 2020; Sharma and Negi, 2022). An unpublished work by Arora (2008) reported the natural occurrence of *O. sinensis* in India in a few locations, including Brahamkot, Chhipalakedar, Chhipalakot, Ghwardhap and Najari Ultapara regions in the Dharchula region of the state of Uttarakhand, Arunachal Pradesh, Sikkim and locations along the Indo-Tibetan and Indo-Nepal borders, including Burfu, Darti, Milam, Mapa, Lashpa,

Ralam and Tola in the Johar Valley, and Chhipalakedar, Golfa Bona, and Nagindhura in the basement of Panchachuli in Munsyari, has also reported its occurrence in recent years. In the interior mountain regions, it is also known by the local names 'Yarsha Gamboo', 'Keera jhar', and 'Keera ghas'. The term 'Yarsa Gamboo' translates to 'fungus cum larvae'. In several works of literature, 'Gonba' or 'Gumba' or 'Gunba' have been substituted for 'Gamboo' (Shrestha et al., 2010; Shrestha and Bawa, 2015). This highly valued mushroom is trafficked from the border town of Dharchula, through Nepal and then to markets outside of India. There is a significant diversity in caterpillar fungus production over the Tibetan Plateau and many other Himalayan locations. The caterpillar fungus's estimated global annual output ranged between 84.2–182.5 tonnes (Belwal et al., 2018).

5. Ecology

The mouth cavity, the insect cuticles, or spiracles are the entry points that the caterpillar fungus *O. sinensis* uses to infect its insect larvae (Fig. 1). Enzymatically, the entrance and proliferation of hyphae serve to breach the insect's immune system by the active secretion of a wide variety of biomolecules. These biomolecules include adenosine, cordycepin, cordycepic acid, dideoxyadenosine, hydroxyethyl adenosine, and others (Baral, 2017). Insect-lethal extracellular enzymes, such as 2-carboxymethyl-4-(30-hydroxybutyl) furan, cepharosporolides, cordycepin, cyclic peptides, dipicolinic acid, ophiocordin, pyridine-2,6-dicarboxylic acid and others, have been reported to cause larvae to be digested by chemo-heterotrophic digestion (Paterson, 2008). After infecting the lymphatic fluid of an arthropod, the infection manifests as a mycelium body and ascocarp before developing into a sclerotium in the shape of a caterpillar. Eventually, fungal spores were discharged from the whole *O. sinensis* fruiting body (Fig. 1). These ascospores disseminate in the soil, penetrate deeper into the ground and initiate a new infection cycle. When the stroma of the mushroom grows above the ground and separates from the sclerotium, it is also picked for medicinal purposes. The host insect's larvae feed on plant roots and caudexes for three to four years or beyond, during their whole underground larval period. The situational changes and richness of fungi associated with *O. sinensis* in the soil regulate the fungus-larvae connection, which influences the occurrence and topographical dispersal of *O. sinensis*. Those affected by the fungus typically perish during winters. The stroma emerges in the following year's spring or summer. Excitingly, the cuticularised insect epidermis protects the host from excessive desiccation caused by solar UV radiation and other possible natural predators. *Ophiocordyceps* has its indigenous clock; interfering with the host and resetting it to match its own could increase its virulence. The manipulation of the circadian clock is proven by the fact that infected moths undoubtedly move from nocturnal to diurnal feeding and by the synchronisation of the modified biting behaviour to a particular time of day. It has evolved to high extreme latitude habitat, including low oxygen, freezing temperature, less pressure and direct ultraviolet (UV) rays. iTRAQ-based comparative proteomic studies of demonstrated ecological variables would lead to the production of reactive oxygen species (ROS) at distinct growth stages, demonstrating the regulatory role of ROS (Tong et al., 2021). In comparison to other closely related, low-latitude fungi, the peroxidase gene repertoire of *O. sinensis* is significantly increased. In addition, the shift in ROS distribution caused by the SOD-1 mutation destroyed light-induced perithecial polarization, indicating that ROS catabolism is implicated in physical arousal (Tong and Guo, 2022). According to one theory, reactive oxygen species may mediate ecological mechanisms for the development of the fruiting body.

6. Nutritional Properties and Pharmacological Values

Entomopathogenic macrofungi are usually called 'functional meals' since they offer several health benefits and contain specific nutrients. Mushrooms are nutrient-dense foods and an excellent source of nutrients due to their vast array of critical components, including proteins, polysaccharides, phenols, sterols, enzymes, minerals, vitamins, dietary fibre, low calories, fats, flavonoids, sterols and toxic metals (Fig. 2). Nutraceutical fungi are medicinal mushrooms used for their therapeutic properties, often in the form of infusions, concentrations, or particles. *O. sinensis* is utilised in Chinese and Tibetan medicine because of its bioactive properties. Extreme environmental stress endows these remarkable herbs with unique bioactive compounds of tremendous medical value, which have been exploited commercially in Oriental medicine worldwide for millennia. However, this fungus' bioactive peptides have not yet been characterised. Due to the recent increase in interest in *Ophiocordyceps* for pharmacological, nutritional and gastronomic biotechnological processes, it is required to do additional research to establish the potential of this medicinal fungus (Yue et al., 2022).

O. sinensis offers a broad spectrum of biological activity relevant to the development of pharmaceuticals. Recent work focuses on its spermatogenesis and medicinal applications (Xie et al., 2021). In addition, *O. sinensis*-related products are renowned nutraceuticals consumed as edible mushrooms or sold in various forms, including extracts, fermented powder and tinctures (Kontogiannatos et al., 2021). These products are renowned for their anti-bacterial, anti-fungal, anti-inflammatory, antimalarial, antioxidant, antitumour, antiviral, hypoglycemic, hypolipidemic, and neuroprotective properties by deterring chronic kidney inflammation, lupus erythematosus and asthma, nourishing the lungs and kidneys, reducing bleeding, limiting phlegm, pro-sexual by enhancing the contractile ability of the atrium and ventricle, stabilising the blood-sugar metabolism and immuno-protective properties (Xu et al., 2016; Rathi et al., 2021; Du et al., 2021) (Table 1).

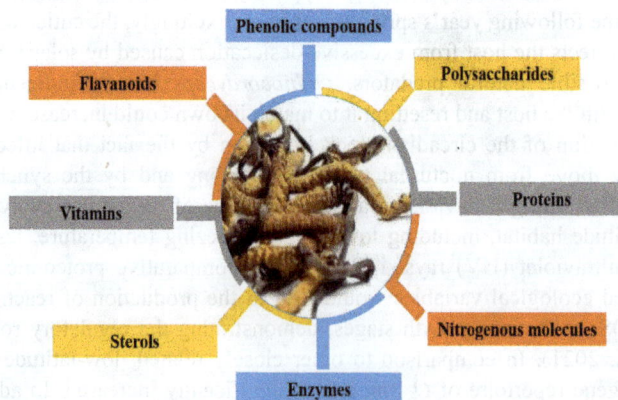

Fig. 2. Nutritional properties of *O. sinensis.*

Table 1. Therapeutic effects of *O. sinensis*.

S. No.	Therapeutic Effects	Mechanisms	References
1.	Anti-microbial	Activated and enhanced the IL-12 and IFN-α expression and acrophage phagocytic activities.	Joshi et al., 2019
2.	Anti-inflammatory	Suppressive effect on neutrophil SO anion production and elastase production; inhibited the activity of the p65, Akt, and MAPK cascades signalling paths and promoting the activation of PP2A in response to ceramide; inhibited tumour necrosis factor-α (TNF-α), IL-1, IL-10, and iNOS expression.	Yang et al., 2021
3.	Anti-cancer	Upregulated interferon-γ, tumor necrosis factor-α, and U937 leukemic cells were induced to develop macrophages, and to express superficial antigens; enhancing E-cadherin up-regulation; accumulated microtubule-associated protein 1A/1B light chain 3 (MAP1LC3) in HCT116 colon cancer cells by blocking the PI3K-Akt-mTOR signalling path and stimulating the AMPK–mTORC1–ULK1 (uncoordinated-51 like kinase 1) signaling network.	Chen et al., 1997
4.	Anti-diabetics	Promoted renal potassium-ATPase (NKA) expression and lowers collagen type-1 deposition and glomerular mesangial matrix buildup while regulating β-cell survival.	Kan et al., 2012
5.	Anti-tumour	Activated caspases-3, 9 and 8, Bid cleavage, 3-NC downregulation of anti-apoptotic Inhibitor of Apoptosis Proteins (IAPs), Bcl-2 expression level, and overexpression of Bax protein with reduced telomerase activity by decreasing hTERT transcriptional activity; anti-metastatic efficacy by inhibiting HGF-accelerated tumour invasion and metastasis.	Wang et al., 2012; Nakamura et al., 2015
6.	Cardio-protective	Reduced glutathione disulfide/glutathione disulfide + glutathione and malondialdehyde/superoxide dismutase + malondialdehyde.	Phull et al., 2022
7.	Immuno-modulator	Suppressed the production of pro-inflammatory molecules production in activated BALF cells; increased cytokine synthesis and TNF-α as well as increase in CD11b expression; stimulate macrophages by elevating of IL-12, iNOS, and TNF-α levels; encouraged the synthesis of nitric oxide in RAW 264.7 macrophages by activating the MAPK and (PI3K) pathway/PKB/Akt signalling pathways; increased the production of nitric oxide, IL-6, IL-113 and TNF-α production by stimulating the p38, Janus kinase 2, and NF-B signalling paths.	Kuo et al., 2001; Kuo et al., 2007; Bi et al., 2020; Liu et al., 2015; Fan et al., 2021; Rong et al., 2021

Table 1 contd. ...

...Table 1 contd.

S. No.	Therapeutic Effects	Mechanisms	References
8.	Liver-protective	Controlled total bilirubin, serum creatinine, serum urea and total cholesterol level; increased the expression of B-cell lymphoma-2, mitofusin-2, PGC1α transcriptional cofactor and downregulated the expression of B-cell lymphoma-2-associated X, cleaved caspase-3 and -9, and reactive oxygen species; in cyclophosphamide-induced liver injury, inhibited the activation of TLR9 and the expression of cyclooxygenase-2, IL-1β, iNOS and TNF-α, while enhancing the production of glutathione peroxidases (GPx).	Fung et al., 2017; Fan et al., 2018
9.	Nephro-protective	Inhibited TGF-β1/Smad pathway by the expression of alpha-smooth muscle actin and ferroptosis suppressor protein 1; inflammation in the glomeruli of the kidney in mesangial cells regulated with extracellular signal-related kinases 1/2/PKB pathways by altered expression of platelet-derived growth factor homodimer BB-induced intercellular ROS in mesangial cells by decreasing tumour necrosis factor-α, (TNF-α), TNF-α receptor 1, and monocyte chemoattractant protein-1 (MCP-1/CCL2).	Wang et al., 2014; Wang et al., 2015a, b
10.	Osteo-protective	Inhibited RANKL-induced osteoclast differentiation by NF-kβ	Ashraf et al., 2020; Phull et al., 2022
11.	Pulmonary-protective	Reduced formation of ROS and rebalanced the MMP-9/TIMP-1; inhibited the TGF-β-1; activation of effectors like white blood cells and epithelial cells, which control activation of the A2AR and A2BR that induce pro-inflammation and/or tissue restoration.	Nakamura et al., 2015; Zhou et al., 2009
12.	Spermatogenesis	Elevated plasma testosterone levels through PKA and PKC signalling pathways.	Xu et al., 2016
13.	Prebiotic effect	Activation of the extracellular signal-regulated kinase 1 signalling network and inhibition of the JNKs and p38 signalling; increasing the number of immunoglobulin A-secreting cells, secretory IgA cells, and goblet cells in the gut and boosting the expression of IFN-γ, IL-2, IL-12 p40, TNF-α, T box transcriptional factor and GATA-binding transcription factor 3 in intestine; structure of the gut microbiome is modulated, with an increase in *Bacteroidetes* and a decrease in *Firmicutes* and *Verrucomicrobia*; induced the release of cytokines and the transcription factors associated to the activation of TLRs and the NF-κB signalling pathway.	Ying et al., 2019; Chen et al., 2019; Chen et al., 2021

Researchers have found a lot of bioactive compounds that could be used as nutraceuticals. In most cases, the entire fungus and its secondary metabolites are devoured. This species complex is said to possess superior nutritional qualities due to its sweet and mildly astringent flavour and its fatty nature. *Ophiocordyceps* could be considered one of the most important mushrooms, as it is rich in numerous compounds with potential nutraceutical benefits (Ashraf et al., 2020). The approximate proportions of total moisture, sugar, amino acid, fat, fibre and ash contents of various species is 7.18%, 7.48%, 21.46%, 1.80%, 6.40% and 55.68%, respectively (Rakhee et al., 2016). The fruiting bodies contain 59.8% protein, 5.7% moisture, 5.1% ash, 8.8% fat and 29.1% carbohydrates, whereas mycelial biomass consists of 39.5% protein, 13.1% moisture, 5.7% ash, 2.2% fat and 39.6% carbohydrates (Ashraf et al., 2020). Mycelium and fruiting bodies, on the other hand, contain 14–69 mg/g of amino acids, respectively. The content of carbohydrates extracted by pressurised liquid extraction (PLE) including both instinctual and cultivated *O. sinensis* was compared and the concentration levels of 10 sugars, such as arabinose, fructose, glucose, galactose, mannitol, mannose, rhamnose, ribose and sorbitol was determined by gas-chromatography-mass spectrophotometer (Guan et al., 2010). Mannitol (also known as cordycepic acid) is a polyol that has been extensively exploited for the treatment of a wide variety of disorders. The variance in mannitol concentrations among three biological samples examined, ranging between 9.85–11.51%, can be attributed to different growth conditions (Wang et al., 2015b).

Numerous enzymes have been linked to the pathogenesis of fungi, but little is known regarding their characteristics, modes of action, regulation, cellular localisation and production levels. *O. sinensis* produces and releases (internally and externally) a wide variety of secondary chemicals with the ability to break the larval cuticle and devour the larvae with excellent efficiency. These secondary metabolites, secreted in both solid surface media and axenic growth environments, confer greater therapeutic properties on the fungus. Extracellular polysaccharides (exopolysaccharides) are the most abundant compounds in fungal cultures. The fungus is known for its therapeutic applications as the 'Himalayan Viagra' due to its extraordinary capacity to boost men's sexual desire. In addition to these uses, it has also been looked at to see if it can be used to make medicines with a lot of therapeutic value (Lo et al., 2013; Cannon et al., 2009).

6.1 Polysaccharides

O. sinensis produces polysaccharides, superoxide disproportionation enzymes and sugar derivatives with various pharmaceutical properties, such as antioxidant, anti-inflammatory, immunomodulatory and probiotic activities (Yan et al., 2014). Numerous journals have published comparative studies of polysaccharides from fruiting bodies and mycelial material using liquid extraction, chromatography and purification (Cheong et al., 2016). More than 17 polysaccharides and sugar derivatives have been structurally characterised (β-glucans, β-mannans, Cordysinocan, cyclofurans, CS-F30, CS-F10, CT-4N, CS-81002, CS-Pp, CS-PS, CME-1, CSP-1, CPS1, CPS2, EPS, PS-A, SCP-I, mannoglucan and D-mannose). In addition, antitumour adenosine derivatives, ophiocordin, and L-tryptophan were identified (Cheong et al., 2016; Bi et al., 2020; Qi et al., 2020; Yuan et al., 2022). Researchers have shown that the mycelial carbohydrate CME-1 protects RAW264.7 cells against peroxide pressure.

This protection is achieved by inhibiting the sphingomyelinases and lowering ceramide amounts (Yang et al., 2011), consequently demonstrating antioxidant and anti-sclerotic activity. Shen et al. (2011) found that APS and CSP-1 protect PC12 pheochromocytoma cells of the rat adrenal medulla from hydrogen peroxide-oxidative losses by enhancing the functions of catalase, GPx, and SOD and reducing the lactate dehydrogenase, intracellular calcium-ions, malondialdehyde and ROS contents. In short, oxidative stress damage is linked with the growth and development of the large majority of disorders and the antioxidant capabilities of their polysaccharides are undeniably advantageous for the treatment of diseases (Yuan et al., 2022). The polysaccharides produced by *O. sinensis* can influence the immunological action of macrophages by reducing the manifestation of pro-inflammatory proteins and the inflammatory reaction by activating many pathways of NF-κB, and boosting the regulation of TNF-α, IL-12, inducible NO synthase and MAPK signalling pathways. Furthermore, numerous studies have shown that the immunostimulatory effect is linked to suppressing the STAT3 signalling pathway (Yuan et al., 2022).

6.2 Nitrogenous Molecules

One of the most important active elements in the caterpiller mushroom is nitrogenous compounds (Das et al., 2021; Rathi et al., 2021; Phull et al., 2022), while adenosine and cordycepin (3'-deoxyadenosine) are employed as quality inspection markers (Nakamura et al., 2015). Interestingly, the amount of cordycepin is 0.97 and 0.36% and adenosine is 0.18 and 0.06% in the mycelium and fruit bodies, respectively (Tuli et al., 2017). Numerous nucleosides and nucleobases have been reported as bioactive pharmacological ingredients with various medical applications, such as immune modulators and antioxidants. Adenosine and cordycepin are essential components that demonstrate the authenticity of *O. sinensis*. The genes encoding glucose–methanol–choline oxidoreductase and telomerase reverse transcriptase are linked to the cordycepin production in *C. militaris* (Zheng et al., 2015). Furthermore, it has been reported that cordycepin prevents the development of lung carcinoma cells in rats and colon carcinoma cells in humans, by stimulating ADORA3 on cancer cells (Yoshikawa et al., 2011). In addition to adenosine and cordycepin, a few more nucleosides and nucleotide bases were commonly found in the aqueous extracts of the stroma and endosclerotium of *O. sinensis*. Among these were 2-nicotinic acid, 20-dideoxyadenosine, 3,6-di(4-hydroxy)benzyl-2,5-dioxopiperazine, adenine, adenosine, caffeine, cordycepin, cytosine, cytidine, deoxyuridine, guanine, guanosine, hypoxanthine, inosine, N-(2'-hydroxy-tetracosanoyl)-2-amino-1,3,4-trihydroxy-octadec-8eene(tetracosanamide), N6-(2-hydroxyethyl) adenosine 3-isopropyl-6-isobutyl-2,5-Dioxopiperazine, thymine, thymidine, uracil and uridine (Lo et al., 2013; Belwal et al., 2018; Jiraungkoorskul and Jiraungkoorskul, 2016; Yuan et al., 2022b). According to reports, the nucleoside content of wild and cultivated *O. sinensis* may differ. Numerous additional nucleosides have been discovered, such as cordysinins A, B, C, D and E, as well as unique structures of 2'-3'-dideoxyadenosine, 6-hydroxyethyladenosine, cordycepin triphosphate, deoxyuridines, and deoxyguanidine. Adenosine and 30-deoxyadenosine are widely exploited therapeutic components with numerous pharmacological effects, including immunomodulatory, antioxidant, etc. Reported nucleotides decrease urethral inflammation, enhance blood circulation and cerebral function and their most significant effect is boosting human immunity (Lo et al., 2013; Belwal et al., 2018; Yang et al., 2011).

6.3 Proteins

In addition to the aforementioned primary components, *O. sinensis* contains proteins, peptides, glycoproteins, polyamines, enzymes, isozymes, essential amino acids and a few unique cyclic dipeptides (Das et al., 2021). In another study, from a liquid culture of *Ophiocordyceps*, a novel cyclodipeptide named cordycedipeptide A and cordyceamides A and B dipeptides were recovered and structures were resolved using 1D and 2D NMR technologies. This demonstrates cytotoxic activity against human linages A375, L929, and Hela cells (Liu et al., 2015; Rakhee et al., 2016). The mycelia powders of *O. sinensis* grown in the laboratory are harmless and safe, as no significant changes in body biomass and antiserum properties were seen in the mouse. In contrast, treated groups demonstrated an upsurge in nourishment, organ mass, organ growth and hematological properties including haemoglobin and lymphocytes. The nontoxic impact of *O. sinensis* was further verified by the histopathology of important organs (Meena et al., 2013). Cordyceps has been used effectively as an antitumour herb in Chinese medicine due to its apoptosis activation and anti-metastasis activities. It has also been found to have anticancer effects on B16–BL6 cells, CW-2 cells, LLC cells and HT1080 cells by enhancing A3-R and which is also followed by activation of a serine/threonine protein kinase called Glycogen Synthase Kinase-3b (GSK-3beta), and cyclin D1 suppression by using G-protein-coupled receptor-radioligand binding assay (Shin et al., 2021; Wang et al., 2017; Yoshikawa et al., 2011). The fruiting bodies are rich in several amino acids, including 1-(5-hydroxymethyl-2-furyl)-βcarboline (perlolyrine), 1-acetyl-α-carbolinearginine, 1-methylpyrimidine-2,4-dione, cadaverine, cordymin, cyclo(L-Pro3-L-Val), cyclo(L-Phe-L-Pro), cyclo(L-Pro-L-Tyr), cyclo-(Thr-Leu) 1,3-diamino propane, flazin, glutamic acid, histidine, L-tryptophan, L-tryptophan, L-arginine, lysine, oxyvaline, perlolyrine, phenylalanine, proline, putrescine, spermidine, spermine, threonine and valine (Yang et al., 2011; Li et al., 2020). Moreover, fatty acid analysis reveals that approximately 70% of the fat content comprises unsaturated fatty acids. A comparison of natural and cultured amino acids showed that glutamic acid is the most prevalent amino acid in *Ophiocordyceps*, preceded by arginine in natural collections and aspartic acid and leucine in cultivated samples (Wang et al., 2015a).

6.4 Other Compounds

Ophiocordyceps contains an abundance of bioactive components, including carotenoids, flavonoids and minerals (Al, Ca, Cr, Cu, Fe, Ga, K, Mg, Mn, Na, Ni, Pi, Se, Si, Sr, Ti, V, Z, Zn and Zr), phenolic compounds, volatile oils, vitamins (B_1, B_2, B_{12}, E and K), and diverse forms of sterols, such as (17R)-17-methylincisterol, â-sitosterol, â-sitosterol 3-O-acetate;daucosterol, 22,23-dihydroergosteryl-3-O-α-D-glucopyranoside, cholesterol, campesterol, dihydrobrassicasterol [D5-ergosterol], fungisterol [D7-ergosterol], stigmasterol, stigmasterol 3- etc. (Elkhateeb and Daba, 2020). β-sitosterol, campesterol and cholesterol, including free and esterified ergosterol in wild *Ophiocordyceps* were detected using GC-MS analysis (Yang et al., 2009). These mycosterols play an important role in treating breast cancer, colorectal cancer, pancreatic cancer, prostatic carcinoma and their bioactivities in the determination of treatment strategies of *O. sinensis*. Polyamine compounds like 1,3-diamino propane, cadaverine, putrescine, spermidine and spermine are also reported as therapeutic molecules from *O. sinensis* (Bhetwal et al., 2021). Free fatty acids, like docosanoic acid, lauric acid, lignoceric acid, linoleic acid, myristic acid, oleic acid, palmitic acid, palmitoleic acid, pentadecanoic acid and stearic acid have also been identified (Yang et al., 2009). They also reported four isoflavones, namely daidzein orobol, 32,42,7-trihydroxyisoflavone, genistein and glycitein from the caterpillar mushroom. Numerous scientists have identified many phenolics and acids, including 2-deoxy-D-ribono-1,4-lactone, 3-hydroxy-2-methyl-4-pyrone, 3,4-dihydroxyacetophenone, 4-hydroxyacetophenone, acetovanillone, furancarboxylic acid, methyl p-hydroxyphenylacetate, p-hydroxybenzoic acid, syringicaci, p-methoxybenzoic acid, p-hydroxyphenylacetic acid, p-methoxyphenol, protocatechuic acid, salicylic acid and vanillic acid (Shrestha et al., 2012). The structures of some chemical constituents of *C. sinensis* are shown in Table 2.

The medical and economic usefulness of *O. sinensis* has been widely recognised and investigated around the globe. According to the *Materia Medica* of the People's Republic of China, *O. sinensis* is also helpful in treating impotence, spermatorrhea and knee and waist pain. In addition, modern medicine has revealed that it exhibits various qualities, including antitumour, anti-hyperglycemia (Paterson, 2008; Joshi et al., 2019) and a significant increase in the immunity of the human body (Cheong et al., 2016). Recent studies have revealed many biological metabolites in *O. sinensis*, including adenine, cordycepic acid, phenols and vitamin B (riboflavin). Transcriptomics and proteomics analysis of the sequential and vital growth phases of *O. sinensis* were done and the antioxidant proteins/peptides, such as catalase, mitochondrial peroxiredoxin PRX1, peroxidase, peroxiredoxin and superoxide dismutase, were examined (Tong and Guo, 2022). Most reported chemical compounds and their health-promoting and healing activities are depicted in Table 3.

7. Commercial Cultivation

O. sinensis is an ascomycete whose life cycle includes the conidium and ascospore phases, commonly known as the anamorph and teleomorph phases. In 2011, Chinese caterpillar fungus alone produced 41% of revenue generated as compared to 62 species of non-timber forest products governed by the Department of Forests in Nepal (Shrestha and Bawa, 2015). Cultivating *O. sinensis* is identical to that of other filamentous fungi, including strain screening, strain activation, intermediate culture and batch culture. Liquid fermentation and solid-state fermentation are two culturing methods that can be used, depending on the physicochemical characteristics of the media (Liu et al., 2021). Temperature, relative humidity, oxygen and nutritional components are crucial to the optimal cultivation and production of valuable products during the culture phase (Elkhateeb and Daba, 2020; Das et al., 2021; Kontogiannatos et al., 2021). Additionally, if a small amount of potassium dihydrogen ortho-phosphate and magnesium sulphate is introduced, the mycelia will grow more quickly and efficiently (Dong and Yao, 2005). There are now numerous patented technologies for isolating fungal strains from *O. sinensis*. Diverse preservation methods have also been devised, with vacuum cryo-preservation being the most effective or simulating natural

Table 2. Structures of some chemical constituents of *C. sinensis* responsible for various pharmacological roles.

Adenine	Guanine
Adenosine	Inosine
Cytidine	Uridine
Cytosine	Thymidine
Uracil	Glucose
Hypoxanthine	Ribose

Table 2 contd. ...

...Table 2 contd.

Cordycepin	Cordycepic acid	Guanosine	Rhamnose
Galactose	Mannose	Xylose	Arabinose
Cordycedipeptide A	Fructose	Mannitol	Sorbose

Ergosterol	CPS-2: A-(1 →4)-D-glucose and A-(1 → 3)-D-mannose
22,23-dihydroergosteryl-3-O-B-D-glucopyranoside (B)	Cordyceamides A and B (a) R₁=OH, R₂=H (b) R₁=OH, R₂=OH
Ergosteryl-3-O-B-D-glucopyranoside (A)	5α,6α-Epoxy-24(R)-methylcholesta-7,22-Dien-3β-Ol (D)
H1-A	5α,8α-Epidioxy-24(R)-methylcholesta-6,22-Dien-3β-D-glucopyranoside (C)

Table 3. Chemical classes and their health benefits described in *O. sinensis*.

Chemical Classes	Compounds	Health Benefits	References
Polysaccharides	PS-A, CSP-1, CS-F30 and CS-F10	It shows anti-diabetic, hypoglycemic and hypolipidemic effect.	Yuan et al., 2022
	Polysaccharide fraction from CS (PSCS)	It shows anti-leukemic and anticancer activity.	Chen et al., 1997
	Mannoglucan	It shows cytotoxic activity.	Wu et al., 2007
	D-galactose	It shows adaptogenic activity.	Liu et al., 2015
	Exo-polysaccharides	It shows immunomodulatory and antitumour effects.	Kuo et al., 2007; Dong and Yao, 2005
	Polysaccharide CPS-2	Liver protective activity.	Wang et al., 2012
	Polysaccharide CPS-F	Kidney protective activity.	Wang et al., 2015b
	Exo-polysaccharides (Fr. A)	Protection against immunosuppressive effects.	Chen et al., 2019
	Acid polysaccharide fraction (APSF)	Anti-inflammatory, antioxidant, immunomodulatory effect and prevention of cardiac arrhythmia.	Liu et al., 2015
	Polysaccharide OSP	Immunomodulatory effect.	Liu et al., 2021
	Cordyceps caterpillar polysaccharide (CCP)	Immunomodulatory effect.	Li et al., 2021b
	Polysaccharide HSWP-2a	Immunomodulatory effect.	Rong et al., 2021
	Polysaccharide CME-1	Anti-inflammatory activity.	Sheu et al., 2018
	Heteropolysaccharides (PS-A)	Hypercholestrolemia.	Shrestha et al., 2012
	Cordyglucans	Antitumour effect.	Meena et al., 2013
	Polysaccharide CSP-1	Antidiabetic activity.	Paterson and Lima, 2015; Yuan et al., 2022; Shrestha et al., 2012

	Polysaccharide NCSP-50	Prebiotics.	Yuan et al., 2022
	Polysaccharide CSPS-1; Polysaccharide MHP-1; Polysaccharide P-4	Anticancer activity.	Xiang et al., 2019
Nucleosides	Adenosine	Anti-inflammatory molecule suppressing the production of cytokine storm, protecting against organ damage and repairing damaged tissue from acute lung diseases, like abdominal and lung diseases and chronic obstructive pulmonary disease (COPD) exhibiting fibrosis, pharmacokinetic effects, and cardioprotection; acts in anti-bacterial, antiviral, anti-neoplastic and immune repair and recovery mechanisms.	Sheth et al., 2014; Zhou et al., 2009; Du et al., 2021
	Guanosine	Immunomodulatory activities.	Wang et al. 2015b
	Cordycepin (3'-Deoxyadenosine)	It has analgestic, anticancer, anti-leukemic, antitumour, immunomodulatory, pharmacokinetic and steroidogenesis stimulating properties; to protect mitochondrial function, cellular metabolism and organs from oxidative stress, cordycepin has an antioxidant activity on reactive oxygen species and can delay, decrease and avoid the oxidative damage of active oxygen/oxygen-free radicals; reduce the level of fat, lower lipoprotein, and triglycerides in the blood.	Paterson, 2008; Deshmukh et al., 2020; Das et al., 2021; Yoshikawa et al., 2011; Nakamura et al., 2015; Zheng et al., 2015; Du et al., 2021; Wang et al., 2015a
	Ergosterol	Anticancer effects, anti-immunomodulatory, anti-metastatic, antimicrobial and cytotoxicity.	Paterson, 2008; Yang et al., 2009; Bhetwal et al., 2021; Shin et al., 2021
	β-sitosterol	Anticancer effects.	
	5a,8a-Epidioxy-24(R)-methylcholesta-6,22-dien-3b-Dglucopyranoside; and 5,6-Epoxy-24(R)-methylcholesta7,22-dien-3b-ol	Anticancer effects.	Bok et al., 1999

Table 3 contd. ...

...Table 3 contd.

Chemical Classes	Compounds	Health Benefits	References
	5a,8a-Epidioxy-22E-ergosta-6,22- dien-3b-ol; 5a,8a-Epidioxy-22E-ergosta6,9(11),22-trien-3b-ol; and 5a,6a-Epoxy-5-ergosta-7,22-dien3b-ol	Prevent the formation of cancel cells.	Matsuda et al., 2009
Proteins	Protein constituents	Hypotensive and vasorelaxant activities.	Liu et al., 2015; Wang et al., 2012; Lo et al., 2013; Jiraungkoorskul and Jiraungkoorskul, 2016
	CSAP anti-bacterial protein	Bactericidal activity.	Zheng et al., 2006
	Cordymin (peptide)	Anti-inflammatory and antinociceptive activities.	Qian et al., 2012
	Cordycedipeptide A (cyclodipeptide)	Cytotoxicity of cell lines (A375, Hela, L929) at the concentration of 50%.	Jia et al., 2005
	Aurantiamide acetate and Cordyceamides A and B	Cytotoxicity of cell lines (A375, Hela, L929).	Jia et al., 2009
Enzymes	Catalase, glutathione peroxidase, and superoxide dismutase	Adaptogenic activity.	Bhetwal et al., 2021
Amino acids, zinc, vitamins also with other trace elements	–	Improve sexual function.	Elkhateeb and Daba, 2020; Das et al., 2021

circumstances for strain preservation. Fang and colleagues tested the conservation of *O. sinensis* strains under natural circumstances and in a freezer, respectively. It is important to consider that the quantity of ascospores per ascus of *O. sinensis* differs from region to region and maturity stages, as does the size and number of septa of each ascospore (Shrestha et al., 2010). Several articles have described the properties and appearance of cultured conidia of isolated *H. sinensis* taken from *O. sinensis* in different places (Lo et al., 2013; Fung et al., 2017; Rong et al., 2021). *O. sinensis* has attracted considerable scientific and economic interest because of its abundance of bioactive compounds that are helpful to human health and the relative simplicity of laboratory cultivation. However, the commercial-scale production of *O. sinensis* has room for improvement, primarily through strain screening, biological degeneration of cultures and substrate enhancement. In particular, culture degeneration, often marked by abnormal fruit body formation and less sporulation, causes significant economic losses and makes investors and potential producers less likely to put money into this exciting business (Kontogiannatos et al., 2021). Mycelia and ascomata of *O. sinensis* can be produced *in situ* even without living and decomposing insect tissues. Numerous techniques for producing mycelia and fruit bodies of *O. sinensis* are described in the relevant works (Dong and Yao, 2005; Guan et al., 2010). Concerning fruit body growth, metabolite biosynthesis and the pathogenicity of *O. sinensis* fungus to target insects, the economic utilisation of *O. sinensis* still has room for development (Kontogiannatos et al., 2021). Strain degeneration is crucial because it significantly impacts *O. sinensis* productivity, resulting in severe economic damage on a commercial scale. *O. sinensis* degeneration is influenced by several intrinsic/ genetic and cultivation-related variables (Zhang et al., 2012).

8. Marketing

The surge in its attractiveness has resulted in the overexploitation of biological species, contributing to their scarcity. The reduction of *O. sinensis* populations is commonly attributed to overharvesting, which will accelerate as climate change reduces the species' range. Several sources say that the main area where the Tibetan Plateau is spread out could shrink by up to 39% over the next 50 years. This poses a direct threat to production levels and suggests that production areas could change, which would be bad for harvesters (Shrestha et al., 2010; Zhou et al., 2012; Winkler, 2011). The Himalayan range may experience diverse climate change effects, depending on the situation. It has been anticipated that the distribution region in Nepal will expand anywhere from a little to 80% (Shrestha and Bawa, 2015). In addition, researchers predicted culturing this organism on an artificial substrate and by the mid-1990s, *O. sinensis* produced in a laboratory setting was being sold globally. After the 1993 Global Athletic Championship games in Stuttgart, Germany, when Chinese athletes were rumoured to have supplemented their diets with caterpillar mushroom and turtle blood, they set numerous long distances running records. *O. sinensis* has been utilised as a nutritional complement and an aphrodisiacal supplement, but after a few decades of opening up its economy, China became part of the global marketplace (Winkler, 2011). This scenario increases global demand for *O. sinensis*, which is one of the world's richest biological resources due to its aphrodisiac and revitalising characteristics. The price of caterpillar fungus is pretty high and its regional trade each year is a billion-dollar enterprise (Negi et al., 2020). In China, a dried and cleaned piece of mushroom can cost as much as $50. Between 1999 and 2004, the price of caterpillar mushrooms in Kumaon increased by 1256% and was deemed to have generated significant demand for organic gold (Pouliot et al., 2018). In terms of weight, caterpillar fungus is more valuable than gold. In Nepal, 1 kg of caterpillars was believed to be worth 30,000–60,000 Nepali rupees; in India, approximately a lakh. The increased commerce and demand for this fungus are sometimes attributed to its exaggerated reputation as Himalayan herbal Viagra (Shrestha and Bawa, 2015). Caterpillar fungus is mainly sold directly between Kathmandu and China's Himalayan megacities. Independent merchants make profit or ration distributions to collectors early, either personally or through intermediaries, based on a mutual agreement, to stimulate value creation. The rural-to-urban transfer of this resource is made possible by the developing unofficial integrity market links, in which the buyer and seller agree on prices (Pouliot et al., 2018). The high-quality items might cost up to $60,000 per kg in the United States (Bibi et al., 2020; 2021). Numerous products derived from caterpillar mushrooms are currently available on global marketplaces and e-commerce websites (Table 4). There is a list of things available on online purchasing platforms and their costs.

9. Future Prospects

O. sinensis is a therapeutic edible fungus that can, without a doubt, be added to common foods, superfoods, functional food ingredients and nutraceuticals. However, extensive scientific studies on constituent analysis are required to evaluate the nutritional quality and bioactivity. It needs *in vivo* confirmation to prove a therapeutic potential of the products and to find and describe novel biomolecules to linkage their bioactivity to value-added products. For large-scale extraction of bioactive substances, it is necessary to use efficient artificial growing techniques to yield wet or dry biomass. For efficient, cost-effective downstream processing and sustainable manufacturing using *in vitro* technologies, it is required to conduct an extensive study on proven scientific concepts. To develop healthy food supplements and nutraceuticals, based on irrefutable scientific facts, there is a pressing need for a potential expansion of the mushroom position in the food and pharmaceutical industries.

Deeper molecular investigations are required to completely comprehend the related pathways. This exhaustive review of studies exposing the molecular basis of fungal infectivity advances our understanding of *Ophiocordyceps* biology, which might also aid in the pharmaceutical exploitation of bioactive chemicals of fungal origin. Evidence for the precise detection of the infection process could help decipher the essential properties this fungus exploits throughout pathogenesis. In addition, disclosing the crimes committed by fungus against the human immune system is a significant concern. Indeed, this would provide the much-needed insight into the functional components of the insect infection process. Another factor that must be considered is biological control and its associated effect on natural enemies other than the targeted agricultural insect pests. The impact should be at least less than that of chemical insecticides. However, it should be kept in mind that not every agricultural area can be treated similarly. It should be determined which species of *Ophiocordyceps* are most effective for minimizing the damage.

Table 4. Products available in Indian markets.

Name of Product	Purpose	Cost (USD)
Cordyceps capsules - 650 mg by Fair & Pure	Dietary supplement	$20.679/120 capsules
Cordyceps sinensis - Vegan by Fair & Pure	Dietary supplement	$17.30/100 g
Kala health Cordyceps by Kala Health Inc., USA	Dietary supplement	$97.95/box (600 capsules)
Mdrive Elite performance by Dream Brands, Inc., USA	Dietary supplement	$79.99/box (90 capsules)
Now Foods Cordyceps by Now Foods Inc., USA	Dietary supplement	$12.34/750 mg
Cordyceps capsules by Host Defense Inc., USA	Dietary supplement	$44.96/pack (120 capsules)
Cordyceps mushroom extract powder by Real Mushrooms, Canada	Dietary supplement	$0.50/g
New China *Cordyceps sinensis* whitening cream by Smile, China	Skin care	$17.65/unit
YanWo drink with Cordyceps extract and rose bud by QiYun B.V, UK	Beverage for antiaging and smooth skin	$60.22/unit
Mountain Fresh Cordyceps homeopathic cream by Mountain Fresh, UK	Skin care	$108.92/unit
Cordyceps organic mushroom by Full and Fill Bio, Thailand	Sexual tonic, reduces sugar and fat levels in the blood	$3/unit
Cordyceps soap by Full and Fill Bio, Thailand	Skin care	$9.79/unit
Le'JOYva Gourmet Arabica black coffee by Le'JOYva Inc., USA	Beverage	$29.99/pack
DXN Cordyceps instant coffee	Beverage	$28.99/pack
CordyMax CS-4 by Pharmanex, Inc., USA	Dietary supplement	$53.30/120 capsules
Organo Gold Red Tea by Organo Gold Enterprises Inc., Canada	Beverage	$33/box
Cordyceps sinensis capsules by Tonicology, LLC	Immune booster	$78.87/350 mg
Urban Platter Cordyceps mushroom extract powder by Urban Platter	Immune booster	$10/50 g (50 g/1.76oz)
Mystique Hills organic *Cordyceps sinensis* powder (Caterpillar Fungus) by Mystique Hills-Organic Living	Health care	$34.28/100 g
Cordyceps sinensis Yarsagumba capsules Iriss Herbals, India	Immune booster	$85.71/60 capsules (500 mg each)
Cordyceps sinensis extract (200 g) by G & E Nutrition	Health care	$138.34/200 g
A Grade IC PRO *Cordyceps sinensis* by IC Pro, India	Clinical	$12857.14/kg
Cordyceps extract, by Herbal Creative, India	Clinical	$22.85/g
Yarsagumba by Aditya Agni Trading Co., India	Clinical	$28571.43/kg

Table 4 contd. ...

...Table 4 contd.

Name of Product	Purpose	Cost (USD)
Cordyceps mushroom extract powder - 60 g by Real Mushrooms, Canada	Dietary supplement	$1.8/g
Cordy Gold Cordyceps *militaris* mushroom 50 g by Itaara Foodtech Private Limited, India	Dietary supplement	$1.2/g
Organic *Cordyceps sinensis* extract capsules, 500 mg (60 Count) by Dr. Bham Lab	Dietary supplement	$1.2/g
Cordyceps sinensis 600 mg 120 caps by Swanson, India	Dietary supplement	$1/g
Ultra Cordyceps plus capsule - 60 capsules by Tonga Herbals	Dietary supplement	$1/g
Cordyceps 500 mg - 60 veg. capsule by Biotrex Nutraceuticals	Dietary supplement	$0.8/g
Synergised Cordyceps Forte (30 veg. capsules) by Plants Med Laboratories Pvt. Ltd.	Dietary supplement	$1/g
Cordyceps 100 capsules by Tianshi Biological Development Co. Ltd, China	Dietary supplement	$1/g
Cordyceps CS-4 750 mg - 60 vegetarian capsules by MRM	Dietary supplement	$2/g

10. Conclusion

Ophiocordyceps is one of the essential medicinal mushrooms used in Chinese and Tibetan medicine, and it has impressive health implications. This would include increased physical endurance for enhanced performance, anticancer capabilities and lung and kidney protection. As a result of its extraordinary therapeutic properties, this fungus is in plentiful supply as a health care product, earning its collectors enormous sums. The importance of the Eastern medical system is further demonstrated by its popularity and demand. In addition, bio-guided extraction and linking the bioactive constituents to a variety of bioactivities offer several study options. Although it has substantial pharmacological effects, its interaction with the biological body at the molecular level is poorly understood. A detailed scientific examination is required to comprehend the caterpillar mushroom properly.

References

Araújo, J.P.M. and Hughesm D.P. (2016). Diversity of entomopathogenic fungi. Which groups conquered the insect body? *Adv. Genet.*, 94: 1–39.

Arora, D. (2008). The houses that Matsutake built. *Econ. Bot.*, 62(3): 278–290.

Ashraf, S.A., Elkhalifa, A.E.O., Siddiqui, A.J., Patel, M., Awadelkareem, A.M., Snoussi, M., Ashraf, M.S., Adnan, M. and Hadi, S. (2020). Cordycepin for health and wellbeing: A potent bioactive metabolite of an entomopathogenic medicinal fungus Cordyceps with its nutraceutical and therapeutic potential. *Molecules*, 25(12): 2735.

Baldrian, P., Větrovský, T., Lepinay, C. and Kohout, P. (2021). High-throughput sequencing view on the magnitude of global fungal diversity. *Fungal Divers*, 19: 1–9.

Baral, B. (2017). Entomopathogenicity and biological attributes of Himalayan treasured fungus *Ophiocordyceps sinensis* (Yarsagumba). *J. Fungi*, 3(1): 4.

Belwal, T., Bhatt, I.D., Kashyap, D., Sak, K., Tuli, H.S., Pathak, R., Rawal, R.S. and Shashidhar, G.M. (2018). *Ophiocordyceps sinensis.* pp. 527–537. *In*: Nabavi, S.M. and Silva, A.S. (eds.). *Nonvitamin and Nonmineral Nutritional Supplements*, Elsevier, New York, USA.

Bhetwal, S., Subrata, C., Samrat, Robin, R., Meenakshi, R. and Seweta, S. (2021). *Cordyceps sinensis*: Peculiar caterpillar mushroom, salutary in its medicinal and restorative capabilities. *Pharma Inno. J.*, 10(4): 1045–1054.

Bi, S., Huang, W., Chen, S., Huang, C., Li, C., Guo, Z., Yang, J., Zhu, J., Song, L. and Yu, R. (2020). *Cordyceps militaris* Polysaccharide converts immunosuppressive macrophages into M1-like phenotype and activates T lymphocytes by inhibiting the PD-L1/PD-1 axis between TAMs and T lymphocytes. *Int. J. Biol. Macromol.*, 150: 261–280.

Bibi, S., Wang, Y.B., Tang, D.X., Kamal, M.A. and Yu, H. (2021). Prospects for discovering the secondary metabolites of *Cordyceps sensulato* by the integrated strategy. *Med. Chem.*, 17(2): 97–120.

Bibi, Shabana, Yuan-Bing Wang, De-Xiang Bi, S., Huang, W., Chen, S., Huang, C., Li, C., Guo, Z., Yang, J., Zhu, J., Song, L. and Yu, R. (2020). *Cordyceps militaris* polysaccharide converts immunosuppressive macrophages into M1-like phenotype and activates T lymphocytes by inhibiting the PD-L1/PD-1 axis between TAMs and T lymphocytes. *Int. J. Biol. Macromol.*, 150: 261–280.

Bok, J.W., Lermer, L., Chilton, J., Klingeman, H.G. and Towers, G.N. (1999). Antitumour sterols from the mycelia of *Cordyceps sinensis. Phytochem.*, 51(7): 891–898.

Cannon, P.F., Hywel-Jones, N.L., Maczey, N., Norbu, L., Samdup, T. and Lhendup, P. (2009). Steps towards sustainable harvest of *Ophiocordyceps sinensis* in Bhutan. *Biodivers. Conserv.*, 18(9): 2263–2281.

Chen, S., Wang, J., Fang, Q., Dong, N. and Nie, S. (2019). Polysaccharide from natural *Cordyceps sinensis* ameliorated intestinal injury and enhanced antioxidant activity in immunosuppressed mice. *Food Hydrocoll.*, 89: 661–667.

Chen, S., Wang, J., Fang, Q., Dong, N., Fang, Q., Cui, S.W. and Nie, S. (2021). A polysaccharide from natural *Cordyceps sinensis* regulates the intestinal immunity and gut microbiota in mice with cyclophosphamide-induced intestinal injury. *Food Funct.*, 12(14): 6271–6282.

Chen, Y.J., Shiao, M.S., Lee, S.S. and Wang, S.Y. (1997). Effect of *Cordyceps sinensis* on the proliferation and differentiation of human leukemic U937 cells. *Life Sci.*, 60(25): 2349–2359.

Cheong, K.L., Meng, L.Z., Chen, X.Q., Wang, L.Y., Wu, D.T., Zhao, J. and Li, S.P. (2016). Structural elucidation, chain conformation and immunomodulatory activity of glucogalactomannan from cultured *Cordyceps sinensis* fungus UM01. *J. Funct. Foods*, 25: 174–185.

Das, G., Shin, H.S., Leyva-Gómez, G., Prado-Audelo, M.L.D., Cortes, H., Singh, Y.D., Panda, M.K., Mishra, A.P., Nigam, M., Saklani, S. and Chaturi, P.K. (2021). Cordyceps spp.: A review on its immune-stimulatory and other biological potentials. *Front. Pharmacol.*, 2250.

Deshmukh, L., Sharma, A.K. and Sandhu, S.S. (2020). Contrive himalayan soft gold Cordyceps species: A lineage of Eumycota bestowing tremendous pharmacological and therapeutic potential. *Curr. Pharmacol. Rep.*, 6(4): 155–166.

Dong, C.H. and Yao, Y.J. (2005). Nutritional requirements of mycelial growth of *Cordyceps sinensis* in submerged culture. *J. Appl. Microbiol.*, 99(3): 483–492.

Du, J., Kan, W., Bao, H., Jia, Y., Yang, J. and Jia, H. (2021). Interactions between Adenosine receptors and Cordycepin (3'-Deoxyadenosine) from *Cordyceps militaris*: Possible pharmacological mechanisms for protection of the brain and the amelioration of Covid-19 pneumoniam. *J. Biotechnol. Biomed.*, 4(2): 26–32.

Elkhateeb, W.A. and Daba, G. (2020). Review: The endless nutritional and pharmaceutical benefits of the himalayan gold, *Cordyceps*; current knowledge and prospective potentials. *Asian. J. Nat. Prod. Biochem.*, 18(2): 70–77.

Fan, H.B., Zheng, Q.W., Han, Q., Zou, Y., Liu, Y.L., Guo, L.Q. and Lin, J.F. (2021). Effect and mechanism of a novel *Cordyceps militaris* immunomodulatory protein on the differentiation of macrophages. *Food Biosci.*, 43: 101268.

Fan, S., Huang, X., Wang, S., Li, C., Zhang, Z., Xie, M. and Nie, S. (2018). Combinatorial usage of fungal polysaccharides from *Cordyceps sinensis* and *Ganoderma atrum* ameliorate drug-induced liver injury in mice. *Food Chem. Toxicol.*, 119: 66–72.

Fung, S.Y., Lee, S.S., Tan, N.H. and Pailoor, J. (2017). Safety assessment of cultivated fruiting body of *Ophiocordyceps sinensis* evaluated through subacute toxicity in rats. *J. Ethnopharmacol.*, 206: 236–244.

Guan, J., Yang, F.Q. and Li, S.P. (2010). Evaluation of carbohydrates in natural and cultured *Cordyceps* by pressurised liquid extraction and gas chromatography coupled with mass spectrometry. *Molecules*, 15(6): 4227–4241.

Jia, J.M., Ma, X.C., Wu, C.F., Wu, L.J. and Hu, G.S. (2005). Cordycedipeptide A, a new cyclodipeptide from the culture liquid of *Cordyceps sinensis* (B ERK.) S ACC. *Chem. Pharm. Bull.*, 53(5): 582–583.

Jia, J.M., Tao, H.H. and Feng, B.M. (2009). Cordyceamides A and B from the culture liquid of *Cordyceps sinensis* (B ERK.) S ACC. *Chem. Pharm. Bull.*, 57(1): 99–101.

Jiraungkoorskul, K. and Jiraungkoorskul, W. (2016). Review of naturopathy of medical mushroom, *Ophiocordyceps sinensis*, in sexual dysfunction. *Pharmacogn. Rev.*, 10(19): 1.

Joshi, M., Anand, S., Shamsher, K. and Sukhdev, S. (2019). Anticancer, Antibacterial and antioxidant activities of *Cordyceps militaris*. *Indian J. Exp. Biol.*, 57(1): 15–20.

Kan, W.C., Wang, H.Y., Chien, C.C., Li, S.L., Chen, Y.C., Chang, L.H., Cheng, C.H., Tsai, W.C., Hwang, J.C., Su, S.B. and Huang, L.H. (2012). Effects of extract from solid-state fermented *Cordyceps sinensis* on type 2 diabetes mellitus. *Evid.-based Complement Alternat. Med.*, 743107: 1–10.

Kontogiannatos, D., Koutrotsios, G., Xekalaki, S. and Zervakis, G.I. (2021). Biomass and *Cordycepin* production by the medicinal mushroom *Cordyceps militaris*—A review of various aspects and recent trends towards the exploitation of a valuable fungus. *J. Fungi*, 7(11): 986.

Kumar, S., Sharma, V.P. and Kamal, S. (2016). A review on insect-fungus interactions with special emphasis on *Cordyceps* spp. *Mush. Res.*, 24(1): 1–8.

Kuo, M.C., Chang, C.Y., Cheng, T.L. and Wu, M.J. (2007). Immunomodulatory effect of exopolysaccharides from submerged cultured *Cordyceps sinensis*: Enhancement of cytokine synthesis, CD11b expression, and phagocytosis. *Appl. Microbiol. Biotechnol.*, 75(4): 769–775.

Kuo, Y.C., Tsai, W.J., Wang, J.Y., Chang, S.C., Lin, C.Y. and Shiao, M.S. (2001). Regulation of bronchoalveolar lavage fluids cell function by the immunomodulatory agents from *Cordyceps sinensis*. *Life Sci.*, 68(9): 1067–1082.

Li, H., Tian, Y., Menolli Jr, N., Ye, L., Karunarathna, S.C., Perez-Moreno, J., Rahman, M.M., Rashid, M.H., Phengsintham, P., Rizal, L. and Kasuya, T. (2021a). Reviewing the world's edible mushroom species: A new evidence-based classification system. *Compr. Rev. Food Sci. Food Saf.*, 20(2): 1982–2014.

Li, L.F., But, G.W.C., Zhang, Q.W., Liu, M., Chen, M.M., Wen, X., Wu, H.Y., Cheng, H.Y., Puno, P.T., Zhang, J.X. and Fung, H.Y. (2021b). A specific and bioactive polysaccharide marker for *Cordyceps*. *Carbohydr. Polym.*, 269: 118343.

Li, L.Q., Song, A.X., Yin, J.Y., Siu, K.C., Wong, W.T. and Wu, J.Y. (2020). Anti-inflammation activity of exopolysaccharides produced by a medicinal fungus *Cordyceps sinensis* Cs-HK1 in cell and animal models. *Int. J. Biol. Macromol.*, 149: 1042–1050.

Liu, Y., Li, Q.Z. and Zhou, X.W. (2021). Immunostimulatory effects of the intracellular polysaccharides isolated from liquid culture of *Ophiocordyceps sinensis* (Ascomycetes) on RAW264. 7 cells via the MAPK and PI3K/Akt signalling pathways. *J. Ethnopharmacol.*, 275: 114130.

Liu, Y., Wang, J., Wang, W., Zhang, H., Zhang, X. and Han, C. (2015). The chemical constituents and pharmacological actions of *Cordyceps sinensis*. *Evid.-based Complement Alternat. Med.*, 575063.

Lo, H.C., Hsieh, C., Lin, F.Y. and Hsu, T.H. (2013). A systematic review of the mysterious caterpillar fungus *Ophiocordyceps sinensis* in DongChongXiaCao (Dōng Chóng Xià Cǎo) and related bioactive ingredients. *J. Tradit. Complement. Med.*, 3(1): 16–32.

Lücking, R., Aime, M.C., Robbertse, B., Miller, A.N., Aoki, T., Ariyawansa, H.A., Cardinali, G., Crous, P.W., Druzhinina, I.S., Geiser, D.M. and Hawksworth, D.L. (2021). Fungal taxonomy and sequence-based nomenclature. *Nature Microbiology*, 6(5): 540–548.

Matsuda, H., Akaki, J., Nakamura, S., Okazaki, Y., Kojima, H., Tamesada, M. and Yoshikawa, M. (2009). Apoptosis-inducing effects of sterols from the dried powder of cultured mycelium of *Cordyceps sinensis*. *Chem. Pharm. Bull.*, 57(4): 411–414.

Meena, H., Singh, K.P., Negi, P.S. and Ahmed, Z. (2013). Sub-acute toxicity of cultured mycelia of Himalayan entomogenous fungus *Cordyceps sinensis* (Berk.) Sacc. in rats. *Indian J. Exp. Biol.*, 51(5): 381–387.

Nakamura, K., Shinozuka, K. and Yoshikawa, N. (2015). Anticancer and anti-metastatic effects of Cordycepin, an active component of *Cordyceps sinensis. J. Pharmacol. Sci.*, 127(1): 53–56.

Negi, V.S., Rana, S.K., Giri, L. and Rawal, R.S. (2020). *Caterpillar Fungus in the Himalaya, Current Understanding and Future Possibilities.* GB Pant National Institute of Himalayan Environment, Kosi-Katarmal, Almora, Uttarakhand, India.

Niazi, A.R. and Ghafoor, A. (2021). Different ways to exploit mushrooms: A review. *All Life*, 14(1): 450–460.

Paterson, R.R.M. (2008). Cordyceps—A traditional Chinese medicine and another fungal therapeutic biofactory? *Phytochem.*, 69(7): 1469–1495.

Paterson, R.R.M. and Lima, N. (2015). Failed PCR of Ganoderma-type specimens affects nomenclature. *Phytochem.*, 114: 16–17.

Phull, A.R., Ahmed, M. and Park, H.J. (2022). *Cordyceps militaris* as a bio-functional food source: Pharmacological potential, anti-inflammatory actions and related molecular mechanisms. *Microorganisms*, 10(2): 405.

Pouliot, M., Pyakurel, D. and Smith-Hall, C. (2018). High altitude organic gold: The production network for *Ophiocordyceps sinensis* from far-western Nepal. *J. Ethnopharmacol.*, 218: 59–68.

Qi, W., Zhou, X., Wang, J., Zhang, K., Zhou, Y., Chen, S., Nie, S. and Xie, M. (2020). *Cordyceps sinensis* polysaccharide inhibits colon cancer cells growth by inducing apoptosis and autophagy flux blockage via mTOR signaling. *Carbohydr. Polym.*, 237: 116113.

Qian, G.M., Pan, G.F. and Guo, J.Y. (2012). Anti-inflammatory and antinociceptive effects of Cordymin, a peptide purified from the medicinal mushroom *Cordyceps sinensis. Nat. Prod. Rep.*, 26(24): 2358–2362.

Qin, Q.L., Zhou, G.L., Zhang, H., Meng, Q., Zhang, J.H., Wang, H.T., Miao, L. and Li, X. (2018). Obstacles and approaches in artificial cultivation of Chinese *Cordyceps. Mycology*, 9(1): 7–9.

Rakhee, S.N., Bhargava, K., Misra, K. and Singh, V.K. (2016). Phytochemical and proteomic analysis of a high altitude medicinal mushroom *Cordyceps sinensis. Journal of Proteins and Proteomics*, 7(3): 187–197.

Rathi, A., Singh, C., Pathak, P., Chaudhary, N. and Vyas, D.A. (2021). Thematic approach on Cordyceps. pp. 139–162. *In*: Poonam, D. (ed.). *Recent Trends Mushroom Biology*, Global Books Organisation, India.

Rong, L., Li, G., Zhang, Y., Xiao, Y., Qiao, Y., Yang, M., Wei, L., Bi, H. and Gao, T. (2021). Structure and immunomodulatory activity of a water-soluble α-glucan from *Hirsutella sinensis* mycelia. *Int. J. Biol. Macromol.*, 189: 857–868.

Sharma, P.K. and Negi, C.S. (2022). Impact of livestock grazing pressure and above-ground vegetational changes on yield of caterpillar Mushroom *Ophiocordyceps sinensis. Indian J. Ecol.*, 49(1): 80–86.

Shen, W., Song, D., Wu, J. and Zhang, W. (2011). Protective effect of a polysaccharide isolated from a cultivated Cordyceps mycelia on hydrogen peroxide-induced oxidative damage in PC12 cells. *Phytotherapy Res.*, 25(5): 675–680.

Sheth, S., Brito, R., Mukherjea, D., Rybak, L.P. and Ramkumar, V. (2014). Adenosine receptors: Expression, function and regulation. *Int. J. Mol. Sci.*, 15(2): 2024–2052.

Sheu, J.R., Chen, Z.C., Hsu, M.J., Wang, S.H., Jung, K.W., Wu, W.F., Pan, S.H., Teng, R.D., Yang, C.H. and Hsieh, C.Y. (2018). CME-1, a novel polysaccharide, suppresses iNOS expression in lipopolysaccharide-stimulated macrophages through ceramide-initiated protein phosphatase 2A activation. *J. Cell. Mol. Med.*, 22(2): 999–1013.

Shin, M.K., Sasaki, F., Ki, D.W., Win, N.N., Morita, H. and Hayakawa, Y. (2021). Anti-metastatic effects of ergosterol peroxide from the entomopathogenic fungus *Ophiocordyceps gracilioides* on 4T1 breast cancer cells. *J. Nat. Med.*, 75(4): 824–832.

Shrestha, B., Zhang, W., Zhang, Y. and Liu, X. (2010). What is the Chinese caterpillar fungus *Ophiocordyceps sinensis* (Ophiocordycipitaceae)? *Mycology*, 1(4): 228–236.

Shrestha, S., Shrestha, B., Park, J.H., Lee, D.Y., Cho, J.G. and Baek, N.I. (2012). Chemical constituents of Yarsagumba (*Ophiocordyceps sinensis* (Berk.) Sung et al.), a valued traditional Himalayan medicine. *Nepal J. Sci. Technol.*, 13(1): 43–58.

Shrestha, U.B. and Bawa, K.S. (2015). Harvesters' perceptions of population status and conservation of Chinese caterpillar fungus in the Dolpa region of Nepal. *Reg. Environ. Change*, 15(8): 1731–1741.

Smiderle, F.R., Baggio, C.H., Borato, D.G., Santana-Filho, A.P., Sassaki, G.L., Iacomini, M. and Van Griensven, L.J. (2014). Anti-inflammatory properties of the medicinal mushroom *Cordyceps militaris* might be related to its linear $(1\rightarrow 3)$-β-D-glucan. *PloS ONE*, 9(10): e110266.

Smith-Hall, C. and Bennike, R.B. (2022). Understanding the sustainability of Chinese caterpillar fungus harvesting: The need for better data. *Biodivers. Cons.*, 1–5.

Sung, G.H., Hywel-Jones, N.L., Sung, J.M., Luangsa-Ard, J.J., Shrestha, B. and Spatafora, J.W. (2007). Phylogenetic classification of Cordyceps and the clavicipitaceous fungi. *Studies in Mycology*, 57: 5–59.

Tong, X. and Guo, J. (2022). High throughput identification of the potential antioxidant peptides in *Ophiocordyceps sinensis*. *Molecules*, 27(2): 438.

Tong, X., Wang, F., Zhang, H., Bai, J., Dong, Q., Yue, P., Jiang, X., Li, X., Wang, L. and Guo, J. (2021). iTRAQ-based comparative proteome analyses of different growth stages revealing the regulatory role of reactive oxygen species in the fruiting body development of *Ophiocordyceps sinensis*. *Peer J.*, 9: e10940.

Tuli, H.S., Kashyap, D. and Sharma, A.K. (2017). Cordycepin: A metabolite with promising therapeutic potential. pp. 761–782: *In*: Mérillon, J.M. and Ramawat, K.G. (eds.). *Fungal Metabolites*, Springer, New York, USA.

Wang, B.S., Lee, C.P., Chen, Z.T., Yu, H.M. and Duh, P.D. (2012). Comparison of the hepatoprotective activity between cultured *Cordyceps militaris* and natural *Cordyceps sinensis*. *J. Funct. Foods*, 4(2): 489–495.

Wang, J., Kan, L., Nie, S., Chen, H., Cui, S.W., Phillips, A.O., Phillips, G.O., Li, Y. and Xie, M. (2015a). A comparison of chemical composition, bioactive components and antioxidant activity of natural and cultured *Cordyceps sinensis*. *LWT – Food Sci. Technol.*, 63(1): 2–7.

Wang, J., Liu, R., Liu, B., Yang, Y., Xie, J. and Zhu, N. (2017). Systems pharmacology-based strategy to screen new adjuvant for hepatitis B vaccine from traditional Chinese medicine *Ophiocordyceps sinensis*. *Sci. Rep.*, 7(1): 1–10.

Wang, X.L. and Yao, Y.J. (2011). Host insect species of *Ophiocordyceps sinensis*: A review. *ZooKeys*, 127: 43.

Wang, Y., Liu, D., Zhao, H., Jiang, H., Luo, C., Wang, M. and Yin, H. (2014). *Cordyceps sinensis* polysaccharide CPS-2 protects human mesangial cells from PDGF-BB-induced proliferation through the PDGF/ERK and TGF-β1/Smad pathways. *Mol. Cell. Endocrinol.*, 382(2): 979–988.

Wang, Y., Wang, Y., Liu, D., Wang, W., Zhao, H., Wang, M. and Yin, H. (2015b). *Cordyceps sinensis* polysaccharide inhibits PDGF-BB-induced inflammation and ROS production in human mesangial cells. *Carbohydr. Polym.*, 125: 135–145.

Winkler, D. (2011). Caterpillar fungus (*Ophiocordyceps sinensis*) production and sustainability on the Tibetan. *Asian Med.*, 5(2): 291–316.

Wu, Y., Hu, N., Pan, Y., Zhou, L. and Zhou, X. (2007). Isolation and characterisation of a mannoglucan from edible *Cordyceps sinensis* mycelium. *Carbohydr. Res.*, 342(6): 870–875.

Xiang, T., Xia, C., Shen, J. and Wang, C. (2019). Research progress on antitumour active ingredients and antitumour mechanism of *Cordyceps militaris*. *Hans. J. Food Nut. Sci.*, 8(4): 258–266.

Xie, F., Su, Q., He, L., Uwitugabiye, V., Xia, Y., Zhou, G. and Chen, Z. (2021). *Transcriptome Analysis of Ophiocordyceps sinensis Under Low-Temperature Treatment. Research Square.* https://Doi.org/10.21203/rs.3.rs-860809/v1.

Xu, J., Huang, Y., Chen, X.X., Zheng, S.C., Chen, P. and Mo, M.H. (2016). The mechanisms of pharmacological activities of *Ophiocordyceps sinensis* fungi. *Phytother. Res.*, 30(10): 1572–1583.

Yan, J.K., Wang, W.Q. and Wu, J.Y. (2014). Recent advances in *Cordyceps sinensis* polysaccharides: Mycelial fermentation, isolation, structure, and bioactivities: A review. *J. Funct. Foods*, 6: 33–47.

Yang, F.Q., Feng, K., Zhao, J. and Li, S.P. (2009). Analysis of sterols and fatty acids in natural and cultured Cordyceps by one-step derivatisation followed with gas chromatography–mass spectrometry. *J. Pharm. Biomed. Anal. J.*, 49(5): 1172–1178.

Yang, M.L., Kuo, P.C., Hwang, T.L. and Wu, T.S. (2011). Anti-inflammatory principles from *Cordyceps sinensis. J. Nat. Prod.*, 74(9): 1996–2000.

Yang, Y., Xiao, Y., Yu, G., Wen, T., Deng, C., Meng, J. and Lu, Z. (2021). *Ophiocordyceps aphrophoridarum* sp. nov., a new entomopathogenic species from Guizhou, China. *Biodivers. Data J.*, 9.

Ying, M., Yu, Q., Zheng, B., Wang, H., Wang, J., Chen, S., Gu, Y., Nie, S. and Xie, M. (2019). Cultured *Cordyceps sinensis* polysaccharides attenuate cyclophosphamide-induced intestinal barrier injury in mice. *J. Funct. Foods*, 62: 103523.

Yoshikawa, N., Nishiuchi, A., Kubo, E., Yamaguchi, Y., Kunitomo, M., Kagota, S., Shinozuka, K. and Nakamura, K. (2011). *Cordyceps sinensis* acts as an adenosine A3 receptor aganist on mouse melanoma and lung carcinoma cells, and human fibrosarcoma and colon carcinoma cells. *Pharmacol. Pharm.*, 2(4): 266–270.

Yuan, Q., Xie, F., Tan, J., Yuan, Y., Mei, H., Zheng, Y. and Sheng, R. (2022). Extraction, structure and pharmacological effects of the polysaccharides from *Cordyceps sinensis*: A review. *J. Funct. Foods*, 89: 104909.

Yue, P., Zhang, H., Tong, X., Peng, T., Tang, P., Gao, T. and Guo, J. (2022). Genome-wide identification and expression profiling of thes MAPK, MAPKK, and MAPKKK gene families in *Ophiocordyceps sinensis. Gene*, 807: 145930.

Zhang, Y., Li, E., Wang, C., Li, Y. and Liu, X. (2012). *Ophiocordyceps sinensis*, the flagship fungus of China: terminology, life strategy and ecology. *Mycology*, 3(1): 2–10.

Zheng, H., Maoqing, Y., Liqiu, X., Wenjuan, T., Liang, L. and Guolin, Z. (2006). Purification and characterisation of an antibacterial protein from the cultured mycelia of *Cordyceps sinensis. Wuhan Univ. J. Nat. Sci.*, 11(3): 709–714.

Zheng, Z.L., Qiu, X.H. and Han, R.C. (2015). Identification of the genes involved in the fruiting body production and cordycepin formation of *Cordyceps militaris* fungus. *Mycobiol.*, 43(1): 37–42.

Zhou, X.W., Li, L.J. and Tian, E.W. (2014). Advances in research of the artificial cultivation of *Ophiocordyceps sinensis* in China. *Crit. Rev. Biotechnol.*, 34(3): 233–243.

Zhou, X.W., Su, K.Q. and Zhang, Y.M. (2012). Applied modern biotechnology for cultivation of Ganoderma and development of their products. *Applied Microbiology and Biotechnology*, 93(3): 941–963.

Zhou, Y., Schneider, D.J. and Blackburn, M.R. (2009). Adenosine signaling and the regulation of chronic lung disease. *Pharmacol. Ther.*, 123(1): 105–116.

CHAPTER 18

Oyster Mushroom
(*Pleurotus ostreatus*)

Parthasarathy Seethapathy,[1,*] *Praveen Thangaraj,*[2] *Anu Pandita,*[3]
Subbiah Sankaralingam[4] *and Deepu Pandita*[5,*]

1. Introduction

There is an incredible variety of species of fungi, each with its unique morphology, developmental patterns, ecological niche, nutritional strategy and a wide range of habitats, such as soil, water, air, animals, and plants, as well as environments with extreme conditions, such as temperatures, metals and salts concentrations (Lücking et al., 2021). Insects are considered the most incredible group of organisms; however, the global fungal species richness ranges between 2.2 (additive method) and 3.8 million, possibly the largest and most diverse group of living things, after insects (global ratio method). Considering 120,000 officially recognised species, at least 8%, and at worst, just 3%, have been named. Appropriate statistical and evolutionary methodologies to examine the continuously expanding quantity of environmental sequencing data are crucial to improving estimates (Hawksworth and Lücking, 2017). It has been precisely estimated that there are between 1.5–5.1 million different species of fungi in global ecosystems (Lücking et al., 2021). Fungi are an essential and varied contributor to the biodiversity found in various habitats. These creatures are composed of a wide variety of fungi belonging to all the major fungal groups and function as enemies and allies. Numerous mycologists, who continue to work on the classification and biodiversity of fungi, have generated differing measures of the prevalence of fungal species. Estimations range between 0.1–9.9 million. The process

[1] Department of Plant Pathology, Amrita School of Agricultural Sciences, Amrita Vishwa Vidyapeetham, Coimbatore, 642109, Tamil Nadu, India.
[2] Department of Plant Pathology, Tamil Nadu Agricultural University, Coimbatore, 641003, Tamil Nadu, India.
[3] Vatsalya Clinic, Krishna Nagar, New Delhi, 110051, India.
[4] PG and Research Department of Botany, Saraswathi Narayanan College, Madurai, 625022, India.
[5] Government Department of School Education, Jammu, Jammu and Kashmir, 180001, India.
* Corresponding authors: spsarathyagri@gmail.com, deepupandita@gmail.com

of discovering new fungal taxonomy is dynamic. This is because there are several natural ecosystems and an extensive range of environments whose fungi diversity has yet to be studied (Gautam et al., 2022). Macrofungi such as mushrooms, play a significant part in the environment by performing various biological functions. These macrofungi decompose dead organic material and develop mutually beneficial partnerships. They can also be edible to living creatures, such as humans and animals. Recycling nutrients, rotting dead matter and mushrooms save soil nutrients, but through mutualistic or pathogenic relationships, certain organisms can be suppressed from overwhelming excessively. Since ancient times, people worldwide have regarded mushrooms as among the most delicious of all consumables. Their exceptionally exquisite flavour captured the attention of luxury-obsessed Roman aristocrats over 2,000 years ago, just as it continues to do so now. Since mushroom use was reported 18,700 years ago during Stone Age in Spain, 6,000 years ago in China and 4,600 years ago in Egypt; numerous societies around the globe have long regarded mushrooms as medicinal and nutritive food (Li et al., 2021). Even modern experts are uncertain about how our ancestors discovered the first edible mushroom species. Despite this, people believed it applied to wild plants and other life forms that were harvested or acquired for livelihood. Li et al. (2020) recognised 2,189 edible mushrooms, among which 2,006 could be eaten securely and an additional 183 needed certain type of preprocessing or were related with anaphylactic reactions in certain individuals. According to the recent data, there are roughly 3,283 species of edible mushrooms, which account for around 20% of all mushroom species described to date. There are about 700 medicinal mushrooms that are thought to be effective. These mushrooms possess the pharmacological properties of 2,000 different species of safe mushrooms. Numerous nations have cultivated 10 of these mushrooms (Baldrian et al., 2021). In the past 60 years, the worldwide production of edible mushrooms has increased thirty times. According to the FAO, China has the most significant volume of mushroom production and anticipates that mushroom cultivation will continue to expand at an accelerating rate in the nearby years. Numerous species belonging to the genera *Agaricus*, *Auricularia*, *Flammulina*, *Lentinula*, *Pleurotus* and *Volvariella* are among the various edible saprophytic mushrooms of substantial social and economic significance belonging to Basidiomycetes. Despite this, approximately 100 species are economically grown, and only 10 of these have been marketed on a global scale: *Agaricus bisporus* (J.E. Lange) Imbach, *Auricularia auricula-judae* (Bull.) Quél, *A. nigricans*, *Flammulina velutipes* (Curtis) Singer, *Lentinula edodes* (Berk.) Pegler, *Pleurotus cornucopiae*, *P. eryngii*, *P. nebrodensis*, *P. ostreatus* (Jacq.) P. Kumm. and *Volvariella volvacea* (Bull.) Singer (Pérez-Moreno et al., 2021). Among a widespread type of edible mushroom is *Pleurotus ostreatus*, sometimes known as the oyster fungus, pearl oyster mushroom, or hiratake. It was initially produced in Germany as a method of survival during World War I, and it is currently commercially produced worldwide to offer nutrition. It is similar to the king oyster mushroom, which is cultivated similarly. The biological decomposition of organic matter is another potential application for oyster mushrooms in the industrial environment. Almost five to six *Pleurotus* species are cultivated in approximately 27% of global mushroom production and utilised in healthcare, pharmaceuticals, the food industry and brewing; they are regarded as a rich protein supplement since they carry all of the essential amino acids, fibre and very little fat. Similarly, they deliver sizeable quantities of vitamins and other bioactive substances, such as unsaturated fats, phenolics, tocopherols and carotenoid pigments. *P. ostreatus* is among the most critical edible mushrooms in terms of quantity of output. It is a heterotrophic fungus with bioactive chemicals derived primarily from phenols. In recent decades, its use has increased due to the putative health benefits associated with its intake. In present agriculture, yields rarely increase twofold, even though mushroom demand has been exceptionally high.

2. History and Significance

The Latin term for edible oyster mushrooms is *Pleurotus*. As a result of their widespread distribution, *Pleurotus* mushroom species are frequently confused with one another and the identities of the group's unique species are always shrouded in mystery. *Pleurotus ostreatus*, one of the numerous species of edible mushrooms, was discovered, by Flack in 1917, growing on tree stumps and logs of wood in Germany. Flack was the first person to experiment with this mushroom. Additionally, Tsao and Hau improved growth techniques in the United States of America. In addition to other regions of continental Europe, the oyster mushroom, also known as *P. ostreatus*, was initially discovered in Great Britain and Ireland. The species name *ostreatus* relates to an oyster, whereas the generic term *Pleurotus* refers to a side ear. Further, Block et al. (1959) cultivated *P. ostreatus* under laboratory conditions for the first-time using sawdust as the colonising substrate. The scientists experimented by mixing oatmeal and sawdust for cultivation and reported the best results on eucalyptus sawdust followed by pine sawdust. In this perspective, Basidiomycetes mushrooms of the genus *Pleurotus* can provide a significant contribution since they possess the potential to create with minimal bioresource consumption and can contribute to the creation of products with added value. Furthermore, *P. ostreatus* are rapidly growing, short-lived fungi that can be collected in temperate and tropical locations with minimum investment and technical proficiency. Moreover, *P. ostreatus* may utilise numerous food wastes and byproducts as growth sources due to their enzymatic combinations, which include phenol oxidases and peroxidases, which efficiently degrade lignocellulose-rich substrates. Through this process, *P. ostreatus* produce fungal biomass rich in essential amino acids, branching β-glucans, minerals, micronutrients and low-molecular-weight bioactive substances.

3. Taxonomy

In mushroom taxonomy, the conception of morphological species predominates; most mushrooms are classed based on morphological traits. Furthermore, the morphological characteristics of fungal species are inconsistent and variable due to their considerable dependence on weather, growth medium, and environmental circumstances. Based on physical attributes, various taxonomists have reached divergent judgments regarding the phylogenetic classification of the same taxon. The taxonomy of *P. ostreatus* was scientifically first described by Dutch naturalist Nikolaus Joseph Freiherr Von Jacquin during 1775 and named *Agaricus ostreatus* (Syn: *Pleurotus opuntiae* (Dur. and Lev.) Sacc.), according to the early stage of fungal taxonomy. Further, German mycologist Paul Kummer defined and renamed the genus *Agaricus* to *Pleurotus* during 1871. Later, Kummer called a new genus as *P. ostreatus* (Jacq.) Kummer, belonging to the Pleurotaceae family (Agaricales, Agaricomycetes), as an accepted generic name for oyster mushroom ((Pat O'Reilly, 2022). The morphology of the basidiocarp is referenced in both the Latin name and the common name of the mushroom. The English common name, oyster, comes from the Latin *ostreatus*, which refers to the structure of the oyster-shaped cap, when the mushroom develops on decaying wood, similar to that of the identified bivalve. The Latin word *pleurotus*, which means 'fungus projecting outwards' or 'sideways', refers to the stipe's growth in a perpendicular direction to the cap. There is a possibility that the smooth texture of the basidiocarp is where the oyster comparison arises. The species *P. ostreatus* is referred to by the common name 'grey oyster' or 'pearl oyster' mushroom. However, the oyster mushroom tastes delicious. Since it typically develops in clumps weighing one pound or more, with five to 10 fruiting bodies per clump, it is an excellent fungus to consume. Saccardo developed a scientific organisation based on the relative position of the cap and stipe. Notably, the taxa of *Pleurotus* are divided into three major categories: excentric, dimidiate and resupinate (Venturella et al., 2015). This grey to brown cap features a semicircular to mussel-shaped pileus, a medial to the eccentric stipe, a white to grey spore print and a monomitic mycelial architecture. The cap is kidney-or oyster-shaped epigeous, when observed upwards, with a short stem that tapers downward from one side and slopes downhill from the margin to the stem. Pileus lingulate to spatulate, convex, then conchate to flabellate with 4–20 cm. The cuticle is velvety, dull and smooth, and its hue ranges from cream-beige to lilac-blackish. Lamellae are white to cream-coloured and decurrent, rudimentary, laterally attached stipe; hymenophoral trama slightly irregular, thin-walled, 2.5×4–2 µm in diameter. The basidiospores are broadly elliptic, smooth and hyaline, measuring 6.2–8×4–6.5 µm in forests and parks, solitary to gregarious, on stumps, fallen trunks and thriving on broad-leaved branches (Venturella et al., 2015). The upper surface is smooth and white or ivory-coloured. The white flesh is delicate and spongy, and the gills are white.

4. Diversity

The genus *Pleurotus* includes 30 species and sub-specific taxa of *P. ostreatus* (Venturella et al., 2015). *Pleurotus* is one of the wild mushrooms that is hunted most frequently, although it can also be grown on hay and other substrates. The odour is similar to that of bitter almonds due to benzaldehyde, which has a bittersweet quality. Further, the research employed mating tests to identify inter-sterility groupings among *Pleurotus* collections from Europe and America. Within the North American *Pleurotus* collections, there were three inter-sterility groupings. The 13 species of *Pleurotus* found in Europe have been categorised into eight inter-sterility groups. In a separate piece of research, researchers identified six clusters of inter-sterility that contain a total of seven *Pleurotus* species native to several continents. In another investigation, mating compatibility tests revealed the existence of five inter-sterility mushrooms (*P. cornucopiae, P. cystidiosus, P. ostreatus, P. pulmonarius* and *P. salmoneostramineus*) and seven independent inter-incompatible mushrooms (*P. calyptratus, P. corticatus, P. eryngii, P. dryinus, P. smithi, P. nebrodensis* and *P. ulmarius*) out of mushrooms studied from Asia. Based on the idea of biological species, there is currently a debate regarding the taxonomic status of five groups of seven species that are incompatible with one another (Bao et al., 2004). The *P. ostreatus* complex includes *P. ostreatus, P. ostreatus* var. *columbinus, P. djamor, P. flabellatus,* and *P. ostreatus* var. *columbinus*. The research by Li et al. (2017) demonstrates the evolutionary studies of *P. ostreatus* species complexes for commercial production. Seven distinct *P. ostreatus* lineages were discovered from 284 specimens with varied commercialised names collected from various fungal spawn-preservation institutes, organisations and wild isolations. Among four possible barcode genes employed, the rpb2 locus has been the most useful biomarker for *P. ostreatus* lineage identification. In addition, many *Pleurotus* lineages are commercially cultivated and have considerable economic value, including *P. citrinopieatus* (golden oyster mushroom), *P. cornucopiae* (branched oyster mushroom), *P. cystidiosus* (abalone oyster), *P. djamor* (pink oyster mushroom), *P. eryngii* (king oyster), *P. ostreatus* (pearl oyster mushroom), *P. pulmonarius* (phoenix mushroom) and *P. sajor-caju* (Indian oyster) (Zhang et al., 2017).

5. Ecology

The *Pleurotus* spp. is white spore oyster mushroom belonging to the category Basidiomycetes and consists of a large number of recognised as 'white rot fungi' because these yield white mycelium and are primarily fostered on non-composted lignocellulosic media by vibrantly producing laccase, ligninase and manganese peroxidase (oxidative enzymes) as well as cellulase, xylanase, and tannase (hydrolytic enzymes). Unlike other mushrooms, *Pleurotus* species require a shorter period of growth and grow better during autumn and summer in the tropical and sub-tropical regions, through which a variety of *Pleurotus* species are economically grown and have high commercial value. It can be produced inexpensively in simple structures with higher production through broad substrate use, sporelessness, wide temperatures and chemical stability, and biological remediation of the environment. This genus cleaves lignin, cellulose and hemicellulose from green and dried woods in suitable conditions, unlike brown rot fungi, which only cleave cellulose and hemicellulose (Machado et al., 2016). Secreted laccases are continually synthesised in minute amounts by this Basidiomycete mushroom and the lignocellulolytic enzymes are regulated by numerous fermentation-related parameters, including substrate concentration, acid levels, temperature, humidity and aeration rate, etc. It is rarely affected by pests and pathogens (Bellettini et al., 2019). *P. ostreatus* releases manganese peroxidases, the activity of which increases through the vegetative stage and reduces after sporocarp expansion (Velázquez-Cedeno et al., 2002). Laccase production is often more significant during substrate colonisation than during reproductive phases. Like laccase, significant amounts of manganese peroxidase enzyme are produced during colonisation (Savoie et al., 2007). Throughout primordial production and fructification, manganese peroxidase levels decrease and a subsequent surge is found during the post-harvest period. This genus has over 200 species that are dispersed globally and thus can be found in ecosystems, preferably decaying tree stumps and twigs, but can be produced in a variety of temperate and tropical environments. The oyster mushroom deteriorates lignin rather than polysaccharides. Furthermore, the bioactive components (mostly polysaccharides) of *Pleurotus* species include antibacterial, antibiotic, anticancer, hypocholesterolemic and immunomodulatory properties. The relationships between *Pleurotus* species, trees, animals and other microbes are diverse. Additionally, *Pleurotus* species are nematophagous and derive sustenance from devouring nematodes. Hyphae enable this with sticky knobs or droplets that connect to passing worms and produce nematicide chemicals (Venturella et al., 2015).

6. Nutritional Properties

The high protein content of mushrooms is preferable to that of cereals and is comparable to other plant-based protein sources. Mushrooms contain numerous non-digestible polysaccharides, including β-glucans, chitin, disaccharides, raffinose and complex carbohydrates. Mushrooms are the healthiest, safest, most nutritious foods and are vital to human well-being. Despite this, mushrooms have been consumed by people of all ages for the past decade due to their superior nutritional value compared to fruits and vegetables. Despite their nutritional value, they also possess medicinal properties and other health-promoting benefits and effects. Since the beginning of civilisation, humans from across the globe have utilised a variety of *Pleurotus* species (Kalač et al., 2009). Not only are mushrooms well-known for their taste and consistency, but also for their nutritional and chemical benefits. According to the Global Hunger Index (2019) findings, many nations across Africa and Asia are dealing with a significant issue of hunger and malnutrition. The Sustainable Development Goals (SDGs) established by the United Nations require all of the world's governments to put an end to hunger, achieve food security and improve nutrition by the year 2030. This is especially important for less fortunate and more vulnerable members of society, such as infants. According to the most recent statistics, 8.9% of the world's population, or approximately 690 million people, suffer from malnutrition. The number of undernourished people is highest in Asia with 381 million, followed by Africa with 250 million and Latin America and the Caribbean with 48 million (WHO, 2020). Mushrooms grown in a controlled environment have increased protein, mineral content, vitamin K, vitamin B, vitamin D and even, occasionally, vitamins A and C, Fe, Zn, Cu, Se and some bioactive mycochemicals, and despite having a low-fat and Na content (Yildiz et al., 1998; Lavelli et al., 2018). Mushrooms regulate protein deficit and complement cereal grains in people who do not consume animal proteins (either because of a lack of availability or due to religious belief). In contrast to the majority of other plant-based protein sources, the protein found in mushrooms contains all nine essential amino acids (EAAs). This is in contrast to most other plant-based protein sources, which typically lack one or two EAAs. Mushrooms have a high concentration of branched-chain amino acids (BCAA), customarily found solely in animal protein sources. According to the nutritional reports, *P. ostreatus* mushroom can potentially serve as a source of novel proteins. There have been discoveries of novel proteins with different capabilities, such as lectins, ligninolytic enzymes, proteolytic enzymes and hydrophobins. In addition, the protein in *P. ostreatus* appears to be higher than that of several popular legumes, such as peanuts, soybeans and protein-rich vegetables, as some *Pleurotus* species contain high levels of protein as well as balanced essential and non-essential amino acid molecules. Also, the fresh *P. ostreatus* mushroom basidiocarp contains up to 85–90% and about 1% of different macromolecules, which may be a source of dietary fibre (Wang et al., 2001). Tolera and Abera (2017) found 28.85% of the protein in fresh *P. ostreatus* mushroom. Using nitrogen-rich inputs along with a supplemental source of nitrogen in the media can potentially increase the amount of protein in mushrooms. The overall amino acid content of food is a reliable predictor of the complete quality of food, including the value of mushrooms. It is also well known that a small number of amino acids enhance the flavour of edible mushrooms, making them delightful. When combined with vitamins C and E, some amino acids, such as tryptophan, cysteine, alanine and glycine display a synergistic effect that enhances the antioxidant activity of the vitamins. The species of the genus *Pleurotus* are regarded as an excellent source of protein and amino acid content. In addition to that, *Pleurotus* species have a high concentration of gamma-aminobutyric acid (GABA), as well as ornithine. GABA is a non-essential amino acid for proper brain activity and mental skills. In addition, muscular proteins containing GABA are utilised in the therapy of wasting muscles that might occur as a result of sickness or post-operative treatment (Lavelli et al., 2018).

On the basis of the efficient usage of agro-waste substrates, a rich supply of minerals and their therapeutic value is gaining popularity today (Sibel Yildiz et al., 2002). In the present modern, technologically-advanced world, research is being conducted on the nutritional content and identification of various bioactive compounds, including steroids, alkaloids, terpenoids, phenols and lectins from the mycelium and fruiting body of *P. ostreatus* to combat deficiencies and diseases (Lindequist et al., 2005). Bonatti (2004) reported an 88.1 g/100 g moisture content for *P. ostreatus* grown in a banana fibre trial, while Reis et al. (2012) recorded 89.2 g/100 g in a commercially grown mushroom. The mushroom aroma has no nutritional value, yet it may stimulate a particular flavour. The odour of *P. ostreatus* stimulated the formation of 1-octanol, 3-octanol, 3-octanon, 1-caprynol-3-ol, 1-octynol-3-ol, 2-octynol-3-ol and 1-caprynol-3-on in large quantities (Deepalakshmi and Sankaran, 2014). The odour produced by *P. ostreatus* primarily depends on the availability of peptides and nucleotides (Bernas et al., 2006). The phenolic component of *P. ostreatus* consists of numerous varieties, such as vanillic acid, myricetin, naringin, homogentic acid, 5-O-caffeoylquinic acid, chrysin, routine, gentisic acid, gallic acid, protocatechuic acid, p-coumaric acid, caffeic acid, tannic acid, syringic acid and cinnamic acid (Rahimah et al., 2019). In addition, it is an excellent source of protein, a low-calorie diet with minimal fat in the form of ergosterol, a complement to the need for lipids and a source of fats, calories, minerals and vitamins (Table 1).

Mushrooms are a low-calorie food with a higher therapeutic benefit for diabetes patients to combat alimentary ulceration and promote weight loss. As a result of the existence of non-starch polysaccharides and larger number of bioactive elements, mushrooms have the potential to be used as a source of dietary fibre. In contrast, the mushroom stem included a more significant proportion of insoluble dietary fibre than the pileus did in every instance. On the other hand, mushrooms contain a total of

Table 1: Proximate macro-nutrient of *P. ostreatus*.

Nutrients	Content/100g Dry Weight
Moisture (g/100 g)	85–90
Crude protein (g/100 g)	20.5–40
Crude fat (g/100 g)	1.2–1.9
Carbohydrate	37–69.7
Total dietary fibre (g/100 g)	20–37
Ash (g/100 g)	7.9–10
Energy value (kcal/100 g)	356–374

intrinsic non-digestible polysaccharides that make up their total dietary fibre (TDF). Chitin makes up the majority of this fibre content in *P. ostreatus*. Glucans found in mushrooms can either be soluble or insoluble, depending on the type of dietary fibre they make up. The chemical structure and orientation play a significant role in determining how soluble they are in the water (Deepalakshmi and Sankaran, 2014). *P. ostreatus* consists of glucan with various forms of glycosidic connections, including linear (1→3)-α-glucans and branched (1→3), (1→6)-β-glucans. Intriguingly, the dehydrated fruiting bodies of *P. ostreatus* demonstrate the production of a particular β-glucan known as pleuran (a polysaccharide anticancer agent with a molecular weight ranging from 600,000 to 700,000 kDa). The antioxydative potential of pleuran is associated with its capacity to activate enzymes that neutralise free radicals: glutathione peroxidase, superoxide dismutase and catalase (Synytsya et al., 2008). In particular, the dietary fibre content of *P. ostreatus* has already been reported to be between 6–8% for soluble fibre and 22–37% for insoluble fibre (Kalac, 2009). The mannitol and trehalose content of the studied samples were as follows: 2.73 and 23.5 g/100 g respectively for *P. ostreatus* (Fernandes et al., 2015).

Although *Pleurotus* mushrooms are low in fat, the generation of necessary fatty acids contributes to maintaining human health. According to various findings, the total fat content of the analysed basidiocarps was significantly lower than the values reported in the past studies. The percentages of total saturated fatty acids (SFA), monounsaturated fatty acids (MUFA) and polyunsaturated fatty acids (PUFA) in the fatty acids' composition of *P. ostreatus* were found to be 20.2%, 10.8% and 69.1%, respectively, with linoleic acid (C18:2n–6c; 68.1%) being the predominant fatty acid (Fernandes et al., 2015). The fat content came in at 5.97 and 6.32 g/100 g for *P. ostreatus* cultured in bananas and paddy straw, respectively (Bonatti, 2004). The values that are described for the amount of protein present are pretty variable, as follows: 19.9–34.7 g/100 g in samples that were cultivated in wheat straw accompanied by beets (Manzi et al., 1999); the protein content of 21 g/100 g in *P. ostreatus* that was produced in wheat straw (Sopanrao et al., 2010) and 9.62 g/100 g in a specimen that was grown on coffee husks (Fan et al., 2000). *P. ostreatus* produced the highest quantities of monounsaturated fatty acids like oleic acid (363 µg/g dried mushroom) and polyunsaturated fatty acids like linoleic acid (533 µg/g dried mushroom) among the essential fatty acids (Hossain et al., 2007). In addition, the minimal lipid content of dried basidiocarps is mainly attributable to n-3 fatty acids, like linoleic acid and arachidonic acid. This linoleic acid is a predecessor of the attractive aroma of dehydrated mushrooms. Using a profile that was more comparable to that of the control, Reis et al. (2012) described the presence of four vitamers of α-, β-, γ- and δ-tocopherols at 42.4 µg/100 g, in a commercial sample of *P. ostreatus*. The estimated composition of organic acids in *P. ostreatus* includes citric acid (2.25 g/100 g), oxalic acid (0.30 g/100 g), fumaric acid (0.12 g/100 g) and quinic acid (0.10 g/100g) (Fernandes et al., 2015). *Pleurotus* species have an extraordinarily high concentration of folic acid (B_9). Folic acid is one of those vitamins that the body cannot create on its own and thus has to get from food. The growth of *P. ostreatus* on a mixed substrate consisting of soybean and wheat straw demonstrated the highest levels of folic acid (0.052 mg/100 g), thiamin, riboflavin, and vitamin C content (Sopanrao et al., 2010).

7. Pharmacological Values

Because these species contain many compounds that exhibit crucial pharmacological and therapeutic properties, they have been used as medicinal mushrooms for a long time. There are lectins with immunomodulatory, antiproliferative and anti-tumour properties; phenolic compounds with antioxidant properties and polysaccharides (polysaccharopeptides and polysaccharide proteins) with immune-enhancing and anticancer properties. Each of these classes of molecules possesses a unique set of characteristics (Table 2). In rats with colitis, beta-glucans isolated from *Pleurotus* sp. were shown to have anti-inflammatory effects (Jagadish et al., 2008), but *P. ostreatus* was shown to prevent leukocyte migration to wounded tissues caused by acetic acid (Jedinak and Sliva, 2008). Similarly, the extract of *P. florida* suppressed inflammation. In addition, the fungus *Pleurotus* has been shown to possess haematological, antiviral, anticancer, antibacterial, hypocholesterolemic, immunomodulatory and antioxidant properties (Jayakumar et al., 2008).

Lavi et al. (2006) found anti-proliferative and pro-apoptotic effects of an aqueous hot-water-soluble fraction of low-molecular-weight α-glucan extracted from the edible *P. ostreatus* on colon cancer HT-29 cells. Similarly, basidiocarps of *P. ostreatus* were extracted with hot water, precipitated with ethanol and fractionated using DEAE-cellulose ion-exchange chromatography and sepharose CL-6B gel filtration chromatography to yield a new water-soluble polysaccharide (POPS-1) with anti-tumour potential (Tong et al., 2009). Further, Jedinak and Sliva (2008) identified *P. ostreatus*, which inhibited the proliferation of several breast and colon cancer cells without impacting the proliferation of epithelial mammary and normal colon cells and was identified as the most potent compound. Flow cytometry demonstrated that the suppression of cancer cell growth by *P. ostreatus* was connected with the arrest of the cell cycle during the G0/G1 phase. In addition, *P. ostreatus* promoted the expression of the tumour suppressor p53 and the cyclin-dependent kinase inhibitor p21, while inhibiting the phosphorylation of the retinoblastoma Rb protein in cells. In addition, *P. ostreatus* up-regulated production of p21. They inhibited Rb phosphorylation in lines, indicating that it inhibits the growth of breast and colon cancer cells through a p53-dependent and independent mechanism. Maity et al. (2011) structurally identified the repeating unit of water-soluble polysaccharides using hydrolysis, methylation, oxidation, degradation and NMR analysis. They documented the stimulation of macrophages with various doses of the heteroglycan with d-glucose and d-galactose in a molar ratio of nearly 7:1, isolated from *P. ostreatus*. Biochemical analysis has been done on the total phenolic content, radical scavenging activity on 2,2-diphenyl-1-picryl-hydrazyl, ascorbic acid content and ion exchange activity on Fe2+ of the fungi, such as *H. ulmarius*, *P. citrinopileatus*, *P. djamor*, *P. eryngii*, *P. flabellatus*, *P. florida*, *P. sajor-caju* including *P. ostreatus*, to reveal its antioxidant potential (Mishra et al., 2013).

There are modest lipids in *Pleurotus* species, making them excellent sources of fatty acids, such as linoleic acid, arachidonic acid and oleic acid. Previous studies have indicated that certain species of *Pleurotus* are promising candidates for anti-inflammatory and hypocholesterolemic effects in the diets of humans (Ganesan and Xu, 2018). According to the findings of Schneider et al. (2011), the *P. ostreatus* diet prevented the buildup of low-density lipoproteins and very-low-density lipoprotein.

Table 2. Therapeutic effects.

S. No.	Therapeutic Effects	Mechanisms	References
1.	Antioxidant activity	Inhibits solid sarcoma 180 tumour development which when implanted in mice, helps in treatment of acute HSV-1 with beneficial effect on the respiratory tract.	Facchini et al., 2014; Urbancikova et al., 2020
		Iron chelating, reducing and scavenging abilities.	Yang et al., 2002
		Improves the catalase gene expression and decreases the incidence of free radical-induced protein oxidation in aged rats.	Jayakumar et al., 2011
2.	Anti-tumour activity	Anti-tumour activity on mice bearing sarcoma S-180 and hepatoma H-22.	Wang et al., 2000
		Water extract of *P. ostreatus* mycelium exhibited significant cytotoxic potential by inducing apoptosis of human carcinoma cells.	Egra et al., 2019
3.	Anticancer activity	Proteoglycans act as immunomodulators in mammals.	Borchers et al., 2008
		Protein extract mevinolin is a good source of potential cancer therapeutic agents.	Wu et al., 2011; Mitra et al., 2013
		Cytotoxicity and induced apoptosis in human androgen-independent prostate cancer PC-3 cells.	Gu and Sivam, 2006
		Inhibition in proliferation of MCF-7 human breast cancer cells.	Martin and Brophy, 2010
4.	Anti-atherosclerotic activity	Ergothioneine, lovastatin and chrysin compounds involved in anti-atherosclerotic activity.	Abidin et al., 2017
5.	Anti-proliferative activity	Induces pro-apoptosis on HT-29 colon cancer cells.	Lavi et al., 2006
6.	Antiviral activity	Inhibits the replication of hepatitis C virus and prevents entry into peripheral blood cells and hepatoma HepG2 cells.	El-Fakharany et al., 2010
7.	Anti-protective activity	Has a protective effect on inflammatory pathologies in diabetic rats.	Jayasuriya et al., 2020
8.	Genoprotective and nephroprotective activity	Suppresses DNA damage in artificially mutated *Drosophila* sp.	Taira et al., 2005
		Reduces the cadmium level in renal tissues and restores DNA fragmentation in rats.	Dkhil, 2020
9.	Antidiabetic activity	Decline in the blood-glucose levels in hyperglycemic conditions; even used for normal amaryl treatments.	Ghaly et al., 2011; Adebayo and Oloke, 2017
		Improves hyperlipidemia and impaired kidney functions.	Ravi et al., 2013
10.	Anti-bacterial activity	Involves cell membrane lysis, inhibition of protein synthesis, proteolytic enzymes and microbial adhesions.	Cowan, 1999

It considerably lowered the levels of cholesterol in humans. Initially, *P. ostreatus* has a hypocholesterolemic impact on rodents with regular cholesterolemia or hypercholesterolemia as well as inherited cholesterol abnormalities (Ozdín and Galbavý, 1998). In addition, they concluded that β-glucans in mushrooms are dispersed between the soluble and insoluble fractions of the diet. The recent review by Dicks and Ellinger (2020), *P. ostreatus* contains around twofold as many β-glucans as *A. bisporus*; these β-glucans are a dietary fibre gaining attractiveness for its role in preventing insulin resistance, hypertension, dyslipidemia and fatness. The aptitude of β-glucans from biological source to produce very viscous solutions in the gastrointestinal region, as well as their fermentability, is believed to play a crucial part in their proven health benefits (Sari et al., 2017). Further bioactive chemicals discovered in *P. ostreatus* may further promote cardiometabolic fitness. Mevinolin, also called lovastatin, inhibits the enzyme 3-hydroxy-3-methylglutaryl coenzyme A (HMG-COA) reductase and is probably involved in hypolipidemic properties by inhibiting cholesterol production in the human body (Abidin et al., 2017). A lovastatin from *P. ostreatus* was reported as an additive targeting lymphocyte. It possesses immunomodulatory, tioxidative and fibrinolytic activities and may protect against atherosclerosis (Sałata et al., 2017). Moreover, it has been demonstrated that numerous bioactive peptides derived from *in vitro* digestion of *P. ostreatus* inhibit the angiotensin-converting enzyme. This may lower blood pressure (BP), given that the same effects occur *in vivo*.

The discovery of ostreatin, a putative ribotoxin-like from the basidiocarp of *P. ostreatus*, as well as its catalytic and molecular characterisation, primarily focus on a group of unique enzymes that presumably play key roles in the homeostasis of Basidiomycetes mushrooms. Ostreatin is a monomeric selective ribonuclease (15 kDa) that could be utilised as a particular element for investigating human ribosomopathies. Ostrein may be able to contribute to the process of assigning possible biotechnological uses in the fields of crop protection against diseases and pests, and cytotoxic effects. In addition, this toxin exhibits snatching endonuclease activity on plasmids by adding and removing metallic magnesium and zinc ions (Landi et al., 2020).

Researchers have found a lot of bioactive compounds including beta-carotene, polysaccharides, phenolic flavonoids, glycoproteins, lectins, a number of lipid components, steroids, ergothioneine, terpenoids, vitamin C and selenium (Rahimah et al., 2019) with therapeutic value from *P. ostreatus*, that could be used as nutraceuticals (Table 3). In most cases, the entire mushroom and its secondary metabolites are devoured. This species complex is said to possess superior nutritional qualities due to its sweet and mildly astringent flavour, and its fatty nature. Mohamed and Farghaly (2014) explained the difference in chemical properties and active metabolites between fresh and dried *P. ostreatus*. They identified 107 distinct compounds among the chemicals tested and suggested using fresh mushroom fruiting bodies as a reservoir of aromatics and dried mushrooms as a potent antioxidant. Numerous enzymes have been linked to the pathogenesis of fungi, but little is known regarding their characteristics, modes of action, regulation, cellular localisation and production levels. *Pleurotus* species are considered a significant source of dietary fibre and contain many essential minerals and polyphenols, which contribute to their antioxidant nature and capacity to suppress free radicals (Golak-Siwulska et al., 2018). Even if the human body is a balance of oxidants and antioxidants, dietary antioxidants are still required to minimise

Table 3. Chemical classes and their health benefits.

Chemical Classes	Compounds	Health Benefits	References
Polysaccharides	Polysaccharide (β-glucans)	Antioxidant activity	Mitra et al., 2013
	Lectin	Anti-tumour activity	Wang et al., 2000
	Proteoglycans	Anticancer agents	Borchers et al., 2008
	Pleuran	Induce gene expression of cytokines	Solomko et al., 1992; Golak-Siwulska et al., 2018
	Selenium polysaccharide (Se-POP-21)	Antioxidant and anti-tumour activities	Zhang et al., 2021
Proteins	Protein extract mevinolin	Anticancer activities	Wu et al., 2011
	Ergothioneine, lovastatin and chrysin	Anti-atherosclerotic activities	Abidin et al., 2017
	ubiquitin (protein)	Anti viral activities	Wang et al., 2000
Enzymes	Laccase	Inhibits hepatitis C virus entry into peripheral blood cells and hepatoma cells	El-Fakharany et al., 2010

peroxidation from the atmosphere (Adebayo et al., 2012). *P. ostreatus* is associated with high moisture capacity and low caloric content of 1510 kJ kg^{-1} edible portions, rendering it appropriate for incorporation into calorie-controlled regimes (Jaworska and Bernás, 2009). The pileus of *P. ostreatus* was primarily regarded for its flavour and its nutritional composition, particularly in vegan diets.

Polysaccharides derived from *Pleurotus* mushroom are the most potent, widely dispersed and well-researched bioactive substances responsible for the suppression of illnesses and deficits. The structurally varied set of macromolecules is linked to a β-glucose backbone that constitutes mushroom polysaccharides (Nakashima et al., 2018). Glucans from polysaccharides are abundant in the mushroom fungal cell wall, primarily in the form of cellulose, which acts as a powerful anticancer polysaccharide, but chitin and chitosan lack polysaccharide properties (Chaitanya et al., 2019; Pandya et al., 2018). *P. ostreatus* polysaccharides are believed to have natural immune-stimulatory characteristics with substantial clinical and medicinal uses (El-Deeb et al., 2019). Diverse polysaccharides derived from the *Pleurotus* mushroom are recognised as strong anticancer agents. Polysaccharides extracted from *P. ostreatus* have been demonstrated to trigger and establish a favourable role for NK-cell cytotoxic effects in the activation and generation of an innate immune system against breast and lung cancer cells. Additionally, *P. ostreatus* is abundant in phenolic compounds, which may elucidate the increased antioxidant effects observed in extracts (Piskov et al., 2020). The phenol is the primary chemical believed to be responsible for the antioxidant properties of white oyster mushrooms. Its hydroxyl content is highly correlated with its antioxidant capacity. The previous study has demonstrated that phenol from *P. ostreatus* prevents liver, lung and kidney damage (Kane et al., 2022). Phenol has the ability to chelate metals and can function as a reduction agent for hydrogen donors and singlet oxygen quenchers.

8. Commercial Cultivation

Recent increases in both health consciousness and the demand for meat substitutes have contributed to a worldwide surge in mushroom consumption. Mushroom cultivation is one of the most efficient and rapid biological processes in terms of the biotransformation of resources into nutrient-rich foods. Because of its low production costs and great yielding capacity, the *Pleurotus* mushroom sector in Asia has experienced a tremendous expansion in recent years. The cultivation of oyster mushrooms depends on various conditions, which may have independent or synergistic impacts during growth and reproduction. The cultivation procedure consists of preparing the substrate, casing and exposing the species to temperature shocks. Chemical constituents, the C:N ratio, minerals, water activity, pH, humidity, sources of ammonia, particle density, the quantity of starter culture, surfactant, antimicrobial agents and the involvement of microbe interactions are recognised as physical, chemical and biological considerations associated with mushroom cultivation. The significant environmental parameters include temperature, moisture, brightness and the underlying substrate's atmospheric component, such as O_2 and CO_2 concentrations. Mushroom growing is an effective strategy for reducing environmental problems. The most commonly utilised substrates in commercial production are those made of broadleaves, hardwoods, sawdust and straws, with various supplements added. The synthetic substrates have to undergo a pretreatment process in a sterile setting, primarily so that any impurities may be removed (Raman et al., 2021).

The cultivation of mushrooms has become an indispensable cottage industry for the community empowerment programme. Despite the apparent ease of large-scale growing, selecting the proper medium and spawning duration can be difficult. Due to the necessity of waste disposal, the oyster mushroom develops on various by-products and garbage from agricultural and agro-industrial processes, which provide an environmental advantage (Cohen et al., 2002). The oyster mushroom is a macro-basidiomycetous fungus that produces a characteristic fruiting body and utilises agro-waste substrates to produce degrading enzymes (Piškur et al., 2011). Naturally, *P. ostreatus* decomposes the complex organic materials (agro-waste substrate) on which it develops to produce a nutrient-rich product for human use. Typically, the substrates colonised by fungi are byproducts or wastes from industry, households and agriculture (Tesfaw et al., 2015). In light of the enormous variety of domestic wastes (food waste, compost, napkins, sawdust, chopped office papers, cardboards and plant fibres) and agricultural by-products {banana dried pseudostem, biogas residual slurry manure, coffee pulp, corn trashes, cotton wastes, fruit wastes, jute wastes, leaf of hazelnut, olive mill waste, pine needles, palm leaves, paddy straw, peanut hull, pulses husk, sugarcane leaves or biomass (trash, bagasse and pith), thatch grass, tomato tuff and wheat straw} were utilised for cultivation of oyster mushrooms (Bellettini et al., 2019). Further, some research on the production of *P. ostreatus* on various substrates including beech, chestnut, coconut coir and fibre, corncobs, linden, oak, potato crop stubbles, peanut haulm, poplar, walnut, etc., nothing is known about the optimal substrate for this mushroom (Mihai et al., 2022). The production of *P. ostreatus* on such waste by-products could be one of the remedies for transforming the inedible pollutants into edible protein-rich biomass with a high economic value as it would be rich in nutrients and has strong antioxidant content. The commercial cultivation technology of *P. ostreatus* mushroom is very interesting and simple. Practically, each cropping period of mushroom production takes 45 days under controlled conditions and hence there can be eight crops per year. *P. ostreatus* can grow at a moderate temperature at 20–30°C with a frequent relative humidity of 70–85% for a period of six to eight months per year. In addition, *P. ostreatus* can also be cultivated during summer by maintaining extra humidity for its growth in hilly areas. Again, for the production of *P. ostreatus*, Lechner and Albertó (2011) reported that the ultimate relative humidity (w/w) of the growing substrate was adjusted to 74% to account for the moisture content. Alananbeh et al. (2014) analysed three spawning rates and discovered that the production, biological effectiveness and cumulative fruiting bodies rose as the proportion of producing *P. ostreatus* enhanced. El-Batal et al. (2015) revealed that the combination of Tween®80 (0.02% (v/v)) results in better yield potential of enzyme production in *P. ostreatus* under solid-state fermentation since there is an indication that all these spreads over the surface agents lead to significantly greater oxygen permeability and extracellular enzyme transit through the cellular membrane of mushrooms (Silva et al., 2007). Growth inhibition ensued when *P. ostreatus* was cultured in hydrolysed bagasse without adding nitrogen and the addition of small amounts of nitrogen sources produced nitrogen levels more than or equal to 1.5% on a dry weight basis. A low nitrogen concentration can enhance ligninolytic enzyme activity, while a high nitrogen concentration inhibits it. Neelam et al. (2011) revealed that ammonium chloride stimulates mycelial growth and development in *P. ostreatus* more effectively than calcium nitrate and sodium nitrate mainly due to nitrate ions, which have indeed been allegedly involved in the inhibitory action of certain mushrooms and may be difficult to move throughout the fungal outer layer, where they can encourage growth. Hoa and Wang (2015) reported that the ammonium chloride concentration enhanced mycelium growth in *P. ostreatus* from 0.03% to 0.09%. Owaid et al. (2015), in their research on composting cardboard waste products to yield *P. ostreatus*, observed that the small contours of the substrate wheat straw have a significant influence on oyster mushroom development. Compared to large pieces of cardboard, this increased decayed wheat straw and extensive mycelia biomass comprised due to the increased substrate permeability for colony growth. Therefore, groups were developed on this material. Heavy metals, such as manganese, mercury, arsenic, zinc, lead and copper can give the dish an unpleasant taste and cause personal poisoning. Oyster mushroom production is a highly profitable crop with high demand and supply patterns and exports.

9. Conclusion

Because of its economic (food), environmental (bioconversion intermediaries) and therapeutic properties (antioxidant and compound sources), the *P. ostreatus* mushroom contains naturally occurring, biologically active substances which benefit humans' health. One of the primary goals of chemoprevention is to bolster the body's natural defence mechanisms by consuming substances of natural origin that are already present in the diet. Biologically active chemicals found in oyster mushrooms' fruiting bodies have been shown in many scientific studies to protect against the onset of diseases associated with modern civilisation and aid in treating such conditions. It has been demonstrated that bioactive compounds found in *P. ostreatus*, primarily ß-glucan and lectin, are effective in the treatment of cancer as well as states of physical and mental fatigue; pleuran is effective in the treatment of cardiovascular and liver diseases, and serotonin is effective in the treatment of neurosis. The magnificent effects of using *P. ostreatus* in the treatment of advanced stages of cancer, frequently with enormous metastatis and in its tremendous help as a supportive component in chemo- and radiotherapy of various cancers, should be emphasised. This mushroom can penetrate a diverse range of lignocellulosic substrates and other organic waste from the agricultural, forestry and food fields. The production of this mushroom is an efficient alternative method for the production of great sources of food, pharmaceuticals and nutraceuticals. *P. ostreatus* has a shorter growth time (in comparison to other edible mushrooms); the substrate that is used needs to be pasteurised; it has high profitability (converting a higher proportion of the growing medium to basidiocarps) and diseases and pests are less prone to attack it than other edible mushrooms. These are the advantages of growing *P. ostreatus*.

References

Abidin, M.H.Z., Abdullah, N. and Abidin, N.Z. (2017). Therapeutic properties of *Pleurotus* species (oyster mushrooms) for atherosclerosis: A review. *International Journal of Food Properties*, 20(6): 1251–1261.

Adebayo, E.A. and Oloke, J.K. (2017.) Oyster Mushroom (*Pleurotus* species); A natural functional food. *Journal of Microbiology, Biotechnology and Food Sciences*, 7(3): 254–264

Adebayo, E.A., Oloke, J.K., Ayandele, A.A. and Adegunlola, C.O. (2020). Phytochemical, antioxidant and antimicrobial assay of mushroom metabolite from *Pleurotus pulmonarius*-LAU 09 (JF736658). *Journal of Microbiology and Biotechnology Research*, 2(2): 366–374.

Alananbeh, K.M., Bouqellah, N.A. and Al Kaff, N.S. (2014). Cultivation of oyster mushroom *Pleurotus ostreatus* on date-palm leaves mixed with other agro-wastes in Saudi Arabia. *Saudi Journal of Biological Sciences*, 21(6): 616–625.

Baldrian, P., Větrovský, T., Lepinay, C. and Kohout, P. (2021). High-throughput sequencing view on the magnitude of global fungal diversity. *Fungal Divers*, 19: 1–9.

Bao, D., Kinugasa, S. and Kitamoto, Y. (2004). The biological species of oyster mushrooms (*Pleurotus* spp.) from Asia based on mating compatibility tests. *Journal of Wood Science*, 50(2): 162–168.

Bellettini, M.B., Fiorda, F.A., Maieves, H.A., Teixeira, G.L., Ávila, S., Hornung, P.S., Júnior, A.M. and Ribani, R.H. (2019). Factors affecting mushroom *Pleurotus* species. *Saudi Journal of Biological Sciences*, 26(4): 633–646.

Bernaś, E., Jaworska, G. and Lisiewska, Z. (2006). Edible mushrooms as a source of valuable nutritive constituents. *Acta Scientiarum Polonorum Technologia Alimentaria*, 5(1): 5–20.

Block, S.S., Tsaq, G. and Han, L. (1959). Experiments in the cultivation of *Pleurotus ostreatus*. *Mushroom Science*, 4: 309–325.

Bonatti, M., Karnopp, P., Soares, H.M. and Furlan, S.A. (2004). Evaluation of *Pleurotus ostreatus* and *Pleurotus sajor-caju* nutritional characteristics when cultivated in different lignocellulosic wastes. *Food Chemistry*, 88(3): 425–428.

Borchers, A.T., Krishnamurthy, A., Keen, C.L., Meyers, F.J. and Gershwin, M.E. (2008). The immunobiology of mushrooms. *Experimental Biology and Medicine*, 233(3): 259–276.

Chaitanya, M.V.N.L., Jose, A., Ramalingam, P., Mandal, S.C. and Kumar, P.N. (2019). Multi-targeting cytotoxic drug leads from mushrooms. *Asian Pacific Journal of Tropical Medicine*, 12(12): 531.

Cohen, R., Persky, L. and Hadar, Y. (2002). Biotechnological applications and potential of wood-degrading mushrooms of the genus *Pleurotus*. *Applied Microbiology and Biotechnology*, 58(5): 582–594.

Cowan, M.M. (1999). Plant products as antimicrobial agents. *Clinical Microbiology Reviews*, 12: 564–582.

Deepalakshmi, K. and Sankaran, M. (2014). *Pleurotus ostreatus*: An oyster mushroom with nutritional and medicinal properties. *Journal of Biochemical Technology*, 5(2): 718–726.

Dicks, L. and Ellinger, S. (2020). Effect of the intake of oyster mushrooms (*Pleurotus ostreatus*) on cardiometabolic parameters—A systematic review of clinical trials. *Nutrients*, 12(4): 1134.

Dkhil, M.A., Diab, M.S., Lokman, M.S., El-Sayed, H., Bauomy, A.A., Al-Shaebi, E.M. and Al-Quraishy, S. (2020). Nephroprotective effect of *Pleurotus ostreatus* extract against cadmium chloride toxicity in rats. *Anais da Academia Brasileira de Ciências*, 92.

Egra, S., Kusuma, I.W., Arung, E.T. and Kuspradini, H. (2019). The potential of white-oyster mushroom (*Pleurotus ostreatus*) as antimicrobial and natural antioxidant. *Biofarmasi. J. Nat. Prod. Biochem.*, 17(1): 14–20.

El-Batal, A.I., El-Kenawy, N.M., Yassin, A.S. and Amin, M.A. (2015). Laccase production by *Pleurotus ostreatus* and its application in synthesis of gold nanoparticles. *Biotechnology Reports*, 5: 31–39.

El-Deeb, N.M., El-Adawi, H.I., EL-wahab, A.E.A., Haddad, A.M., El Enshasy, H.A., He, Y.W. and Davis, K.R. (2019). Modulation of NKG2D, KIR2DL and cytokine production by *Pleurotus ostreatus* glucan enhances natural killer cell cytotoxicity toward cancer cells. *Frontiers in Cell and Developmental Biology*, 7: 165.

El-Fakharany, E., Haroun, B., Ng, T. and Redwan, E.R. (2010). Oyster mushroom laccase inhibits hepatitis C virus entry into peripheral blood cells and hepatoma cells. *Protein and Peptide Letters*, 17(8): 1031–1039.

Facchini, J.M., Alves, E.P., Aguilera, C., Gern, R.M.M., Silveira, M.L.L., Wisbeck, E. and Furlan, S.A. (2014). Anti-tumour activity of *Pleurotus ostreatus* polysaccharide fractions on Ehrlich tumour and sarcoma 180. *International Journal of Biological Macromolecules*, 68: 72–77.

Fan, L., Soccol, C.R. and Pandey, A. (2000). Produção de cogumelo comestível Pleurotus em casca de café e avaliação do grau de detoxificação do substrato. In Symposium search of coffee from Brazil, 1, (pp. 687–670). Minas Gerais, Brasil.

Fernandes, Â., Barros, L., Martins, A., Herbert, P. and Ferreira, I.C. (2015). Nutritional characterisation of *Pleurotus ostreatus* (Jacq. ex Fr.) P. Kumm., produced, using paper scraps as substrate. *Food Chemistry*, 169: 396–400.

Flack, R. 1917. Uber dle walkulter des austernpilzes (*Agaricus ostreatus*) out laubholzstubben. *Forst-sagdwes*, 49: 159–165.

Ganesan, K. and Xu, B. (2018). Anti-obesity effects of medicinal and edible mushrooms. *Molecules*, 23(11): 2880.

Gautam, A.K., Verma, R.K., Avasthi, S., Bohra, Y., Devadatha, B., Niranjan, M. and Suwannarach, N. (2022). Current insight into traditional and modern methods in fungal diversity estimates. *Journal of Fungi*, 8(3): 226.

Ghaly, I.S., Ahmed, E.S., Booles, H.F., Farag, I.M. and Nada, S.A. (2011). Evaluation of antihyperglycemic action of oyster mushroom (*Pleurotus ostreatus*) and its effect on DNA damage, chromosome aberrations and sperm abnormalities in streptozotocin-induced diabetic rats. *Global Veterinaria*, 7(6): 532–544.

Global Hunger Index. (2019). Global hunger index. Global Hunger Index-Peer-Reviewed Annual Publication Designed to Comprehensively Measure and Track Hunger at the Global, Regional, and Country Levels. https://www. globalhungerindex. org/india.html.

Golak-Siwulska, I., Kałużewicz, A., Spiżewski, T., Siwulski, M. and Sobieralski, K. (2018). Bioactive compounds and medicinal properties of oyster mushrooms (sp.). *Folia Horticulturae*, 30(2): 191–201.

Gu, Y.H. and Sivam, G. (2006). Cytotoxic effect of oyster mushroom *Pleurotus ostreatus* on human androgen-independent prostate cancer PC-3 cells. *Journal of Medicinal Food*, 9(2): 196–204.

Hawksworth, D.L. and Lücking, R. (2017). Fungal diversity revisited: 2.2 to 3.8 million species. *Microbiol. Spectrum*, 5(4): FUNK-0052-2016.

Hoa, H.T. and Wang, C.L. (2015). The effects of temperature and nutritional conditions on mycelium growth of two oyster mushrooms (*Pleurotus ostreatus* and *Pleurotus cystidiosus*). *Mycobiology*, 43(1): 14–23.

Hossain, M.S., Alam, N., Amin, S.R., Basunia, M.A. and Rahman, A. (2007). Essential fatty acid contents of *Pleurotus ostreatus*, *Ganoderma lucidum* and *Agaricus bisporus*. *Bangladesh J. Mushroom*, 1: 1–7.

Jagadish, L.K., Shenbhagaraman, R., Venkatakrishnan, V. and Kaviyarasan, V. (2008). Studies on the phytochemical, antioxidant and antimicrobial properties of three indigenous *Pleurotus* species. *Journal of Molecular Biology and Biotechnology*, 1(1): 20–29.

Jaworska, G. and Bernaś, E. (2009). Qualitative changes in *Pleurotus ostreatus* (Jacq.: Fr.) Kumm. mushrooms resulting from different methods of preliminary processing and periods of frozen storage. *Journal of the Science of Food and Agriculture*, 89(6): 1066–1075.

Jayakumar, T., Sakthivel, M., Thomas, P.A. and Geraldine, P. (2008). *Pleurotus ostreatus*, an oyster mushroom, decreases the oxidative stress induced by carbon tetrachloride in rat kidneys, heart and brain. *Chemico-biological Interactions*, 176(2-3): 08–120.

Jayakumar, T., Thomas, P.A., Sheu, J.R. and Geraldine, P. (2011). *In-vitro* and *in-vivo* antioxidant effects of the oyster mushroom *Pleurotus ostreatus*. *Food Research International*, 44(4): 851–861.

Jayasuriya, W.J.A., Handunnetti, S.M., Wanigatunge, C.A., Fernando, G.H., Abeytunga, D.T.U. and Suresh, T.S. (2020). Anti-inflammatory activity of *Pleurotus ostreatus*, a culinary medicinal mushroom, in Wistar rats. *Evidence-based Complementary and Alternative Medicine*, 2020.

Jedinak, A. and Sliva, D. (2008). *Pleurotus ostreatus* inhibits proliferation of human breast and colon cancer cells through p53-dependent as well as p53-independent pathway. *International Journal of Oncology*, 33(6): 1307–1313.

Kalač, P. (2009). Chemical composition and nutritional value of European species of wild growing mushrooms: A review. *Food Chemistry*, 113(1): 9–16.

Kane, F.C., Kimassoum, D., Brice, S.F., Paul, M.F. and Mbacham, W.F. (2022). Antioxidant property of a dietary supplement of *Moringa oleifera* leaves and *Pleurotus ostreatus* in wistar rats subjected to forced swimming endurance test. *Food and Nutrition Sciences*, 13(5): 493–503.

Landi, N., Ragucci, S., Russo, R., Valletta, M., Pizzo, E., Ferreras, J.M. and Di Maro, A. (2020). The ribotoxin-like protein ostreatin from *Pleurotus ostreatus* fruiting bodies: Confirmation of a novel ribonuclease family expressed in Basidiomycetes. *International Journal of Biological Macromolecules*, 161: 1329–1336.

Lavelli, V., Proserpio, C., Gallotti, F., Laureati, M. and Pagliarini, E. (2018). Circular reuse of bio-resources: The role of *Pleurotus sajor-caju* in the development of functional foods. *Food & Function*, 9(3): 1353–1372.

Lavi, I., Friesem, D., Geresh, S., Hadar, Y. and Schwartz, B. (2006). An aqueous polysaccharide extract from the edible mushroom *Pleurotus ostreatus* induces anti-proliferative and pro-apoptotic effects on HT-29 colon cancer cells. *Cancer Letters*, 244(1): 61–70.

Lechner, B.E. and Albertó, E. (2011). Search for new naturally occurring strains of *Pleurotus* to improve yields. *Pleurotus albidus* as a novel proposed species for mushroom production. *Revista Iberoamericana de Micologia*, 28(4): 148–154.

Li, H., Tian, Y., Menolli Jr, N., Ye, L., Karunarathna, S.C., Perez-Moreno, J., Rahman, M.M., Rashid, M.H., Phengsintham, P., Rizal, L. and Kasuya, T. (2021). Reviewing the world's edible mushroom species: A new evidence-based classification system. *Comprehensive Reviews in Food Science and Food Safety*, 20(2): 1982–2014.

Li, J., He, X., Liu, X.B., Yang, Z.L. and Zhao, Z.W. (2017). Species clarification of oyster mushrooms in China and their DNA barcoding. *Mycological Progress*, 16(3): 191–203.

Lindequist, U., Niedermeyer, T.H.J. and Jülich, W.D. (2005). The pharmacological potential of mushrooms. *Evid-based Complement Altern Med.*, 2: 285–299.

Lücking, R., Aime, M.C., Robbertse, B., Miller, A.N., Aoki, T., Ariyawansa, H.A., Cardinali, G., Crous, P.W., Druzhinina, I.S., Geiser, D.M. and Hawksworth, D.L. (2021). Fungal taxonomy and sequence-based nomenclature. *Nature Microbiology*, 6(5): 540–548.

Machado, A.R.G., Teixeira, M.F.S., de Souza Kirsch, L., Campelo, M.D.C.L. and de Aguiar Oliveira, I.M. (2016). Nutritional value and proteases of *Lentinus citrinus* produced by solid state fermentation of lignocellulosic waste from tropical region. *Saudi Journal of Biological Sciences*, 23(5): 621–627.

Maity, K.K., Patra, S., Dey, B., Bhunia, S.K., Mandal, S., Das, D., Majumdar, D.K., Maiti, S., Maiti, T.K. and Islam, S.S. (2011). A heteropolysaccharide from aqueous extract of an edible mushroom, *Pleurotus ostreatus* cultivar: Structural and biological studies. *Carbohydrate Research*, 346(2): 366–372.

Manzi, P., Gambelli, L., Marconi, S., Vivanti, V. and Pizzoferrato, L. (1999). Nutrients in edible mushrooms: An inter-species comparative study. *Food Chemistry*, 65(4): 477–482.

Martin, K.R. and Brophy, S.K. (2010). Commonly consumed and specialty dietary mushrooms reduce cellular proliferation in MCF-7 human breast cancer cells. *Experimental Biology and Medicine*, 235(11): 1306–1314.

Mihai, R.A., Melo Heras, E.J., Florescu, L.I. and Catana, R.D. (2022). The edible grey oyster fungi *Pleurotus ostreatus* (Jacq. ex Fr.) P. Kumm a potent waste consumer, a biofriendly species with antioxidant activity depending on the growth substrate. *Journal of Fungi*, 8(3): 274.

Mishra, K.K., Pal, R.S., Arunkumar, R., Chandrashekara, C., Jain, S.K. and Bhatt, J.C. (2013). Antioxidant properties of different edible mushroom species and increased bioconversion efficiency of *Pleurotus eryngii* using locally available casing materials. *Food Chemistry*, 138(2-3): 1557–1563.

Mitra, P., Khatua, S. and Acharya, K. (2013). Free radical scavenging and NOS activation properties of water soluble crude polysaccharide from *Pleurotus ostreatus*. *Asian J. Pharm. Clin. Res.*, 6(3): 67–70.

Mohamed, E.M. and Farghaly, F.A. (2014). Bioactive compounds of fresh and dried *Pleurotus ostreatus* mushroom. *International Journal of Biotechnology for Wellness Industries*, 3(1): 4–14.

Nakashima, A., Yamada, K., Iwata, O., Sugimoto, R., Atsuji, K., Ogawa, T., Ishibashi-Ohgo, N. and Suzuki, K. (2018). β-Glucan in foods and its physiological functions. *Journal of Nutritional Science and Vitaminology*, 64(1): 8–17.

Neelam, S., Chennupati, S. and Singh, S. (2011). Comparative studies on growth parameters and physio-chemical analysis of *Pleurotus ostreatus* and *Pleurotus florida*. *Asian Journal of Plant Science & Research*, 18(3): 201–207.

Niazi, A.R. and Ghafoor, A. (2021). Different ways to exploit mushrooms: A review. *All Life*, 14(1): 450–460.

Owaid, M.N., Abed, A.M. and Nassar, B.M. (2015). Recycling cardboard wastes to produce blue oyster mushroom *Pleurotus ostreatus* in Iraq. *Emirates Journal of Food and Agriculture*, 537–541.

Ozdín, L. and Galbavý, Š. (1998). Dose-and time-dependent hypocholesterolemic effect of oyster mushroom (*Pleurotus ostreatus*) in rats. *Nutrition*, 14(3): 282–286.

Pandya, U., Dhuldhaj, U. and Sahay, N.S. (2018). Bioactive mushroom polysaccharides as anti-tumour: An overview. *Natural Product Research*, 33(18): 2668–2680.

Pat O'Reilly. (2022). Fascinated by fungi. First Nature, pp. 1–450.

Pérez-Moreno, J., Mortimer, P.E., Xu, J., Karunarathna, S.C. and Li, H. (2021). Global perspectives on the ecological, cultural and socioeconomic relevance of wild edible fungi. *Studies in Fungi*, 6(1): 408–424.

Piskov, S., Timchenko, L., Grimm, W.D., Rzhepakovsky, I., Avanesyan, S., Sizonenko, M. and Kurchenko, V. (2020). Effects of various drying methods on some physico-chemical properties and the antioxidant profile and ACE inhibition activity of oyster mushrooms (*Pleurotus ostreatus*). *Foods*, 9(2): 160.

Piškur, B., Bajc, M., Robek, R., Humar, M., Sinjur, I., Kadunc, A., Oven, P., Rep, G., Petkovšek, S.A.S., Kraigher, H. and Jurc, D. (2011). Influence of *Pleurotus ostreatus* inoculation on wood degradation and fungal colonisation. *Bioresource Technology*, 102(22): 10611–10617.

Rahimah, S.B., Djunaedi, D.D., Soeroto, A.Y. and Bisri, T. (2019). The phytochemical screening, total phenolic contents and antioxidant activities in vitro of white oyster mushroom (*Pleurotus ostreatus*) preparations: Open access. *Macedonian Journal of Medical Sciences*, 7(15): 2404.

Raman, J., Jang, K.Y., Oh, Y.L., Oh, M., Im, J.H., Lakshmanan, H. and Sabaratnam, V. (2021). Cultivation and nutritional value of prominent *Pleurotus sajor-caju*: An overview. *Mycobiology*, 49(1): 1–14.

Ravi, B., Renitta, R.E., Prabha, M.L., Issac, R. and Naidu, S. (2013). Evaluation of antidiabetic potential of oyster mushroom (*Pleurotus ostreatus*) in alloxan-induced diabetic mice. *Immunopharmacology and Immunotoxicology*, 35(1): 101–109.

Reis, F.S., Barros, L., Martins, A. and Ferreira, I.C. (2012). Chemical composition and nutritional value of the most widely appreciated cultivated mushrooms: An inter-species comparative study. *Food and Chemical Toxicology*, 50(2): 191–197.

Sałata, A., Lemieszek, M. and Parzymies, M. (2017). The nutritional and health properties of an oyster mushroom (*Pleurotus ostreatus* (Jacq. Fr) P. Kumm.). *Acta Sci. Pol. Hortorum Cultus,* 17: 185–197.

Sari, M., Prange, A., Lelley, J.I. and Hambitzer, R. (2017). Screening of beta-glucan contents in commercially cultivated and wild growing mushrooms. *Food Chemistry,* 216: 45–51.

Savoie, J.M., Salmones, D. and Mata, G. (2007). Hydrogen peroxide concentration measured in cultivation substrates during growth and fruiting of the mushrooms *Agaricus bisporus* and *Pleurotus sajor-caju. Journal of the Science of Food and Agriculture,* 87(7): 1337–1344.

Schneider, I., Kressel, G., Meyer, A., Krings, U., Berger, R.G. and Hahn, A. (2011). Lipid lowering effects of oyster mushroom (*Pleurotus ostreatus*) in humans. *Journal of Functional Foods,* 3(1): 17–24.

Silva, E.G., Dias, E.S., Siqueira, F.G. and Schwan, R.F. (2007). Chemical analysis of fructification bodies of *Pleurotus sajor-caju* cultivated in several nitrogen concentrations. *Food Science and Technology,* 27: 72–75.

Solomko, E.F. (1992). The physiology-biochemical properties and biosynthetic activities of higher Basidiomycetes mushroom *Pleurotus ostreatus* (Jacq.: Fr.) Kumm. *In: Submerged Culture: Dr. Sci. Submerged Culture,* 49–63.

Sopanrao, P.S., Abrar, A.S., Manoharrao, T.S. and Vaseem, B.M.M. (2010). Nutritional value of *Pleurotus ostreatus* (Jacq: Fr) Kumm cultivated on different lignocellulosic agro-wastes. *Innovative Romanian Food Biotechnology,* 7: 66–76.

Synytsya, A., Mickova, K., Jablonsky, I., Sluková, M. and Copikova, J. (2008). Mushrooms of genus *Pleurotus* as a source of dietary fibres and glucans for food supplements. *Czech J. Food Sci.,* 26(6): 441–446.

Taira, K., Miyashita, Y., Okamoto, K., Arimoto, S., Takahashi, E. and Negishi, T. (2005). Novel anti-mutagenic factors derived from the edible mushroom *Agrocybe cylindracea. Mutation Research/Genetic Toxicology and Environmental Mutagenesis,* 586(2): 115–123.

Tesfaw, A., Tadesse, A. and Kiros, G. (2015). Optimidation of oyster (*Pleurotus ostreatus*) mushroom cultivation using locally available substrates and materials in Debre Berhan, Ethiopia. *Journal of Applied Biology and Biotechnology,* 3(1): 0–2.

Tolera, K.D. and Abera, S. (2017). Nutritional quality of Oyster Mushroom (*Pleurotus ostreatus*) as affected by osmotic pretreatments and drying methods. *Food Science and Nutrition,* 5(5): 989–996.

Tong, H., Xia, F., Feng, K., Sun, G., Gao, X., Sun, L., Jiang, R., Tian, D. and Sun, X. (2009). Structural characterisation and *in vitro* anti-tumour activity of a novel polysaccharide isolated from the fruiting bodies of *Pleurotus ostreatus. Bioresource Technology,* 100(4): 1682–1686.

Urbancikova, I., Hudackova, D., Majtan, J., Rennerova, Z., Banovcin, P. and Jesenak, M. (2020). Efficacy of pleuran (β-glucan from *Pleurotus ostreatus*) in the management of herpes simplex virus type 1 infection. *Evidence-based Complementary and Alternative Medicine,* 2020.

Velázquez-Cedeño, M.A., Mata, G. and Savoie, J.M. (2002). Waste-reducing cultivation of *Pleurotus ostreatus* and *Pleurotus pulmonarius* on coffee pulp: Changes in the production of some lignocellulolytic enzymes. *World Journal of Microbiology and Biotechnology,* 18(3): 201–207.

Venturella, G., Gargano, M. and Compagno, R. (2015). The genus *Pleurotus* in Italy. *Flora Mediterranea,* 25: 143–155.

Wang, D., Sakoda, A. and Suzuki, M. (2001). Biological efficiency and nutritional value of *Pleurotus ostreatus* cultivated on spent beer grain. *Bioresource Technology,* 78(3): 293–300.

Wang, H., Gao, J. and Ng, T.B. (2000). A new lectin with highly potent antihepatoma and antisarcoma activities from the oyster mushroom *Pleurotus ostreatus. Biochemical and Biophysical Research Communications,* 275(3): 810–816.

World Health Organization (WHO). (2020). As more go hungry and malnutrition persists, achieving Zero Hunger by 2030 in doubt, UN report warns. https://www.who.int/, 13 Jul 2020.

Wu, J.Y., Chen, C.H., Chang, W.H., Chung, K.T., Liu, Y.W., Lu, F.J. and Chen, C.H. (2011). Anticancer effects of protein extracts from *Calvatia lilacina, Pleurotus ostreatus* and *Volvariella volvacea. Evidence-based Complementary and Alternative Medicine,* 2011.

Yang, J.H., Lin, H.C. and Mau, J.L. (2002). Antioxidant properties of several commercial mushrooms. *Food Chemistry,* 77(2): 229–235.

Yildiz, A., Karakaplan, M. and Aydin, F. (1998). Studies on *Pleurotus ostreatus* (Jacq. ex Fr.) Kum. var. salignus (Pers. ex Fr.) Konr. et Maubl.: cultivation, proximate composition, organic and mineral composition of carpophores. *Food Chemistry*, 61(1-2): 127–130.

Yildiz, S., Yildiz, Ü.C., Gezer, E.D. and Temiz, A. (2002). Some lignocellulosic wastes used as raw material in cultivation of the *Pleurotus ostreatus* culture mushroom. *Process Biochemistry*, 38(3): 301–306.

Zhang, X., Wang, L., Ma, F., Yang, J. and Su, M. (2017). Effects of arbuscular mycorrhizal fungi inoculation on carbon and nitrogen distribution and grain yield and nutritional quality in rice (*Oryza sativa* L.). *Journal of the Science of Food and Agriculture*, 97(9): 2919–2925.

Zhang, Z., Zhang, Y., Liu, H., Wang, J., Wang, D., Deng, Z., Li, T., He, Y., Yang, Y. and Zhong, S. (2021). A water-soluble selenium-enriched polysaccharides produced by *Pleurotus ostreatus*: Purification, characterisation, antioxidant and anti-tumour activities *in vitro*. *International Journal of Biological Macromolecules*, 168: 356–370.

CHAPTER 19

Sheep Polypore
(Polyporus confluens)

Ikbal Hasan,[1] *Tuyelee Das,*[1] *Mimosa Ghorai,*[1] *Abdel Rahman Al-Tawaha,*[2] *Ercan Bursal,*[3] *Mallappa Kumara Swamy,*[4] *Potshangbam Nongdam,*[5] *Mahipal S Shekhawat,*[6] *Devendra Kumar Pandey*[7] *and Abhijit Dey*[1,*]

1. Introduction

P. confluens is a very useful fungi species common to Scandinavian countries, Mexico, U.S.A, Canada, China, India, Bangladesh, Nepal, Thailand, Myanmar, Japan, Korea, Bulgaria and Russia and belongs to the family Albatrellaceae (Dahlberg, 2019). The vernacular names for this species are Semmel Purling (German), Políporo confluente (Spanish), Brödticka (Swedish), Krásnoporka (Czech), and Ningyotake (Japan) (Kotl. and Pouzar, 1957; Mizuno and Sugiyama, 1995). The colour of the top of the cap varies with age; sometimes the cap appears white, whitish-grey or brown (Harris, 2014). Usually found in coniferous forests, this fungus is sometimes called fused polypore due to its large population size. It establishes an ectomycorrhizal association with conifers. In their young stage, this species is edible, but in their older state, they become bitter and unappetising (Dahlberg, 2019). Approximately 80% of the global population is dependent upon herbal medicine. People have used plants as a source of traditional medicine since ancient times (Rouhi-Boroujeni et al., 2015). A number

[1] Department of Life Science, Presidency University, Kolkata-700073.
[2] Department of Biological Sciences, Al-Hussein Bin Talal University, Maan, Jordon.
[3] Department of Biochemistry, Mus Alparslan University, Turkey.
[4] Department of Biotechnology, East West First Grade College of Science (Bangalore University), Bengaluru, Karnataka, India.
[5] Department of Biotechnology, Manipur University, Canchipur, Imphal-795003, Manipur, India.
[6] Plant Biotechnology Unit, KM Government Institute for Postgraduate Studies and Research, Puducherry, India.
[7] Department of Biotechnology, Lovely Professional University, Punjab, India.
* Corresponding author: abhijit.bs@presiuniv.ac.in

of traditional Chinese and Japanese medicines contain this fungus. The fruit body of this fungus contains grifolinone, aurovertin and albatrellin, which have been shown to have anticancer, cytotoxic, acute neurotoxic, anti-inflammatory and antimicrobial effects (Keller et al., 2002). *P. confluens* is well-known in Ayurveda and in Nepali and Chinese traditional medicine for its anticancer properties due to the presence of the substance called grifolin (Panda et al., 2021). Chinese, Nordic European people and Indian population frequently use *P. confluens* as a medicinal mushroom (Yu et al., 2017, Kumar et al., 2017; Keller et al., 2002). We discuss the presence and use of *P. confluens* in the food and nutraceutical industries, as well as its acceptance alongside its use as a pharmaceutical product.

1.1 Taxonomic Classification

Kingdom: Fungi
Division: Basidiomycota
Class: Agaricomycetes
Order: Russulale
Family: Polyporaceae
Genus: *Albatrellus* Grey
Species: *Polyporus confluens* (*Albatrellus confluens*)

2. Geographical Distribution

P. confluens grows in several continents in the world and is considered one of the most beautiful mushroom species (Fig. 1). In terms of distribution, *P. confluens* is widely distributed throughout most of Europe and is also found in North America. The species *P. confluens* is abundant throughout the Nordic region, primarily in Norway and Sweden. *P. confluens* is usually found in mountainous regions of North America, such as the Rocky Mountains, Appalachian Mountains and Mexican mountain areas (Dahlberg, 2019). This species is widely distributed also in Asian countries, such as Japan, China, Korea, India, Indonesia, Bangladesh and Myanmar (Yu et al., 2017; Kumar et al., 2017; Keller et al., 2002).

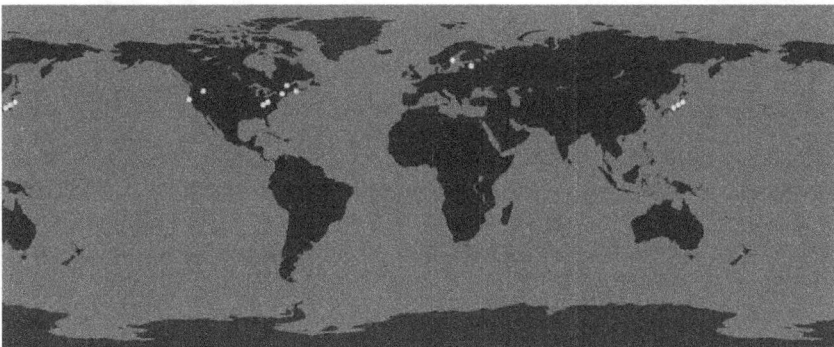

Fig. 1. Geographical distribution of *Polyporus confluens* (*Albatrellus confluens*) (https://www.gbif.org/species/5248892).

3. Mycobiota as Nutraceuticals and Food: Historical Aspect

P. confluens was the first cultivated mushroom by people around AD 600. After that, nearabout AD 800–900, *Flammulina velutipes* was cultivated in China, where the practice of mushroom cultivation has been followed for thousands of years. The Kyusuye tribe of Japan was the first mushroom cultivator in Japan and offered a shiitake mushroom to the Japanese Emperor Chuai (Rahman and Choudhury, 2012). Since the early stages of civilisation, the people of the African desert have been using macrofungi for food and medicinal purposes. The people of ancient Egypt used fungi for their religious rituals (El Enshasy et al., 2013). The first data on the initiation of mushroom cultivation in Europe was found in the reign of Louis XIV in France (1637–1715). The history of Indian mushroom cultivation is very recent. In 1886, some specimen species of mushrooms were first grown for the Horticulture Society of India by N.W. Newton. Dr. B.C. Roy carried out a chemical analysis of local mushroom species in Calcutta Medical College between 1886 and 1887 (Singh and Mishra, 2008).

4. *Polyporus confluens* as Food among Populations

Humans have used mushrooms as food and medicine since ancient times. The existence of mushrooms on earth goes back to the Cretaceous period; it is believed that primitive humans consumed mushrooms before human evolution. In Islam, mushrooms are regarded as a food from *Jannat*. The Greeks considered mushrooms to be a royal dish. Egyptians referred to them as 'food for Pharaohs'. Kotowski (2019) explains that mushrooms were considered an elixir of life in Chinese folklore. Bulgarians prepare delicious pickles, using *P. confluens* (Stoyneva-Gärtner and Uzunov, 2015). In Yunnan province of China, mushrooms are sold as food ingredients, including *P. confluens* (Yu et al., 2020). The Jaunsari tribe and Nepali population of Chakrata cantonment in Dehradun district of Uttarakhand consume *P. confluens* as a food item (Kumar et al., 2017). *P. confluens* is also used in Bangladeshi as a food item and for medicinal purposes (Kabir, 1999). The indigenous people of China have been using mushrooms in their diets since ancient times due to the abundance of higher-order fungi in the country (Orita et al., 2017).

5. Medicinal Properties

5.1 Anticancer Activity

Ethanol extract from the fruiting body of *P. confluens* exhibits considerable anti-tumour activity. Toyopearl HW-65F gel column filtered extracts of the fungi had high anti-tumour activity with 65–99% inhibition of tumour growth, regression and mortality (Mizuno et al., 1992). Grifolin, a secondary metabolite isolated from the fruit body of *P. confluens*, shows an inhibitory effect on SW480, CNE1, MCF7, K562, and HeLa cells. Grifolin causes the death-associated protein kinase 1 (DAPK1) gene to be upregulated in nasopharyngeal carcinoma cells CNE1 via caspase-mediated apoptosis, thus causing apoptosis (Luo et al., 2011). Grifolin extract from *P. confluens* can act as an anticancer agent against gastric tumour cells by inhibiting the ERK1/2 pathway and inducing apoptosis (Wu and Li, 2017). Albaconol, another biologically active compound isolated from *P. confluens*, also induced apoptosis with a high concentration in RAW264.7 cells (Liu et al., 2008). Compared with albaconol, a neoalbaconol derivative of albaconol suppresses cell proliferation, migration and invasion of VEGF-induced human umbilical vascular endothelial cells (Yu et al., 2017). Albaconol inhibits DNA topoisomerase II. It stabilises and increases the effectiveness of the enzyme's cleavage complex, but it does not inhibit topoisomerase I. Albaconol inhibits tumour cell growth by inhibiting DNA topoisomerase-mediated DNA cleavage (Qing et al., 2004). Auerovertin B inhibits cell proliferation in breast cancer cell lines by activating apoptosis and arresting the G0/G1 phase of the cell cycle. Aurovertin B inhibits breast cancer cell proliferation by inhibiting ATP synthesis in carcinoma cells (Huang et al., 2008). In 2015, Yu et al., discovered that neoalbaconol promotes necroptosis pathways through the autocrine secretion of TNF-α by the RIPK/NF-κB regulated signalling pathways and the production of ROS through RIPK3 signakling.

5.2 Anti-inflammatory Activity

Inflammation in the human body is considered a crucial and complex response to harmful stimuli, such as irritation, pathogens, or damaged cells (Elsayed et al., 2014). Various inflammatory biomarkers, such as (IL-1β), IL-6; IL-8; nuclear factor-κB (NF-κB) and tumour necrosis factor (TNF-α); are produced by pro-inflammatory cells, such as monocytes, macrophages, or other host cells in response to harmful stimuli (Levine et al., 2011). In the mushroom, biomolecules exhibit immunomodulating effects, including mitogenicity and activation of several types of immune cells. In previous research, albaconol was found in high concentrations in *P. confluens* (Zhi-Hui et al., 2001). Several studies on dendritic cell stimulators have shown that high concentrations of albaconol suppress lipopolysaccharide-induced proinflammatory cytokinin production. Based on the available data, albaconol inhibits DCs-induced antigen-specific CD4+ T cell responses. Albaconol significantly disrupts the function of CDs, resulting in the suppression of NF-κB activation (Liu et al., 2008a). In another study, albaconol was found to inhibit the LPS-induced inflammatory biomarkers, IL-1β, IL-6, TNF-α, and NO production by macrophages through the upregulation of SOCS1 expression and the suppression of NF-κB activation (Liu et al., 2008b).

5.3 Antimicrobial Activity

Grifolin from *P. confluens* is an effective antibacterial agent. Grifolin acts as antiparasitic and cytotoxic agent against *Trypanosoma cruzi* and *Leishmania species*. It is a known antibacterial agent against Gram-positive bacteria, like *Staphylococcus aureus*. Dichloromethane extracts of *P. confluens* show antifungal activity against *Bulgaria inquinans* (black bulgar), *Cladosporium cucumerinum* (effective pathogen of cucumbers) and *Candida albicans* (human gut flora). *P. confluens* provide larvicidal activity against *Agaricus xanthoderma* (Keller et al., 2002). Confluenines A–F, a derivative of L-isoleucine, has been isolated from *P. confluens*. This compound acts as an antimicrobial agent against *Staphylococcus aureus* (Zhang et al., 2018a).

Table 1. Lists of metabolites and its medicinal properties present in *P. confluens.*

Compound Name	Class	Medicinal Properties	References
8β-hydroxy-18-norcleistAnth-4	Cleistanthane-type diterpenes	–	Zhou et al., 2009
3α,5α,8β-trihydricxycleistanth-13	Cleistanthane-type diterpenes	–	Zhou et al., 2009
12b,15b,22R,23S,24S)-22,25Epoxy-12,15,23-trihydroxyergost-4,6,8 (14)-trien-3-one	Ergostane-type steroid	Cytotoxic	Guo et al., 2015
5-(20,30-Epoxy-30,30dimethylpropoxy)-7-methoxy-6methylphthalide	Phthalide derivatives	Cytotoxic	Guo et al., 2015
(20)-(Z)-5-(30-Hydroxymethyl-30methylallyloxy)-7-methoxy-6methylphthalide	Phthalide derivatives	Cytotoxic	Guo et al., 2015
5-(30,30-Dimethylallyloxy)-7hydroxy-6-methylphthalid	Phthalide derivatives	Cytotoxic	Guo et al., 2015
Albatrellin	Meroterpenoid	Cytotoxic activity against HepG2 human lung carcinoma cells	Yang et al., 2008
Aurovertin E	Polyene pyrone	Neurotoxic	Wang et al., 2005
Aurovertin B	–	Neurotoxic, anti-breast cancer	Yang et al., 2004
Conflamide A-I	Nitrogen-containing heterocyclic compounds	–	Zhang et al., 2018
Grifolin	–	Cytotoxic	Qing et al., 2004
Grifolinone C	Phenol	–	Yang et al., 2008
Ponfluenines A–F	L-isoleucine derivatives	Antimicrobial	Zhong and Xiao, 2009
Sulfurenic acid	–	Dopamine D_2 receptor agonistic activity	Zjawiony, 2004

6. Conclusion

Mushrooms have been regarded and appreciated for centuries as cures for various types of degenerative diseases. In light of the diverse ecological zones and habitats of mushrooms across the world, they are considered the next generation's nutraceutical food. Mushrooms are a complete health food that contain protein, dietary fibre, vitamins and minerals, making them an appropriate food for individuals of all ages. In recent years, new information has become available regarding their chemical composition, nutritional value and therapeutic properties. Iimmense nutraceutical properties of *P. confluens* have been documented and verified. There are many reasons for this, including the presence of bioactive molecules, such as aurovertin B, Conflamide A and I, Albatredines A and B, Grifolin, Albaconol, Sulfurenic acid, Grifolinone C, Albatrellin, Pyrazinol, Pmeheterone, etc. (Zhong and Xiao, 2009). Therapeutically, these molecules are highly effective. So overall, we can say that we do not yet understand the range of physiological activities governed by *P. confluens*. The field of genomics, proteomics, secondary metabolite secretion and their involvement in biochemical pathways governing their therapeutic roles require extensive research. It is necessary to investigate the precise mechanisms by which these therapeutic actions work in order to provide the general public with a clear understanding of them.

Acknowledgment

Authors are extremely thankful to UGC, Government of India, for financial assistance and to Presidency University-FRPDF fund, Kolkata for providing the necessary research facilities.

References

Dahlberg, A. (2019). e.T122090225A122090858. *Albatrellus confluens.* https://dx.Doi.org/10.2305/IUCN.UK.2019-3.RLTS.T122090225A1 2090858.en; accessed on 4 February 2022.

El Enshasy, H., Elsayed, E.A., Aziz, R. and Wadaan, M.A. (2013). Mushrooms and truffles: *Historica biofactories* for complementary medicine in Africa and in the Middle East. *Evidence-based Complementary and Alternative Medicine,* 2013: 1–10.

Elsayed, Elsayed, A., El Enshasy, Hesham, Wadaan, Mohammad, A.M. and Aziz, Ramlan. (2014). Mushrooms: A Potential natural source of anti-inflammatory compounds for medical applications. *Mediators of Inflammation,* 2014: 1–15. Doi:10.1155/2014/805841.

Guo, H., Li, Z.H., Feng, T. and Liu, J.K. (2015). One new ergostane-type steroid and three new phthalide derivatives from cultures of the Basidiomycete *Albatrellus confluens. J. Asian Nat. Prod. Res.,* 17(2): 107–113.

Harris, J.C. (2014). *Pocket Guide to Mushrooms.* Bloomsbury Publishing.

Huang, T.C., Chang, H.Y., Hsu, C.H., Kuo, W.H., Chang, K.J. and Juan, H.F. (2008). Targeting therapy for breast carcinoma by ATP synthase inhibitor aurovertin B. *J. Proteome Res.,* 7(4): 1433–1444.

Kabir, Y. (1999). Nutritious food mushrooms: Problems and prospects of production and consumption in Bangladesh. *South Asian J. Nutr.,* 1(1&2): 57–67.

Keller, C., Maillard, M., Keller, J. and Hostettmann, K. (2002). Screening of European fungi for antibacterial, antifungal, larvicidal, molluscicidal, antioxidant and free-radical scavenging activities and subsequent isolation of bioactive compounds. *Pharm. Biol.,* 40(7): 518–525.

Kotl and Pouzar. (1957). *In: Česká Mykol.,* 11(3): 154.

Kotowski, M. (2019). History of mushroom consumption and its impact on traditional view on mycobiota—An example from Poland. *MB,* 4(3): 1–13. https://Doi.org/10.21608/mb.2019.61290.

Kumar, M., Harsh, N.S.K., Prasad, R. and Pandey, V.V. (2017). An ethnomycological survey of Jaunsar, Chakrata, Dehradun, India. *J. Threat. Taxa.,* 9(9): 10717–10725.

Levine, B., Mizushima, N. and Virgin, H.W. (2011). Autophagy in immunity and inflammation. *Nature,* 469(7330): 323–335.

Liu, Q., Shu, X., Sun, A., Sun, Q., Zhang, C., An, H., Liu, J. and Cao, X. (2008a). Plant-derived small molecule albaconol suppresses LPS-triggered pro-inflammatory cytokine production and antigen presentation of dendritic cells by impairing NF-κB activation. *Int. Immunopharmacol.,* 8(8): 1103–1111.

Liu, Q., Shu, X., Wang, L., Sun, A., Liu, J. and Cao, X. (2008b). Albaconol, a plant-derived small molecule, inhibits macrophage function by suppressing NF-κB activation and enhancing SOCS1 expression. *Cell. Mol. Immunol.*, 5(4): 271–278.

Luo, X.J., Li, L.L., Deng, Q.P., Yu, X.F., Yang, L.F., Luo, F.J., Xiao, L.B., Chen, X.Y., Ye, M., Liu, J.K. and Cao, Y. (2011). Grifolin, a potent anti-tumour natural product upregulates death-associated protein kinase 1 DAPK1 via p53 in nasopharyngeal carcinoma cells. *Eur. J. Cancer*, 47(2): 316–325.

Mizuno, T. and Sugiyama, K. (1995). Ningyotake, *Polyporus confluens*: Biologically active substances. *Food Rev. Int.*, 11(1): 179–184.

Mizuno, T., Ando, M., Sugie, R., Ito, H., Shimura, K., Sumiya, T. and Matsuura, A. (1992). Anti-tumour activity of some polysaccharides isolated from an edible mushroom, Ningyotake, the fruiting body, and the cultured mycelium of Polyporous confluens. *Biosci. Biotechnol. Biochem.*, 56(1): 34–41.

Orita, M., Nakashima, K., Taira, Y., Fukuda, T., Fukushima, Y., Kudo, T., Endo, Y., Yamashita, S. and Takamura, N. (2017). Radiocesium concentrations in wild mushrooms after the accident at the Fukushima Daiichi Nuclear Power Station: Follow-up study in Kawauchi village. *Sci. Rep.*, 7(1): 1–7.

Panda, M.K., Paul, M., Singdevsachan, S.K., Tayung, K., Das, S.K. and Thatoi, H. (2021). Promising anticancer therapeutics from mushrooms: Current findings and future perceptions. *Curr. Pharm. Biotechnol.*, 22(9): 1164–1191.

Qing, C., Liu, M.H., Yangi, W.M., Zhang, Y.L., Wang, L. and Liu, J.K. (2004). Effects of albaconol from the Basidiomycete *Albatrellus confluens* on DNA topoisomerase II-mediated DNA cleavage and relaxation. *Planta Med.*, 70(09): 792–796.

Rahman, T. and Choudhury, M.B.K. (2012). Shiitake mushroom: A tool of medicine. *Bangladesh J. Med. Biochem.*, 5(1): 24–32.

Rouhi-Boroujeni, H., Gharipour, M., Mohammadizadeh, F., Ahmadi, S. and Rafieian-Kopaei, M. (2015). Systematic review on safety and drug interaction of herbal therapy in hyperlipidemia: A guide for internist. *Acta Biomed. Ateneo Parmense.*, 86(2): 130–6.

Singh, R.P. and Mishra, K.K. (2008). *Mushroom Cultivation, Mushroom Research and Training Centre*. G.B. Pant University of Agriculture and Technology, Pantnagar-263 145 (Uttarakhand).

Stoyneva-Gärtner, M.P. and Uzunov, B.A. (2015). An ethnobiological glance on globalization impact on the traditional use of algae and fungi as food in Bulgaria. *Nutr. Food Sci.*, 5(5): 1.

Wang, F., Luo, D.Q. and Liu, J.K. (2005). Aurovertin E, a new polyene pyrone from the Basidiomycete. *J. Antibiot.*, 58(6): 412–415.

Wu, Z. and Li, Y. (2017). Grifolin exhibits anticancer activity by inhibiting the development and invasion of gastric tumour cells. *Oncotarget*, 8(13): 21454.

Yang, W.M., Liu, J.K., Ding, Z.H., Shi, Y., Wu, W.L. and Chen, Z.H. (2004). The antioxidative effect of albaconol from *Albatrellus confluens*. *Chin. J. Trad. West Med.*, 5: 1864–1866.

Yang, X.L., Qin, C., Wang, F., Dong, Z.J. and Liu, J.K. (2008). A new meroterpenoid pigment from the Basidiomycete *Albatrellus confluens*. *Chem. Biodivers.*, 5(3): 484–489.

Yu, F., Guerin-Laguette, A. and Wang, Y. (2020). Edible mushrooms and their cultural importance in yunnan, China. *In*: *Mushrooms, Humans and Nature in a Changing World, Springer Sci. Rev.*, 163–204. Doi: 10.1007/978-3-030-37378-8_6.

Yu, X., Deng, Q., Li, W., Xiao, L., Luo, X., Liu, X., Yang, L., Peng, S., Ding, Z., Feng, T. and Zhou, J. (2015). Neoalbaconol induces cell death through necroptosis by regulating RIPK-dependent autocrine TNFα and ROS production. *Oncotarget*, 6(4): 1995.

Yu, X., Li, W., Deng, Q., You, S., Liu, H., Peng, S., Liu, X., Lu, J., Luo, X., Yang, L. and Tang, M. (2017). Neoalbaconol inhibits angiogenesis and tumor growth by suppressing EGFR-mediated VEGF production. *Mol. Carcinog.*, 56(5): 1414–1426.

Zhang, S., Huang, Y., He, S., Chen, H., Li, Z., Wu, B., Zuo, J., Feng, T. and Liu, J. (2018a). Albatredines A and B, a pair of epimers with unusual natural heterocyclic skeletons from edible mushroom *Albatrellus confluens*. *RSC Adv.*, 8(42): 23914–23918.

Zhang, S., Huang, Y., He, S., Chen, H., Wu, B., Li, S., Zhao, Z., Li, Z., Wang, X., Zuo, J. and Feng, T. (2018b). Heterocyclic compounds from the mushroom *Albatrellus confluens* and their inhibitions against lipopolysaccharides-induced B lymphocyte cell proliferation. *J. Org. Chem.*, 83(17): 10158–10165.

Zhang, S.B., Huang, Y., Chen, H.P., Li, Z.H., Wu, B., Feng, T. and Liu, J.K. (2018c). Confluenines A–F, N-oxidised l-isoleucine derivatives from the edible mushroom *Albatrellus confluens. Tetrahedron Lett.*, 59(34): 3262–3266.

Zhi-Hui, D., Ze-Jun, D. and Ji-Kai, L. (2001). Albaconol, A novel prenylated resorcinol (= Benzene-1, 3-diol) from Basidiomycetes *Albatrellus confluens. Helv. Chim. Acta*, 84(1): 259–262.

Zhong, J.J. and Xiao, J.H. (2009). Secondary metabolites from higher fungi: Discovery, bioactivity, and bioproduction. pp. 79–150. *In*: Zhong, J.J., Bai, F.W. and Zhang, W. (eds.). *Biotechnology in China I, Advances in Biochemical Engineering/Biotechnology*, vol. 113. Springer, Berlin, Heidelberg. https://Doi.org/10.1007/10_2008_26.

Zhou, Z.Y., Liu, R., Jiang, M.Y., Zhang, L., Niu, Y., Zhu, Y.C., Dong, Z.J. and Liu, J.K. (2009). Two new cleistanthane diterpenes and a new isocoumarine from cultures of the Basidiomycete *Albatrellus confluens. Chem. Pharm. Bull.*, 57(9): 975–978.

Zjawiony, J.K. (2004). Biologically active compounds from Aphyllophorales (polypore) fungi. *J. Nat. Prod.*, 67(2): 300–310.

CHAPTER 20

Turkey Tail (*Trametes versicolor*)

Carlos Gregorio Barreras-Urbina,[1,2] *Francisco Rodríguez-Félix,*[2]
Tomás Jesús Madera-Santana,[1] *Eneida Azaret Montaño-Grijalva,*[2]
Cielo Estefanía Figueroa-Enríquez,[2] *Frida Lourdes*
García Larez,[2,3] *Danya Elizabeth Estrella-Osuna*[2]
and *José Agustín Tapia-Hernández*[2,*]

1. Introduction

Nowadays, the medicinal mushrooms, such as those which are edible and usually live on wood, such Phelinus, Coriolus and Ganoderma, are used as a natural source for the bioactive compounds. These compounds could maintain the human health due to their ability to present several benefits in decreasing and preventing certain diseases. Their benefits are antidiabetic properties, antitumour activity, immunomodulating properties, cardiovascular, antimicrobial, hepatoprotective and antioxidative properties (Kıvrak et al., 2020). Due to its shape and several colours, the *T. versicolor* is usually called 'Turkey tail' because it is similar to the colours of wild turkey. This fungus normally grows in temperate climate as in North America, Europe, UK and Asia (Habtemariam, 2020). In eastern countries, is called as Yun-Zhi (China) or Kawaratake (Japan) (Tišma et al., 2021). The physical appearance of *T. versicolor* is seen in Fig. 1.

The habitat of *T. versicolor* is several hardwood trees as oak and prunus; also some conifers as fir and pine where the basidium appears on trunks and stubs during the year. The fungus presents a morphological complex structure with different

[1] Centro de Investigación en Alimentación y Desarrollo, A.C., Coordinación de Tecnología de Alimentos de Origen Vegetal, Carretera Gustavo Enrique Astiazarán Rosas Núm. 46. La Victoria, C.P. 83304, Hermosillo, Sonora, Mexico.

[2] Departamento de Investigación y Posgrado en Alimentos (DIPA). Blvd. Luis Encinas y Rosales S/N, Col. Centro, Hermosillo, Sonora, México. C.P. 83000.

[3] Posgrado en Sustentabilidad, Universidad de Sonora, Blvd. Luis Encinas y Rosales, S/N, Colonia Centro, 83000 Hermosillo, Sonora, Mexico.

* Corresponding author: Joseagustin.tapia@unison.mx

Fig. 1. The physical appearance of *T. versicolor* (Zhang, 2019).

forms. The *T. versicolor* prefers to consume the ligning than celullose from the wood, causing 'white riot' (Tišma et al., 2021). Unlike edible mushrooms (*L. edodes*), *T. versicolor* is used mainly for production of metabolites and its applications are in (1) environmental: degradation of dyes, such as bisphenol A (Hongyan et al., 2019), laccase production (Xu et al., 2020) and (2) in health as its antioxidant, anti-proliferative and antimicrobial effect (Rašeta et al., 2020; Bains and Chawla, 2020), antidiabetic (Yang et al., 2012).

T. versicolor is produced for use in the chemical and pharmaceutical industry. There are several techniques for its cultivation, among which submerged fermentation (SmF) and solid-state fermentation (SSF) are commonly adopted. SmF involves liquid media which are nutrients that are dissolved or suspended and are accessible for the microorganisms and, also, the nutrient media for inoculum production which differs from the bioproduct media. Morphology and the physiology are the main factors that are necessary to explain and predict the design process and bioreactor performance. On the other hand, *T. versicolor* is cultivated on inert or non-inert solid substrates. Among the latter, lignocellulosic biomass is the most common substrate used for both inoculum and bioproduct production. In some cases, the SSF product is a biotransformed lignocellulosic biomass that can be applied to biofuel production, biobased product production or animal feed.

Due to its industrial importance and beneficial (Znidarsic, 2001; Lizardi-Jiménez and Hernández-Martínez, 2017; Diorio et al., 2021) effects on health, this chapter aims to present the taxonomy, structure and compounds that have been isolated with therapeutic effect from *T. versicolor*.

2. Structure and Taxonomy of *Trametes versicolor*

T. versicolor, one of the most easily identifiable polypores and a common wood-rotting species on dead hardwoods, is a multicolored mushroom recognisable throughout the world (Hobbs, 2004). The upper surface of the mushroom shows typical concentric zones in different colours. Flesh is 1–3 mm thick, leathery texture and the lid is brown or in darker brownish rust, sometimes blackish in the older specimens. It may have areas with green algae growing on them, thus making the mushroom appear green. It commonly grows in layers of tiles. The cap is flat, up to 8 × 5 × 0.5–1 cm, often

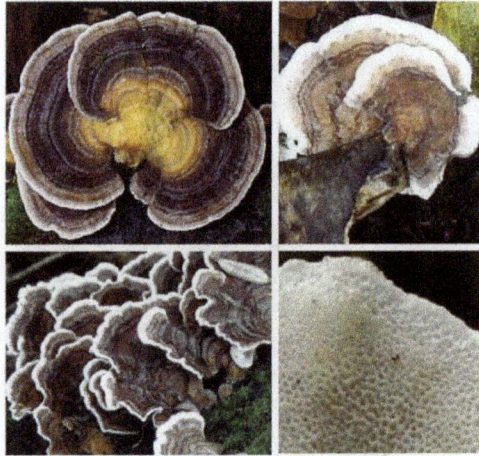

Fig. 2. Structure of *T. versicolor* (Habtemariam, 2020).

Table 1. Taxonomic tree of *T. versicolor*.

Phylum	Basidiomycota
Class	Agaricomycetes
Subclass	Agaricomycetidae
Order	Polyporales
Family	Polyporaceae
Genus	Trametes
Species	Versicolor

triangular or round, with patches of fine hair. The porous surface has white to light brown pores all around and with age becomes twisted and labyrinthine. There are 2–5 pores per millimetre. The structure of *T. versicolor* is seen in Fig. 2.

T. versicolor or *C. versicolor* is a taxonomic member of the phylum Basidiomycota, order Polyporales, and family Polyporaceae (Hsu et al., 2013). Table 1 shows the taxonomy of *T. versicolor*.

3. Chemical Composition of *Trametes versicolor*

T. versicolor is a fungus that contains various components in its structure, both internal and external, and these are mainly polysaccharides, minerals and vitamins. In the case of polysaccharides, they can be classified into two types: intrapolysaccharides (IPTV) and extrapolysaccharides (EPTV) of *T. versicolor*. Table 2 shows the monosaccharide content in IPTV and EPTV. The IPTV is composed of seven different monosaccharides including ribose, rhamnose, arabinose, xylose, galactose, mannose and glucose with a mass percentage of 0.19%, 0.22%, 0.38%, 1.56%, 14.00%, 25.98% and 57.67%, respectively, while the EPTV is composed of ribose, rhamnose, arabinose, xylose, galactose, glucose and mannose with a mass percentage composition of 0.24%, 0.27%, 0.71%, 7.77%, 19.99%, 34.15% and 36.87%, respectively (Huang et al., 2020). In

Table 2. Monosaccharide content in intrapolysaccharides (IPTV) and extrapolysaccharides (EPTV) of *T. versicolor.*

Monosaccharide Composition	IPTV (mass %)	EPTV (mass %)
Ribose	0.19	0.24
Rhamnose	0.22	0.27
Arabinose	0.38	0.71
Xylose	1.56	7.77
Galactose	14.00	19.99
Mannose	25.98	36.87
Glucose	57.67	34.15

Table 3. Trace elements present in *T. versicolor.*

Minerals	Concentration (μgkg^{-1}) Akgul et al., 2017
Fe	154.34 ± 6.71
Mg	133.54 ± 8.65
Zn	15.68 ± 0.94
Cu	8.94 ± 0.54
Na	214.6 ± 5.09
Ca	161.62 ± 1.98
Vitamins	**Concentration* (mg/100 g)** Ivanova et al., 2015
B_1 (thiamine)	0.390 ± 0.005
B_2 (riboflavin)	1.080 ± 0.020
B_3 (niacin)	10.67 ± 0.77
B_9 (folic acid)	0.56 ± 0.02

* Reported concentration in the mycelium.

addition, it contains trace elements of vitamins and minerals. Table 3 shows the type and concentration of vitamin (complex B, B_1, B_2, B_3, B_9) and minerals in *T. versicolor.*

4. Nutraceutical and Bioactive Properties of *Trametes versicolor*

The fruiting body of *T. versicolor*, such as of other mushrooms, has medicinal and nutritional properties. The macromolecules that are present in mushrooms are carbohydrates, proteins, minerals and lipids. This variety is known to contain several secondary metabolites with potential for pharmacological activity and belongs to small molecular weight compounds. Previous studies have isolated new molecules derived from terpenoids (piroaxane sesquiterpenes) with possible therapeutic properties (Fig. 3) (Wang et al., 2015).

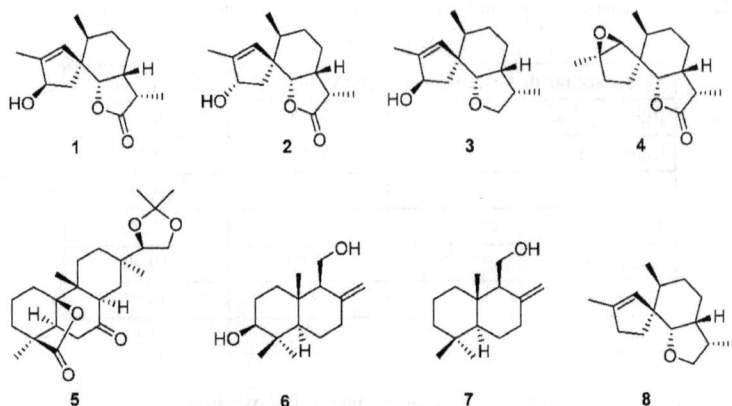

Fig. 3. Terpenoids from *T. versicolor* (Wang et al., 2015).

Table 4. *T. versicolor*-based drugs and health product approved by CFDA in China (Zhang, 2019).

Drug Name	Number of Manufacturers	CFDA Approval Number
Coriolus versicolor Gantai granules	109	Z23022161, Z22020288, Z23020110, Z36020495
Gansukang capsules	35	Z23021143, Z20055670, Z20054485, Z23022214
Polystictus Glycopeptide capsules	23	H45021256, H50021714, H50021713, H50021712
Polystictus glycopeptide	6	H32025831, H32025759, H50020928, H13023841
Coriolus versicolor Gantai capsules	4	Z20026733, Z20026732, Z20026735, Z20026734
Coriolus versicolor Gantai granules	4	Z20027003, Z20027001, Z20027006, Z20027002
Polystictus Glycopeptide oral solution	3	H20184045, H20184065, H50020933
Coriolus versicolor capsules	2	Z32021221, Z21020484
Polystictus Glycopeptide tablets	2	H50020934, H50020935
Posaverptidum	1	Z10980087
Posaverptidum capsules	1	Z20163011
Posaverptidum granules	1	Z20174012
Shen Qi Coriolus versicolor granules	1	B20070004
Yikang Versicolor capsules	1	No. 0516

There exist 13 types of *T. versicolor*-based drugs and one *T. versicolor*-based health product as authorised by the China State Food and Drug Administration (SFDA) for clinical and commercial use. Table 4 shows drugs isolated form *T. versicolors* and approved by CFDA of China.

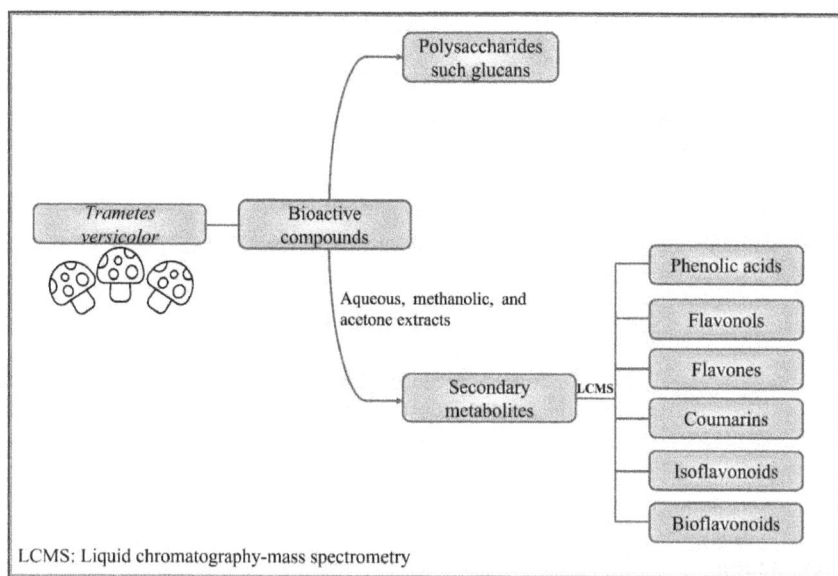

Fig. 4. Bioactive compounds of the fungus *T. versicolor.*

4.1 Antioxidant Capacity

T. versicolor is a fungus with interesting characteristics, among which are its great variety of bioactive compounds (Fig. 4). These compounds have various bioactive properties, such as antioxidant capacity. Recent studies have focused on determining the antioxidant capacity of the fungus. It has been mentioned that in addition to the content of phenolic compounds, it has polysaccharides, such as beta-glucans and some polymers of D-glucose together with glucuronic acids, arabinose, fucose, galactose, mannose, xylose and galactose (Puia et al., 2018).

The antioxidant activity of the fungus *T. versicolor* has been compared with other fungi as in the study by Agkul et al. (2017), where they determined the antioxidant activity of *Trametes versicolor* and *Auricularia auricula* and found that the fungus *Trametes versicolor* presented higher antioxidant capacity when evaluated by the DPPH free radical and compared to the ascorbic acid standard. In addition, they managed to find that the effectiveness of the DPPH radical in ethanol extracts depends directly on its concentration. This type of mushroom can be consumed taking care of the quantities and, in turn, they can be an alternative of antioxidants for the organism as clarified in the research by Agkul et al. (2017).

Janjusević et al. (2018) developed a study where the authors determined that the ethanolic extract had antioxidant capacity. The authors have conducted studies over the years to determine the presence of compounds, such as flavonoids catechin and quercetin which possess biological properties. Therefore, like other studies, the fungus *T. versicolor* has antioxidant capacity and is considered a natural antioxidant source in food supplements with health benefits.

T. versicolor and *S. subtomentosum* were studied by Raseta et al. (2020) to determine the antioxidant capacity of both fungi. The authors of this study found that the antioxidant capacity is greater in the aqueous extract than in the ethanol extract for *T. versicolor*, with findings like IC50 = 5.6 μg/mL, IC50 = 0.6 μg/mL for the DPPH radical and the OH radical, respectively. In addition, both fungi have other biological properties, such as anti-proliferative. The authors mention that *T. versicolor* may be recommended as a natural antioxidant source in the future.

Tel-Cayan et al. (2021) studied extracts of various fungi, among which were the methanolic extracts of *T. versicolor*, where the authors found that these have antioxidant activity and can be used as preventive and therapeutic agents in diseases related to oxidative stress.

4.2 Antimicrobial Capacity

T. versicolor is currently important because of its broad spectrum of application both in the food and pharmaceutical industries. Some fungi, such as *T. versicolor* have effective antimicrobial properties against Gram-positive and Gram-negative microorganisms, due to the presence of a wide variety of bioactive compounds extracted from the fungus. The mechanism of action of the antimicrobial activity lies in the hydroxyl groups of the components, which produce oxidised compounds that prevent the interaction of the membrane with some enzymes that bind to it (Bains and Chawla, 2020).

Studies similar to the ones mentioned above have been carried out with the aim of determining the antimicrobial capacity of extracts from different fungi. Canli et al. (2019) obtained extracts of the fungus *T. versicolor* using ethanol, where they found the same results as other studies. The ethanol extract presents bioactive compounds that have low to very high antimicrobial capacity. The authors of this study suggested identifying each bioactive compound present in the ethanolic extract in order to learn what the main causes of antimicrobial activity are.

Orzali et al. (2020) carried out studies where they tested the extracts of *T. versicolor* against bacteria that develop in the tomato seed, causing losses. The extracts of *T. versicolor* showed that they have the ability to inhibit bacterial growth, that is, reduce the concentration of pathogenic microorganisms by up to 90%. Although there is a great diversity of natural compounds with bioactive capacities, the fungus *T. versicolor* has been shown to contain natural components within its extracts, capable of achieving beneficial activities for health and for various industries, such as food.

4.3 Chronic Degenerative Diseases

4.3.1 Anticancer Activity

The anticancer activity of polysaccharide compounds, such as β-glucans obtained from the fungus *T. versicolor*, was proven in several studies, such as those reported in a literature review by Bains et al. (2021) where they explain that these polysaccharides have anticancer properties. Some polysaccharopeptides act as regulators of immune function, activating macrophages, T and B lymphocytes, as well as the formation of antibodies and in general, activating and strengthening the immune system. This proves that these types of components obtained from *T. versicolor* help in conventional cancer treatments to reduce the adverse effects. However, the bioactive components from fungi have potential applications in human health and have to be studied in depth.

4.3.2 Antidiabetics

Daveci et al. (2021) reported a study where they analysed the biological properties of extracts from various fungi, among which they studied *T. versicolor* with its ethanolic and methanolic extract. The authors reported that these extracts have an effect on the inhibition of the enzymes α-amylase and α-glucosidase, which are related to type 2 diabetes. By inhibiting the enzymes, the release of monosaccharides of complex carbohydrates occurs and which can be absorbed, thus delaying the glucose absorption which slows the rapid rise in blood glucose (Daveci et al., 2021). This indicates the high potential that the bioactive compounds of different extracts have for the benefit of human health, not only of *T. versicolor*, but also of other fungi, according to the studies mentioned above.

4.3.3 Antiobesogenic

Compounds from extracts of the fungus *T. versicolor* have had an antidiabetic approach in recent years. Compounds, such as ternatin, which is a highly methylated cyclic heptapeptide, have been seen to have an effect on type 2 diabetes and obesity since it inhibits the accumulation of fat in the body (Bains et al., 2021). This compound reduces the mRNA levels of the protein CCAAT which is a sterol-regulatory element and enhancer-binding protein-1c (SREBP-1c) when it is in an early stage of preadipocyte differentiation. These results, in the suppression of C receptors proliferation of peroxisomes, at the end is represented by the reduction of adipocyte fatty acid binding protein, lipoprotein lipase, fatty acid synthase and acetylCoA carboxylase with reduction in cellular lipids (Bains et al., 2021).

4.3.4 Cardiovascular Diseases

T. versicolor promotes human health since the compounds extracted from its extracts have various health benefits, such as in heart diseases and related to these are hypercholesterolemia, obesity, hepatic steatosis, among others. Huang et al. (2020) carried out a study where they mentioned that the effect of polysaccharide extracts on cardiovascular diseases and those mentioned above has not yet been clearly studied at present. Therefore, the authors carried out an intracellular and extracellular characterisation of the structure of *T. versicolor* polysaccharides, where they found that both improved serum lipid profiles in mice with high-fat diets. These types of studies suggest an improvement in cardiovascular health, as well as in the liver, due to the high consumption of fat in the diet of the mice studied.

4.3.5 Immunomodulatory Properties

The role of immunomodulatory properties of *T. versicolor* through the effect of polysaccharopeptides (PSP) has been studied. The predominantly pro-inflammatory cytokines induce the immunomodulatory effect of PSP. The induction of IL-12 also shows the ability of PSP to induce the production of cytokines associated with TNF-α. The IL-12, a cytokine capable of enhancing NK and CD8+ T cell cytotoxic activities and their expression of TNF-α, is related to a T helper (Th1). In antitumoural effect of PSP, the most used model to study is *in vitro* culture of human leukemia HL-60 cells, which have demonstrated reduced proliferation by disruption of their cell cycle, induction of apoptosis and sensitisation to various chemotherapeutics,

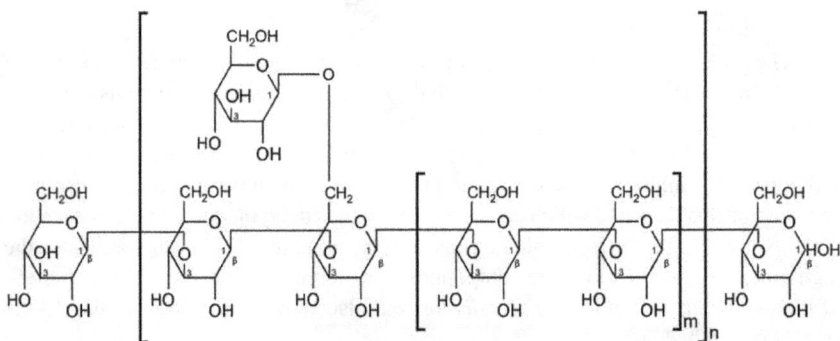

Fig. 5. Partial structure of polysaccharopeptide (PSP) contains a β-$(1\rightarrow3)$-d-glucan (Sale et al., 2017).

such as camptothecin, doxorubicin and etoposide (Sale et al., 2017). The structure of polysaccharopeptide (PSP) is shown in Fig. 5.

5. Uses of *Trametes versicolor* in Functional Food

The use of bioactive compounds and different types of plant extracts for their application in functional foods has been studied in recent decades and the fungus *T. versicolor* has not been the exception. The components extracted from this fungus present different biological activities demonstrated through years of study and reported in scientific information over time. This has opened the doors to the use of these components for the development of functional foods. Bains et al. (2021) presented a section in their review article where they report about the potential uses of these extracts in functional foods, as well as the benefits already proven in health. This is an interesting area of opportunity to produce functional foods that help prevent health problems in the population as well as improve the already compromised health of some people.

6. Conclusion

From extracts obtained from *T. versicolor*, it is possible to induce different biological activities, such as antioxidant and antimicrobial derived from the compounds present, such as phenolics, carotenoids and others. In addition, the isolation of compounds, such as PSP, has applications as immunomodulators and with the potential to be evaluated in different diseases, such as diabetes and cancer. However, it is a fungus that is rarely consumed as an ingredient or food supplement, unlike others (*L. edodes*). So, more studies in this area should be carried out.

References

Agkul, H., Sevindik, M., Coban, C., Alli, H. and Selamoglu, Z. (2017). New approaches in traditional and complementary alternative medicine practices: *Auricularia auricula* and *Trametes versicolor*. *J. Tradit. Med. Clin. Natur.*, 6(2): 239.

Bains, A. and Chawla, P. (2020). *In vitro* bioactivity, antimicrobial and anti-inflammatory efficacy of modified solvent evaporation assisted *Trametes versicolor* extract. *3 Biotech*, 10(9): 1–11.

Bains, A., Chawla, P., Kaur, S., Najda, A., Fogarasi, M. and Fogarasi, S. (2021). Bioactives from Mushroom: Health attributes and food industry applications. *Materials*, 14(24): 7640.

Canli, K., Benek, A., Şenturan, M., Akata, İ. and Altuner, E.M. (2019). *In vitro* antimicrobial activity of *Morchella esculenta* and *Trametes versicolor. Mantar Dergisi*, 10(3): 28–33.

Daveci, E., Çayan, F., Tel-Çayan, G. and Duru, M.E. (2021). Inhibitory activities of medicinal mushrooms on α-amylase and α-glucosidase-enzymes related to type 2 diabetes. *South African Journal of Botany*, 137: 19–23.

Diorio, L.A., Fréchou, D.S. and Levin, L.N. (2021). Removal of dyes by immobilisation of *Trametes versicolor* in a solid-state micro-fermentation system. *Revista Argentina de Microbiología*, 53(1): 3–10.

Habtemariam, S. (2020). *Trametes versicolor* (Synn. *Coriolus versicolor*) polysaccharides in cancer therapy: Targets and efficacy. *Biomedicines*, 8(5): 135.

Hobbs, C. (2004). Medicinal value of turkey tail fungus *Trametes versicolor* (L.: Fr.) Pilát (Aphyllophoromycetideae). A literature review. *International Journal of Medicinal Mushrooms*, 6(3).

Hongyan, L., Zexiong, Z., Shiwei, X., He, X., Yinian, Z., Haiyun, L. and Zhongsheng, Y. (2019). Study on transformation and degradation of bisphenol A by *Trametes versicolor* laccase and simulation of molecular docking. *Chemosphere*, 224: 743–750.

Huang, Z., Zhang, M., Wang, Y., Zhang, S. and Jiang, X. (2020). Extracellular and intracellular polysaccharide extracts of *Trametes versicolor* improve lipid profiles via serum regulation of lipid-regulating enzymes in hyperlipidemic mice. *Current Microbiology*, 77(11): 3526–3537.

Hsu, W.K., Hsu, T.H., Lin, F.Y., Cheng, Y.K. and Yang, J.P.W. (2013). Separation, purification, and α-glucosidase inhibition of polysaccharides from *Coriolus versicolor* LH1 mycelia, *Carbohydrate Polymers*, 92(1): 297–306.

Ivanova, T.S., Bisko, N.A., Titova, L.O. and Megalinska, G.P. (2015). Vitamin content in medicinal Mushrooms Schizophyllum commune and *Trametes versicolor* cultivated on breadcrumb BBK 30.16 R43 M59: 154.

Janjušević, L., Pejin, B., Kaišarević, S., Gorjanović, S., Pastor, F., Tešanović, K. and Karaman, M. (2018). *Trametes versicolor* ethanol extract, a promising candidate for health-promoting food supplement. *Natural Product Research*, 32(8): 963–967.

Kıvrak, İ., Kivrak, S. and Karababa, E. (2020). Assessment of bioactive compounds and antioxidant activity of Turkey tail medicinal mushroom *Trametes versicolor* (Agaricomycetes). *International Journal of Medicinal Mushrooms*, 22(6).

Lizardi-Jiménez, M.A. and Hernández-Martínez, R. (2017). Solid state fermentation (SSF): Diversity of applications to valorize waste and biomass. *3 Biotech.*, 7(1): 1–9.

Orzali, L., Valente, M.T., Scala, V., Loreti, S. and Pucci, N. (2020). Antibacterial activity of essential oils and *Trametes versicolor* extract against Clavibacter michiganensis subsp. michiganensis and *Ralstonia solanacearum* for seed treatment and development of a rapid *in vivo* assay. *Antibiotics*, 9(9): 628.

Puia, I.C., Aida, P.U.I.A., Chedea, V.S., Leopold, N., Bocsan, I.C. and Buzoianu, A.D. (2018). Characterisation of *Trametes versicolor*: Medicinal mushroom with important health benefits. *Notulae Botanicae Horti Agrobotanici Cluj-Napoca*, 46(2): 343–349.

Rašeta, M., Popović, M., Knežević, P., Šibul, F., Kaišarević, S. and Karaman, M. (2020). Bioactive phenolic compounds of two medicinal mushroom species *Trametes versicolor* and *Stereum subtomentosum* as antioxidant and anti-proliferative agents. *Chemistry & Biodiversity*, 17(12): e2000683.

Saleh, M.H., Rashedi, I. and Keating, A. (2017). Immunomodulatory properties of *Coriolus versicolor*: The role of polysaccharopeptide. *Frontiers in Immunology*, 8: 1087.

Tel-Çayan, G., Çayan, F., Deveci, E. and Duru, M.E. (2021). Phenolic profile, antioxidant and cholinesterase inhibitory activities of four Trametes species: *T. bicolor, T. pubescens, T. suaveolens*, and *T. versicolor. Journal of Food Measurement and Characterization*, 15(5): 4608–4616.

Tišma, M., Žnidaršič-Plazl, P., Šelo, G., Tolj, I., Šperanda, M., Bucić-Kojić, A. and Planinić, M. (2021). *Trametes versicolor* in lignocellulose-based bioeconomy: State of the art, challenges and opportunities. *Bioresource Technology*, 330: 124997.

Wang, S.R., Zhang, L., Chen, H.P., Li, Z.H., Dong, Z.J., Wei, K. and Liu, J.K. (2015). Four new Spiroaxane sesquiterpenes and one new rosenonolactone derivative from cultures of Basidiomycete *Trametes versicolor*. *Fitoterapia*, 105: 127–131.

Xu, L., Sun, K., Wang, F., Zhao, L., Hu, J., Ma, H. and Ding, Z. (2020). Laccase production by *Trametes versicolor* in solid-state fermentation using tea residues as substrate and its application in dye decolourisation. *Journal of Environmental Management*, 270: 110904.

Yang, J.P., Hsu, T., Lin, F., Hsu, W. and Chen, Y. (2012). Potential antidiabetic activity of extracellular polysaccharides in submerged fermentation culture of *Coriolus versicolor* LH1. *Carbohydrate Polymers*, 90(1): 174–180.

Zhang, L. (2019). *Progress in Molecular Biology and Translational Science: Glycans and Glycosaminoglycans as Clinical Biomarkers and Therapeutics-Part B*. Academic Press.

Znidarsic, P. (2001). The morphology of filamentous fungi in submerged cultivations as a bioprocess parameter. *Food Technol. Biotechnol.*, 39: 237–252.

CHAPTER 21

Snow Ear (*Tremella fuciformis*)

Kaksha Sankhe and *Tabassum Khan**

1. Introduction

Man has been hunting mushrooms for food since ancient times. Because of their nutritional value, distinct flavour and palatability, they were the only source of food thousands of years ago (Lan et al., 2021). The edible mushrooms have a high nutritional value because they are high in fibres, proteins, vitamins and minerals (K, Na, Mn, P, Mg, Fe, Ca, and Mg) and low in fat and cholesterol (Kalač, 2013). Mushrooms are important in vegetarian diets because they provide all the essential amino acids. They have higher protein content than most vegetables and edible mushrooms contain a variety of bioactive compounds that have numerous health benefits. Mushrooms have high moisture content, ranging at 80–95 g/100 gm. They are also a good source of protein (200–250 g/kg dry weight), with the most important being valine, leucine, glutamic acid and aspartic acid (Oloke, 2015). *Tremella fuciformis* (*T. fuciformis*), a white jelly-like edible mushroom, also known as a crown of fungi, snow fungus and snow ear, has been widely farmed and used as food and medicine in China. *T. fuciformis* was extensively used to fight ageing and as a tonic against weakness. In China, *T. fuciformis* was known as Yiner or Baimur. It is a member of the Tremellaceae family and the Tremellales order, which belongs to the heterobasidiomycetes class. *T. fuciformis* fruiting bodies are gelatinous and pale yellow or white in appearance (Newman and Cragg, 2016). They are high in polysaccharides, dietary fibres and proteins, but low in lipids, according to their nutritional value. *T. fuciformis* has a moisture content of $11.4 \pm 0.5\%$, an ash content of $3.4 \pm 0.2\%$, carbohydrate content of 76.6% and a protein content of $5.7 \pm 0.2\%$. Vitamins, minerals and trace elements are also present in *T. fuciformis* (Shahrajabian et al., 2022). Polysaccharides (TFPS) are a prominent component of this. Neutral polysaccharides, acidic polysaccharides, acidic oligosaccharides, exopolysaccharides and cell wall polysaccharides are the five forms of TFPS. Mannose, glucose, xylose, fucose, rhamnose, galactose, arabinose,

Department of Pharmaceutical Chemistry, SVKM's Dr. Bhanuben Nanavati College of Pharmacy, Mumbai, Maharashtra, India.
* Corresponding author: tabassum.khan@bncp.ac.in

galacturonic acid, ribose and glucuronic acid are among the 10 monosaccharides found in these TFPS (Edri et al., 2021). Mannose is one of these compounds that plays an important role in bioactivities (Khondka, 2009). Hypoglycaemic, antioxidant, anti-aging, enhancing immunological response and anti-tumour bioactivities of TFPS have been documented in the literature. In 2002, the Chinese Food and Drug Administration approved TFPS enteric-coated capsules. This medication is prescribed to patients who have leukopenia because of radiotherapy or chemotherapy (Xu et al., 2021).

2. Extraction and Purification of TFPS

TFPS extraction is an important process that affects yield, chemical structure, quality and bioactivity (Chen et al., 2018). The most typical TFPS extraction process involves pulverised fruiting bodies being stirred for several hours in hot water (Chen et al., 2018). Although this extraction process is simple to use, it requires a long extraction time, a big amount of solvent and high temperatures (Chen, 2010). TFPS can be extracted in several ways. In the aqueous extraction process, three methods were used: microwave-assisted extraction (Chen et al., 2012) (750 W power, for 60s and liquid/solid ratio of 20:1), enzymatic extraction of TFPS (Ma et al., 2021a) (0.7% pectinase enzyme, 50°C, 60 min and liquid/solid ratio of 60:1), sonicated assisted extraction (Zou and Hou, 2017) (85°C, 2 hours, 6 W/cm^2 and liquid/solid ratio of 46:1). In the alkaline extraction of TFPS, 0.7 M NaOH was used along with liquid and solid at 90:1 for 3 hours (Wu et al., 2007). 10 g of *T. fuciformis* powder was mixed with 300 mL of PEG aqueous solution in organic solvents and ultrasonic-assisted extraction was performed. The supernatant was separated by filtration or centrifugation after an extraction method is chosen. The extracted compound was then concentrated to a quarter of its original volume and precipitated at 40°C for 24 hours with three volumes of anhydrous ethanol. The precipitate was collected, using centrifugation and then low temperature dried or freeze-dried to obtain crude TFPS (Ma et al., 2021c). Typically, crude TFPS is purified using ion-exchange chromatography, with acid PS and neutral PS eluted with water and a varying concentration of salt (NaCl). The separation principle is associated with the balance of molecular weight effects and ion exchange. Furthermore, TFPS can be separated, based on their molecular weight using the gel chromatography technique. A higher molecular weight enables faster separation with the aid of column chromatography (Li et al., 2018).

3. Structural Features of TFPS

TFPS is a mixture of different polysaccharides with molecular weights ranging at 5.82×10^5 Da to 3.74×10^6 Da, because of different experimental conditions yielding different polysaccharide fractions (Zou and Hou, 2017). The greater the molecular weight of TFPS, the stronger the intermolecular hydrogen bond interaction force and the higher the viscosity of its aqueous solution. These can be used to improve food lubrication and skin sensation in cosmetic applications. TFPS, on the other hand, has a low molecular weight (1000–10,000 Da) and can cross multiple cell membranes to reach the target organ for activity (Du et al., 2014). In neutral conditions, TFPS forms a weak gel structure that crosslinks with proteins, strengthening protein hydrogen bonds and increasing solution viscosity (Zou and Hou, 2017). Yui et al., discovered the structure of TFPS in 1995 and it consists of a linear main backbone or chain of

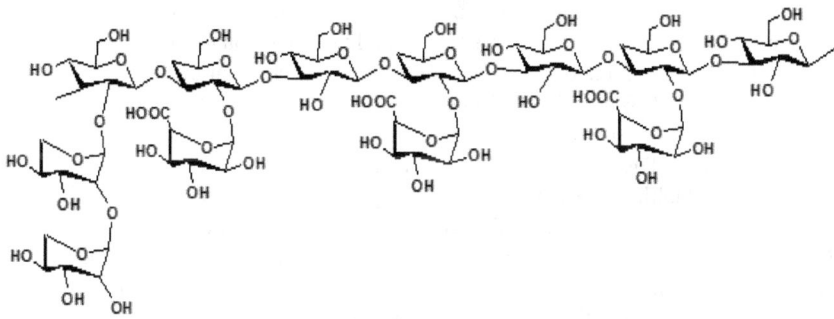

Fig. 1. Chemical structure of TFPS.

α-D-mannose (1→3-linked-α-D-mannose) along with α-D-fucose, β-D-xylobiose and β-D-gluconic acid which is linked to the C-2 of the backbone of mannose (Fig. 1). As side chains, they also contain galactose, glucose and glucuronic acid. The chain is made up of six mannose residues and has a left-handed triple helix structure. These mannose residues, together with three side chains, form a repeating unit with a central axis of 2.42 nm (Yui et al., 2006). Chemical modifications of TFPS, such as alkylation, acetylation, sulfation, carboxymethylation and phosphorylation provide new agents with effective therapeutic potential by improving solubility, pharmacokinetics and bioavailability. The more hydrophilic groups, such as hydroxyl and acetyl groups, there are in TFPS, the better its solubility. Acetyl groups can inhibit free radicals and have an effective scavenging effect (Ma et al., 2021b). Du et al., reported acetylated TAPA1 (mannose and xylose in a ratio of 5.00:4.00, 1.35×10^6 Da) and TFPS showed significant nitric oxide production than deacylated TAPA2 in mice (Du et al., 2014). Besides that, sulphate, as an electron-withdrawing group in TFPS, has a scavenging effect while increasing the water solubility and thus enhancing bioactivities (Zhao et al., 2013). Furthermore, catechin-TFPS conjugates have a better scavenging effect against TFPS (41.67%) and carboxymethylated TFPS has moisture-preserving and antioxidant activity (Liu et al., 2016).

4. Biological Activities

Chen Bin et al., extracted polysaccharides from *T. fuciformis* in boiling distilled water for four hours at 100°C, with a solvent-to-raw-material ratio of 5 (v/v). They investigated the antioxidant and anti-tumour properties of extracted polysaccharides. *T. fuciformis* polysaccharides exhibited effective superoxide anion radical scavenging activity, which increased from 17.1% to 86.9% with increasing TFPS concentration. TFPS at 0.2 mg/mL inhibited hydroxyl radicals more effectively (79.8%). The hydroxyl radical scavenging effect was proportional to TFPS concentration. They conducted *in-vivo* studies for anti-tumour activity and included powdered *T. fuciformis* in the diets of animals, which were found to have cancer-preventive effects and a restriction of tumour metastasis. TFPS have increased its activity from 73.4% to 92.1% on the increasing concentration of TFPS (Chen, 2010). Hongyu Xiao et al., investigated the activity of TFPS on ulcerative colitis (UC) using a dextran sodium sulphate (DSS)-induced BALB/c mouse model and lipopolysaccharides (LPS)-stimulated Caco-2 cell model. TFPS (100 or 200 mg/kg) and 5-aminosalicyclic acid (5-APA) supplementation significantly reversed pathological weight loss, swelling and colon shortening compared to the DSS group of mice. In comparison to LPS-induced Caco-2 cells, TFPS at 25, 50 and 100 µg/mL reduced the expression of pro-inflammatory cytokines TNF-, IL-6 and IL-10 (Xiao et al., 2022). Lingrong Wen et al., extracted TFPS with hot water and tested it for anti-photoaging effects in a UV-irritated animal assay for 30 days. They discovered that TFPS significantly reduces water and collagen loss in skin and inhibits the increasing level of glycosaminoglycans in response to UV irradiation (Wen et al., 2016). Zhili Sheng et al., investigated the effects of fermented *T. fuciformis*/blueberry (FTB) on body weights, food intake and lipid profiles in a metabolically healthy obese (MHO) model of rats. After FTB treatment, the abundance of bacteria associated with obesity prevention (*prevotella, blautia, allobaculum*, and *parabacteroides*) increased, while pathogenic bacteria abundance decreased (Sheng et al., 2022). Ben Niu et al., prepared TFPS microcapsules with maltodextrin and whey protein as a carrier to reduce moisture absorption and caking strength of the prepared microcapsules. They examined the hypoglycemic activity of TFPS microcapsules (200 mg/kg) in comparison to metformin (200 mg/kg). After two weeks of treatment, they found that fasting glucose levels in TFPS-treated mice were 8.18 mmol/L and standard metformin sugar levels were 10.82 mmol/L, compared to diabetic-induced mouse groups (17.32–20.12 mmol/L). The insulin receptor relies heavily on the PI3K/Akt signalling pathway. PI3K is a kinase and any dysfunction in it causes glucose and lipid metabolism. Akt phosphorylation, on the other hand, increases glucose absorption by promoting the GLUT4 to the plasma membrane. The expression of the PI3K and Akt genes increased in mice treated with TFPS, promoting the recovery of islet cell injury (Niu et al., 2022). Zhou et al., discovered TFPS could slow the progress of Ehrlich as cites carcinoma and block cancer cell DNA synthesis in mice model. Nonetheless, this has no activity *in vitro*, showing that TFPS cannot directly hinder or destroy cells of cancer, suggesting that it could function as a biological response modulator by improving the immunological function of the organism (Zhou et al., 1987). After treating RAW 264.7 cells with LPS, Ruan et al., discovered that up-regulation of miR-15 and later up-regulation of NFB stimulation occurred in the cells. The result was reversed when TFPS was employed to treat RAW 264.7 cells. IL-6 and TNF-α expression levels in RAW 264.7 cells dropped in a TFPS concentration-dependent way when TFPS concentration reached 200 µg/mL (Ruan et al., 2018). Also, TFPS reduced the severity of sepsis in mice by suppressing aberrant CD4⁺CD25 elevated regulation of T cells. TFPS can suppress the growth of CD4⁺CD25 Treg CD4⁺ T cells via lowering the IL-4 and IL-10 and enhancing IFN-γ production when administered to treat mice with pseudomonas aeruginosa infection and burns (Zhou et al., 2018). Kum Ju Park et al., examined neuroprotective activity of *T. fuciformis*. The hot water extract was evaluated to monitor its tendency to NGF in PC12h cells. The greatest neurite length of every cell was assessed after 48 hrs of treatment of *T. fuciformis* hot-water extract at concentration ranging at 0.1–1 µg/mL. Cells treated with test compounds had a mean neurite size of 69.51 ± 2.3 µm and 77.84 ± 0.92 µm. These findings indicate that *T. fuciformis* could be employed as a preventative measure in neurodegenerative diseases, like Alzheimer disease (Park et al., 2007). *T. fuciformis* is being studied in Korea in an eight-week prospective randomised double-blind placebo-controlled clinical trial (NCT02377024) to see how it affects cognitive biomarkers of cognitive functions. The main goal of this study was to detect the cognitive enhancement safety and effect of *T. fuciformis* at two dosages —TF 600 mg and TF 1200 mg, in age groups ranging between 40–65-years old, using clinical and cognitive indicators (psychometer speed, attention and memory) as well as brain imaging for healthy adults suffering from subjective cognitive impairment 'Effects of *Tremella Fuciformis* on Improvement in Cognitive-Bio-Markers of Cognitive Functions (Full Text View—ClinicalTrials.Gov' 2022).

5. Applications of *T. fuciformis*

TFMS are organic ingredients that are safe and non-toxic and can be used for a long term in humans through oral administration. It means that TFMS is suitable for the routine health needs of the general population who are not in good health and could be widely used in medicine, food and chemical industry (green cosmetics material) (Zhang et al., 2019). Because of its excellent viscosity and uniformity, TFPS can be used as a great thickener, food fortifier and emulsifier in the food industry to make nutritious foods while minimising the use of industrial ingredients. Pseudoplasticity will be observed in TFMS dilute solution (> 5%). If monovalent cations (sodium, potassium ions) are added to the solution, the viscosity goes up. This property could be used in conjunction with a gelling food additive to improve the gelling impact (Yang et al., 2019). For example, by combining the gel properties of TFMS and lotus seed starch, a *Tremella* lotus seed cake with a distinct flavour can be created. Furthermore, the accumulation of TFPS in raw meat can improve water retention and elasticity of myofibrillar protein (Wang et al., 2019). The addition of TFPS can keep the bread moist, thus further boosting the bread's shelf-life (Strop, 2014). TFMS induces proteins to cross-link for achieving binder effects on protein-rich foods, such as yogurt and sausages (Lan et al., 2021). Hongjic Yuan et al., extracted TFPS with hot water and purified it with DEAE sepharose fast-flow ion-exchange chromatography to produce TFPS 1. They compared the water retention properties of TFPS and TFPS 1 to glycerine. They discovered that glycerol had a higher moisture absorption rate than TFPS 1 and TFPS because it contained more hydroxyl groups per unit mass than the other two test samples. The moisture absorption rate of TFPS was lower than that of purified TFPS 1 due to impurities. TFPS 1 demonstrated better water retention due to the presence of uronic acid and hydroxyl groups, which can form hydrogen bonds with water and thus retain water. In addition, the flexible chain of TFPS intertwined with each other to form a spatial network in the solution to improve water retention. As a result, they can be used as a supplement in the food, cosmetics and pharmaceutical industry (Yuan et al., 2022). Jian Zhang et al., compared the emulsifying properties of *T. fuciformis* to lotus seed (LTS), gum arabic (GA) and purple sweet potato (PSP) in oil/water emulsions. They assessed their emulsions' emulsifying properties, such as zeta potential, shear viscosity, emulsifying activity, mean droplet size and emulsifying stability. TFPS demonstrated excellent emulsifying stability and activity (100%) after 21 days at 21°C. When exposed to 100°C for 20 min., TFS emulsions proven effective in reducing droplet size that was supreme to GA, LTS and PSP. TFPS was found to be a better emulsifier than GA and LTS in zeta potential test due to its higher magnitude. The lowest droplet size was found in TFPS emulsions at pH 10, accompanied by pH 3 and 6.5. At 4% concentration, the non-Newtonian shear thinning behaviour of all samples remained consistent, while TFPS had the highest viscosities of all. During freeze thaw cycles, the cream index of 4% TFPS was also the greatest. As a result, TFPS can be used in the food industry as a thickener and emulsifier (Zhang et al., 2019).

6. Conclusion

T. fuciformis has found widespread use in a variety of industries, including cosmetics, food and pharmaceuticals. The excellent chemical (gelation, emulsification) and physiochemical characteristics of TFMS deliver high stability in liquid dosage forms. The viscosity of TFMS allows for close contact with the protein, increasing lubrication and elasticity of the preparation. As a result, it is useful to produce a variety of products. *T. fuciformis* has also shown tremendous progress in research of the use of TFMS as an antioxidant, immunomodulatory, hypoglycaemic and other properties. *T. fuciformis* is a non-toxic, inexpensive and safe natural ingredient that could be used in the treatment and prevention of a variety of diseases and can be consumed on a regular basis as a source of nutrition in the form of a supplement by people.

References

Chen, Bin. (2010). Optimisation of extraction of *Tremella fuciformis* polysaccharides and its antioxidant and anti-tumour activities *in vitro*. *Carbohydrate Polymers*, 81(2): 420–424. Doi:10.1016/J.CARBPOL.2010.02.039.

Chen, Guangjing, Fan Bu, Xuhui Chen, Changfeng Li, Shasha Wang and Jianquan Kan. (2018). Ultrasonic extraction, structural characterisation, physicochemical properties and antioxidant activities of polysaccharides from bamboo shoots (*Chimonobambusa quadrangularis*) processing by-products. *Int. J. Biol. Macromol.*, 112(June): 656–666. Doi:10.1016/J.IJBIOMAC.2018.02.013.

Chen, Guangjing, Kewei Chen, Renfeng Zhang, Xiaolong Chen, Peng Hu and Jianquan Kan. (2018). Polysaccharides from bamboo shoots processing by-products: New insight into extraction and characterisation. *Food Chemistry*, 245(April): 1113–1123. Doi:10.1016/J. FOODCHEM.2017.11.059.

Chen, Yuzhen, Lei Zhao, Benguo Liu and Sasa Zuo. (2012). Application of response surface methodology to optimize microwave-assisted extraction of polysaccharide from *Tremella*. *Physics Procedia*, 24. Elsevier BV: 429–433. Doi:10.1016/J.PHPRO.2012.02.063.

Du, Xiuju, Jingsong Zhang, Zhiwei Lv, Libin Ye, Yan Yang and Qingjiu Tang. (2014). Chemical modification of an acidic polysaccharide (TAPA1) from *Tremella Aurantialba* and potential biological activities. *Food Chemistry*, 143: 336–340. Doi:10.1016/J.FOODCHEM.2013.07.137.

Edri, Eitan, Nina Armon, Ehud Greenberg, Shlomit Moshe-Tsurel, Danielle Lubotzky, Tommaso Salzillo, Ilana Perelshtein, Maria Tkachev, Olga Girshevitz and Hagay Shpaisman. (2021). Laser printing of multilayered alternately conducting and insulating microstructures. *ACS Applied Materials and Interfaces*, 13(30): 36416–36425. Doi:10.1021/ACSAMI.1C06204.

Effects of *Tremella fuciformis* on Improvement in Cognitive-Bio-markers of Cognitive Functions - Full Text View - ClinicalTrials.Gov., 2022; accessed on April 30. https://clinicaltrials.gov/ct2/show/ NCT02377024.

Kalač, Pavel. (2013). A review of chemical composition and nutritional value of wild-growing and cultivated Mushrooms. *Journal of the Science of Food and Agriculture*, 93(2): 209–218. Doi:10.1002/JSFA.5960.

Khondka, Proma. (2009). Composition and partial structure characterisation of *Tremella* polysaccharides. *Mycobiology*, 37(4): 286. Doi:10.4489/MYCO.2009.37.4.286.

Lan, Xuyue, Yihan Wang, Sha Deng, Jiayuan Zhao, Ling Wang, Kai Yao and Dongying Jia. (2021). Physicochemical and rheological properties of *Tremella fuciformis* polysaccharide fractions by ethanol precipitation. *CyTA—Journal of Food*, 19(1): 645–655. Doi:10.1080/19476337.2021.1 950212.

Li, Yujuan, Fangxue Xu, Mengmeng Zheng, Xiaozhi Xi, Xiaowei Cui and Chunchao Han. (2018). Maca polysaccharides: A review of compositions, isolation, therapeutics and prospects. *Int. J. Biol. Macromol.*, 111(May): 894–902. Doi:10.1016/J.IJBIOMAC.2018.01.059.

Liu, Jun, Chen guang Meng, Ye hua Yan, Ya na Shan, Juan Kan and Chang Hai Jin. (2016). Structure, physical property and antioxidant activity of catechin grafted *Tremella fuciformis* polysaccharide. *International Journal of Biological Macromolecules*, 82(January): 719–724. Doi:10.1016/J. IJBIOMAC.2015.11.027.

Ma, Xia, Meng Yang, Yan He, Chuntao Zhai and Chengliang Li. (2021a). A review on the production, structure, bioactivities and applicationsof *Tremella* polysaccharides. *International Journal of Immunopathology and Pharmacology*, 35, SAGE Publications. Doi:10.1177/20587384211000541.

Ma, Xia, Meng Yang, Yan He, Chuntao Zhai and Chengliang Li. (2021b). A review on the production, structure, bioactivities and applications of *Tremella* polysaccharides. *International Journal of Immunopathology and Pharmacology*, 35. Doi:10.1177/20587384211000541.

Ma, Xia, Meng Yang, Yan He, Chuntao Zhai and Chengliang Li. (2021c). A review on the production, structure, bioactivities and applications of *Tremella* polysaccharides. *International Journal of Immunopathology and Pharmacology*, 35(April). SAGE Publications. Doi:10.1177/20587384211000541.

Newman, David J. and Gordon M. Cragg. (2016). Natural products as sources of new drugs from 1981 to 2014. *Journal of Natural Products American Chemical Society*, 79(3): 629–661. Doi:10.1021/ ACS.JNATPROD.5B01055/SUPPL_FILE/NP5B01055_SI_002.DOCX.

Niu, B., Feng, S., Xuan, S. and Shao, P. (2022). Food Research International and Undefined 2021–2022. Moisture and Caking-resistant *Tremella fuciformis* Polysaccharides Microcapsules with Hypoglycemic Activity, Elsevier; accessed on April 30. https://www.sciencedirect.com/science/ article/pii/S0963996921003197.

Oloke, J.K. (2015). Effectiveness of immunotherapies from oyster Mushroom (*Pleurotus* species) in the management of immunocompromised patients. *International Journal of Immunology*, 3(2), Science Publishing Group: 8. Doi:10.11648/J.IJI.S.2015030201.12.

Park, Kum Ju, Sang-Yun Lee, Hyun-Su Kim, Matsumi Yamazaki, Kenzo Chiba and Hyo-Cheol Ha. (2007). The neuroprotective and neurotrophic effects of *Tremella fuciformis* in PC12h cells. *Mycobiology*, 35(1), (KAMJE): 11. Doi:10.4489/MYCO.2007.35.1.011.

Ruan, Yang, Hong Li, Lianmei Pu, Tao Shen and Zening Jin. (2018). *Tremella fuciformis* polysaccharides attenuate oxidative stress and inflammation in macrophages through MiR-155. *Analytical Cellular Pathology* (Amsterdam), 2018, Hindawi Limited. Doi:10.1155/2018/5762371.

Shahrajabian, Mohamad Hesam, Wenli Sun, Qi Cheng and Mehdi Khoshkharam. (2022). Exploring the quality of foods from ancient china based on traditional Chinese medicine. *Functional Foods and Nutraceuticals in Metabolic and Non-communicable Diseases*, January. Academic Press, 87–105. Doi:10.1016/B978-0-12-819815-5.00048-3.

Sheng, Z., Yu, L., Li, X., Zhao, Y., Dai, W. and Chang, S.K. (2022). The anti-obesity effect of fermented *tremella*/blueberry and its potential mechanisms in metabolically healthy obese rats. *Journal of Functional and Undefined* 2021–2022. Elsevier; accessed on April 30. https://www. sciencedirect.com/science/article/pii/S1756464621003194.

Strop, Pavel. (2014). Versatility of microbial transglutaminase. *Bioconjugate Chemistry*, 25(5): 855–862. Doi:10.1021/BC500099V/ASSET/IMAGES/MEDIUM/BC-2014-00099V_0005.GIF.

Wang, C., Ji, Y., Deng, W. et al. (2019). Preparation of *Tremella* and lotus seed cake. *Food and Fermentation Industries*. https://www.cabdirect.org/cabdirect/abstract/20193242240.

Wen, Lingrong, Qing Gao, Chung wah Ma, Yazhong Ge, Lijun You, Rui Hai Liu, Xiong Fu and Dong Liu. (2016). Effect of polysaccharides from *Tremella Fuciformis* on UV-induced photoaging. *Journal of Functional Foods*, 20(January): 400–410. Doi:10.1016/J.JFF.2015.11.014.

Wu, Qiong, Cheng Zheng, Zheng Xiang Ning and Bao Yang. (2007). Modification of low molecular weight polysaccharides from *Tremella fuciformis* and their antioxidant activity *in vitro*. *International Journal of Molecular Sciences* 8(7): 670–679. Doi:10.3390/I8070670.

Xiao, H., Li, H., Wen, Y., Jiang, D., Zhu, S. and He, X. (2022). *Tremella fuciformis* polysaccharides ameliorated ulcerative colitis via inhibiting inflammation and enhancing intestinal epithelial barrier function. *International Journal of and Undefined, 2021–2022*; accessed on April 30. https://www.sciencedirect.com/science/article/pii/S0141813021006012.

Xu, Yingyin, Liyuan Xie, Zhiyuan Zhang, Weiwei Zhang, Jie Tang, Xiaolan He, Jie Zhou and Weihong Peng. (2021). *Tremella fuciformis* polysaccharides inhibited colonic inflammation in dextran sulfate sodium-treated mice via Foxp3+ T cells, gut microbiota and bacterial metabolites. *Frontiers in Immunology*, 12(April): 948. Doi:10.3389/FIMMU.2021.648162/BIBTEX.

Yang, J., Liu, T., Zhang, S. et al. (2019). Optimisation of microwave-assisted extraction and rheological and gelling properties of polysaccharide from *Tremella fuciformis*. *Food Science*. https://www.cabdirect.org/cabdirect/abstract/20193425721.

Yuan, Hongjie, Lin Dong, Zhiyuan Zhang, Yan He and Xia Ma. (2022). Production, structure and bioactivity of polysaccharide isolated from *Tremella fuciformis*. *Food Science and Human Wellness*, 11(4): 1010–1017. Doi:10.1016/J.FSHW.2022.03.030.

Yui, Toshifumi, Kozo Ogawa, Mariko Kakuta and Akira Misaki. (2006). Chain conformation of a glucurono-xylo-mannan isolated from fruit body of *Tremella fuciformis* Berk, Taylor & Francis Group, 14(2): 255–263. Doi:10.1080/07328309508002068.

Zhang, Jian, Ya Kun Zhang, Yong Liu and Jun Hui Wang. (2019). Emulsifying properties of *Tremella fuciformis*: A novel promising food emulsifier. *International Journal of Food Engineering*, 15(3-4), De Gruyter. Doi:10.1515/IJFE-2018-0217/Machinereadablecitation/RIS.

Zhao, Xiaona, Yuanliang Hu, Deyun Wang, Jianzhu Liu and Liwei Guo. (2013). The comparison of immune-enhancing activity of sulfated polysaccharidses from *Tremella* and *Condonpsis pilosula*. *Carbohydr. Polym.*, 98(1): 438–443. Doi:10.1016/J.CARBPOL.2013.06.043.

Zhou, A.R., Wu, Y.K. and Hou, Y.Y. (1987). Effects of *Tremella* polysaccharide on anti-tumour. *J. Beijing Med. Univ.*, 19(3): 150.

Zhou, Yalin, Xiaoyong Chen, Ruokun Yi, Guijie Li, Peng Sun, Yu Qian and Xin Zhao. (2018). Immunomodulatory effect of *Tremella* polysaccharides against cyclophosphamide-induced immunosuppression in mice. *Molecules*, 23: 239–23. Doi:10.3390/MOLECULES23020239.

Zou, Yu and Xiyan Hou. (2017). Extraction optimisation, composition analysis and antioxidation evaluation of polysaccharides from white jelly Mushroom, *Tremella fuciformis* (Tremellomycetes). *Int J. Med Mushrooms*, 19(12): 1113–1121. Doi:10.1615/INTJMEDMUSHROOMS.2017024590.

CHAPTER 22

Paddy Straw Mushroom (*Volvariella volvacea*)

Dhiraj L Wasule, Anjali M Gaharwar, Prashant R Shingote*
and *Darasing R Rathod*

1. Introduction

Steep rise in population, reduced cultivable land-holding capacity and climate change
cause inadequate quality food supplies, leading to de-treating quality of health which
is a major concern. These serious problems give an alternative to search for nutrient-
rich food and the answer lies in cultivation of nutritive *Volvariella volvacea*, the third
most-consumed mushroom. Simple, low-cost technology can help it to grown in rural
areas when the temperature remains around 30–35°C. Mushroom is a typical fleshy,
spore-bearing fruiting body of a macro fungus, devoid of leaves and chlorophyll-
containing tissues. Presently, in the world, there are 12,000 mushroom species out
of which 2,000 edible species were reported by Beulah et al., in 2013. Recently the
number increased from available 12,000 to approximately 38,000 species.

Paddy straw mushroom (*V. volvacea*) belongs to family Pluteaceae of the
Basidiomycetes (Ahlawat and Arora, 2016). *Volvariella* varieties are found at
global level and include *V. coesiotincta*, *V. gloiocephala*, *V. bombycine*, *V. iranica*,
V. hypopithys, *V. lepiotospora*, *V. sathei*, *V. jamaicensis*, *V. speciosa*, *V. peckii*,
V. surrecta and *V. volvacea*. For edible purposes, *V. volvacea* species are popularly
used. Volva means 'wrapper' that looks as cover over the fruiting body that was
successively named as 'button', 'egg', 'elongation' and 'mature'. The volva on
maturity ruptures to form an umbrella. Mushroom in general is used as a healthy food
as its role is in prevention of arteriosclerosis, chronic hepatitis, hyperlipidemia and
act as an antioxidant, anticancer, antimalarial, and anti-allergic agent. It is also known
as a warm mushroom as it requires high temperature for growth and development.

Dr. Panjabrao Deshmukh Krushi Vidyapeeth, Akola, Maharashtra - 444104.
* Corresponding author: shingoteprashant@pdkv.ac.in

This mushroom is found growing at 30–35°C temperature with relative humidity at 80–90%. Due to its delicious taste, pleasant flavour, higher protein content and less production period compared to other mushrooms, it is ranked the third most important mushroom in the world (Gupta et al., 2016). It is a fast-growing mushroom which is reported to complete its crop cycle within a short period of three to four weeks. Composts for mushroom comprise of lignocellulosic wastes which is made available from different field crops. Easy availability of raw material for compost lowers the production cost of mushroom and helps to make more money. Different crops, like rice straw, sugar cane bagasse, cereal straws, cotton waste and oil palm pericarp can be utilised for mushroom production (Yella et al., 2021). Straw/wastes like paddy straw, cotton wastes or other cellulosic organic waste are used to manufacture commercial produce (Uddin et al., 2012). The only concern is that the substrate should be rich in nutritional composition for quality mushroom production.

2. History

Volvariella volvacea grows in the east and Southeast Asia (Kamaliah et al., 2021). Out of 38,000 reported varieties of mushrooms, only 300 mushroom species are edible and 10 are grown commercially. Mushroom cultivation was first time reported in China in 1822, by Chang (1969). Further reported mushroom cultivation began in the 18th century and was first recorded as Chinese mushroom, popular as Nanhua mushroom, with the name given according to the Nanhua Temple in China. Buddhist monks used to grow this mushroom for their table purpose; later on, the cultivation extended for the royal family as a tribute by 1875. Around 1932 to 1935, it was introduced by Chinese in the tropical and subtropical area of the Philippines, Malaysia and some South Asian countries (Ahlawat et al., 2016). Mushroom cultivation first started in America at Pennsylvania, which came to be known as the 'mushroom capital of the world'. Mushroom has now become a regular cultivated crop of China, India and Southeast Asian countries, like Indonesia, Malaysia and Thailand which incorporate 16% of the total world production (Umor et al., 2020). China (47%) stands at first position in the world as a mushroom producer; USA (11%) comes second and The Netherlands (4%) is at third position. During 1940, Su and Seth cultivated mushroom first time in India. At Coimbatore, the scientific cultivation report of *V. diplasia* through spawn was reported by Thomas in 1943 (Reddy, 2015). Moreover 19 edible species of *Volvariella* have been reported in India among which some important ones are: *V. esculenta*, *V. diplasia* and *V. volvacea*. Coastal states, like Andhra Pradesh, Orissa, Kerala, Tamil Nadu and West Bengal, which are paddy-growing states, also grow mushrooms on a commercial scale. Even states like Punjab and Haryana grow mushrooms as there is agro waste which makes composting easily available (Ahlawat and Tewari, 2007).

3. Environmental Requirement for Growth and Production

V. volvacea mushroom is suitable for growing in tropical as well as subtropical climate (Bao et al., 2013). This mushroom, due to its short duration of production, has been popularised for easy production technology with ample availability of agro wastes (Zikriyani et al., 2018). Mushrooms undergo optimum mycelial growth and fruiting body respectively at temperature range 30–35°C and 28–30°C (Le-Duy-Thang, 2006). The humidity ranges between 70–90% (Biswas and Layak, 2014). Compost pH is one the factors affecting mushroom growth, hence the pH suitable for growth has been determined at 6.5 (Akinyele and Adetuyi, 2005).

3.1 Factors Affecting Growth and Production of Mushroom

1. *Temperature*: As mushroom is a saprophytic crop, climate impacts the growth of the fungus. Temperature is the very first factor that has to be controlled for optimum growth. Each mushroom species has its own optimal temperature. More or less the temperature adversely affects the growth of fruiting body of mushroom. *Volvariella*, as a warm mushroom, requires high temperatures for mycelia and fruiting-body growth.

2. *Relative humidity*: As the mushroom body contains about 80–90% water, it is necessary to maintain humidity in the growth chamber to meet the moisture requirement and to avoid damage to fungi. High humidity leads to less transpiration as the mushroom body favours its growth.

3. *Light*: Mushroom loves shade, indoor conditions and so provide the environment that it needs for growth and production.

4. *Air exchange*: Air exchange is important which make the mushroom breathe. It literally reduces the temperature and thus adjusts the humidity and CO_2 levels around the mushroom.

5. *pH*: The pH of substrate directly impacts the uptake of nutrients form substrate which affects the mycelia growth, the number of fruiting bodies and the production of mushrooms. At higher pH, the growth gets retarded. Hence, if required, lime should be added to substrate to maintain the substrate pH at 7.

6. *Particle size of substrate*: The particle size of substrate matters in the growth of fungi. As a general conclusion, coarser the particle density the greater the time taken for growth of mushrooms. However, when fine substrate used, early growth is observed.

7. *Nutrients*: All kinds of mushroom grow on substrate consisting of cellulose and lignin. However optimum C:N ratio may influence the growth and production of mushroom.

8. *Substrate*: The quality of substrate determines the quality and productivity of mushroom. (Ahlawat et al., 2011).

The general parameters required for the growth of mushrooms is presented in Tables 1 and 2.

Table 1. Environmental requirement for growth and production of *V. volvacea*.

Parameter	Temperature (°C)		Relative Humidity (%)		pH	
	Range	Optimal	Range	Optimal	Range	Optimal
Mycelium	15–42	35 ± 2	50–70	60 ± 5	6–7	6.5
Fruiting	25–30	28 ± 2	80–100	90 ± 5	6–7	6.5

Adapted from Thuc et al. (2020).

Table 2. Different substrate for mushroom production and its biological efficiency.

Substrates	Biological Efficiency (%)
Rice straw	10.2–15.0
Cotton waste	17.7–40.0
Banana leaves	8.6–15.2
Sugarcane trash	13.2
Wheat	12.1
Water hyacinth	8.7

Adapted from Thuc et al. (2020).

3.2 Morphological Characteristics

Mushroom is an umbrella-shaped fungus which is always seen in stories and pictures books. In Nature, whatever the wild types grow, they have the most common umbrella shape. At very early phase of mushroom development, a fungal hypha grows, bears buttons in later stage and on reaching maturity, burst to form the umbrella. *V. volvacea*, the name of paddy straw mushroom, is derived from *volva*, meaning a wrapper that wraps the main fruiting body during the young stage. The fruiting body formation starts with distinct tiny clusters of white hyphal aggregates, called primordial. This primordial at successive stage bears a bulging growth named as 'button', 'egg', 'elongation' before becoming 'mature'. At maturity, hypha bears an umbrella-shaped cap, called pileus. The growth structures of mushroom are described under the respective stages as follows:

1. *Volva*: It appears at stipe of mushroom to resume a cup-like structure. It is more or less similar to a veil shape.

2. *Stipe*: It is a long, rounded structure with soft surface and no annulus. It seems off-white to dull brown in colour. It shows a slightly enlarged structure at the base of the bulb. The stalk (stipe) bears a brownish-grey coloured cup-like structure called *volva* (Chang and Miles, 2004).

3. *Pileus*: The more or less umbrella-shaped or budge-cup-shaped fleshy structure is attached to the stipe. Pileus is around 5–15 cm broad and it varies under different environmental conditions or as per the mushroom species. Colour of pileus varies from dark grey to brown with diameter 8–10 cm. At a younger stage, its cap has an oval shape like egg; later on at maturity it becomes conical and then turns domed or flat.

4. *Gills*: The gills, also known as lamellae, are observed as radial vertical plate on the lower surface of the pileus.

3.3 Physical Characters

1. It is a warm mushroom, suitable to be grown in tropical and subtropical regions.

2. This mushroom grows relatively at high temperatures.

3. It completes its life cycle in approximately four to five weeks under favourable growing conditions (Bao et al., 2013; Biswas, 2014).

4. Life cycle duration may differ as per the climatic conditions of the place where it grows.

5. Mushroom body comprises of the vegetative parts, viz., volva, stipe, pileus and gills.

On the basis of biological characteristics, the mushroom is divided into different developmental stages.

- *Pinhead stage*: A very early stage, where the fruit body is of pinhead size and veil is spotlessly white.

- *Button stage*: Button stage comes after pinhead stage and comprises tiny buttons followed by button stage. At the tiny button stage, the veil is brown from the top and the rest is white. At the later stage of growth, on making longitudinal sectional cuts, it is distinct as per its various parts, like stalk, cap and gills which are generally identified as stipe, pileus, and lamellae respectively. At this stage, it is sold in the market at a good commercial price. The button stage mushrooms have a high protein content of about 25% and good palatability and storage life; hence it is preferred for consumption.

- *Egg stage*: During the egg stage, the pileus thrown out from the veil and volva remains as such later on. It has commercial value in the market.

- *Elongation stage*: When the mushroom volva ruptures exposing the stipe or stalk and the cap, it is termed as elongation stage. The stipe grows to its maximum length.

- *Mature stage*: Maturity is the last stage where the volva is fully expanded and exposed with brownish gills on its lower surface. The pileus is attached to the stipe at the centre and spowns later on burst to release the spores. Basidiospores vary in shapes and colour.

3.4 Methodology for Production of Paddy Straw Mushroom in Indoor Conditions

Methodology or process includes in the production of *V. volvacea* from paddy straw is as follows:

1.1: Platform preparation

A raised platform of about 1m in length and 0.75 m in breadth or appropriate dimension with wooden planks/bamboos be erected, providing bricks or any support at all the four corners, or prepare racks for mushroom production.

1.2: Bed preparation

a. Weigh about a kg of paddy straw into bundles.

b. Soak the paddy straw in water mixed with lime at 3–5% for about two to four hours. Thereafter chop the straws into pieces 2–5 cm long, add 1% calcium carbonate and 1–2% rice bran. Maintain water pH at 13–14. Drain the excess water and take out 10 bundles. Do not add excess water as it inhibits rice straw mycelium growth and water is there at the bottom.

c. Place the bundles over the platform with their butt/tail end on one side.

d. Build the second layer by placing the butt end towards the other direction.

e. Sprinkle a spoonful of coarse gram powder and apply spawn all along the periphery of bundles, about 8–12 cm inside the margin. 500 gm spawn and 150 gm of pigeon pea powder generally is used with each bed comprising an average of 30–40 kg.

f. Build another layer as prepared earlier and spawn the layer.

g. Build up 4–5 layers and spawn as usual.

h. Apply pressure over bed to make it compact as possible and cover it with a transparent polythene sheet.

i. Keep the beds undisturbed for a few days.

1.3: Environment maintenance

Maintain 80–85% humidity and temperature of 30–35°C in the 'mushroom room'. Plastic sheet be removed after seven to eight days of spawning. Now maintain temperature of 28–32°C and relative humidity of about 80%.

1.4: Growth of mushroom

After four to five days of removal of sheets from the bed, mushrooms will start appearing and continue to grow for the next 20 days.

1.5: Harvesting of mushroom

Fruit will appear after five to seven days of spawning. For best storage, the fruiting bodies are harvested while they are still at the egg stage. Mushrooms can be kept in a paper bag and placed for cool preservation in a fridge for a few days. They are then dried out and stored in an airtight container until ready to cook.

4. Nutritional and Sensory Profile of *V. Volvacea*

The trend to view mushrooms not only as food but as nutritive and medicinal fungi for consumptive use is now gaining momentum. Out of 38,000 different varieties of worldwide available mushrooms, 300 mushroom species are found edible and are being exploited for medicinal purposes through research (Roy et al., 2014). *V. volvacea* contains 240 KJ (58 Kcal) of food energy, cholesterol and fat, salt-free and has low alkalinity and is a rich source of carbohydrates, minerals, protein, fibres (chitin) and vitamins (vitamin C, riboflavin, biotin and thiamine) (Senatore, 1990; Adewusi and Oke, 1993). Protein content in mushrooms is as good as found in vegetable and animal sources (Kurtzman, 1976; Purkayastha and Nayak, 1981). *Volvacea* contains fats (5.7%), carbohydrates (56.8%), essential amino acids and unsaturated fatty acids. One cup of *V. volvacea* provides an average of 50.36% selenium, 46.60% sodium, 32.50% iron, 26.89% copper, 11.09% zinc, 15.86% phosphorus with low calorific values (Breene, 1990; Mshandete et al., 2007; Ouzouni et al., 2009). The nutritive values are presented in Tables 3–10. Overall mushroom contains roughly about 90% water. The remainder is made up of 10–40% protein, 3–28% carbohydrate, 8–10% ash, 3–32% fibre, minerals and 2–8% fat (Breene, 1990). Various bioactive molecules also present terpenoids, nucleotides, ergosterol, polysaccharides and glycoproteins. Also present are crude fats, lipids and a low percentage of saponifiable fat (58.8%) (Huang et al., 1985).

Bales (Paddy straw)

(Moisture content : 15-18% for wet

bales)

Lime water treatment (3-5% CaCO3,

pH=13-14, 10-15 min.)

Lime water-soaked rice straw

Excess water removal (3-5min.)   Spawning of yeast (100g)/

Spawning 10 spawn bags

Spawn   Spawning   Bed preparation

160g/bed)1,2   Stimulator

m)

Keep beds in open sunlight for three days

Dry straw layering on straw beds

At 35 ± 2°C temp. and

relative humidity at 60 ± 5%

Growth of mycelial colony

Sprinkling of water, 1-2 litre/ bed

At 28 ± 2°C temp. and

relative humidity of 90 ± 5%

Fruitification

Harvesting of mushroom

Rice straw mushroom

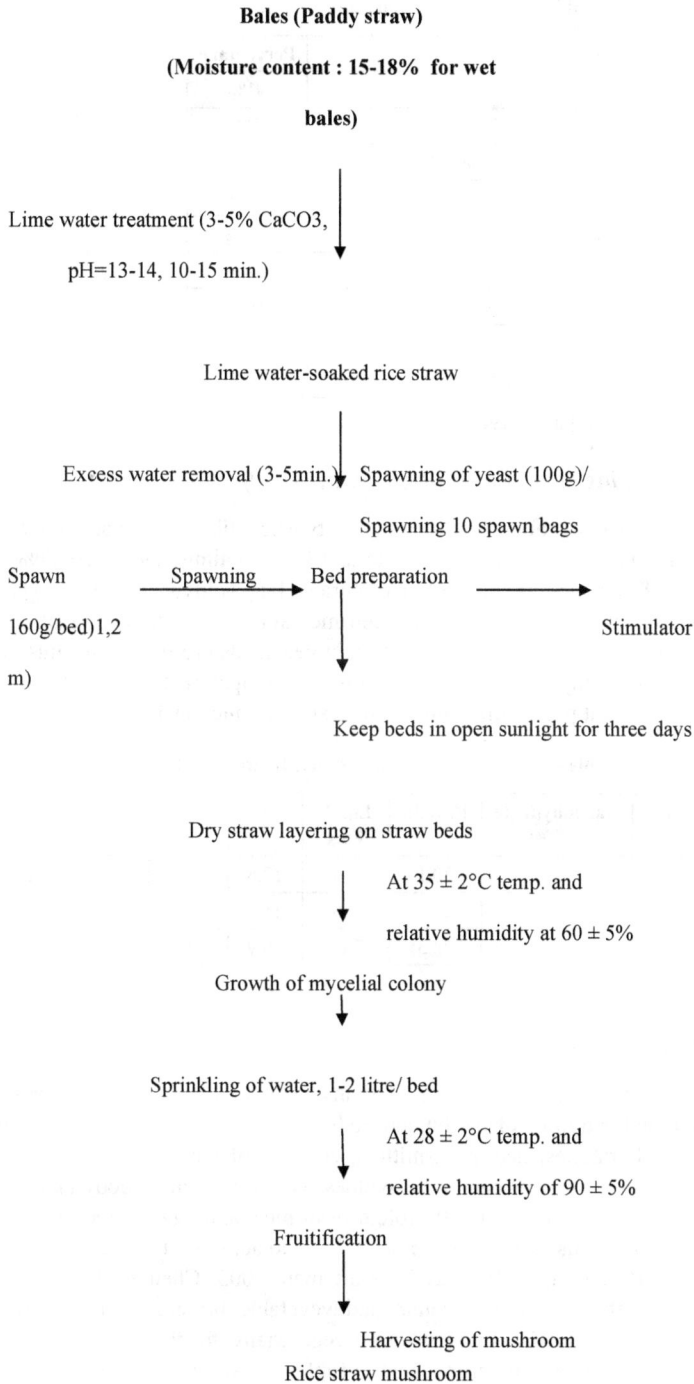

Fig. 1. Flow chart for paddy straw mushroom production (Thuc et al., 2020).

Table 3. Proximate composition of *V. volvacea* mushroom.

Composition	Percentage
Moisture	91%
Dry matter	9%
Total nitrogen	6.5%
Crude protein	28%
Crude fat	3.3%
Crude fibre	9.8%
Ash	10%
Total carbohydrates	50%
Nitrogen-free extract	41%

Adapted from Zahid et al. (2019).

4.1 Fatty Acids (Unsaturated and Saturated)

Mushrooms contain a good percentage of fatty acids, like linoleic acid 69.91%, oleic acid 12.74%, palmitic acid 10.5%, stearic acid 3.47%, palmitoleic acid 0.62%, myristic acid and 0.48% (Huang et al., 1989). Unsaturated fatty acids are an essential part of our diet while saturated fatty acids have a detrimental effects on human health (Holman, 1976). Certain unsaturated fatty acids and linoleic acids present in this mushroom are considered as healthy and rich food components. Comparative nutritional composition studies of different paddy straw mushrooms are presented in Table 4.

Table 4. Nutritional parameters of paddy straw mushrooms (%).

Volvariella Species	Carbohydrate	Protein	Lipid/ Fats	Ash	Fibre	References
V. volvacea	50.9	30.1	6.4	12.6	11.9	Lee and Chang, 1975
V. diplasia	57.4	28.5	2.6	11.5	17.4	Chang and Hayes, 2013
V. bombycina	38.9	28.3	2.72	10.9	24.6	Jagadeesh et al., 2010

Adapted from Ghosh (2000).

4.2 Amino Acids

Amino acids are vital components of living beings and are essential for structural and functional activities of the human body. Amino acids are involved in synthesis of proteins, hormones, neurotransmitters, etc. Out of the 20 amino acids, nine are essential amino acids that cannot be synthesised by the human body and need to be supplemented from food. Animal protein from meat and egg are the best sources of essential amino acids. All the nine essential amino acids are reported to be present in *V. volvacea* (Chang and Miles, 2004; Kurtzman, 2005; Cheung, 2003). Mushroom protein lies between animal protein and vegetable protein in its nutritive value (Kurtzman, 1975). Hence, those who are vegetarians, for them mushrooms are one of importance sources of amino acids (Table 7). Amino acids in the mushroom are equivalent to that of egg proteins as they are rich in lysine and can be used as supplementary food in lysine-deficient diets (Sohi et al., 1990). *V. bombycine* contains tyrosine and cysteine. The important functions of essential amino acids are:

1. *Phenylalanine*: Phenylalanine is a key molecule for synthesis of enzymes and other amino acids. It is a precursor for the neurotransmitters like tyrosine, dopamine, epinephrine and norepinephrine.

2. *Valine*: Valine is important to boost muscle growth as well as in regeneration, tissue repair and also energy production.

3. *Threonine*: This amino acid is an important residue of many proteins, such as tooth enamel, collagen and elastin. It plays a vital role in fat metabolism, prevents building up of fat in liver and builds immunity.

4. *Tryptophan*: Precursor to serotonin, it is a neurotransmitter which regulates your appetite, sleep and to maintains proper nitrogen balance.

5. *Leucine*: It helps in protein synthesis and muscle repair. It produces growth hormones, regulates blood sugar levels and heals wounds.

6. *Lysine*: Lysine amino acid is involved in calcium absorption, biosynthesis of protein and hormones. It also regulates the production of collagen to strengthen the bony structure and elastin maintains the original shape of organs and both together make the skin youthful and stronger.

7. *Methionine*: It regulates metabolism and detoxification and is involved in tissue growth and absorption of minerals, such as zinc and selenium which are vital for human strength.

8. *Isoleucine*: Isoleucine contributes to human health in rapid wound healing, detoxification of nitrogenous wastes, stimulation of immune functions, secretion of several hormones and regulation of blood sugar and energy levels.

9. *Histidine*: Histidine amino acid plays the role of protecting the cover of nerve cells, that is myelin sheaths, and in metabolising histamine in neurotransmitters.

Table 5. Amino acid contents of *Volvariella volvaceae* mushroom (g/100 g of protein).

Amino Acids	Content
Leucine	7.55 g
Lysine	5.20 g
Cystine	0.95 g
Phenylalanine	6.22 g
Tyrosine	4.79 g
Threonine	4.88 g
Tryptophan	14.7 g
Valine	3.77 g
Alanine	7.14 g
Aspartic Acid	12.4 g
Glutamic Acid	27.9 g
Proline	6.60 g

Adapted from Zahid et al. (2019).

Table 6. Comparative composition of amino acids between different spp. of paddy straw mushrooms.

Amino acids	*V. volvacea* (Chang and Miles, 1989)	*V. diplasia* (Chang and Miles, 1989)	*V. bombycina* (Cheung, 2008)	Hen's Egg	FAO/WHO Requirement
Valine	5.4	9.7	3.58	7.3	3.5
Leucine	4.5	5.0	5.01	8.8	6.6
Isoleucine	3.4	7.8	5.41	6.6	2.8
Threonine	3.5	6.0	4.65	5.1	3.5
Methionine	1.1	1.2	0.122	3.1	2.5c
Lysine	7.1	6.1	5.41	6.4	5.8
Phenylalanine	2.6	7.0	6.02	5.8	6.3d
Tryptophan	1.5	1.5	ND	1.6	1.1
Histidine	3.8	4.2	----	2.4	1.9
Cysteine	---	---	1.91	--	---
Tyrosine	---	---	4.58	--	---
Total EAAs	32.9b	48.5b	36.692	47.1	32.8

Adapted from Ghosh (2000).

Table 7. Mineral content of *Volvariella volvacea* mushroom.

Minerals	Percentage
Potassium	52.52%
Oxygen	28.72%
Phosphorus	8.96%
Chlorine	3.57%
Sulphur	2.72%
Magnesium	0.99%
Silicone	0.79%
Calcium	0.62%
Iron	0.38%
Aluminium	0.27%
Zinc	0.12%
Rubidium	0.09%
Copper	0.07%
Molybdenum	0.07%
Manganese	0.05%

Adapted from Zahid et al. (2019).

4.3 Mineral Composition

In *V. volvaceae* mushroom, potassium (3345.21 mg/100 g) is the most abundant mineral and is an ideal source of minerals like potassium, phosphorous, magnesium, calcium, zinc, iron, manganese and copper (Mshandete and Cuff, 2007). Carbonyl compounds and octavalent carbonate alcohols produce the peculiar aroma of *V. volvacea*. Elements, like sulphur, nitrogen, potassium, zinc, phosphorus, iron, nucleotides, amino acids impact auto-oxidation of unsaturated fatty acids and thus the aroma (Grzybowski et al., 1978).

4.4 Vitamins

V. volvacea is rich in vitamins like riboflavin, niacin and folates (Matilla et al., 2001). Niacin is involved in controlling activity of DNA damage responses and signalling events in stress responses, such as apoptosis, that may impact the cancer-risk reduction (Kirkland, 2003). Niacin is also found to be effective in treatment for atherosclerotic cardiovascular disease (Guyton, 1998). However, tamin A and C were not found present in any stages of fruiting body of *V. volvacea* (Milton and Dulay, 2015). This mushroom is rich in vitamin C, thiamine, riboflavin, niacin and biotin (Ahlawat and Singh, 2016; Chang and Hayes, 2013) (Tables 8 and 9).

Table 8. Vitamin contents in *Volvariella volvacea* mushroom.

Vitamins	Content
Vitamin A	0.001 mg/kg
Vitamin D	50.711 mg/kg
Vitamin K	0.006 mg/kg
Vitamin C	48 mg /100g

Zahid et al. (2019).

Table 9. Relative nutritional content in different edible straw mushroom species.

Parameter	*V. volvacea* (Ahlawat and Singh, 2016)	*V. diplasia* (Chang and Hayes, 2013)	*V. bombycina* (Ahlawat and Singh, 2016)
Vitamin D (IU/g)	462.05	---	106.995
Calcium (mg/100 g)	39.74	58.0	25.61
Potassium (%)	4.16	3.353	4.12
Iron (mg/kg)	72.51	177.0	72.50
Copper (mg/kg)	42.55	---	50.20
Zinc (mg/kg)	94.28	---	119.95
Sodium (mg/kg)	345.34	ND	---
Magnesium (%)	0.11	---	0.12

Adapted from Ghosh (2000).

Fibre is a part of a healthy and balanced diet. It stabilises the blood sugar level in diabetic patients by reducing the daily requirements of insulin (Anderson and Ward, 1979). In *V. bombycina*, *V. diplasia* and *V. volvacea* the fibre content is 24.60%, 17.40% and 11.90% respectively.

4.4.1 Bioactive molecules

Various biomolecules, such as carbohydrates, terpenoids, proteins and nucleotides are abundantly found in mushrooms. Huang and Chang (1985) identified ergosterol and provitamin D_2 in mushroom with low amounts of saponifiable fat (58.8%).

4.4.2 Metabolites of V. Volvacea and their Effects

Metabolites promote high antioxidant activities. In *V. volvacea*, metabolites that are present include terpenes, polypeptides and steroids as also different phenolic compounds, such as phenolic acids, tannins and flavonoids. Huang et al. (2012) reported that mushrooms contain high quantities of free phenolic compounds for antioxidant activity. The fruiting body or mushroom in dried straw form is rich in antioxidant enzymes, like superoxide dismutase, catalase, glutathione reductase, glutathione-S-transferase, glutathione peroxidase and peroxidase (Ramkumar et al., 2012). Mushroom is a rich source of protein which contains flammutoxin and volvatoxin and cardio-toxic proteins are known to slow down the respiration in tumour cells. Paddy straw mushroom possesses polysaccharides and has antitumour properties as reported by Zhang et al. (1994). Mushroom has a high quantity of water, which with methanol extracts of paddy straw mushrooms is rich in antioxidative properties that help to prevent cardiovascular diseases, cancer (Cheung et al., 2003); also neurodegenerative diseases and inflammation (Joseph et al., 1999).

Table 10. Profile of soluble sugars in canned *Volvariella volvacea* mushroom (mg/g).

Soluble Sugars	Content
Arabinose	3.19 mg
Fructose	2.26 mg
Glucose	0.81 mg
Myo-inosito	1.20 mg
Mannose	2.40 mg
Ribose	5.07 mg
Sucrose	2.13 mg
Trehalose	5.86 mg
Total	22.92 mg

Adapted from Zahid et al. (2019).

4.4.3 Secondary Metabolites

Kalava and Menon (2012) reported that *V. volvacea* is a good source of polypeptides, terpenes and phenolic compounds like flavonoids, phenolic acids, and tannins that contribute to high antioxidant capacity. *V.* volvacea was found to be rich in antioxidative activities (Cheung et al., 2003). Nhi and Hung (2012) found that antioxidant properties of *V. volvacea* are governed by higher amounts of phenolics. Foods that are rich in antioxidants play an essential role in the prevention of diseases like arthritis, chronic inflammation, cancers and cardiovascular diseases (Ames et al. 1993; Dragsted et al., 1993). Ramkumar et al. (2012) reported the presence of highest levels of reactive oxygen specie scavengers, such as catalase, peroxidase, glutathione reductase and superoxide dismutase in this mushroom. Higher levels of variegatic acid and diboviquinone antioxidative agents were also reported in *V. volvacea* (Kalaiselvan, 2007).

5. Health Benefits of *V. volvacea*

5.1 Nutritional Food

V. volvaceae is a rich source of carbohydrates, proteins, fibre, minerals and vitamins. The protein content in paddy straw mushroom is higher than that found in meat, fish, vegetables, or citrus fruits and satisfies the need for eight essential amino acids required by the human body. This makes it a promising food to fight against malnutrition (Endang Sukara et al., 1985).

5.2 Inhibits Cancer Cell Proliferation

The cardio-toxic proteins, flammutoxin and volvatoxin, slow down the respiration in tumour cells and Beta d-glucan and lectin, thus inhibiting the production and development of tumour cells during leukaemia, sarcoma and colorectal (Kishida et al., 1992).

5.3 Lowers Cholesterol Level

Phenolic acid, present in paddy straw mushroom, prevents lipid peroxidation and decreases blood cholesterol, low-density lipoprotein (LDL) cholesterol and triglycerides level (Cheung et al., 2005).

5.4 Controls Blood Pressure

Various biomolecules, such as carbohydrates, terpenoids, proteins and nucleotides, are abundantly found in mushroom. These bioactive molecules reduce blood pressure and prevent hypertension, thus reducing the risk of cardiovascular diseases (Chiu et al., 1995).

5.5 Control Blood Sugar Level and Manage Diabetes

V. volvaceae is found to reduce blood cholesterol and thus control diabetes. Different polysaccharides, lectins, lactones, β-glucans, alkaloids, terpenoids, sterols and phenolics enable pancreatic cell function to improve glucose tolerance and maintain blood glucose levels (Punitha et al., 2016).

5.6 Build Immunity

V. volvaceae found Fip-vvo protein having immunomodulatory properties and protecting against various chronic diseases by improving immunity. Lectin present in it also boosts immunomodulatory activity (She et al., 1998).

6. Conclusion

Production of *V. volvaceae*, paddy straw mushroom, favouring tropical climatic conditions is one of the commercially grown mushrooms worldwide. The easily available agro waste, like paddy straw in Indian villages, facilitates proper reutilisation of huge biomass and is the easiest way to make this venture money earning. This is not only an economical way of farming but also produces good quality of nutritive food against the land limitation criterion conditioned to provide microclimatic conditions in a sustainable and economical manner within a short period of time and fight against malnutrition. Simple cultivation technology makes its production possible at all levels of farming. Owing to the highly nutritive and taste-bud nourishing recipes for different mushroom foods, it becomes a dish of five-star hotels as also of root-level consumers. Being the only non-animal source of vitamin D, it provides food to malnourished people and helps to maintain the health of people and build the human immune system. This mushroom, being a rich source of antioxidants, vitamins and minerals with more protein than any other vegetable, is found promising in protection against several human illness and diseases for its anti-inflammatory, antidiabetic, cytotoxic, antitumour, anticancer, antibacterial, antiviral, anti-HIV activities and for being an antioxidant and immunity booster. High proportion of polysaccharides leads to considerable antitumour and immunomodulating properties. As per the nutritional and medicinal attributes of this mushroom, it is popular as a novel food in the nutraceutical, pharmaceutical and pharmacological industries.

References

Adewusi, S.R.A., Alofe, F.V., Odeyemi, O., Afolabi, O.A. and Oke, O.L. (1993). Studies on some edible wild mushrooms from Nigeria: 1. Nutritional, teratogenic and toxic considerations. *Plant Foods for Human Nutrition*, 43(2): 115–121.

Ahlawat, O.P. and Arora, B. (2016). *Paddy Straw Mushroom (Volvariella volvacea) Cultivation*. National Research Centre for Mushroom *(ICAR)*: Solan, India, 165–182.

Ahlawat, O.P. and Tewari, R.P. (2007). *Cultivation Technology of Paddy Straw Mushroom (Volvariella volvacea)*, vol. 36, India: National Research Centre for Mushroom.

Ahlawat, O.P., Singh, R. and Kumar, S. (2011). Evaluation of *Volvariella volvacea* strains for yield and diseases/insect-pests resistance using composted substrate of paddy straw and cotton mill wastes. *Indian Journal of Microbiology*, 51(2): 200–205.

Akinyele, B.J. and Adetuyi, F.C. (2005). Effect of agro wastes, pH and temperature variation on the growth of *Volvariella volvacea. African Journal of Biotechnology*, 4(12).

Ames, B.N., Shigena M.K. and Hagen, T.M. (1993). Oxidants, antioxidants and the degenerative diseases of ageing. *Proceedings of the National Academy of Sciences*, USA, 90: 7915–7922.

Anderson, J.W. and Ward, K. (1979). High-carbohydrate, high-fibre diets for insulin-treated men with diabetes mellitus. *The American Journal of Clinical Nutrition*, 32(11): 2312–2321.

Bao, D., Gong, M., Zheng, H., Chen, M., Zhang, L., Wang, H. and Tan, Q. (2013). Sequencing and comparative analysis of the straw mushroom (*Volvariella volvacea*) genome. *PLoS ONE*, 8(3): e58294.

Beulah, H., Margret, A.A. and Nelson, J. (2013). Marvellous medicinal mushrooms. *Int. J. Pharma. Bio. Sci.*, 3(1): 611–615.

Biswas, M.K. (2014). Cultivation of paddy straw mushrooms (*Volvariella volvacea*) in the lateritic zone of West Bengal—A healthy food for rural people. *International Journal of Economic Plants*, 1(1): 23–27.

Biswas, M.K. and Layak, M. (2014). Techniques for increasing the biological efficiency of paddy straw mushroom (*Volvariella volvacea*) in eastern India. *Food Science and Technology*, 2(4): 52–57.

Breene, W.M. (1990). Nutritional and medicinal value of specialty mushrooms. *Journal of Food Protection*, 53(10): 883–894.

Chang, S.T. (1969). A cytological study of spore germination of *Volvariella volvacea*. *Bot. Mag.*, 82: 102–109.

Chang, S.T. and Hayes, W.A. (eds.). (2013). *The Biology and Cultivation of Edible Mushrooms*. Academic Press.

Chang, S.T. and Miles, P.G. (1989). *In: Edible Mushrooms and Their Cultivation*. Boca Raton, Florida: CRC Press Inc., 27–40.

Chang, S.T. and Miles, P.G. (2004). *Mushrooms: Cultivation, Nutritional Value, Medicinal Effect and Environmental Impact*. CRC Press.

Cheung, L.M. and Cheung, P.C. (2005). Mushroom extracts with antioxidant activity against lipid peroxidation. *Food Chemistry*, 89(3): 403–409.

Cheung, L.M., Cheung, P.C. and Ooi, V.E. (2003). Antioxidant activity and total phenolics of edible mushroom extracts. *Food Chemistry*, 81(2): 249–255.

Cheung, P.C.K. (ed.). (2008). *Mushrooms as Functional Foods*. Hoboken, New Jersey: A John Wiley and Sons, INC.

Chiu, K.W., Lam, A.H.W. and Pang, P.K.T. (1995). Cardiovascular active substances from the straw mushroom, *Volvariella volvacea*. *Phytotherapy Research*, 9(2): 93–99.

Dragsted, L.O., Strube, M. and Larsen, J.C. (1993). Cancer-protective factors in fruits and vegetables: Biochemical and biological background. *Pharmacology and Toxicology*, 72: 116–135.

Ghosh, K. (2000). A review on edible straw mushrooms: A source of high nutritional supplement, biologically active diverse structural polysaccharides. *J. Sci. Res.*, 64: 295–304.

Grzybowski, R. (1978). Nutrient properties of the fructification and vegetative mycelium of mushrooms. *Przem. Spo.*, 32(1): 13–16.

Gupta, S., Summuna, B., Gupta, M. and Mantoo, A. (2016). Mushroom cultivation: A means of nutritional security in India. *World*, 3: 6–50.

Guyton, J.R. (1998). Effect of niacin on atherosclerotic cardiovascular disease. *The American Journal of Cardiology*, 82(12): 18U–23U.

Holman, Ralph T. (1976). Essential fatty acids in human nutrition. *Function and Biosynthesis of Lipids*, Springer, Boston, MA, 1977: 515–534.

Huang, B.H., Yung, K.H. and Chang, S.T. (1985). The sterol composition of *Volvariella volvacea* and other edible mushrooms. *Mycologia*, 77(6): 959–963.

Huang, B.H., Yung, K.H. and Chang, S.T. (1989). Fatty acid composition of *Volvariella volvacea* and other edible mushrooms. *Mushroom Science*, 12: 533–540.

Jagadeesh, R., Raaman, N., Periyasamy, K., Hariprasath, L., Thangaraj, R., Srikumar, R. and Ayyappan, R. (2010). Proximate analysis and antibacterial activity of an edible mushroom *Volvariella Bombycina*, centre for advance studies in Botany, University of Madras. *Intl. J.*, 1(3): 110–113.

Joseph, J.A., Shukitt-Hale, B., Denisova, N.A., Bielinski, D., Martin, A., McEwen, J.J. and Bickford, P.C. (1999). Reversals of age-related declines in neuronal signal transduction, cognitive, and motor behavioral deficits with blueberry, spinach, or strawberry dietary supplementation. *Journal of Neuroscience*, 19(18): 8114–8121.

Kalaiselvan, B. (2007). Studies on modern techniques for cultivation of paddy straw mushroom (*Volvariella volvacea*) Sing. on commercial scale, M.Sc. (Agriculture) thesis, Tamil Nadu gricultural University, Coimbatore, pp. 89–95.

Kalava, S.V. and Menon, S.G. (2012). Ameliorative effect of *Volvariella volvacea* aqueous extract (Bulliard Ex Fries) singer on gentamicin induced renal damage. *International Journal of Pharma and Bio Sciences*, 3(3): 105–117.

Kamaliah, N., Salim, S., Abdullah, S., Nobilly, F., Mat, S., Norhisham, A.R. and Azhar, B. (2021). Evaluating the experimental cultivation of edible mushroom, *Volvariella volvacea* underneath tree canopy in tropical agroforestry systems. *Agroforestry Systems*, 1–13.

Kirkland, J.B. (2003). Niacin and carcinogenesis. *Nutrition and Cancer*, 46(2): 110–118.

Kishida, E., Kinoshita, C., Sone, Y. and Misaki, A. (1992). Structures and antitumour activities of polysaccharides isolated from mycelium of *Volvariella volvacea*. *Bioscience, Biotechnology and Biochemistry*, 56(8): 1308–1309.

Kurtzman, R.H. (1975). Mushrooms as a source of food proteins. *Protein Nutritional Quality of Foods and Feeds*, 2: 305–318.

Kurtzman, R.H. (1976). Nutrition of *Pleurotus sapidus*: Effects of lipids. *Mycologia*, 68(2): 286–295.

Kurtzman, R.H. (2005). Mushrooms: Sources for modern Western medicine. *Micologia Aplicada International*, 17(2): 21–33.

Lee, T.F. and Chang, S.T. (1975). Nutritional analysis of *Volvariella volvacea*. *J. Hortic. Soc. China*, 21(1): 13–20.

Matilla, P., Könkö, K., Eurola, M., Pihlava, J.M., Astola, J., Vahteristo, L. and Piironen, V. (2001). Contents of vitamins, mineral elements, and some phenolic compounds in cultivated mushrooms. *Journal of Agricultural and Food Chemistry*, 49(5): 2343–2348.

Milton, R. and Dulay, R. (2015). Nutrient composition and functional activity of different stages in the fruiting body development of philippine paddy straw Mushroom, *Volvariella volvacea* (Bull.:Fr.) Sing. *Advances in Environmental Biology*, 9(22): 54–56.

Mshandete, A.M. and Cuff, J. (2007). Proximate and nutrient composition of three types of indigenous edible wild mushrooms grown in Tanzania and their utilization prospects. *African Journal of Food, Agriculture, Nutrition and Development*, 7(6).

Nhi, N.N.Y. and Hung, P.V. (2012). Nutritional composition and antioxidant capacity of several edible mushrooms grown in South Vietnam.

Ouzouni, P.K., Petridis, D., Koller, W.D. and Riganakos, K.A. (2009). Nutritional value and metal content of wild edible mushrooms collected from West Macedonia and Epirus, Greece. *Food Chemistry*, 115(4): 1575–1580.

Punitha, S.C. and Rajasekaran, M. (2016). Proximate, elemental and GC-MS study of the edible mushroom *Volvariella volvacea* (Bull Ex Fr) Singer. *J. Chem. Pharm. Res.*, 7: 511–518.

Purkayastha, R.P. and Nayak, D. (1981). Analysis of protein patterns of an edible mushroom by gel-electrophoresis and its amino acid composition. *Journal of Food Science and Technology*, 89–91.

Ramkumar, L., Ramanathan, T. and Johnprabagaran, J. (2012). Evaluation of nutrients, trace metals and antioxidant activity in *Volvariella volvacea* (bull. Ex. Fr.), Sing. *Emirates Journal of Food and Agriculture*, 113–119.

Reddy, S.M. (2015). Diversity and applications of mushrooms. *In: Plant Biology and Biotechnology*, pp. 231–261, Springer, New Delhi.

Roy, A., Prasad, P. and Gupta, N. (2014). *Volvariella volvacea*: A macrofungus having nutritional and health potential. *Asian Journal of Pharmacy and Technology*, 4(2): 110–113.

Senatore, F. (1990). Fatty acid and free amino acid content of some mushrooms. *Journal of the Science of Food and Agriculture*, 51(1): 91–96.

She, Q.B., Ng, T.B. and Liu, W.K. (1998). A novel lectin with potent immunomodulatory activity isolated from both fruiting bodies and cultured mycelia of the edible mushroom *Volvariella volvacea*. *Biochemical and Biophysical Research Communications*, 247(1): 106–111.

Sohi, H.S., Nair, M.C. and Balakrishnan, S. (1990). Studies on paddy straw mushroom (*Volvariella* spp.) cultivation in India, present status and future prospectus. *Beneficial Fungi and their Utilisation*, 462–465.

Sukara, E. (1985). *Cara menanam jamur merang*: The cultivation of the paddy straw mushroom. *Bulletin of the British Mycological Society*, 19(2): 129–132.

Su, U.T. and Seth, L.N. (1940). Cultivation of the straw mushroom. *Indian Farming*, 1: 33–333.

Thang, L.D. (2006). *Growing Edible Mushroom*. Ho Chi Minh Agricultural Publisher, 242 p.

Thuc, L.V., Corales, R.G., Sajor, J.T., Truc, N.T.T., Hien, P.H., Ramos, R.E. and Hung, N.V. (2020). Rice-straw Mushroom production. *In: Sustainable Rice Straw Management*, pp. 93–109, Springer, Cham.

Uddin, M.J., Haque, S., Haque, M.E., Bilkis, S. and Biswas, A.K. (2012). Effect of different substrate on growth and yield of button Mushroom. *Journal of Environmental Science and Natural Resources*, 5(2): 177–180.

Umor, N.A., Abdullah, S., Mohamad, A., Ismail, S., Ismail, S.I. and Misran, A. (2020). Challenges and current state-of-art of the *Volvariella volvacea* cultivation using agriculture waste: A brief review. *Advances in Waste Processing Technology*, 145–156.

Yella, V.K., Chadrapati, A., Kuri, A., Miglani, I., Andrews, A.A. and Singh, S. (2021). *Cultivation Technology and Spawn Production of Volvariella volvacea: Paddy Straw Mushroom.*

Zahid, A., Fozia, M.R. and Ahmed, S. (2019). Nutritional and Medicinal significance of paddy straw Mushroom (*Volvariella volvacea*). *Science Letters*, 7(3): 99–103.

Zhang, J., Wang, G., Li, H., Zhuang, C., Mizuno, T., Ito, H. and Li, J. (1994). Anti-tumoor polysaccharides from a Chinese mushroom, 'yuhuangmo', the fruiting body of *Pleurotus citrinopileatus. Bioscience, Biotechnology and Biochemistry*, 58(7): 1195–1201.

Zikriyani, H., Saskiawan, I. and Mangunwardoyo, W. (2018). Utilisation of agricultural waste for cultivation of paddy straw mushrooms (*Volvariella volvacea* (Bull.), Singer, 1951). *Intl. J. Agric. Technol.*, 14(5): 805–814.

CHAPTER 23

China Root
(*Wolfiporia extensa* (Peck) Ginns)

Jeetendra Singh, SK Soni and *Rajiv Ranjan**

1. Introduction

Poria cocos (Syn. *Wolfiporia cocos*), commonly known as '*Tuckahoe*', is a plant prevalent in North America. Most studies have also used the term 'Fuling'. The *Pachyma hoelen* and other East Asian *Pachyma* species have not been thoroughly explored. In spite of Merrill's discovery, a couple of centuries ago, Asia's and North America's *Pachyma Cocos* looked alike but is different. There was a high level of homogeneity in the East Asian sclerotia, commercial cultivars and fruiting bodies samples from *Pachyma hoelen* that differentiated it from *Wolfiporia cocos* and other species in North America. The sclerotia and basidiocarps, which are both unusually developed, were examined on the core of evolutionary relationships and morphological evidence. According to our observations, the commonly grown Fuling *Pachyma hoelen* is not conspecific as *Pachyma hoelen* of East Asia and *Wolfiporia cocos* in North America. In accordance with *Pachyma*, which was validated by Fries, now the genus has priority (ICN, Art. F.3.1) and this name adequately depicts the economically relevant stage of the generic type of nomenclature for algae, fungi and plants because of the amendments to Article 59 of the International Code. Accordingly, we recommend using *Pachyma* instead of *Pachyma hoelen* and *Pachyma cocos* as both are relevant to *Wolfiporia* terminology for use in East Asia as Fuling and in North America as Tuckahoe. Furthermore, *Pachyma pseudococos* is offered as a unique combination. In addition, *Pachyma cocos* appears to be a genus complex which includes three species which are native to North America (Wu et al., 2020).

In the Dikarya subkingdom, thick aggregations of fungi, called sclerotia can sustain adverse environmental circumstances, as well as providing the fungus with a habitat to grow (Willets and Bullock, 1992; Coley-Smith and Cooke, 1971; Smith et al., 2015). Sclerotia is a useful fungus food because it is such a durable fungal

Department of Botany, Faculty of Science, Dayalbagh Educational Institute, Dayalbagh, Agra-5.
* Corresponding author: rajivranjanbt@gmail.com

structure that it contains physiologically active secondary metabolites (Wong and Cheung, 2009; Lau and Abdullah, 2016). In the past, indigenous peoples throughout the entire world have consumed massive underground sclerotia of a few mushroom species (Aguiar and Sousa, 1981; Bandara et al., 2015; Lau et al., 2015; Oso, 1977). Native Americans in North American countries eat Tuckahoe or 'Indian bread', which is made from the hypogenous sclerotia of a fungus species called *Tuckahoe* (Gore, 1881; Weber, 1929). In 1822, the first scientifically credible von Schweinitz described and named the fungal sclerotia as *Schwein. Sclerotium cocos* Fries (1822) acknowledged this name when he created the genus *Pachyma* Fr. Tuckahoe mushroom's binomial *Pachyma cocos* (Schwein.) Fr. is the most often used (Currey and Hanbury, 1860; Gore, 1881; Prilleaux, 1889; Elliott, 1922). The sexual stage of *P. cocos*, on the other hand, has been unknown for nearly a century. Wolf found its whitish resupinate period fruiting body. *Poria Pers. Poriacocos* (Schwein.) F.A. Wolf by Wolf was the general name for all light-coloured and resupinate polypores at the time (Murrill, 1920; 1923). As a result, the sexual stage was given the name *Poria cocos* (Schwein.) F.A. Wolf by Wolf (1922).

Poria cocos have undergone a change in categorisation. This species was given to a new genus, Macrohyporia I Johans & Ryvarden, by Johansen and Ryvarden (1979) and is represented by *M. dictyopora* (Cooke) I Johans & Ryvarden. Gilbertson and Ryvarden (1984), based on the distinctive spore morphology of *M. dictyopora*, said Poria cocos belongs to Wolfiporia Ryvarden and Gilb. genus, which is represented by Wolfiporia Ryvarden and Gilb. *Poria cocos* were classed as Macrohyporia by Ginns and Lowe (1983), putting it in the same category as the teleomorphic name *Daedalea extensa* Peck. As a result, Ginns (1984) adopted Ryvarden and Gilbertson's generic revision and published Ginns to correct the binomial *Wolfiporia extensa* (Peck) species name. Despite this, due to the findings and the medical community's affection for it, Redhead and Ginns (2006) recommend that *Cocos poria* (syn. *Wolfiporia cocos*) be preserved above *Daedalea extensa* (syn.). The Fungi Nomenclature Commission likewise recommended the preservation of *Poria cocos* (Norvell, 2008). In traditional Chinese medicine, *Poria cocos* is also referred as Fulling, a fungal sclerotium that is used for cough relief, inducing diuresis, reducing stress, reducing fever, anti-carcinogenicity, anti-hepatitis B virus, anti-inflammation, anti-metastasis and antitumour hypoglycaemia (Wang et al., 2013; Wu et al., 2019). These qualities need be validated in pharmacological research (Li et al., 2019; Zhang et al., 2018; Sun, 2014; Zhang et al., 2018; Li et al., 2019). In China, sclerotia Fuling remains extensively cultivated (Wang et al., 2013), as well as its derivatives have being distributed to more than 40 nations (Chi et al., 2018). The term '*hoelen*' is derived from Rumphius' posthumous publication of pre-Linnean research literature (Xu et al., 2014; Li et al., 2016) and is also used to describe this renowned therapeutic fungus. Fries (1822) recognised *hoelen* against a little pharmacological activity from China under the name *Pachyma*. In China, *Pachyma hoelen* Fr has being mostly found on pine shrubs, according to Merrill (1917). It was formerly known as *Poria cocos*. He wanted to know who *P. hoelen* was W.A. Murrill was a Chinese Scientist (procured *P. hoelen* a pharmacy) and sent for screening. The Chinese sclerotia samples were identical to those collected at different sites across America, although *Pachyma hoelen* Fries is distinguishable from *Pachyma cocos* Fries, according to Murrill (Merrill, 1917).

Edible and therapeutic mushrooms seem to be natural resources that provide significant health advantages and have long played as a major food and medication throughout the whole globe (Njouonkou et al., 2016; Yahia et al., 2017). These mushrooms have a unique flavour valued not simply because of their flavour and appearance, but also for their nutritional value and are widely available in local stores as a source of protein, carbs, essential amino acids, vitamins and minerals to make for a balanced diet (Cruz et al., 2016; Wang et al., 2016). Many types of mushroom species utilised in daily life have traditionally remained in wild environments based on ethno-mycological expertise, and some attempts have been made to artificially produce the mushrooms on a viable level through genetic engineering With increased knowledge of the need for edible and therapeutic mushrooms, the best quality raw materials are used in limited stream and farmed species have become the primary alternatives for wild species as they can thrive in various environments. As a result of the variety in the quality of mushroom, choice of the right ones has become a major cause of worry for the general population. The study of differences amongst mushrooms in their natural state is critical for quality control and estimation of edible and medicinal properties before they are transformed into finished products. However, mushrooms are complex systems and their multi-component characteristics make it difficult to identify their bioactive ingredients (Pala et al., 2013; Phan et al., 2017). Furthermore, these mushrooms' unique characteristics are linked with not just one or many components, but also the holistic composition's synergistic impact (Phan et al., 2015). With the help of current analytical techniques, the full quality comparison and assessment procedures have been created. In common, the spectroscopic instrumental approach is simple, rapid, effective and inexpensive, making it one of the most popular methods for studying complex biological materials (Márquez et al., 2016; Xiong et al., 2015).

The Polyporaceae saprophytic mushroom *Wolfporia cocos* (F.A. Wolf) Ryvarden and Gilb is well known. This species is a key component in the well-known snack *Tuckahoe pie* and is one of the most well-known materials used in Chinese and other Asian traditional medicines (Yu et al., 2017; Ling et al., 2012). The medicinal portions of *W. cocos* are the inner parts and epidermis, which have been, for a long time, documented in the *Chinese Pharmacopoeia* thats a programme in China. The key active ingredients are as follows: *W. cocos* has anti-tumour, antioxidant, anti-rejection and anti-hyperglycaemic properties. This species is commonly utilised to treat symptoms of gastritis, stomach agony, acute enterotoxin catarrh, dizziness, nausea and emesis (Wang et al., 2013; Tai et al., 1995; Okui et al., 1996). *W. cocos* is an important culinary ingredient and medicinal fungus extensively spread across various provinces of China – Yunnan, Guizhou, Sichuan, Hubei, Anhui, Guangxi and Fujian Yunnan which serve as the typical natural habitat. Because natural *W. cocos* is uncommon in natural conditions, it is highly prized and the nurtured variety has always been in great demand. It grows in the wild and farmed *W. cocos* might be useful for its varying standards of excellence and therapeutic effects as it is grown in distinct growth environments. As a result, the distinction and raw resources from *W. cocos* should be highlighted first as they are associated with the acceptance of specific quality requirements.

2. Taxonomy

Although its scientific name is *Wolfiporia extensive* (Peck) Ginns, in 1984, the synonym *Poriacocos* (Schwein.) A Wolf, F.A. was chosen for this study since it appears in the preponderance of reviewed studies. Naming of this species seems somewhat perplexing and is summarised here.

Sclerotium cocos Schwein was offered as a term for the sclerotia that developed the fungus in 1822. Fries offered the name *Pachyma cocos* the following year, based on Schweintz's description: *P. cocos, oblongum, cortice* dura fibroso-squamoso brunneo (Esteban, 2009; Wolf, 1922). FA Wolf combined the species as *Poria cocos* (Schwein.) F.A. Wolf a century later owing to the genus *Poria's* basidiocarp similarities (Wolf, 1922).

Following the examination of the American specimens of genus *Poria* (which includes *P. cocos*) was renamed *Wolfiporia cocos* (F.A. Wolf) Ryvarden and Gilb 47 and a new genus, *Wolfiporia* (dedicated to F.A. Wolf), was suggested for them. However, like *P. cocos*, which is a synonym of *D. extensive*, *Wolfiporia extensive* (Peck) Ginns19 is a name that is now approved in the *Index Fungous*.

3. Habit and Distribution

As a perennial basidiocarp, *P. cocos* resupinates, forming small patches on wood that gradually combine and spread. The hymenophore is ochraceous to pinkish and has pores, with one to two angular holes per mm. The basidiocarp's flesh is thin and cream to pink in colour (Gilbertson and Ryvarden, 1987).

Between the roots of certain trees, *P. cocos* produces a hypogeal sclerotium, which comes in various shapes and sizes, ranging from rectangular to globose, 10–30 cm long and weighing up to 1 kg. The outside is wrinkled and fibrous, while the interior is white or slightly pinkish. It is resistant and flexible when young (Ying, 1987). Some remnants of *P. cocci's* basidiocarps can sometimes be seen in the sclerotia's outer layer; fresh basidiocarps are occasionally observed on dead trees that have no relationship to the sclerotia. Sclerotia can also be produced by incubating basidiocarp in a humid room (Gilbertson and Ryvarden, 1987).

It causes root rot in trees and shrubs of the genera *Picea*, *Tsuga* and *Pinus* (*Pinusmassoniana, Pinustaiwanensis, Pinusyunnanensis*, and *Pinusthunbergii*) as well as in trees and shrubs of several genera (*Citrus, Diospyros, Melaleuca, Fagus, Magnolia* and *Pinusthunbergii*) (Ginns et al., 1983). It has even been used to find maize (*Zea mays*) roots. Because of its damaging impact on cellulose, *P. cocos* can degrade stored wood (Gujre et al., 2021).

The distribution of *P. cocos* on parasitised trees is inferred from the discovery of sclerotia, cultures grown from plants infected with the fungus and, in rare cases, basidiocarp collections (Hudson, 1968). Eastern Asia, particularly China, eastern Australia, the southern United States and western Canada are home to this species. It is also known as *Macrohyporia cocos* in Africa and Tibet.

4. Morphology of *Wolfiporia extensa*

Pachyma cocos collections were examined by researchers and isolates that were natural and traditional were developed from strains from East Asia (China, Japan) and

North America. Beijing Forestry University's Institute of Microbiology (BJFC) and Beijing Forestry University's Institute of Microbiology, the Herbarium Mycologium, Chinese Academy of Sciences, have voucher specimens in Beijing, China (HMAS). *Pachyma hoelen* has a defined neotype MycoBank entry (Dong 897, HMAS 248370) (Wu et al., 2020). Field notes and dried herbarium specimens are used to create macro-morphological descriptions. Following Dai's instructions, dry sample and slide preparations with cotton blue and Melzer's reagent were used to dye the collections to make microscopic observations and illustrations (2010).

5. Material and Procedures for Growing *Wolfiporia extensa*

5.1 W. extensa Strain

This strain, which produced enormous sclerotia during field culture, was kept on potato dextrose agar (Kitamura et al., 2022).

5.1.1 Bottle for Mushroom Culture

To grow mushrooms, start with a culture container with a screw-on lid which has an air filter. It was divided into three sections: the cap, top section and lower section, all of which were autoclavable and had a volume of 2300 cm^3, a diameter of 15 cm, and a height of 16 cm (Kubo et al., 2006). The three air filters were constructed with commercially accessible materials including fabric, paper and urethane resin. A control was also employed and it was a closed container with the lid covered.

5.1.2 Indoor Cultivation and Inoculum

Mycelia from the strain were inoculated on sawdust rice bran (3:1 v/v) medium with a moisture content of 70% and grown for one month at 30°C.

Three pine logs from *Pinus densiflora* SIEB et ZUCC, each measuring about 5 cm in diameter and 10 cm in length, were soaked in water overnight before being placed in each bottle and sterilised for 30 minutes before cooling at room temperature. The logs were injected with this spawn and the bottles were incubated in the dark at 25°C, which is the optimum sclerotia formation temperature (Kubo et al., 2006).

5.1.3 Harvesting of Sclerotia

The sclerotia that had developed on the pinewood was harvested. The outer layer was removed and dried at 50°C for 18 hours. The yields were determined by using fresh and frozen weight on a wood volume (m3) basis.

5.1.4 CO$_2$ Concentration in the Bottle

A gas detector tube linked to the bottle with silicon tubing was used to monitor CO$_2$ concentration.

5.1.5 Weight Loss as a Result of Wood Decay

To begin with, the dry weight of the initial pine logs was calculated by using the wet weight and mean moisture content of soaked pine logs. After sclerotia developed on pine logs, cultured wood was dried at 105°C for 18 hours and the dry weight was calculated. The weight loss due to wood deterioration was calculated using the difference in the dry weight of pine logs before and after culture.

5.1.6 Based on Weight, the Alkaline Solubility in 1% Aqueous NaOH

The sapwood of dried pine logs was crushed and roughly 1 g of powder was extracted in boiling water with 50 ml of 1% NaOH for one hour. The solution was filtered through a glass filter (1G3) and then rinsed with 150 cc of hot water. After dissolving the residue in 25 mL of 10% CH_3COOH, it was washed with 150 mL of hot water. The glass filter was then dried at 105°C for 18 hours. Based on the weight of rotten wood, the difference between the dried weight of the sample and the filtered residue was used to compute the alkaline solubility.

5.1.7 Wood's pH

The sapwood of dried pine logs was used to make the log chips, which measure roughly 222 cm in length. The wood's pH level was measured after each chip was soaked in 17 cc of distilled water for five days.

5.1.8 Pachymic Acid and Dehydropachymic Acid Analysis

2 g dried sclerotia powder was precisely weighed and extraction was done with 20 ml of CH_3OH for 20 min. under sonication process. Subsequently, centrifuge was done for 10 min. at 3000 rpm. The residue was removed by adopting the same procedure. Both supernatants were mixed and CH_3OH was used to dilute them to 50 ml concentration respectively. HPLC was used to analyse the levels of pachymic acid and dehydropachymic acid. The HPLC system used a 40°C ODS-80TM column (1504.6 I.D. mm) with a 20 ml injection volume. CH_3CN, H_2O and CH_3COOH were the mobile phases (700: 300: 1). The flow rate was 1.0 mL per minute. The summits were discovered.

Pachymic acid is detected at 210 nm, whereas dehydropachymic acid is detected at 240 nm wavelengths. Dry weight amounts of pachymic acid and dehydropachymic acid were estimated after drying powder at 105°C temperature for five hours.

5.2 Edible Aspects of Wolfiporia extensa

The sclerotia that grows from the fungus's mycelium is edible and safe for human consumption (Jin et al., 2019). It has a sweet flavour and a silky texture. *P. cocos* is on the FAO's list of economically important wild fungus (Boa, 2005). It has been recorded as a food source in Nigeria (Smith et al., 2002), probably mistaken with the sclerotia of *Pleurotus* tuber-regium (Rumph. ex Fr.) Singer, as well as in the United States, where roasted sclerotia have been consumed (Weber, 1929). In China, a dessert known as 'Fuling cake' is produced from the sclerotia of *P. cocos*. It's now found in a variety of traditional Chinese medicine formulations. In addition to *P. cocos'* sclerotia, in recent years, China has been more interested in the sclerotia of the *Pleurotus tuber-regium* (Rumph. ex Fr.) Polyporusrhino cerus and Singer.

5.3 Medicinal Aspects

The dried sclerotia of *P. cocos* has been used as a diuretic and sedative in traditional Chinese medicine, while the interior section is utilised as an energising agent (Jia et al., 2016). Polysaccharides have anticancer and immunomodulatory properties (Huang, 2007; Wang et al., 2004; Zhang et al., 2006). Because *P. cocos* polysaccharides have been demonstrated to exhibit substantial anti-suppressive action in human leukemic cells *in vitro* (Muszynska et al., 2018), they might be an option in cancer therapy. Polysaccharides from mycelium have been shown to have antitumour action in sarcoma tumour cells in 180% of cases in mice (Jin et al., 2003). Similarly, the so-called poriatin (a combination of various chemicals isolated with chloroform from *P. cocos*) has been shown to impede DNA production in leukaemia cells (Shingu et al., 1992). The mycelium's glucans have recently been discovered to have an inhibiting impact on cancer cells produced *in vitro* from afflicted breasts (Zhang et al., 2006). Some of the sclerotia's triterpene-type lanostanes have been demonstrated to hinder tumour cell proliferation (Ghosh, 2020). Dehydroebriconic acid inhibits DNA topoisomerase II and prevents cancer cell proliferation in humans (Mizushina et al., 2004), whereas pachimic acid induces apoptosis (cell death) in prostate cancer cells (Mizushina et al., 2004). Poricoic acid has been shown to have antitumour properties.

Li et al. (2018) isolated acid C and 16-deoxyporicoic acid B from sclerotia's exterior area, as well as a protein that may stimulate the immune system in mice. Various triterpene-type lanostanes isolated from sclerotia, such as 3-p-hydroxybenzoyldehydrotumulosic acid, pachemical, and dehydrotumulosic acids, have also been found to have an anti-inflammatory effect in mice (Chen et al., 2019). Other therapeutic properties of *P. cocos* components include antinephritic (Hattori, et al., 1992), antiemetic (Tai et al., 1995), antioxidant (Li et al., 2021), nematic activity (Li et al., 2005), acne and oily skin therapy (Meybeck et al., 1998) and even reduction in the symptoms of hot flashes (Plotnikoff et al., 2005). Some lanostane-type compounds (tumulosic, dehydrotumulosic, 3-0-acetyltumulosic, porichoic A and B) have recently been shown to have anti-inflammatory properties. Experimentally, BC and dihydrolanostan acids have been shown to cure dermatitis in people (Fuchs et al., 2006). The sclerotia of *P. cocos* was used in several recipes for the treatment of the skin of Imperial Chinese Court's women as it helped them to keep their skin clean and prevent pimples, blackheads and wrinkles (Khan et al., 2017). Currently, it may be found in a variety of face creams and other cosmetics.

Acknowledgment

We are thankful to the Director, Dayalbagh Educational Institute, Dayalbagh, Agra for providing kind support and infrastructure.

References

Aguiar, I. de J.A. and Sousa, M.A. de (1981). *Polyporusindigenus I. Araujo & amp, M.A. Sousa, nova espécie da Amazônia, ActaAmazonica*, 11(3): 449–455. https://Doi.org/10.1590/1809-43921981113449.

Bandara, A.R., Rapior, S., Bhat, D.J., Kakumyan, P., Chamyuang, S., Xu, J. and Hyde, K.D. (2015). *Polyporusumbellatus*, an edible-medicinal cultivated mushroom with multiple developed health-care products as food, medicine and cosmetics: A review. *Cryptogamie Mycologie*, 36(1): 3–42. https://Doi.org/10.7872/crym.v36.iss1.2015.3.

Boa, E. (2005). *Los hongossilvestres comestibles: Perspectiva global de suuso e importancia para la población. Productosforestales no madereros, Food & Agriculture Org. (FAO)*.

Chen, B., Zhang, J., Han, J., Zhao, R., Bao, L., Huang, Y. and Liu, H. (2019). Lanostane triterpenoids with glucose-uptake-stimulatory activity from peels of the cultivated edible mushroom *Wolfiporia cocos. Journal of Agricultural and Food Chemistry*, 67(26): 7348–7364.

Chi, X.-L., Yang, G., Ma, S., Cheng, M. and Que, L. (2018). Analysis of characteristics and problems of international trade of Poriacocos in China, *ZhongguoZhong Yao ZaZhi = ZhongguoZhongyaoZazhi = China Journal of Chinese Materia Medica*, 43(1): 191–196. https://Doi.org/10.19540/j.cnki.cjcmm.20171030.004.

Coley-Smith, J.R. and Cooke, R.C. (1971). Survival and germination of fungal sclerotia. *Annual Review of Phytopathology*, 9(1): 65–92. https://Doi.org/10.1146/annurev.py.09.090171.000433.

Cruz, A., Pimentel, L., Rodríguez-Alcalá, L.M., Fernandes, T. and Pintado, M. (2016). Health benefits of edible mushrooms focused on *Coriolus versicolor*: A review. *Journal of Food and Nutrition Research*, 4(12): 773–781.

Currey, F. and Hanbury, D. (1860). VIII. Remarks on Sclerotiumstipitatum, Berk. et Curr., *Pachyma cocos*, Fries, and some similar productions. *Transactions of the Linnean Society of London*, 23(1): 93–97. https://Doi.org/10.1111/j.1096-3642.1860.tb00122.x.

Elliott, J.A. (1922). Some characters of the Southern Tuckahoe. *Mycologia*, 14(4): 222–227. https://Doi.org/10.1080/00275514.1922.12020383.

Esteban, C.I. (2009). *Interés medicinal de Poriacocos (= Wolfiporiaextensa)*, *Revista Iberoamericana de Micología*, 26(2): 103–107. https://Doi.org/10.1016/S1130-1406 (09)70019-1.

Esteban, C.I. (2009). *Revista Iberoamericana de Micología, Rev IberoamMicol*, 26(2): 103–107.

Fuchs, S.M., Heinemann, C., Schliemann-Willers, S., Härtl, H., Fluhr, J.W. and Elsner, P. (2006). Assessment of anti-inflammatory activity of *Poria cocos* in sodium lauryl sulphate-induced irritant contact dermatitis. *Skin Research and Technology*, 12(4): 223–227. https://Doi.org/10.1111/j.0909-752X.2006.00168.x.

Ghosh, S.K., Sanyal, T.A. and Bera, T.A. (2020). Antiproliferative and apoptotic effect of methanolic extract of edible mushroom *Agaricus bisporus* against HeLa, MCF-7 and MDA-MB-231 cell lines of human cancer and chemoprofile by GC-MS. *Plant Cell Biotechnology and Molecular Biology*, 21(39-40): 109–122.

Gilbertson, R.L. and Ryvarden, L. (1987). *North American Polypores*, vol. 2, *Megasporoporia-Wrightoporia. North American Polypores*, vol. 2. *Megasporoporia-Wrightoporia*, 437–885.

Ginns, J. (1984). New names, new combinations and new synonymy in the Corticaceae, Hymenochaetaceae and Polyporaceae. *Mycotaxon*, 21(10-11): 325–333.

Ginns, J. and Lowe, J.L. (1983). *Macrohyporia extensa* and its synonym *Poriacocos. Canadian Journal of Botany*, 61(6): 1672–1679. https://Doi.org/10.1139/b83-180.

Gore, J.H. (1881). Tuckahoe or Indian bread. *Ann. Rep. Smithson. Inst.*, 687–701.

Gujre, N., Soni, A., Rangan, L., Tsang, D.C.W. and Mitra, S. (2021). Sustainable improvement of soil health utilising biochar and arbuscular mycorrhizal fungi: A review. *Environmental Pollution*, 268: 115549. https://Doi.org/10.1016/j.envpol.2020.115549.

Hattori, T., Hayashi, K., Nagao, T., Furuta, K., Ito, M. and Suzuki, Y. (1992). Studies on antinephritic effects of plant components (3): Effect of Pachyman, a main component of Poria Cocos Wolf on original-type anti-GBM Nephritis in rats and its mechanisms. *Japanese Journal of Pharmacology*, 59(1): 89–96. https://Doi.org/10.1254/jjp.59.89.

Huang, Q., Jin, Y., Zhang, L., Cheung, P.C.K. and Kennedy, J.F. (2007). Structure, molecular size and antitumor activities of polysaccharides from Poria cocos mycelia produced in fermenter. *Carbohydrate Polymers*, 70(3): 324–333. https://Doi.org/10.1016/j.carbpol.2007.04.015.

Hudson, H.J. (1968). The ecology of fungi on plant remains above the soil. *New Phytologist*, 67(4): 837–874. https://Doi.org/10.1111/j.1469-8137.1968.tb06399.x.

Jia, X., Ma, L., Li, P., Chen, M. and He, C. (2016). Prospects of Poria cocos polysaccharides: Isolation process, structural features and bioactivities. *Trends in Food Science & Technology*, 54: 52–62. https://Doi.org/10.1016/j.tifs.2016.05.021.

Jin, J., Zhou, R., Xie, J., Ye, H., Liang, X., Zhong, C., Shen, B., Qin, Y., Zhang, S. and Huang, L. (2019). Insights into triterpene acids in fermented mycelia of edible fungus Poria cocos by a comparative study. *Molecules*, 24(7): 1331. https://Doi.org/10.3390/molecules24071331.

Jin, Y., Zhang, L., Zhang, M., Chen, L., Keung Cheung, P.C., Oi, V.E.C. and Lin, Y. (2003). Antitumour activities of heteropolysaccharides of *Poria cocos* mycelia from different strains and culture media. *Carbohydrate Research*, 338(14): 1517–1521. https://Doi.org/10.1016/S0008-6215(03)00198-8.

Johansen, I. and Ryvarden, L. (1979). Studies in the aphyllophorales of Africa. *Transactions of the British Mycological Society*, 72(2): 189–199. https://Doi.org/10.1016/S0007-1536(79)80031-5.

Khan, A.A., Gani, A., Masoodi, F.A., Mushtaq, U. and Naik, A.S. (2017). Structural, rheological, antioxidant, and functional properties of β–glucan extracted from edible mushrooms *Agaricus bisporus*, *Pleurotus ostreatus* and *Coprinus attrimentarius*. *Bioactive Carbohydrates and Dietary Fibre*, 11: 67–74.

Kitamura, M., Muramatsu, N., Yokogawa, T., Kiba, Y. and Suzuki, R. (2022). Fruit body formation and intra-species DNA polymorphism in Japanese *Wolfiporia cocos* strains. *Journal of Natural Medicines*, https://Doi.org/10.1007/s11418-022-01617-2.

Kubo, T., Terabayashi, S., Takeda, S., Sasaki, H., Aburada, M. and Miyamoto, K. (2006). Indoor cultivation and cultural characteristics of *Wolfiporia cocos* sclerotia using Mushroom culture bottles. *Biological and Pharmaceutical Bulletin*, 29(6): 1191–1196. https://Doi.org/10.1248/bpb.29.1191.

Lau, B.F. and Abdullah, N. (2016). Sclerotium-forming Mushrooms as an emerging source of medicinals. *In*: *Mushroom Biotechnology*, pp. 111–136, Elsevier. https://Doi.org/10.1016/B978-0-12-802794-3.00007-2.

Lau, B.F., Abdullah, N., Aminudin, N., Lee, H.B. and Tan, P.J. (2015). Ethnomedicinal uses, pharmacological activities, and cultivation of *Lignosus* spp. (Tiger's Milk mushrooms) in Malaysia—A review. *Journal of Ethnopharmacology*, 169: 441–458. https://Doi.org/10.1016/j.jep.2015.04.042.

Leung, A.Y. and Foster, S. (2003). *Encyclopedia of Common Natural Ingredients Used in Food, Drugs and Cosmetics*, 2nd ed., New Jersey: John Wiley & Sons, Inc.

Li, C.-Y., Liu, L., Zhao, Y.-W., Chen, J.-Y., Sun, X.-Y. and Ouyang, J.-M. (2021). Inhibition of calcium oxalate formation and antioxidant activity of carboxymethylated poria cocos polysaccharides. *Oxidative Medicine and Cellular Longevity*, 1–19. https://Doi.org/10.1155/2021/6653593.

Li, G.H., Shen, Y.M. and Zhang, K.Q. (2005). Nematicidal activity and chemical component of Poria cocos. *Journal of Microbiology*, 43(1): 17–20.

Li, X., He, Y., Zeng, P., Liu, Y., Zhang, M., Hao, C., Wang, H., Lv, Z. and Zhang, L. (2019). Molecular basis for *Poriacocos* mushroom polysaccharide used as an antitumour drug in China. *Journal of Cellular and Molecular Medicine*, 23(1): 4–20. https://Doi.org/10.1111/jcmm.13564.

Li, Y. and Wang, Y. (2018). Differentiation and comparison of *Wolfiporia cocos* raw materials based on multi-spectral information fusion and chemometric methods. *Scientific Reports*, 8(1): 13043. https://Doi.org/10.1038/s41598-018-31264-1.

Li, Y., Zhang, J., Zhao, Y., Liu, H., Wang, Y. and Jin, H. (2016). Exploring geographical differentiation of the hoelen medicinal Mushroom, *Wolfiporia extensa* (Agaricomycetes), using fourier-transform infrared spectroscopy combined with multivariate analysis. *International Journal of Medicinal Mushrooms*, 18(8): 721–731. https://Doi.org/10.1615/IntJMedMushrooms.v18.i8.80.

Ling, Y., Chen, M., Wang, K., Sun, Z., Li, Z., Wu, B. and Huang, C. (2012). Systematic screening and characterization of the major bioactive components of Poria cocos and their metabolites in rats by LC-ESI-MSn. *Biomedical Chromatography*, 26(9): 1109–1117. https://Doi.org/10.1002/bmc.1756.

Márquez, C., López, M.I., Ruisánchez, I. and Callao, M.P. (2016). FT-Raman and NIR spectroscopy data fusion strategy for multivariate qualitative analysis of food fraud. *Talanta*, 161: 80–86. https://Doi.org/10.1016/j.talanta.2016.08.003.

Merrill, E.D. (1917). *An Interpretation of Rumphius's Herbarium Amboinense*, vol. 9. Bureau of Printing.

Meybeck, A. and Bonte, F. (1998). *U.S. Patent No. 5,716,800*. Washington, DC: U.S. Patent and Trademark Office.

Mizushina, Y., Akihisa, T., Ukiya, M., Murakami, C., Kuriyama, I., Xu, X., Yoshida, H. and Sakaguchi, K. (2004). A novel DNA topoisomerase inhibitor: Dehydroebriconic acid, one of the lanostane-type triterpene acids from Poria cocos. *Cancer Science*, 95(4): 354–360. https://Doi.org/10.1111/j.1349-7006.2004.tb03215.x.

Murrill, W.A. (1920). The genus Poria. *Mycologia*, 12: 47–51.

Muszyńska, B., Grzywacz-Kisielewska, A., Kała, K. and Gdula-Argasińska, J. (2018). Anti-inflammatory properties of edible mushrooms: A review. *Food Chemistry*, 243: 373–381. https://Doi.org/10.1016/j.foodchem.2017.09.149.

Njouonkou, A.L., de Crop, E., Mbenmoun, A.M., Kinge, T.R., Biyé, E.H. and Verbeken, A. (2016). Diversity of edible and medicinal mushrooms used in the noun division of the West region of cameroon. *International Journal of Medicinal Mushrooms*, 18(5): 387–396. https://Doi.org/10.1615/IntJMedMushrooms.v18.i5.20.

Norvell, L.L. (2008). Report of the nomenclature committee for fungi: 15. *Mycotaxon*, 110(1): 487–492. https://Doi.org/10.5248/110.487.

Okui, Y., Monta, M., Iizuka, A., Komatsu, Y., Okada, M., Maruno, M. and Niijima, A. (1996). Effects of hoelen on the efferent activity of the gastric vagus nerve in the rat. *Japanese Journal of Pharmacology*, 72(1): 71–73. https://Doi.org/10.1254/jjp.72.71.

Oso, B.A. (1977). *Pleurotus Tuber-Regium* from Nigeria. *Mycologia*, 69(2): 271–279. https://Doi.org/10.1080/00275514.1977.12020058.

Pala, S.A., Wani, A.H. and Bhat, M.Y. (2013). Ethnomycological studies of some wild medicinal and edible Mushrooms in the Kashmir Himalayas (India). *International Journal of Medicinal Mushrooms*, 15(2): 211–220. https://Doi.org/10.1615/IntJMedMushr.v15.i2.100.

Phan, C.-W., David, P. and Sabaratnam, V. (2017). Edible and medicinal Mushrooms: Emerging brain food for the mitigation of neurodegenerative diseases. *Journal of Medicinal Food*, 20(1): 1–10. https://Doi.org/10.1089/jmf.2016.3740.

Phan, C.-W., David, P., Naidu, M., Wong, K.-H. and Sabaratnam, V. (2015). Therapeutic potential of culinary-medicinal mushrooms for the management of neurodegenerative diseases: Diversity, metabolite, and mechanism. *Critical Reviews in Biotechnology*, 35(3): 355–368. https://Doi.org/10.3109/07388551.2014.887649.

Plotnikoff, G.A. (2005). Poria cocos (Schwein.) F.A. Wolf in Japanese traditional herbal medicines: Insights from Kampo case studies and implications for contemporary research. *International Journal of Medicinal Mushrooms*, 447–448. https://Doi.org/10.1615/IntJMedMushrooms.v7.i3.850.

Prilleaux, M. (ed.). (1889). Le *Pachyma Coco se*n France. *Bulletin de La SociétéBotanique de France*, 36(9): 433–436. https://Doi.org/10.1080/00378941.1889.10830499.

Redhead, S.A. and Ginns, J. (2006). (1738) Proposal to conserve the name *Poria cocos* against *Daedalea extensa* (*Basidiomycota*). *TAXON*, 55(4): 1027–1052. https://Doi.org/10.2307/25065702.

Rumphius, G.E. (1750). Herbarium amboinense. *In*: Burmann, J. (ed.). *Amsterdam: Fransicum Changuion and Hermannum Uytwerf.*

Ryvarden, L. and Gilbertson, R.L. (1984). Type studies in the Polyporaceae. 15. Species described by L.O. Overholts, either alone or with J.L. Lowe, *Mycotaxon.*

Shingu, T., Tai, T. and Akahori, A. (1992). A lanostane triterpenoid from *Poria cocos*. *Phytochemistry*, 31(7): 2548–2549. https://Doi.org/10.1016/0031-9422(92)83325-S.

Smith, J., Rowan, N. and Sullivan, R. (2002). *Medicinal Mushrooms: Their Therapeutic Properties and Current Medical Usage with Special Emphasis on Cancer Treatments*, p. 256, London: Cancer Research, UK.

Smith, M.E., Henkel, T.W. and Rollins, J.A. (2015). How many fungi make sclerotia? *Fungal Ecology*, 13: 211–220. https://Doi.org/10.1016/j.funeco.2014.08.010.

Sun, Y. (2014). Biological activities and potential health benefits of polysaccharides from Poriacocos and their derivatives. *International Journal of Biological Macromolecules*, 68: 131–134. https://Doi.org/10.1016/j.ijbiomac.2014.04.010.

Tai, T. (1995). Triterpenes from the surface layer of *Poria cocos*. *Phytochemistry*, 39(5): 1165–1169. https://Doi.org/10.1016/0031-9422(95)00110-S.

Tai, T., Akita, Y., Kinoshita, K., Koyama, K., Takahashi, K. and Watanabe, K. (1995). Anti-emetic principles of *Poria cocos*. *Planta Medica*, 61(06): 527–530. https://Doi.org/10.1055/s-2006-959363.

von Schweinitz, L.D. (1822). *Synopsis fungorum Carolinae superioris secundum observationes Ludovici Davidis de Scweinitz*, Johann Ambrosius Barth.

Wang, R. and Eghbalian, M. (2016). Multi-scale model for damage fluid flow in fractured porous media. *International Journal for Multi-scale Computational Engineering*, 14(4): 367–387. https://Doi.org/10.1615/IntJMultCompEng.2016016951.

Wang, Y., Zhang, L., Li, Y., Hou, X. and Zeng, F. (2004). Correlation of structure to antitumour activities of five derivatives of a β-glucan from Poria cocos sclerotium. *Carbohydrate Research*, 339(15): 2567–2574. https://Doi.org/10.1016/j.carres.2004.08.003.

Wang, Y.-Z., Zhang, J., Zhao, Y.-L., Li, T., Shen, T., Li, J.-Q., Li, W.-Y. and Liu, H.-G. (2013). Mycology, cultivation, traditional uses, phytochemistry and pharmacology of *Wolfiporia cocos* (Schwein.) Ryvarden et Gilb.: A review. *Journal of Ethnopharmacology*, 147(2): 265–276. https://Doi.org/10.1016/j.jep.2013.03.027.

Weber, G.F. (1929). The occurrence of Tuckahoes and *Poria cocos* in florida. *Mycologia*, 21(3): 113–130. https://Doi.org/10.1080/00275514.1929.12016943.

Wolf, F.A. (1926). Tuckahoe on Maize. *J. Elisha Mitchell Sci. Soc.*, 41: 288–290.

Wong, K.-H. and Cheung, P.C.K. (2009). Sclerotia: Emerging functional food derived from Mushrooms. *In: Mushrooms as Functional Foods*, pp. 111–146, John Wiley & Sons, Inc. https://Doi.org/10.1002/9780470367285.ch4.

Wu, F., Li, S.-J., Dong, C.-H., Dai, Y.-C. and Papp, V. (2020). The genus Pachyma (Syn. Wolfiporia) reinstated and species clarification of the cultivated medicinal Mushroom 'Fuling' in China. *Frontiers in Microbiology*, 11. https://Doi.org/10.3389/fmicb.2020.590788.

Wu, F., Zhou, L.-W., Yang, Z.-L., Bau, T., Li, T.-H. and Dai, Y.-C. (2019). Resource diversity of Chinese macrofungi: Edible, medicinal and poisonous species. *Fungal Diversity*, 98(1): 1–76. https://Doi.org/10.1007/s13225-019-00432-7.

Xiong, C., Liu, C., Pan, W., Ma, F., Xiong, C., Qi, L., Chen, F., Lu, X., Yang, J. and Zheng, L. (2015). Non-destructive determination of total polyphenols content and classification of storage periods of Iron Buddha tea using multispectral imaging system. *Food Chemistry*, 176: 130–136. https://Doi.org/10.1016/j.foodchem.2014.12.057.

Xu, Z., Meng, H., Xiong, H. and Bian, Y. (2014). Biological characteristics of teleomorph and optimised *in vitro* fruiting conditions of the hoelen medicinal mushroom, *Wolfiporia extensa* (Higher Basidiomycetes). *International Journal of Medicinal Mushrooms*, 16(5): 421–29. https://Doi.org/10.1615/IntJMedMushrooms.v16.i5.20.

Yahia, E.M., Gutiérrez-Orozco, F. and Moreno-Pérez, M.A. (2017). Identification of phenolic compounds by liquid chromatography-mass spectrometry in seventeen species of wild mushrooms in Central Mexico and determination of their antioxidant activity and bioactive compounds. *Food Chemistry*, 226: 14–22. https://Doi.org/10.1016/j.foodchem.2017.01.044.

Ying, C.C. (1987). *Icons of Medicinal Fungi from China*. Science Press.

Yu, M., Xu, X., Jiang, N., Wei, W., Li, F., He, L. and Luo, X. (2017). Dehydropachymic acid decreases bafilomycin A1 induced β-Amyloid accumulation in PC12 cells. *Journal of Ethnopharmacology*, 198: 167–173. https://Doi.org/10.1016/j.jep.2017.01.007.

Zhang, M., Chiu, L., Cheung, P. and Ooi, V. (2006). Growth-inhibitory effects of a β-glucan from the mycelium of Poria cocos on human breast carcinoma MCF-7 cells: Cell-cycle arrest and apoptosis induction. *Oncology Reports*. https://Doi.org/10.3892/or.15.3.637.

Zhang, W., Chen, L., Li, P., Zhao, J. and Duan, J. (2018). Antidepressant and immunosuppressive activities of two polysaccharides from Poria cocos (Schw.) Wolf. *International Journal of Biological Macromolecules*, 120: 1696–1704. https://Doi.org/10.1016/j.ijbiomac.2018.09.171.

Index

About the Editors

Deepu Pandita, M.Phil.

Senior Lecturer, Government Department of School Education Jammu, Union Territory of Jammu and Kashmir, India.

Deepu Pandita is a Senior Lecturer in Government Department of School Education, Jammu, Union Territory of Jammu and Kashmir, India. She has 20 years of teaching experience and has done her Masters in Botany (M.Sc.) from University of Kashmir, Jammu and Kashmir, India and Master of Philosophy (M.Phil.) in Biotechnology from University of Jammu, Jammu and Kashmir, India. Deepu Pandita has qualified fellowships like JRF-NET and SRF from Council of Scientific & Industrial Research (CSIR), New Delhi India, Biotechnology Fellowship, Government Department of Science and Technology, Jammu & Kashmir, India and IAS-INSA-NASI Summer Research Teacher Fellowship, India. She has presented her research work at important conferences and was awarded Best Oral Presentation Award in an International Conference on Biotechnology for Better Tomorrow, Women Researcher Award and Research Excellence Award by 2 Professional Associations in India. She is a Life-Time Member of various scientific societies. She is a reviewer in numerous journals and editor of several reputed journals. She has published a number of editorials, 32 book chapters (Springer, Elsevier, CRC, etc.), 20 reviews and research articles in various journals of national and international repute like Cells (MDPI) (IF = 7.666), Frontiers in Plant Sciences (IF = 6.627) and Frontiers in Physiology (IF = 4.755) and has currently a number of books in publication.

Anu Pandita, M.Sc.

Dietician, Vatsalya Clinic, New Delhi, India.

Anu Pandita is a Senior Dietician at Vatsalya Clinic, New Delhi, India. Previously she worked as a Lecturer in Bee Enn College of Nursing, Talab Tillo, Jammu, India and as a dietician in Ahinsa Dham Bhagwan Mahavir Charitable Health Centre, New Delhi, India. Anu Pandita has done her M.Sc. Internship and a Course in the Dietetics' Department of PGI, Chandigarh, India. She took a case study at Paediatric Gastroenterology Ward in "Nehru Hospital" PGI, Chandigarh, India on a patient suffering from "Chronic Liver Disease." She has done a Certificate Course in Food & Nutrition as well. She has a number of Trainings, Refresher Courses and workshops to her credit. She is a Life-Time Member of Indian Dietetic Association and Indian

Science Congress Association, Kolkata, India. She has presented her research work at both the national and international conferences. Anu Pandita has published various book chapters with Springer, CRC, etc., and has a number of review and research articles in various journals of national and international repute like Cells (MDPI) (IF = 7.666), and Frontiers in Physiology (IF = 4.755) and has a number of books in publication.

For Product Safety Concerns and Information please contact our EU
representative GPSR@taylorandfrancis.com
Taylor & Francis Verlag GmbH, Kaufingerstraße 24, 80331 München, Germany

www.ingramcontent.com/pod-product-compliance
Lightning Source LLC
Chambersburg PA
CBHW060755220326
41598CB00022B/2445

* 9 7 8 1 0 3 2 3 4 4 5 6 0 *